W9-DEU-536

Basic Electric Circuit Theory

A One-Semester Text

Basic Electric Circuit Theory

A One-Semester Text

I. D. Mayergoyz
University of Maryland
Department of Electrical Engineering
College Park, Maryland

W. Lawson
University of Maryland
Department of Electrical Engineering
College Park, Maryland

ACADEMIC PRESS
An Imprint of Elsevier
San Diego London Boston
New York Sydney Tokyo Toronto

This book is printed on acid-free paper. 

Academic Press
An Imprint of Elsevier
525 B Street, Suite 1900, San Diego, California 92101-4495, USA
200 Wheeler Road, Burlington, MA 01803, USA
http://www.academicpress.com

Academic Press
84 Theobalds Road, London WC1X 8RR, UK
http://www.academicpress.com

Library of Congress Cataloging-in-Publication Data
Mayergoyz, I.D.
Basic electric circuit theory : a one-semester text / I. Mayergoyz, W. Lawson.
p. cm.
Includes index.
ISBN-13: 978-0-12-480865-2 ISBN-10: 0-12-480865-4 (alk. paper)
1. Electric circuits. I. Lawson, W. (Wes) II. Title.
TK454.M395 1996
621.319'2—dc20 96-18904
CIP

ISBN-13: 978-0-12-480865-2
ISBN-10: 0-12-480865-4

PRINTED IN THE UNITED STATES OF AMERICA
06 07 SB 9 8 7

To our wives Deborah and Kathy
with gratitude for their patience and inspiration.

Contents

Preface

You have in your hands an undergraduate text on basic electric circuit theory. As such, it contains no new material for distinction or long remembrance, but it does reflect the current state of instruction in basic circuit theory in electrical engineering (EE) departments in the United States. And this is a state of transition. This transition is brought about by the necessity to introduce new topics into the undergraduate electrical engineering curriculum in order to accommodate the important recent developments in electrical engineering. This can be accomplished only by restructuring classical courses such as basic circuit theory. As a result, there is increasing pressure to find new ways to teach basic circuit theory in a concise manner without compromising the quality and the scope of the exposition of this theory. This text represents an attempt to explore such new approaches.

The most salient feature of this book is that it is designed as a one-semester text on basic circuit theory, and it was used as such in our teaching of the topic at the University of Maryland. Since this is a one-semester text, some traditional topics which are usually presented in other books on electric circuit theory are not covered in this text. These topics include Fourier series, Laplace and Fourier transforms, and their applications to circuit analysis. There exists a tacit consensus that these topics should belong to a course on linear systems and signals. And this is actually the case at many EE departments, where only a one-semester course on basic circuit theory is offered.

Another salient feature of this text is its structure. Here, we deviated substantially from the existing tradition, in which resistive circuits are introduced first and numerous analysis techniques are presented first only for these circuits. In this text, resistors, capacitors, and inductors along with independent sources are introduced from the very beginning and ac steady-state analysis of electric circuits with these basic elements is then developed.

It is known that the ac steady-state equations and the basic equations for resistive electric circuits have identical mathematical structures. As a result, the analysis techniques for ac steady-state and resistive circuits closely parallel one another and are almost identical. The only difference is that in the case of ac steady-state one deals with phasors and impedances, whereas in the case of resistive circuits one deals with instantaneous currents (voltages) and resistances. For this reason, the analysis of resistive circuits can be treated as a particular case of ac steady-state analysis. This is the approach which is adopted in this text.

There are several important reasons for this approach.

First, we believe that EE undergraduate students are well prepared for this style of exposition of the material. They usually have (or should have) sufficient familiarity with the basic circuit elements from a physics course on electricity and magnetism.

Second, this approach allows one to introduce the phasor technique and the notion of impedance at the very beginning of the course and to use them frequently and systematically throughout the course. As our teaching experience suggests, this results in better comprehension and absorption of the phasor technique and the impedance concept by the students. This is crucially important because the notions of phasor and impedance are central and ubiquitous in modern electrical engineering. When the traditional approach to the exposition of electric circuit theory is practiced in the framework of a one-semester course, phasors and impedances are usually introduced toward the end of the course. As a result, students do not have sufficient experience with and exposure to these very important concepts.

Third, the approach adopted here allows one to present numerous analysis techniques (e.g., equivalent transformations of electric circuits, superposition principle, Thevenin's and Norton's theorems, nodal and mesh analysis) in phasor form. In the traditional approach, these techniques are first presented for resistive circuits and subsequently modified for ac steady-state analysis. As a result, this style of exposition requires more time, which is very precious in the framework of a one-semester course.

Finally, the introduction of phasors at the very beginning of the course allows one to use them in the analysis of transients excited by ac sources. This makes the presentation of transients more comprehensive and meaningful. Furthermore, the machinery of phasors paves the road to the introduction of transfer functions, which are then utilized in the analysis of transients, and the discussion of Bode plots and filters.

Another salient feature of the structure of this text is the consolidation of the material concerned with dependent sources and operational amplifiers. In many textbooks, this material is scattered over several chapters, which somewhat undermines its integrity and importance. In this text, this material is consolidated in one chapter where dependent sources are introduced as linear models for semiconductor devices on the basis of small-signal analysis. Then, electric circuits with dependent sources and operational amplifiers are systematically studied.

Finally, we have not completely avoided the temptation to introduce new topics in our textbook. These topics include the use of symmetry in the analysis of electric circuits, the Thevenin theorem for resistive electric circuits with single nonlinear resistors, diode bridge rectifier circuits with *RL* and *RC* loads, the transfer function approach to the analysis of transients in electric circuits, active *RC* filters, and the synthesis of transfer functions by using *RC* operational amplifier circuits. These topics are either not discussed or barely covered in the existing textbooks. We realize that the choice of new topics is always debatable. However, we feel that these topics are of significant educational importance, which prompted our decision to introduce them in this text.

Usually, the basic circuit theory course is the first electrical engineering course taken by undergraduates. For this reason, we believe that it is incumbent upon this course to give students a "taste" of electrical engineering, to kindle their curiosity and enthusiasm about electrical engineering, and to prepare them psychologically for future courses. Probably, the appropriate way to achieve this is to emphasize the connections of electric circuit theory with various areas of electrical engineering. This is exactly what we have tried to accomplish in this text. For instance, we have stressed the connections of basic circuit theory with the area of linear systems and signals when we covered such topics as unit impulse and step responses of linear circuits, the convolution integral technique, the concept of transfer functions and utilization of their poles and zeros in transient analysis, Bode plots, and synthesis of transfer functions by *RC* circuits with operational amplifiers. We have emphasized the connections of basic circuit theory with electronics when we covered dependent sources as linear models for transistors. Finally, we have also stressed that circuit theory has close ties with electromagnetic theory. In basic circuit theory, it is assumed that the values of resistances, inductances, and capacitances are given. The calculation of these quantities is the task of electromagnetic field theory. Furthermore, Kirchhoff's laws, which are treated as basic axioms in circuit theory, can be derived (can be proved) from Maxwell's equations of electromagnetic field theory. More important, by using electromagnetic field theory, the approximate nature of Kirchhoff's laws can be clearly elucidated and the limits of applicability of these laws (and circuit theory) can be established.

There is ongoing discussion concerning the place and role of SPICE (or MicroSim® PSpice®) simulators in a basic circuit theory course. We believe that these circuit simulators should play a complementary role in this course. It is important to emphasize the usefulness of these computer aided design tools and that their effectiveness increases in the hands of "educated consumers." It is equally important to stress that these tools are not a substitute for sound knowledge of electric circuit theory and to provide this knowledge is the ultimate goal of the basic circuit theory course. In other words, we would like to warn against undue invasion of the basic circuit theory course by SPICE and PSpice simulators, an invasion which may compromise the very goals of this course. For this reason, PSpice examples are confined to the final sections of some of the later chapters and a list of PSpice references is relegated to Appendix C.

In undertaking this project, we wanted to produce a student-friendly textbook. We have come to the conclusion that students' interests will be best served by a short book which will closely parallel the presentation of material in class. We have not avoided the discussion of complicated concepts; on the contrary, we have tried to introduce them in a straightforward way and strived to achieve clarity and precision in exposition. We believe that material which is carefully and rigorously presented is better absorbed. From our teaching experience, we have found that there are some topics which are more difficult for students to digest than others. We have observed that the mathematical form of circuit theory is not the major obstacle. Students usually encounter more difficulties in reading connectivity of the electric circuits than in understanding the mathematics of circuit equations. For instance, we have found that it is difficult for students to recognize even simple series and parallel connections if they are masked (obscured) by the drawing of the electric circuit. For this reason, we have made a special effort to explain carefully these "psychologically" difficult topics. It is for the students to judge to what extent we have succeeded.

In writing this book, we have been assisted by several of our students and colleagues. In particular, C. Buehler aided us in the development of the first version of our lecture notes. D. Kerr and Chung Tse helped us in further modifications of our manuscript. Our colleagues, Professors T. Antonsen, N. Goldsman, and C. Striffler read our manuscript and provided us with their suggestions and constructive criticism. We are specially thankful to Professor C. Striffler for using our manuscript in his basic circuit theory class. Mrs. P. Keehn patiently and diligently typed several versions of our manuscript. We are very grateful to our students and colleagues mentioned above for their invaluable help in our work on this book.

Chapter 1

Basic Circuit Variables and Elements

1.1 Introduction

Before we discuss the equations which describe the operation of electric circuits, we must first review a few fundamental physical concepts that will be needed in our study and then define the basic elements which are the building blocks of electric circuits. We begin with basic circuit variables. The reader should have some familiarity with these circuit variables from physics courses on electromagnetism. These variables are the electric charge, electric and displacement currents, voltage and electric potential, electric energy and power, and magnetic flux linkage. Throughout this text we will always use the international system of units for these variables, which is denoted as SI or MKSA for meter-kilogram-second-ampere. The values of the physical quantities will range over many orders of magnitude, so we will make liberal use of the common multiplying factors and abbreviations as given in Table 1.1.

Afterward, we will introduce the concept of a two-terminal element and define the convention for reference directions. It is crucial that the reader adhere to this convention for every circuit problem she or he encounters; failure to do so may result

Table 1.1: Common multiplying factors.

Factor	Prefix name	Symbol	Factor	Prefix name	Symbol
10^{12}	tera	T	10^{-6}	micro	μ
10^{9}	giga	G	10^{-9}	nano	n
10^{6}	mega	M	10^{-12}	pico	p
10^{3}	kilo	k	10^{-15}	femto	f
10^{-3}	milli	m	10^{-18}	atto	a

in many embarrassing sign errors. The basic two-terminal elements which will be discussed below are the resistor, capacitor, inductor, ideal voltage source, and ideal current source. We will explain the operation of these circuit elements and will derive the *terminal relationships* between voltage and current for each element. We will also derive expressions for the energy stored and power dissipated in terms of each element's voltage, current, and physical characteristics. Understanding the terminal relationships for these elements is essential in order to master the material in the following chapters.

1.2 Circuit Variables

1.2.1 Electric Charge

Electric charge is one of the most fundamental quantities in physics. As such, it cannot be defined—it can only be described. Electric charge is a property of particles which manifests itself through what is termed the electromagnetic interaction. This electromagnetic interaction occurs between charged particles at a distance, without actual contact as in the case of mechanical interaction. As a result of the electromagnetic interaction, forces appear. These forces act on charged particles and have two distinct components. The first of these components is due to the instantaneous positions of the charged particles, while the second component is due to both the instantaneous positions and the velocities of the charged particles.

In order to describe the first component of the forces, the notion of the *electric field* is introduced. Each charged particle creates an electric field which is distributed in space and interacts with other charged particles. This interaction is called the Coulomb interaction and is characterized by Coulomb's law. To describe the second component of the interactive forces, the notion of the *magnetic field* is introduced. Magnetic fields are created by moving charged particles and exert forces on moving charged particles. By using the notions of electric and magnetic fields, interaction at a distance can be described.

The circuit notation for electric charge is $q(t)$, implying that the charge may vary with time. The MKSA unit for electric charge is the coulomb, denoted by C:

$$[q] = \text{C}. \tag{1.1}$$

Electric charges can be either positive or negative. For example, electrons are negatively charged particles, while protons are positively charged particles. The sign of the charge can be distinguished through the Coulomb interaction force, by which like charges repel and opposite charges attract. As a unit of electric charge, the coulomb is quite large in comparison to the elementary charges of an electron or a proton. One coulomb in comparison with the charge of one proton is

$$1\,\text{C} = 6.24 \times 10^{18} q_p, \tag{1.2}$$

where q_p is the positive charge of one proton. An electron has a charge equal in magnitude to a proton but opposite in sign, so that it takes approximately 6.24×10^{18} electrons to make -1 C.

The *principle of conservation of electric charge* is a fundamental property of nature. It states that in a closed system the total charge does not change with time. However, equal amounts of positive and negative charges can be simultaneously created or annihilated without disrupting the total balance of electric charge. This creation/annihilation phenomenon occurs only in very high energy systems; in our study of electric circuits we can assume that the total number of both positive and negative charges does not change with time.

1.2.2 Electric and Displacement Currents

By definition, electric charges in motion constitute an *electric current*. To characterize electric current quantitatively, we assume that a net charge $q(t)$ flows through an arbitrary surface S. Then the current through the (open) surface is defined to be the instantaneous time rate of change of the net charge flow through the surface (see Figure 1.1). This can be expressed mathematically as:

$$i(t) = \frac{dq(t)}{dt}. \tag{1.3}$$

By integrating equation (1.3) we can find the total charge through S in terms of electric current:

$$q(t) = q(t_0) + \int_{t_0}^{t} i(\tau)d\tau. \tag{1.4}$$

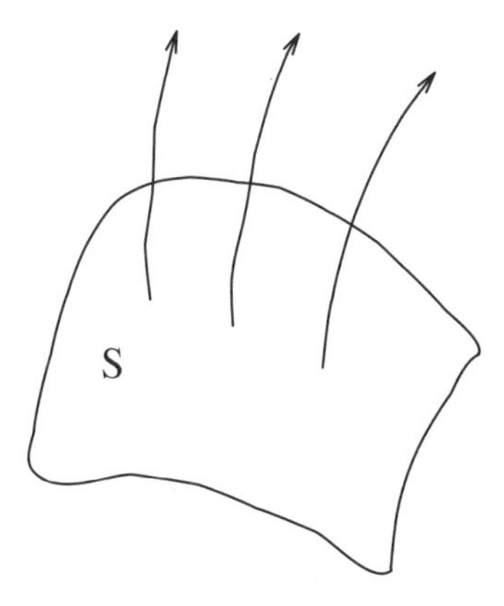

Figure 1.1: Surface S with current flow.

The MKSA unit of electric current is the ampere, which is defined as a coulomb per second:

$$[i] = \mathrm{A} = \mathrm{C/s}. \tag{1.5}$$

An electric current always creates a magnetic field because it is (by definition) the motion of electric charges.

EXAMPLE 1.1 Suppose that under steady-state conditions, 10^{12} electrons flow through a surface S every microsecond. What current does this represent?

According to the previous definition, we have:

$$I = dq/dt = \Delta q/\Delta t = \text{(electron charge)} \times \text{(\#electrons)/(time interval)} \tag{1.6}$$

$$I = -1.602 \times 10^{-19}\ \mathrm{C} \times 10^{12}/10^{-6}\ \mathrm{s} = -0.16\ \mathrm{A}. \tag{1.7}$$

■

In circuit theory, the electric current normally flows in metal wires, so the surfaces we construct to apply the above definition are often ones that simply cut through the wire. These cuts are called cross sections. For convenience, we often just drop the picture of the surface S altogether since the current flowing through the wire is a fairly straightforward concept.

In addition to the previously described electric current, there is another type of current which is not associated with the flow of electric charges. This current is called *displacement current*. Displacement currents occur due to the time variation of electric fields. These currents also create magnetic fields just as the currents due to the motion of electric charges do. As shown below, one important example of displacement current is the current through a capacitor. Displacement currents are also responsible for electromagnetic wave propagation through empty space.

Another very important concept is the *principle of continuity of current*, which states that the total current (electric + displacement) through any closed surface at any instant of time is always zero. If the currents entering the surface are taken with positive signs, and the currents leaving the surface are taken with negative signs, then the principle of continuity of current can be expressed mathematically as follows:

$$\sum_k i_k = 0. \tag{1.8}$$

From the principle of continuity of current, it is evident that displacement currents are continuations (extensions) of currents due to the motion of electric charges. For example, consider the case of a capacitor, shown in Figure 1.2 with the surface S enclosing one of the two plates of the capacitor. The principle of continuity of current asserts that the current entering S from the wire must equal the current leaving S on the right. But the only current leaving S on the right is due to the time variation of

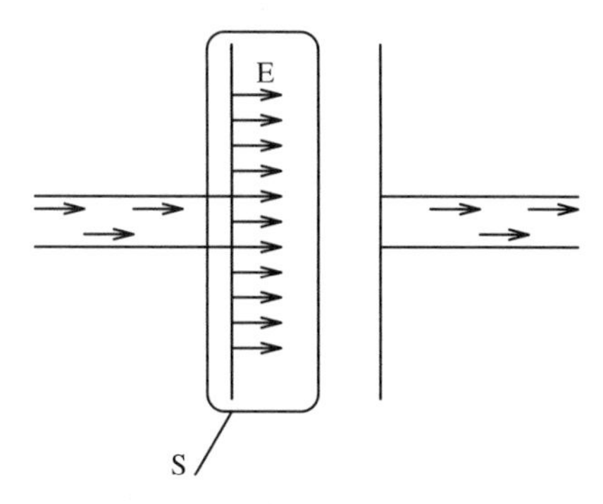

Figure 1.2: Capacitor with displacement current.

the capacitor's electric field, and therefore it is a displacement current. Thus, we can see that the displacement current is actually a continuation (extension) of the current through the wire, which is due to the motion of electric charges.

EXAMPLE 1.2 To illustrate the principle of continuity of current (1.8) consider the following example. There are currents i_1 and i_4 flowing through surface S into the volume in Figure 1.3 and i_2 and i_3 flowing out of the volume through separate wires. If $i_1 = 10$ A, $i_2 = 6$ A, and $i_3 = 7$ A, find i_4.

Formula (1.8) states that $i_1 - i_2 - i_3 + i_4 = 0$, so $i_4 = 6 + 7 - 10 = 3$ A. Note that if i_1 had been 15 A, then i_4 would have been -2 A. This means that, contrary to the picture, current is actually leaving S through the fourth wire. We will talk more about assumed (reference) directions for current later in this chapter. ■

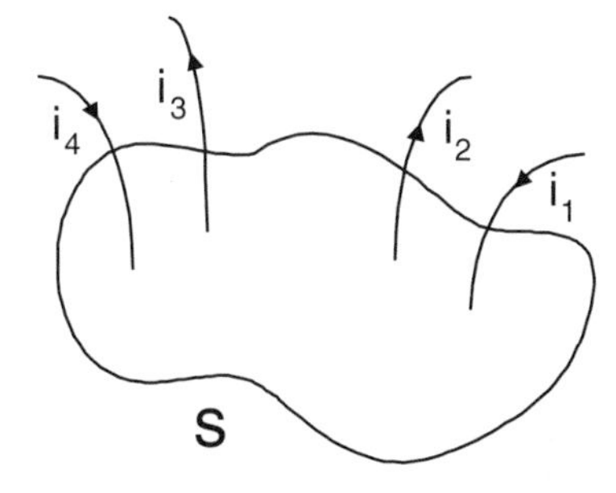

Figure 1.3: Example of conservation of current.

1.2.3 Electric Energy

In order to move electric charges and produce current, work must be performed against interaction forces, resulting in an expenditure of energy. This energy is called *electric energy*. Of course, this is only one of the many forms of energy, which include light energy, mechanical energy, heat energy, chemical energy, etc. However, electric energy has some attractive features which distinguish it from other forms of energy.

Electric energy is relatively cheap and easy to produce, because it can be centrally generated in large quantities at power plants. However, this first property would hardly be an advantage if electric energy were not so easily transmittable over large distances through power transmission lines to almost anywhere it is needed. Electric energy is also extremely versatile—it can be easily converted into other forms of energy, such as light energy or mechanical energy, or even be used to encode and process information.

The mathematical notation for energy (work) is $w(t)$ and the MKSA unit is the joule:

$$[w] = \text{J}. \tag{1.9}$$

1.2.4 Voltage

Voltage is normally discussed as existing between two different points (terminals). Consider two points P and Q in Figure 1.4. By definition, *voltage* (denoted $v(t)$) is the work done on a *unit* positive charge by moving it from point P to point Q.

Therefore,

$$w(t) = qv(t) \tag{1.10}$$

is the work done on moving an arbitrary charge q. From equation (1.10) we find

$$v(t) = \frac{w(t)}{q}. \tag{1.11}$$

The MKSA unit for voltage is the volt, defined as a joule per coulomb:

$$[v] = \text{J/C} = \text{V}. \tag{1.12}$$

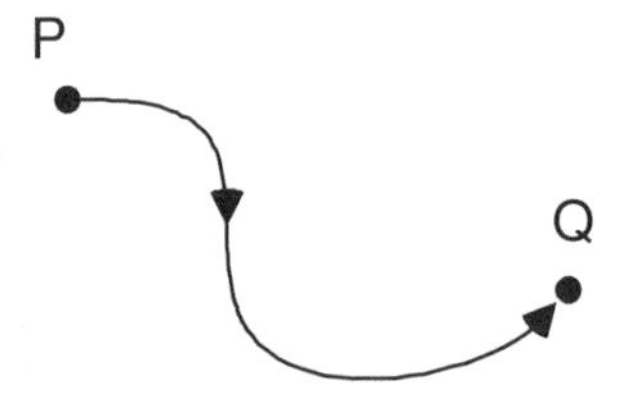

Figure 1.4: Diagram of an arbitrary path between two points.

An important property of voltage is its path independence. For static and quasi-stationary fields, the voltage is independent of the path along which the charge is moved. Thus, voltage will depend only on the position of the two endpoints of the path and can be expressed as a difference of the potentials at each point,

$$v(t) = \phi_Q(t) - \phi_P(t). \tag{1.13}$$

This aspect of voltage will be exploited later in the book when the method of nodal potentials is used for the analysis of electric circuits.

1.2.5 Electric Power

Electric power is defined as the rate of energy expenditure. Mathematically it means that the power $p(t)$ is given by:

$$p(t) = \frac{dw(t)}{dt}. \tag{1.14}$$

The MKSA unit of electric power is the *watt*, defined as a joule per second:

$$[p] = \mathrm{J/s} = \mathrm{W}. \tag{1.15}$$

The expression for power can be integrated with respect to time to find the energy:

$$w(t) = w(t_0) + \int_{t_0}^{t} p(\tau)d\tau. \tag{1.16}$$

A very important consideration to keep in mind is that there are no sources of infinite power. Thus, power is always *finite*. The importance of this fact becomes more evident later when the continuity of voltage across capacitors and the continuity of current through inductors are derived.

Now, by using the expression relating energy and voltage, we can derive the expression for electric power in terms of voltage and current. Consider a two-terminal[1] electric device (shown in Figure 1.5) to which a voltage $v(t)$ is applied and through which a current $i(t)$ flows. Note that the current is pictured in Figure 1.5 as flowing into the positive terminal and also that an equal current must be flowing out of the negative terminal. Then, in an infinitesimally small time period from t to $t + dt$, an infinitesimally small charge dq passes through the device. From the previous relationship between charge and work (equation (1.10)) we have

$$dw = v(t)dq, \tag{1.17}$$

[1] A terminal is a location on a device (usually a metal contact or wire) where connections are generally made to other devices.

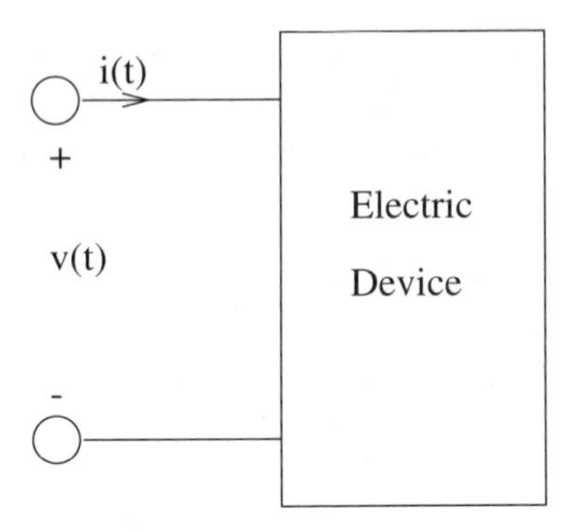

Figure 1.5: Electric device connected to a power source.

where dw is the infinitesimally small amount of work done on moving dq. Dividing both sides of equation (1.17) by dt, we have

$$\frac{dw}{dt} = v(t)\frac{dq}{dt}. \tag{1.18}$$

Recall from equations (1.3) and (1.14) that the time derivative of charge is the current $i(t)$, and the time derivative of work is the power $p(t)$, so

$$p(t) = v(t)i(t). \tag{1.19}$$

This expression for power is especially useful because voltage and current are the two most often encountered circuit variables and are readily measurable.

1.2.6 Flux Linkages

The concept of flux linkages will be crucial when we analyze inductors as circuit elements. To define *flux linkages* we first consider a simple one-turn coil of wire with a current passing through it (see Figure 1.6). As stated previously, the current in the wire creates a magnetic field, which can be represented by magnetic field lines. Recall that the right-hand rule gives the direction of the field lines. These magnetic field lines enclose the electric current and form the magnetic flux Φ which links the coil. This flux is defined mathematically to be the integral of the magnetic flux density $\vec{B}$ over the surface S bounded by the coil: $\Phi = \int_S \vec{B} \cdot d\vec{s}$. When the coil of wire has several closely spaced turns, the flux linkage is defined as

$$\Psi = N\Phi, \tag{1.20}$$

where N is the number of turns and Φ is the magnetic flux linking one turn.

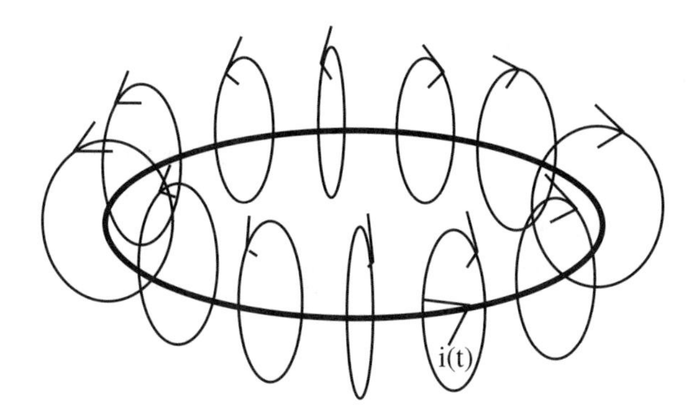

Figure 1.6: A one-turn coil carrying a current $i(t)$ and some resulting field lines.

Flux linkages are very important in the consideration of Faraday's law, which states that the time variation of flux linkage induces voltage. Mathematically, this is expressed as

$$v(t) = \frac{d\Psi}{dt}. \tag{1.21}$$

We can obtain an expression for flux linkages through integration of equation (1.21):

$$\Psi(t) = \Psi(t_0) + \int_{t_0}^{t} v(\tau)d\tau. \tag{1.22}$$

The MKSA unit for flux linkages is the weber:

$$[\Psi] = \text{Wb} = \text{V} \cdot \text{s}. \tag{1.23}$$

A collection of all the standard units which will be used in this book is given in Table 1.2.

Table 1.2: Basic circuit quantities and associated units.

Variable	Symbol	Unit	Notation
Charge	$q(t)$	coulomb	C
Current	$i(t)$	ampere	A
Energy	$w(t)$	joule	J
Voltage	$v(t)$	volt	V
Power	$p(t)$	watt	W
Flux linkage	$\psi(t)$	weber	Wb
Resistance	R	ohms	Ω
Capacitance	C	farads	F
Inductance	L	henries	H

1.3 Reference Directions

Most of the circuit elements that we will encounter in this text will have two terminals and can be represented schematically as in Figure 1.7. Examples of these elements include resistors, capacitors, inductors, and sources.

Each two-terminal element can be completely characterized by the voltage across the terminals of the element and the current through the element. In order to write meaningful equations relating the voltages and currents in electric circuits, it is necessary to assign a direction to the current and a polarity to the voltage for each two-terminal element. These assigned directions are called *reference directions*, and they are used in determining the signs given to the circuit variables when setting up Kirchhoff's equations (which are discussed in the following chapter).

It is important to stress that these reference directions are assigned entirely *arbitrarily*. The actual directions of the currents and the polarities of the voltages in a circuit are not generally known *a priori*.[2] They can be found only by analyzing the circuit, which (as we will soon demonstrate) cannot proceed until the reference directions for each element are assumed. Finally, we note that whereas reference directions for circuit variables are fixed, the actual currents and voltages often reverse their directions and polarities as the state of the circuit evolves with time. For example, in ac circuits (which are introduced in Chapter 3 and figure prominently in electrical engineering), voltages and currents *periodically* alter their actual polarities and directions.

In circuit notation, the reference direction for a current is indicated by an arrow (Figure 1.7a), while the reference polarity for a voltage is shown by using plus and

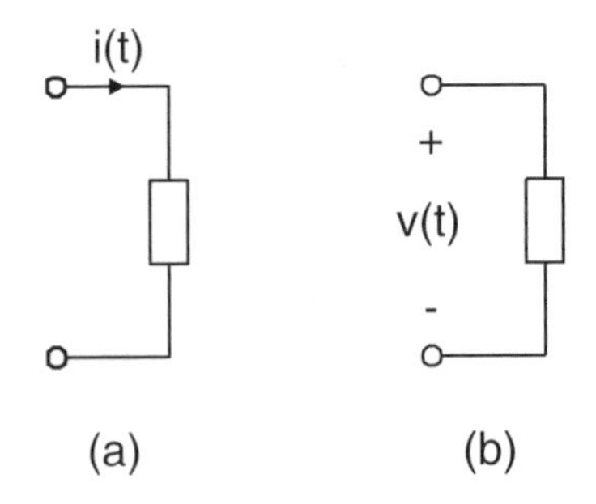

Figure 1.7: Two-terminal element with (a) reference direction for current and (b) reference polarity for voltage.

[2]As one gains familiarity with the analysis of electric circuits, inspection of a particular circuit layout will often seem to suggest the appropriate reference directions. While the student should be encouraged to make an educated guess as to the actual current directions and voltage polarities and to assign the reference directions accordingly, remember that ultimately it is the solution of Kirchhoff's equations that provides the correct answers.

minus signs placed next to element terminals (Figure 1.7b). Although they are chosen arbitrarily, the reference directions and polarities are coordinated for passive circuit elements such as resistors, inductors, and capacitors. This coordination means that the reference direction of the current is always chosen to be from the positive reference terminal to the negative reference terminal. By adopting this *passive sign convention*, we know that a positive value for the power $p(t) = v(t)i(t)$ always means that energy is flowing into the element at that instant in time.

Reference directions are used in order to write Kirchhoff's equations for currents and voltages. First, the arbitrary reference directions and polarities are assigned to all elements in the circuit. Then the circuit equations are set up by using these reference directions. These equations are solved and the signs of the circuit variables are found. If, after solving these equations, a current is found to be positive

$$i(t) > 0, \tag{1.24}$$

then the actual direction of that current at time t coincides with the reference direction. If, on the other hand, the current is negative

$$i(t) < 0, \tag{1.25}$$

then the actual direction of the current at time t is opposite to the reference direction.

The same rule applies to the reference polarity of the voltage. If the voltage is determined to be positive,

$$v(t) > 0, \tag{1.26}$$

then the actual polarity coincides with the reference polarity. If the voltage is found to be negative,

$$v(t) < 0, \tag{1.27}$$

then the actual polarity is opposite to the reference polarity. Thus, the reference directions allow one to write circuit equations, to solve them, and then to find the actual directions and polarities for circuit variables.

1.4 The Resistor

The circuit notation for a **resistor** is shown in Figure 1.8. The resistor is a two-terminal element which impedes the flow of electric current through it by converting some of the electrical energy to heat. The degree to which it impedes the current flow is characterized by *resistance*. Every normal conductor possesses this property to some extent. For an idealized resistor, the value of resistance is independent of the applied voltage and current and is a function only of the conductor geometry and physical properties of the materials from which it is made.

As stated previously, it is very important to know the terminal relationship between current and voltage for each of the circuit elements. For a resistor, this

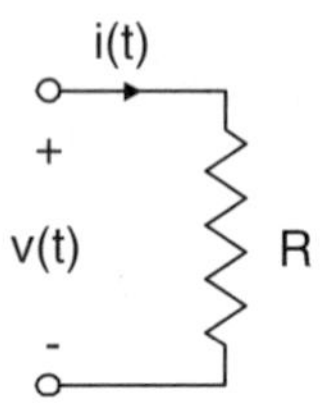

Figure 1.8: Circuit notation for a resistor.

relationship is given by Ohm's law:

$$v(t) = Ri(t). \tag{1.28}$$

Thus, this law states that the voltage drop across the resistor is directly proportional to the current through it. The MKSA unit for resistance is the ohm, denoted by Ω. The ohm is by definition a volt per ampere:

$$[R] = \mathrm{V/A} = \Omega. \tag{1.29}$$

In effect, resistance is a coefficient of proportionality between voltage and current, as indicated in Figure 1.9. Solving equation (1.28) for current, we have the equivalent relationship

$$i(t) = \frac{v(t)}{R}. \tag{1.30}$$

A resistor can also be characterized by a quantity called *conductance* G, defined as the reciprocal of the resistance:

$$G = \frac{1}{R}. \tag{1.31}$$

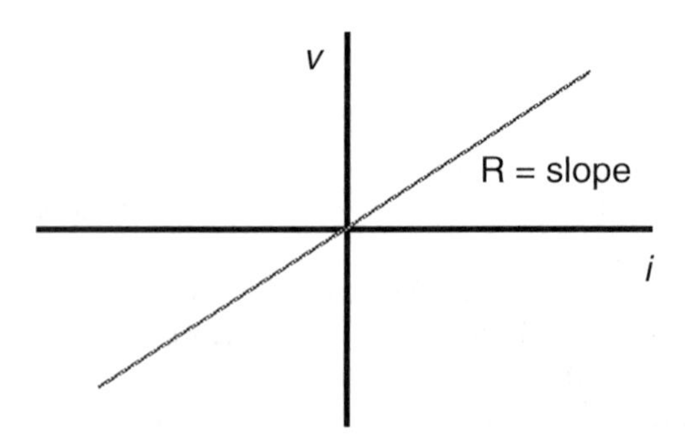

Figure 1.9: The voltage-current relationship for an ideal linear resistor.

In terms of conductance, equation (1.30) can be rewritten as

$$i(t) = Gv(t), \tag{1.32}$$

which is sometimes a more useful form of Ohm's law, depending on the particular application. The unit for conductance is the mho (ohm spelled backward). The mho is defined as an ampere per volt and is also sometimes referred to as a siemens (S):

$$[G] = \mathrm{A/V} = \mho. \tag{1.33}$$

We now recall that instantaneous power was derived earlier as $p(t) = v(t)i(t)$. From this equation and equation (1.28) we see that power can be expressed as:

$$p(t) = Ri^2(t). \tag{1.34}$$

The same formula can be rewritten in the form:

$$p(t) = Gv^2(t), \tag{1.35}$$

or equivalently as

$$p(t) = \frac{v^2(t)}{R}. \tag{1.36}$$

As shown by these equations, the instantaneous power in a resistor is always positive. This means that the resistor always consumes power—it can never give it back to the source. Therefore, the resistor is an energy-dissipating element. Because of this fact, resistors are often used to model irreversible losses of electric energy.

In circuit theory, it is customarily assumed that the values for resistances are given. However, when solving real-world problems, these values are not always known, and they must be computed (or measured) by the engineer in order to undertake a circuit analysis. The problem of the calculation of resistance belongs to electromagnetic field theory. However, in the case of a conductor of uniform cross section (shown in Figure 1.10), the following simple formula is valid:

$$R = \frac{L}{\sigma A}, \tag{1.37}$$

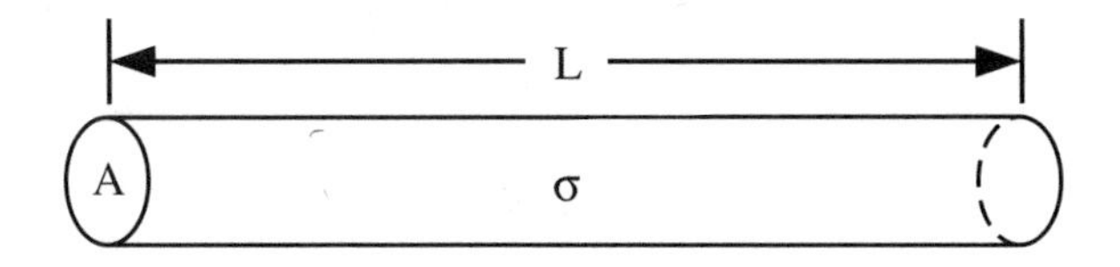

Figure 1.10: A uniform cross-sectional area conductor.

where L is the length of the conductor, A is the cross-sectional area, and σ is a physical constant called conductivity which depends on the material composition and temperature. From this equation we see that the resistance is directly proportional to length and inversely proportional to cross-sectional area. Practically, this means that longer wires have larger resistances while thicker wires have smaller resistances. However, if the cross section is not uniform it is very difficult to calculate the resistance, and we must invoke electromagnetic field theory to solve this problem.

Standard off-the-shelf resistor values range from the milliohm level up to the multi-megohm level. Several different methods of fabrication are required to span such a large range. In addition to the nominal resistance value, resistors are usually categorized by type of construction, power handling capability, and the tolerance of the resistance value. Fabrication materials for resistors include wire-wound metal film, carbon film, and carbon composition. The latter type are by far the most common in low-power packages. By varying the carbon concentration, a wide range of resistance values can be achieved in the same geometry. For example, standard 1/4 watt resistors come in cylindrical casings 0.635 cm long and 0.229 cm in diameter, while 2 watt resistors are 1.746 cm long and 0.794 cm in diameter. Standard tolerances for resistance values are $\pm 10\%$, $\pm 5\%$, and $\pm 1\%$ (tighter tolerance means higher cost). The carbon composition resistors come in standard values from 1.0 up to 9.1 in each decade from one ohm up to the multi-megohm level. The standard values as well as a color code scheme for identifying the values can be found in *Electronic Components and Measurements* by B. D. Wedlock and J. K. Roberge (Prentice-Hall, 1969). The basic idea is that each standard resistor has a resistance about 10% (or less) larger than the one below it. If one makes resistors with 10% tolerance, then in principle all values of resistances will be fabricated.

EXAMPLE 1.3 What is the resistance of a cylindrical carbon composition resistor that is $L = 1$ cm long, has a radius of $r = 1$ mm, and has a conductivity of $\sigma = 2 \times 10^4$ ℧/m?

From equation (1.37) we have:

$$R = L/\sigma A = L/\sigma \pi r^2 = 0.01/(2 \times 10^4 \times \pi \times (10^{-3})^2) = 0.16\ \Omega. \quad (1.38)$$

■

EXAMPLE 1.4 As an application of a resistor, consider a heater which is connected to some fixed voltage source (see Figure 1.11). The power of this heater (i.e., how much heat is produced) is given by the expression

$$p_{\text{heater}} = \frac{v^2(t)}{R_{\text{heater}}}, \quad (1.39)$$

and it is obvious that a smaller resistance is desired for a greater heating effect. Such heaters are common in many household items including toasters, incandescent

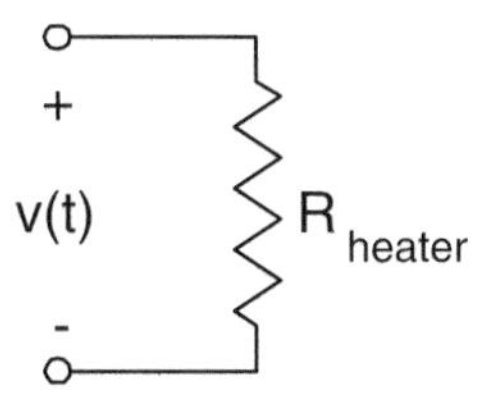

Figure 1.11: Simple diagram of a heater.

lightbulbs, electric ovens, electric clothes dryers, electric water heaters and furnaces, and hair dryers. ■

EXAMPLE 1.5 Consider the transmission of electric power through power lines (Figure 1.12). In this case, there is a fixed amount of power intended for transmission:

$$p_{\text{trans}} = v(t)i(t). \tag{1.40}$$

There is also some power that is lost due to the resistance R_{wire} of the transmission line. This power is given by the expression:

$$p_{\text{lost}} = R_{\text{wire}} i^2(t). \tag{1.41}$$

Therefore, to minimize the power loss, we have two alternatives:

1. Reduce R_{wire}, which, according to (1.37), requires an increase in the cross-sectional area A. This makes for a very expensive alternative.
2. Make the current $i(t)$ as small as possible. Since p_{trans} is fixed, the voltage $v(t)$ must be high in order to transmit the same power with the desired small current.

The latter approach is the basic idea behind high-voltage power transmission lines. ■

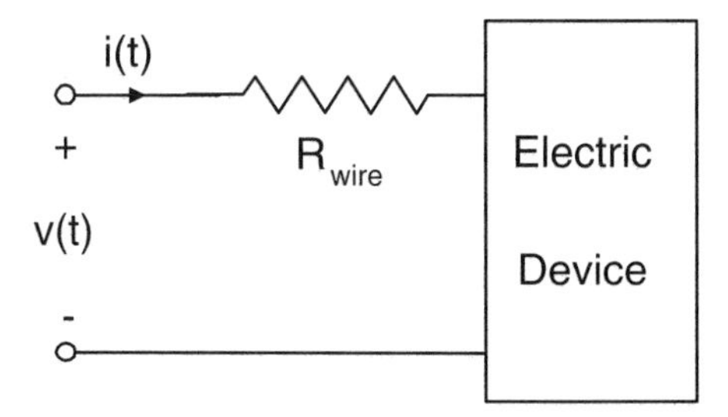

Figure 1.12: Simple diagram of power transmission.

1.5 The Inductor

An **inductor** is a two-terminal element which stores energy in magnetic fields and is characterized by *inductance*. The circuit notation for an inductor is given in Figure 1.13. In practice, an inductor is a coil of wire consisting of many closely spaced turns. In order to enhance the inductance of the coil, an iron or ferrite core is usually inserted because of the high magnetic permeability of these substances. Inductance, denoted by L, is by definition the ratio of total magnetic flux linking the inductor turns to the current flowing through the inductor. This can be written mathematically as follows:

$$L = \frac{\Psi(t)}{i(t)}. \tag{1.42}$$

However, the inductance does not depend on either the flux linkage or the current—it depends solely on the design of the inductor itself. A general formula for inductance is too complicated for this text; however, it can be shown that inductance is determined by the square of the number of turns, the geometric dimensions of the inductor, and the magnetic permeability of the core. In effect, inductance is a coefficient of proportionality between flux linkage and current:

$$\Psi(t) = Li(t). \tag{1.43}$$

The MKSA unit of inductance is the henry,

$$[L] = \mathrm{H} = \frac{\mathrm{Wb}}{\mathrm{A}} = \frac{\mathrm{V} \cdot \mathrm{s}}{\mathrm{A}} = \Omega \cdot \mathrm{s}. \tag{1.44}$$

To determine the relationship between current and voltage in the inductor, we use Faraday's law:

$$v(t) = \frac{d\Psi}{dt}. \tag{1.45}$$

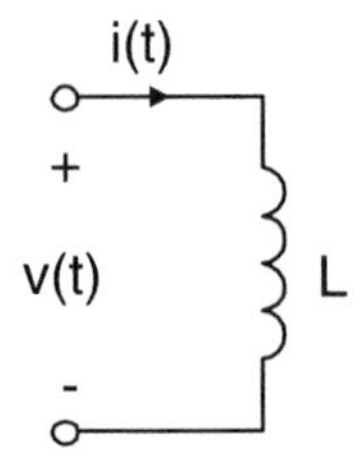

Figure 1.13: Circuit notation for an inductor.

By substituting (1.43) into (1.45) and using the product rule for differentiation, we find:

$$v(t) = \frac{dL}{dt}i(t) + L\frac{di(t)}{dt}. \tag{1.46}$$

In this text, inductance is assumed to be constant (time invariant). Consequently, the last equation leads to the following terminal relationship:

$$v(t) = L\frac{di(t)}{dt}. \tag{1.47}$$

In the special case of dc (direct current), there is no time variation of $i(t)$, and consequently, there is no voltage induced because the flux linkage is constant.

Knowing the voltage across the inductor, we can derive an expression for current through it by integrating equation (1.47):

$$i(t) = i(t_0) + \frac{1}{L}\int_{t_0}^{t} v(\tau)d\tau. \tag{1.48}$$

To derive an expression for power delivered to the inductor, we use the expression for instantaneous power in terms of voltage and current and equation (1.47):

$$p(t) = v(t)i(t) = Li(t)\frac{di(t)}{dt}. \tag{1.49}$$

From this equation we can conclude that the current through an inductor must be a continuous function of time. To draw this conclusion, we must recall an earlier statement that there are no sources of infinite power. Because power can never be infinite, $di(t)/dt$ is never infinite as well, which means that $i(t)$ must always be differentiable. And from calculus we know that differentiability implies continuity. Mathematically, this means that the value of the current immediately before an instant of time t_0 must equal the value of the current immediately after t_0:

$$i(t_{0+}) = i(t_{0-}). \tag{1.50}$$

This is the principle of continuity of electric current through an inductor. This principle will be extensively used throughout the text.

By using the chain rule for differentiation in equation (1.49), power can also be expressed as:

$$p(t) = \frac{L}{2}\frac{d(i^2(t))}{dt}. \tag{1.51}$$

Thus, we can conclude that power is positive if the square of the current increases with time and that it is negative if the square of the current decreases with time. Mathematically, if

$$\frac{d(i^2(t))}{dt} > 0, \tag{1.52}$$

then the absolute value of the current is monotonically increasing with time and the power is positive. However, if

$$\frac{d(i^2(t))}{dt} < 0, \tag{1.53}$$

then the absolute value of the current is monotonically decreasing with time and the power is negative.

From the above discussion we conclude that, unlike the resistor, the inductor is an energy storage element and can give energy back as well as absorb and store it. If the absolute value of the current is increasing with time, then power is being consumed by the inductor and energy is being stored in the inductor's magnetic field. If the absolute value of the current is decreasing with time, then power is given back at the expense of the energy previously stored in the magnetic field. This suggests a practical application of inductors for energy storage. However, since inductors are made from wires, practical inductors often have a resistance associated with them and this leads to some energy dissipation into heat. We will usually ignore the problem by just assuming that the inductor wire has $\sigma \rightarrow \infty$. This is actually the case in superconducting materials, when the materials are cooled to the point where they offer no resistance to current flow. The energy storage application is especially attractive when superconducting wires are used, since there is no power loss associated with the circulating current in the inductor.

Recalling that power is the time derivative of energy, and using this in equation (1.51), we can determine the energy stored in the magnetic field of the inductor by integration. Assuming that at zero time we have

$$w(0) = i(0) = 0, \tag{1.54}$$

we find through integration of (1.51) that the energy stored in the inductor is

$$w(t) = \frac{Li^2(t)}{2}. \tag{1.55}$$

EXAMPLE 1.6 The current through a 10 mH inductor is given (in amperes) by $i(t) = 2t^2e^{-t/10}$ when $t > 0$. Find (a) the voltage across the inductor, (b) the power absorbed by the inductor, and (c) the energy stored in the inductor.

(a)
$$\begin{aligned} v(t) &= L\frac{di}{dt} = .04te^{-t/10} - .002t^2e^{-t/10} \text{ V} \\ &= .04te^{-t/10}(1 - t/20) \text{ V}. \end{aligned}$$

(b)
$$\begin{aligned} p(t) &= v(t)i(t) = .04te^{-t/10}(1 - t/20)2t^2e^{-t/10} \text{ W} \\ &= .08t^3e^{-t/5}(1 - t/20) \text{ W}. \end{aligned}$$

(c)
$$w(t) = \frac{Li^2(t)}{2} = .02t^4e^{-t/5} \text{ J}.$$

■

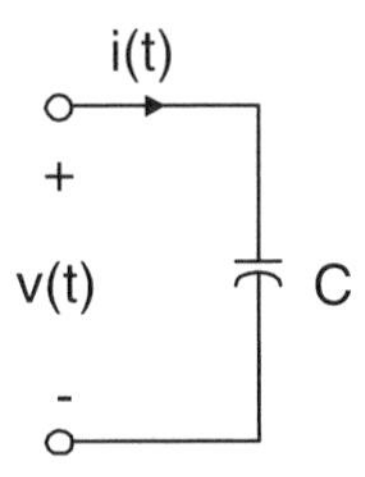

Figure 1.14: Circuit notation for a capacitor.

1.6 The Capacitor

A **capacitor** is a two-terminal element which stores energy in the electric field and is characterized by *capacitance*, denoted C. Its circuit notation is shown in Figure 1.14.

A capacitor is formed by two *conductors* separated by some distance. Sometimes a dielectric material may be placed between the conductors to provide a means of separation and to enhance the capacitance. Both conductors of the capacitor possess charges of equal magnitude but opposite sign (see Figure 1.15):

$$q_1 = -q_2 = q. \tag{1.56}$$

These charges create an electric field between the two conductors. We wish to relate the voltage across the capacitor to the charge of the capacitor. Recall that voltage is usually specified as the potential difference between two points. However, since each conductor under static (or quasistatic) conditions has the same potential at every point, we do not need to specify two points to determine voltage—we can just speak of the voltage between the two conductors. We now define *capacitance* as the ratio between the charge and this voltage:

$$C = \frac{q(t)}{v(t)}. \tag{1.57}$$

Although the capacitance is the ratio of charge to voltage, it does not depend on either of these two variables. Capacitance is determined only by the geometric

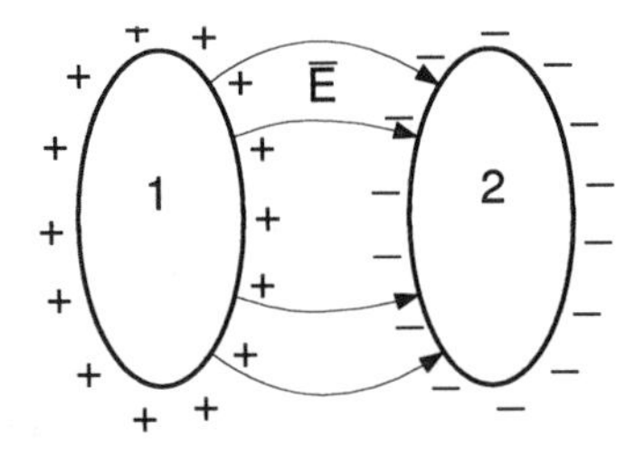

Figure 1.15: Two conductors with the corresponding electric field.

dimensions of the conductors, the separation distance between the conductors, and the permittivity of the dielectric material between the conductors. In this respect, the capacitance serves as a constant of proportionality between charge and voltage, much as inductance is between flux linkages and current:

$$q(t) = Cv(t). \tag{1.58}$$

Physically, capacitance determines the ability of the capacitor to accumulate charge. Thus, the larger the capacitance, the greater the charge a capacitor can store for the same applied voltage.

The MKSA unit for capacitance is the farad, defined as a coulomb per volt:

$$[C] = \mathrm{F} = \frac{\mathrm{C}}{\mathrm{V}} = \frac{\mathrm{A} \cdot \mathrm{s}}{\mathrm{V}} = \frac{\mathrm{s}}{\Omega}. \tag{1.59}$$

A farad is a huge unit of capacitance and in practice we usually use micro-, nano-, and picofarads.

In order to obtain the terminal relationship between voltage and current in a capacitor, we use the product rule to differentiate equation (1.58) with respect to time:

$$i(t) = \frac{dq}{dt} = C\frac{dv(t)}{dt} + \frac{dC}{dt}v(t). \tag{1.60}$$

Assuming capacitance is not a function of time (its derivative is equal to zero), we arrive at the terminal relationship:

$$i(t) = C\frac{dv(t)}{dt}. \tag{1.61}$$

From this equation we clearly see that ac voltage (alternating voltage) can cause a current through a capacitor because this voltage is time varying. However, dc voltage cannot cause a current because it is a constant function of time. This is consistent with the notion that the current through a capacitor is a displacement current which exists only when the electric field in the capacitor varies with time.

We can derive an expression for voltage by integrating equation (1.61):

$$v(t) = v(t_0) + \frac{1}{C}\int_{t_0}^{t} i(\tau)d\tau. \tag{1.62}$$

We also find the expression for instantaneous power delivered to a capacitor:

$$p(t) = v(t)i(t) = Cv(t)\frac{dv(t)}{dt}. \tag{1.63}$$

This last equation shows that the voltage across a capacitor must be a continuous function of time because power can never be infinite. Indeed, the voltage must be differentiable and differentiability implies continuity. From this we find that the voltage immediately before any instant of time t_0 must equal the voltage immediately after this instant of time:

$$v(t_{0+}) = v(t_{0-}). \tag{1.64}$$

This is the principle of continuity of voltage across a capacitor, a principle which will be extensively used throughout the text.

By using the chain rule for differentiation in equation (1.63), we derive another expression for power:

$$p(t) = \frac{C}{2}\frac{d(v^2(t))}{dt}. \tag{1.65}$$

From this equation we see that the power is positive when the square of the voltage is monotonically increasing,

$$\frac{d(v^2(t))}{dt} > 0, \tag{1.66}$$

and it is negative when the square of the voltage is monotonically decreasing,

$$\frac{d(v^2(t))}{dt} < 0. \tag{1.67}$$

This suggests that the capacitor is also an energy storage element; however, it stores energy in the electric field. When the power is positive, energy is consumed and stored in the electric field, and when the power is negative, the energy is given back from the electric field. This also suggests that capacitors can be utilized for energy storage, as is the case for many pulse power applications. It is important to stress that energy storage in capacitors is not usually associated with energy loss as in the case of inductors. This is because dielectrics are much closer to ideal than are conducting wires.

To find the energy stored in the capacitor in terms of voltage, we integrate equation (1.65) by using the (assumed) initial conditions

$$w(0) = v(0) = 0 \tag{1.68}$$

and derive that

$$w(t) = \frac{Cv^2(t)}{2}. \tag{1.69}$$

The capacitance is usually a difficult quantity to calculate, and this is done by using electromagnetic field theory. However, there are some widely used expressions for a few standard configurations. Consider the capacitor consisting of two parallel plates separated by a dielectric slab of permittivity $\epsilon = \epsilon_r \epsilon_0$, where $\epsilon_0 = 8.854 \times 10^{-12}$ F/m is the permittivity of vacuum and ϵ_r is the relative dielectric constant. If the plate area is A and the plate separation is d, the capacitance is given by

$$C = \frac{\epsilon A}{d}. \tag{1.70}$$

Standard off-the-shelf capacitance values range from the picofarad level up to the thousands of microfarads level. The parallel plate configuration cannot be used to span the entire range. There is a practical limit to how large the plate area A can be and a physical limit to how large the permittivity ϵ can be. Likewise, if you make the separation distance d too small, the capacitor will not be able to sustain high voltage. In addition to the nominal capacitance value, capacitors are usually categorized by type of construction, voltage handling capability, and the tolerance of the capacitance value. Fabrication methods for capacitors include parallel plates with ceramic or mica dielectrics, electrolytic, and rolled paper/foil. Standard tolerances for capacitance values are $\pm 10\%$, $\pm 5\%$, and $\pm 1\%$.

EXAMPLE 1.7 What is the capacitance of a system of two circular disks of radius $r = 1$ cm separated by a 1 mm thick piece of alumina ($\epsilon_r = 6$)?

From equation (1.70) we have:

$$\begin{aligned} C &= \epsilon A/d = \epsilon \pi r^2/d = 6 \times 8.854 \times 10^{-12} \times \pi \times (10^{-2})^2/10^{-3} \qquad (1.71)\\ &= 16.68 \text{ pF}. \end{aligned}$$

■

EXAMPLE 1.8 The current through a 1 μF capacitor is shown in Figure 1.16. Assuming that the initial capacitor voltage is $v(0) = 0$, plot the instantaneous power and electric energy stored in the capacitor.

The current indicated in the figure can be expressed as:

$$i(t)\ (\text{mA}) = \begin{cases} 2 & 0 \le t < 2 \text{ ms}, \\ 0 & 2 \le t < 3 \text{ ms}, \\ -2 & 3 \le t < 4 \text{ ms}, \\ t-6 & 4 \le t < 6 \text{ ms}, \\ 2 & t \ge 6 \text{ ms}. \end{cases} \tag{1.72}$$

We can obtain the capacitor voltage by integrating [see equation (1.62)]:

$$v(t)\ (\text{V}) = \begin{cases} 2t & 0 \le t < 2 \text{ ms}, \\ 4 & 2 \le t < 3 \text{ ms}, \\ 4-2(t-3) & 3 \le t < 4 \text{ ms}, \\ (t-6)^2/2 & 4 \le t < 6 \text{ ms}, \\ 2(t-6) & t \ge 6 \text{ ms}. \end{cases} \tag{1.73}$$

The power is found by using (1.72) and (1.73) in equation (1.19):

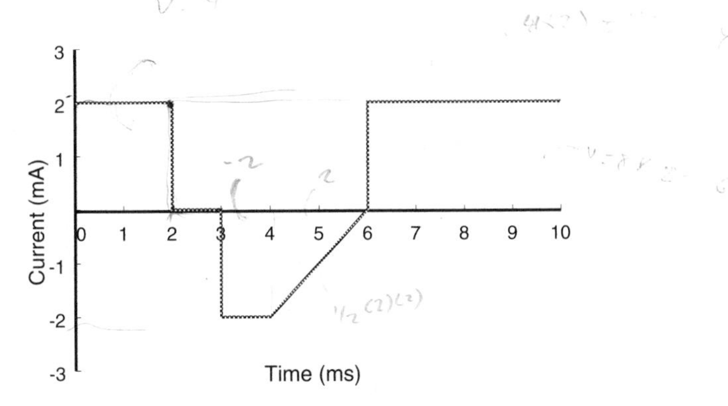

Figure 1.16: The capacitor current.

$$p(t)\,(\text{mW}) = \begin{cases} 4t & 0 \le t < 2 \text{ ms}, \\ 0 & 2 \le t < 3 \text{ ms}, \\ -2(10-2t) & 3 \le t < 4 \text{ ms}, \\ (t-6)^3/2 & 4 \le t < 6 \text{ ms}, \\ 4t-24 & t \ge 6 \text{ ms} \end{cases} \tag{1.74}$$

and is plotted in Figure 1.17a. The total stored energy comes by integrating the power:

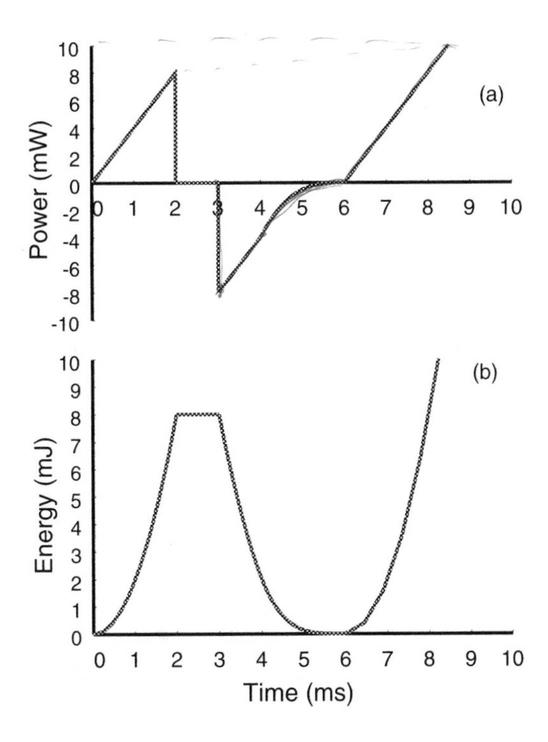

Figure 1.17: Physical quantities for the capacitor: (a) power and (b) energy.

$$w(t)(\mu\text{J}) = \begin{cases} 2t^2 & 0 \le t < 2\text{ ms}, \\ 8 & 2 \le t < 3\text{ ms}, \\ (10-2t)^2/2 & 3 \le t < 4\text{ ms}, \\ (t-6)^4/8 & 4 \le t < 6\text{ ms}, \\ (4t-24)^2/8 & t \ge 6\text{ ms}. \end{cases} \tag{1.75}$$

This result for the energy is shown in Figure 1.17b. ■

1.7 Ideal Independent Voltage and Current Sources

By definition, an **ideal voltage source** is a two-terminal element with the property that the *voltage* across the terminals is specified at every instant in time. This voltage does not depend on the current through the source. That is, any current in any direction could possibly flow through the source. This current will be determined solely by the circuit elements connected to this source and it can be computed according to the rules which we will learn in the next chapter. Depending on the actual direction of the current through the source, the voltage source can either provide power or absorb it. The circuit notation for an ideal voltage source is given in Figure 1.18a. If the specified voltage is a constant, we often refer to the source as a battery and use the symbol shown in Figure 1.18b.

An **ideal current source** is by definition a two-terminal element with the property that the *current* flowing through the device is specified at every instant in time. This current does not depend on the voltage across the source. The voltage will be determined solely by the circuit elements which are connected to this source and it can be computed according to the rules which we will learn in the next chapter. Depending on the actual polarity of the voltage across the source, the current source can either provide or absorb power. The circuit notation for an ideal current source is given in Figure 1.19.

All sources fall into two main categories: nonperiodic and periodic. The non-periodic sources will be treated in Chapter 7 when we study transient analysis. For

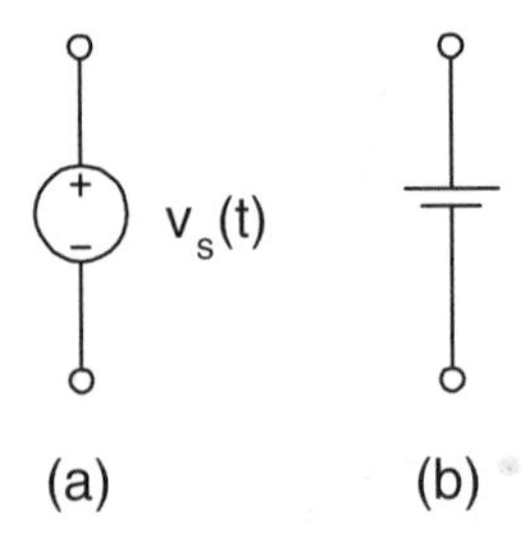

Figure 1.18: Circuit notation for an ideal voltage source: (a) general symbol and (b) symbol for a battery.

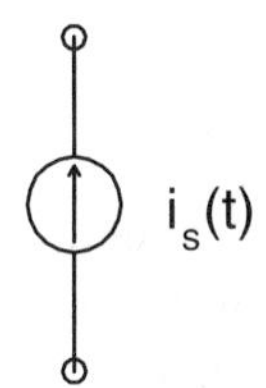

Figure 1.19: Circuit notation for an ideal current source.

example, such a source could be described by exponential or polynomial time dependences. Examples of four different periodic source functions are given in Figure 1.20. The ac or sinusoidal source will be the key source that we analyze in detail starting in Chapter 3. The dc source can be treated as a special case of the ac source when the period approaches infinity (and the frequency approaches zero). Finally, the more complicated periodic sources like the ramp (sawtooth) source and the pulse (square wave) are often encountered in applications.

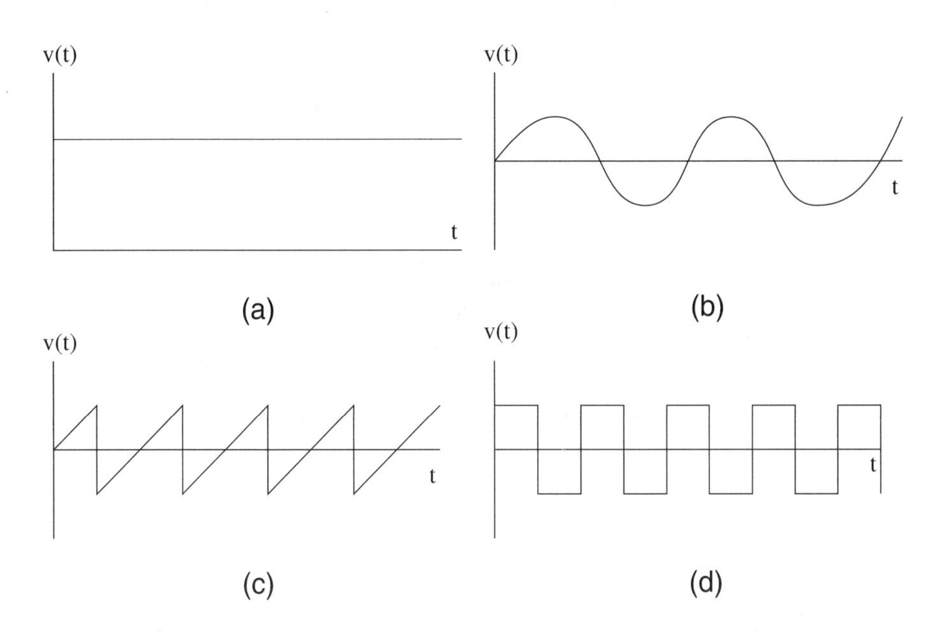

Figure 1.20: (a) dc source, (b) ac source, (c) ramp source, and (d) pulse source.

1.8 Summary

In this chapter we have first discussed the key circuit variables from an electromagnetic theory point of view. The convention for assigning the assumed (reference) directions for voltages and currents in passive two-terminal elements has also been presented.

We have then introduced the three principal passive two-terminal elements (the resistor, the inductor, and the capacitor) and discussed the connections between the circuit variables for each. There is a linear dependence between two of the circuit variables for each element, with the proportionality constant being determined by the geometry and the material composition of the element. For the resistor, the ratio of voltage to current is the resistance R. For the inductor, the inductance L is the ratio of flux linkages to the current. For the capacitor, the ratio of the charge to the voltage is the capacitance C.

Of primary importance are the terminal relationships which describe the connections between voltage and current for each element. If we examine the terminal relationships for capacitors and inductors from a purely mathematical standpoint, we can see that these equations reveal some duality. That is, the equation for the inductor is mathematically identical to that of the capacitor, with the roles of voltage and current being reversed (and interchanging C and L). This parallelism is useful to note and may prove helpful in remembering these equations.

Furthermore, we have introduced the ideal independent voltage and current sources. We have also briefly mentioned the types of source functions that we will investigate in the upcoming chapters. Because a change in current through an ideal voltage source doesn't affect the voltage, we often say that an ideal voltage source has zero internal resistance. Likewise, we say that an ideal current source has infinite internal resistance or zero internal conductance because a change in the applied voltage does not affect the current.

Finally, we point out that all five elements that have been studied so far are *ideal* elements. Our treatment in this chapter of resistors, inductors, capacitors, and sources ignores some small parameters and other deviations from ideal behavior which occur for real-world circuit elements. For example, some capacitors and resistors have measurable inductances and most inductors have some resistance. Furthermore, many of the devices exhibit nonlinear effects when some parameters become large. Nonideal circuit elements and their models as well as nonlinear resistors will be treated in Chapter 5.

1.9 Problems

1. Convert the following derived units to the four basic SI units (m, kg, s, A): (a) joules, (b) coulombs, (c) volts, (d) webers.

2. Express the following quantities in terms of multiplying factors: (a) 2.54×10^{11} electrons, (b) -5.63×10^{-14} C, (c) 0.0035 Wb, (d) 0.00000915 A.

3. Find the number of electrons required to generate a total charge of: (a) -1.6×10^{-7} C, (b) -1.17×10^{-17} C, (c) 12 μC, (d) -3.73 pC.

4. If a $-3\ \mu$A current flows through a surface S, how many electrons pass through S in (a) 1 s, (b) 3 μs, (c) 53.4 fs?

5. If the current flowing through a surface S has the time variation shown in Figure P1-5, plot the time dependence of net charge accumulating on the other side of S. Assume zero initial charge.

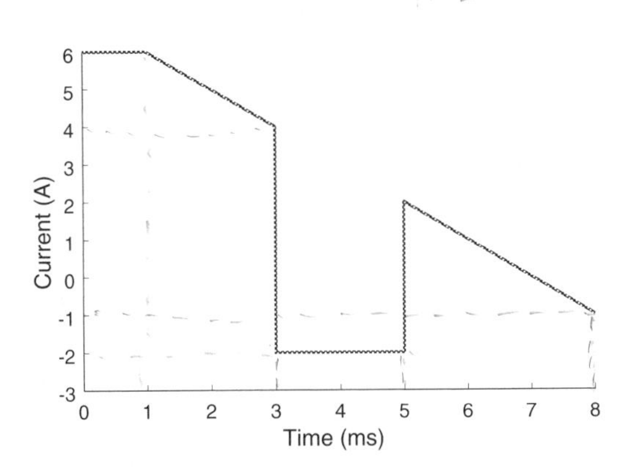

Figure P1-5

6. The total charge in a volume V as a function of time is shown in Figure P1-6. Plot the time variation of electric current leaving V.

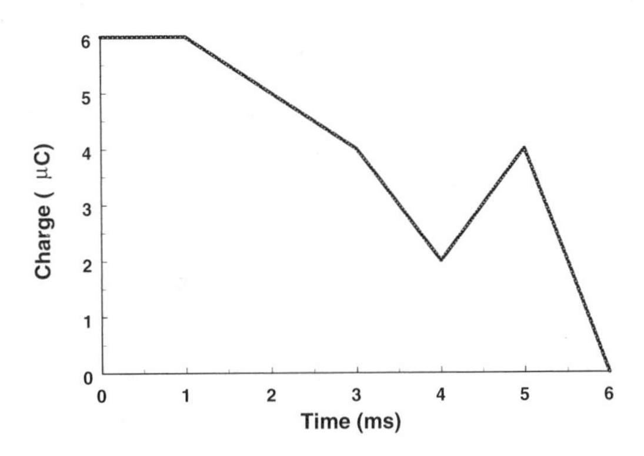

Figure P1-6

7. The time dependences of the voltage across and the current through a two-terminal electric device are shown in Figure P1-7. Plot the power flowing into the device and the total stored energy as a function of time. Assume $w(0) = 0$.

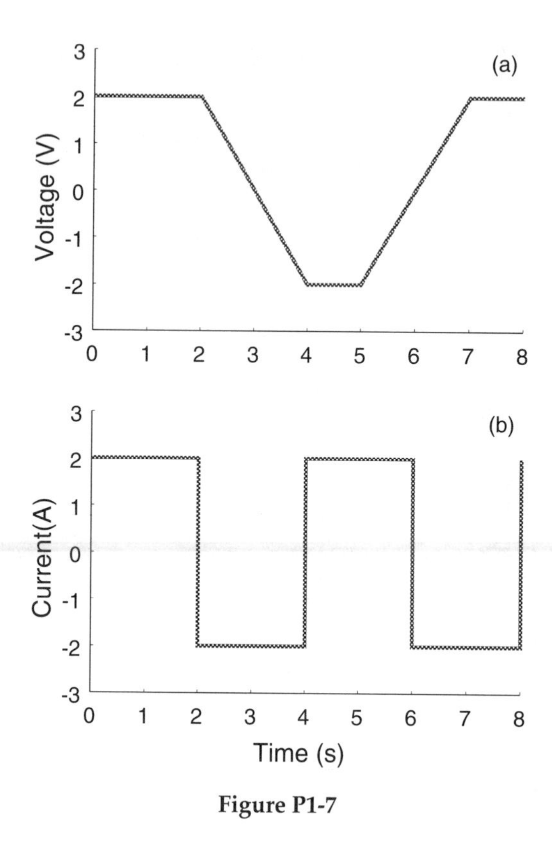

Figure P1-7

8. The voltage developed across a 12-turn coil of wire is shown in Figure P1-8. Plot the time variation of the magnetic flux linking one turn of the device. Assume zero initial flux.

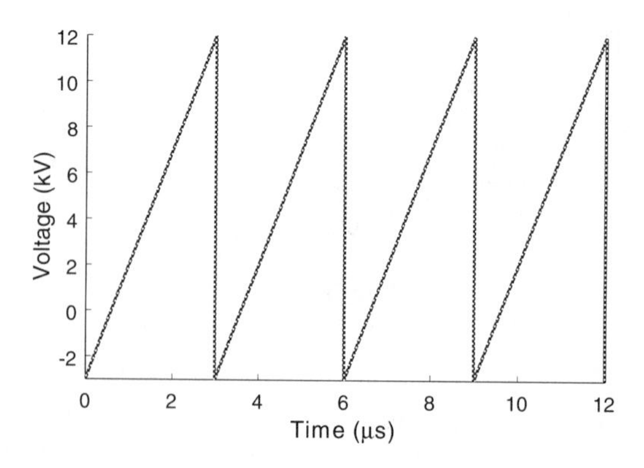

Figure P1-8

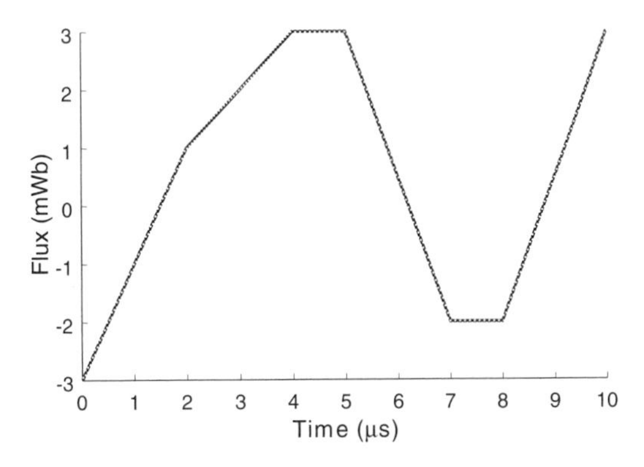

Figure P1-9

9. The total magnetic flux linking a circuit is shown in Figure P1-9 as a function of time. Plot the resulting voltage developed across the device.

10. The current through a circuit is given by $i(t) = (3 + t)$ A. Determine the flux (at time t) that must be linked through the device if the power consumed is a constant 1 W. Assume zero initial flux at $t = 0$.

11. If the current through a closed surface is given by $i(t) = 1 + 5t^2 + \cos(3t)$ A, what is the expression for the charge accumulating in the volume? Assume $q(0) = 0$.

12. If the voltage across a two-terminal device is $v(t) = te^{-t}$ V and the current through it is $i(t) = (1 - t)e^{-t}$ A, plot the energy stored in the device as a function of time (for $t > 0$). Assume $w(0) = 0$.

13. Calculate the nominal resistance of the following automotive hardware (which run off a 12 V battery): (a) a 12 W dome light, (b) a 50 W horn, and (c) a 300 W headlight.

14. To reach a new customer, the local electric company must transport five miles a current of $i(t) = 200\cos(377t)$ amperes. What is the required diameter of a copper wire that would keep the average power transmission losses below 2 kW? The conductivity of copper is $\sigma = 5.8 \times 10^7\ (\Omega\text{m})^{-1}$.

15. Assume you have a carbon composition resistor with a conductivity of $\sigma = 2.0 \times 10^3 (\Omega\text{m})^{-1}$. If you use this material to make 2 cm long resistors, what is the required diameter to produce a resistance of: (a) 4 mΩ, (b) 1 Ω, (c) 1 kΩ, and (d) 1 MΩ? Which of these designs do you think are practical ones? How might you actually construct the impractical designs, if any?

16. Calculate the resistance value for each of the following cases of a circular-cross-section wire:

Resistor	Conductance $(\Omega\text{m})^{-1}$	Length (cm)	Radius (mm)
a	5	1	2
b	2×10^3	2	1
c	5.8×10^7	100	.003

17. Find the average power absorbed by a 20 Ω resistor when a current of $i(t) = 3/4\sin(3t)$ A is applied.

18. A 10 μH inductor is driven by a current $i(t) = 30\cos(500t)$ mA. Calculate the power supplied to the inductor.

19. The voltage across a 2 mH inductor is shown in Figure P1-19. Plot the current flowing through the inductor assuming an initial current of 10 mA.

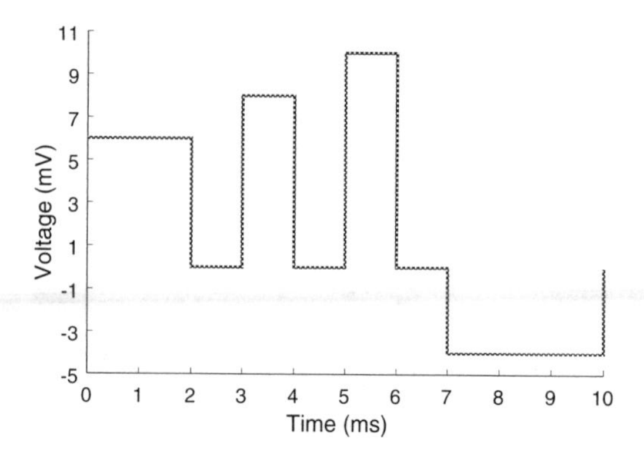

Figure P1-19

20. The current flowing through a 39 μH inductor is shown in Figure P1-20. Plot the power in the inductor as a function of time.

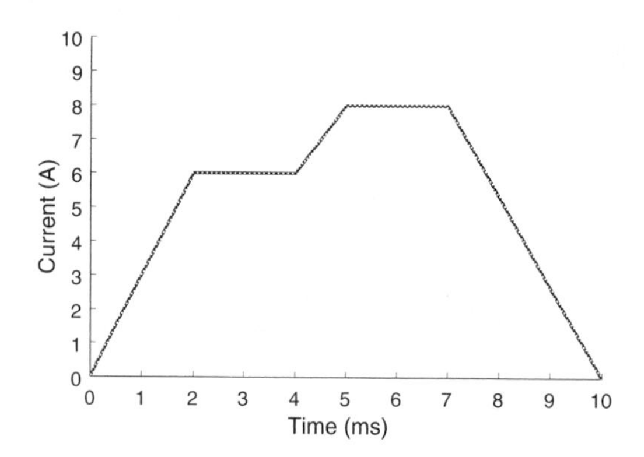

Figure P1-20

21. The current through a 91 μH inductor is given by $i(t) = 7e^{-t/20}$ A. Find expressions for the flux linked by the inductor and the voltage across the inductor. What is the total charge that passes through the inductor from $t = 0$ to $t = \infty$? Assume $q(0) = 0$.

22. Calculate the capacitance for each of the following cases of a parallel plate capacitor:

Capacitor	Relative dielectric constant ε_r	Plate gap (mm)	Radius (cm)
a	1	2	1
b	2.3	.1	4
c	32.4	.003	10

23. The voltage across a 10 μF capacitor is shown in Figure P1-23. Plot the power supplied to the capacitor as a function of time.

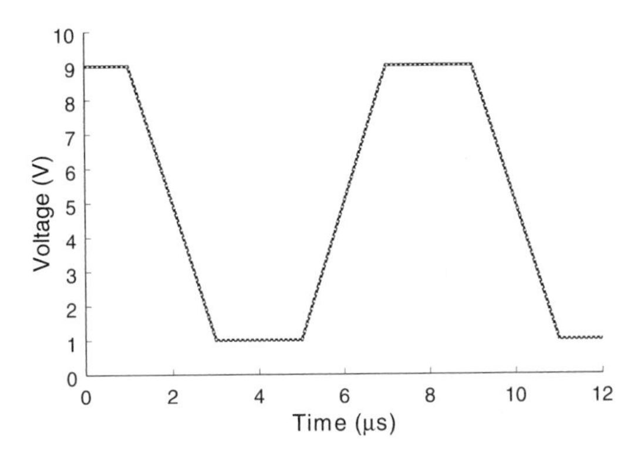

Figure P1-23

24. A 47 pF capacitor is driven by a voltage $v(t) = 1 + 6e^{-2t}$ V. Calculate the power supplied to the capacitor.

25. The current through a 20 μF capacitor is given by $i(t) = 2e^{-3t}$ A for $t > 0$. Find expressions for the voltage, accumulated charge, instantaneous power, and electric energy stored in the capacitor. Assume zero initial charge.

26. The voltage across a 5 nF capacitor is given by $v(t) = 1 + t + 3t^2$ V for $t > 0$. Find expressions for the current, accumulated charge, instantaneous power, and electric energy stored in the capacitor.

27. The voltage across a 5 mH inductor is given by $v(t) = 3t^2 + 5t^4$ V for $t > 0$. Find expressions for the current, magnetic flux, instantaneous power, and electric energy stored in the inductor. Let $i(0) = 0$.

28. The current through a 33 μH inductor is given by $i(t) = 20t\cos(30t)$ mA for $t > 0$. Find expressions for the voltage, magnetic flux, instantaneous power, and electric energy stored in the inductor.

29. The current through a 1 kΩ resistor is given by $i(t) = 2e^{-t}\sin(10t)$ mA for $t > 0$. Find expressions for the voltage across the resistor and the instantaneous power dissipated by the resistor. What is the maximum power dissipated in the resistor?

30. The voltage across a 1.4 Ω resistor is given by $v(t) = e^{-t/10} - e^{-t/100}$ V for $t > 0$. Find expressions for the current through the resistor and the instantaneous power dissipated by the resistor. Plot the power dissipated in the resistor.

Chapter 2

Kirchhoff's Laws

2.1 Introduction

What exactly is an electric circuit? So far circuit variables and circuit elements have been discussed, but we still have not defined precisely what an electric circuit is. Basically, an electric circuit is an interconnection of circuit elements. In this definition two things are important: the circuit elements themselves and the way in which they are connected. In circuit theory there exist two types of relationships between currents and voltages that correspond to the two facets of electric circuits. In the previous chapter, we derived the terminal relations which are determined by the physical nature of the circuit elements. In this chapter we will present the rules that govern the ramifications of the circuit interconnections. These relationships are sometimes referred to as *topological* ones. Before presenting these rules we will first have to characterize the topology (layout) of the circuit and introduce the terminology related to various aspects of the circuit. The notion of the *graph* of the electric circuit will be introduced as a way to describe the circuit interconnections. We will then present the two fundamental laws of circuit analysis, known as *Kirchhoff's laws*, that will be used in conjunction with the terminal relations to determine the voltages and currents everywhere in a circuit. Afterward, we will demonstrate a method for assembling the proper number and types of equations that will guarantee unique solutions for electric circuit problems. We will defer the discussion of the techniques needed to solve these systems of equations to the following chapters. The chapter will close with the discussion of resistive circuits which contain only sources and resistors.

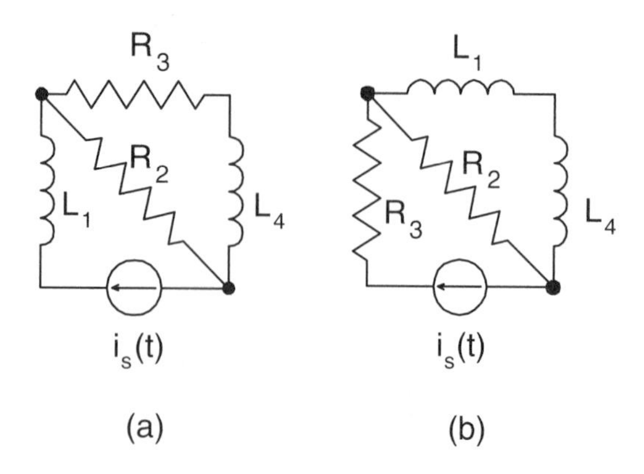

Figure 2.1: Two example circuits with identical elements.

2.2 Circuit Topology

Consider the following two examples of electric circuits, shown in Figure 2.1. Although they are both made from the same set of elements they are different as circuits. It is their interconnections that make the difference. As we will soon see, these differences will give rise to different voltages and currents in these two circuits. This example emphasizes the importance of both aspects of the definition of an electric circuit.

In order to characterize circuit interconnections, we must examine the definitions and rules of circuit topology. There are several important terms that we will use in our discussion of electric circuit topology. These terms are defined in the following paragraphs.

A *node* of an electric circuit is a point where two or more elements are connected together. For both circuits in Figure 2.1, there are four nodes. An example of a node is the point where the inductor L_1 and resistors R_2 and R_3 come together. A node where only two elements are connected together will be called a trivial node. It will be clear from the subsequent discussion that trivial nodes can be easily excluded from the circuit analysis.

A *branch* of an electric circuit is a two-terminal element. There are five branches in each circuit of Figure 2.1, which represent the individual circuit elements.

A *loop* is defined as a set of branches that form a closed path with the property that each node is encountered only once as the loop is traced. For planar circuits (i.e., circuits which can be drawn without any intersections) we can define *meshes*. They form a subset of loops and have the property that they do not enclose any branches. There are only three different loops and two different meshes in Figure 2.1. The leftmost mesh of circuit (a) includes the inductor L_1, resistor R_2, and the current source. The other mesh traces through the inductor L_4 and the two resistors. The last

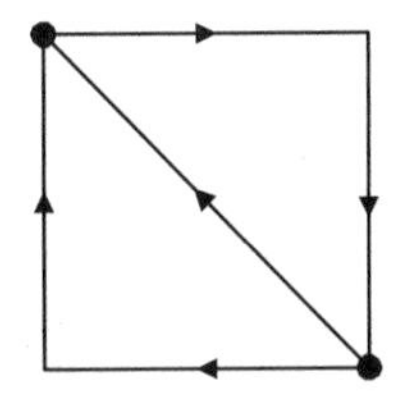

Figure 2.2: Oriented graph of previous circuits.

loop traces the outermost elements of the circuit and is not a mesh because it encloses the resistor R_2.

To characterize the connectivity of a circuit, we introduce the notion of a *circuit graph*. In a circuit graph we ignore the physical nature of the elements and represent only their interconnections. To achieve this, each branch is shown as a line. If we introduce directions which coincide with the reference directions of currents, we arrive at what is called a *directed* or *oriented graph*. Figure 2.2 shows the oriented graph that corresponds to the previous circuits of Figure 2.1.

A *graph tree* of an electric circuit is a subset of the branches of the graph with the property that all nodes of the circuit are connected together, but there are no closed loops formed by these branches. You can obtain a graph tree from the graph of a circuit by removing branches until there are no closed loops. The graph trees are not unique; two examples which correspond to the circuits of Figure 2.1 are given in Figure 2.3.

A *cut set* is a subset of the branches of a circuit with the following two properties: (1) if you remove all the branches of the cut set, the circuit will be divided into two disconnected parts; (2) if you then place any one of the removed branches back into the circuit, it is no longer disconnected. Cut sets are also not unique. For circuit (b) in Figure 2.1, one cut set includes the two resistors and the inductor L_4 while another cut set contains the inductor L_1, the resistor R_2, and the current source.

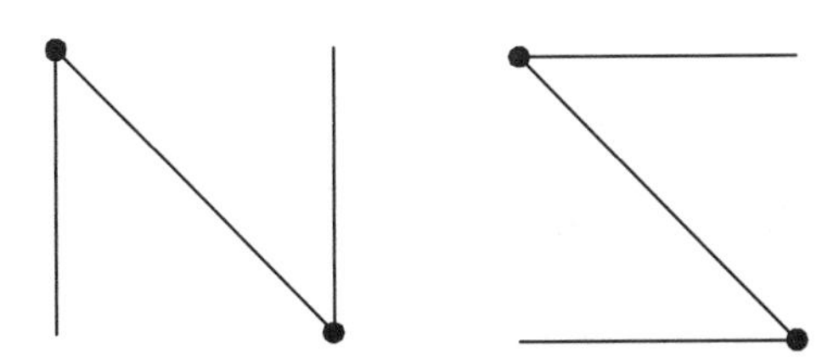

Figure 2.3: Two examples of graph trees.

2.3 Kirchhoff's Laws

2.3.1 Kirchhoff's Current Law

Kirchhoff's current law is generally abbreviated as KCL. KCL states that *the algebraic sum of electric currents at any node of an electric circuit is equal to zero at every instant of time*. The term "algebraic sum" implies that some currents are taken with positive signs while others are taken with negative signs. It is by convention that we assign positive signs to the currents entering the node and negative signs to the currents leaving the node. Expressed mathematically, KCL is written as

$$\sum_k i_k(t) = 0. \tag{2.1}$$

In circuit theory, KCL is considered to be an axiom or postulate. It is to be accepted on faith and considered as a mathematical generalization of numerous observations and experimental data. However, it should be noted that KCL can be proved by using electromagnetic field theory, namely, the principle of continuity of electric current which was described in the last chapter.

As an example, consider the node shown in Figure 2.4. The currents i_1 and i_3 are entering the node while the remaining currents are leaving the node. An application of equation (2.1) results in the expression: $i_1 - i_2 + i_3 - i_4 - i_5 = 0$.

With the aid of the principle of continuity of electric current, we can also see that KCL applies to cut sets of a circuit. In this case, KCL would state that the algebraic sum of currents for any cut set is equal to zero. However, in most cases discussed in this text it will be sufficient to consider only KCL for nodes.

EXAMPLE 2.1 Consider the directed graph shown in Figure 2.5. KCL for node A requires that $i_1 + i_2 - i_4 = 0$, KCL for node B requires that $-i_1 - i_3 = 0$, KCL for node C yields $i_3 + i_5 - i_2 = 0$, and KCL for node D yields $i_6 + i_8 - i_5 = 0$. Consider the surface S_1 which cuts the circuit into two distinct parts via the fourth, sixth, and eighth branches of the circuit. These three branches constitute a cut set.

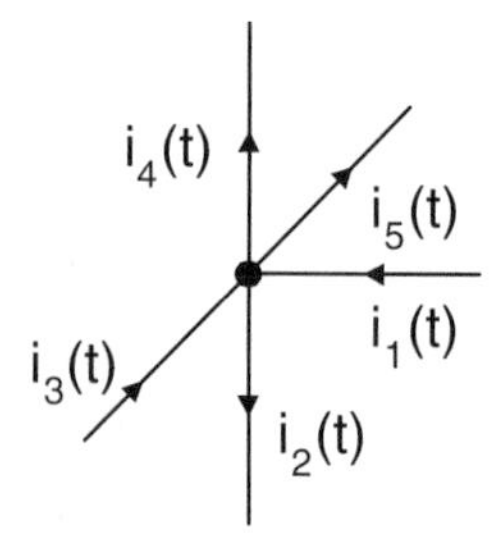

Figure 2.4: A node with currents entering and leaving.

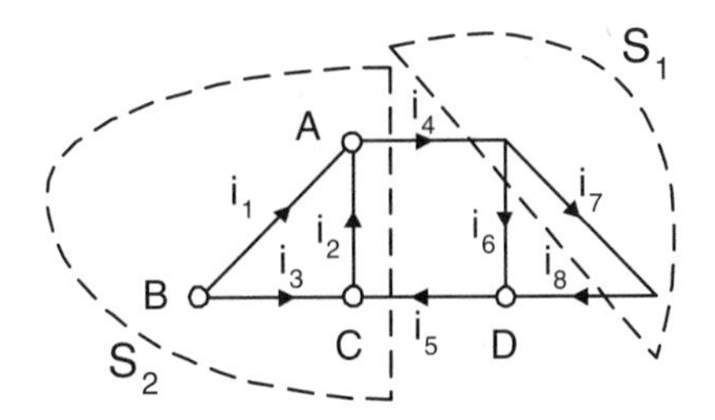

Figure 2.5: KCL example.

The corresponding KCL equation is $i_4 - i_6 - i_8 = 0$. The minus signs in front of the latter two terms arise because the currents from those branches are leaving the volume inside S_1. KCL yields $i_5 - i_4 = 0$ for the cut set generated by the surface S_2 indicated in the figure. There are two interesting points to note here. First, we can arrive at the KCL equation for a cut set by summing the KCL equations for the nodes inside the cut set. For the surface S_2, we sum the KCL equations for nodes A, B, and C and get $i_5 - i_4 = 0$ as claimed. Second, note that the negative sum of the two cut set equations is equal to the KCL equation for node D (the only node which is not contained within a cut set). This example clearly demonstrates that the cut set equations follow from the node equations (and vice versa) and leads us to the notion of independent equations, which will be discussed later in this chapter. ■

2.3.2 Kirchhoff's Voltage Law

Kirchhoff's voltage law is the second fundamental axiom of circuit theory and is often abbreviated as KVL. KVL is applied to voltages in loops and states that *the algebraic sum of branch voltages around any loop of an electric circuit is equal to zero at every instant of time*. Mathematically, it can be written as follows:

$$\sum_k v_k(t) = 0. \tag{2.2}$$

To actually write KVL equations, we shall need the following rule for voltage polarities. We start by introducing reference voltage polarities for each branch. (Remember that these reference voltages are coordinated with the reference currents for the passive elements.) Then, we trace each loop in an arbitrary direction (it is helpful to trace every loop in the same direction—so let us always agree to go clockwise in this text). If, while tracing through the loop, we enter the "plus" terminal of an element and exit its "minus" terminal, then we take the voltage across this element with a positive sign. If, on the other hand, we enter the negative terminal while tracing the loop, we take the corresponding voltage with a negative sign.

To demonstrate the above rule, consider the example circuit shown in Figure 2.6 with the loops already labeled (keep in mind that these are not the only possible loops

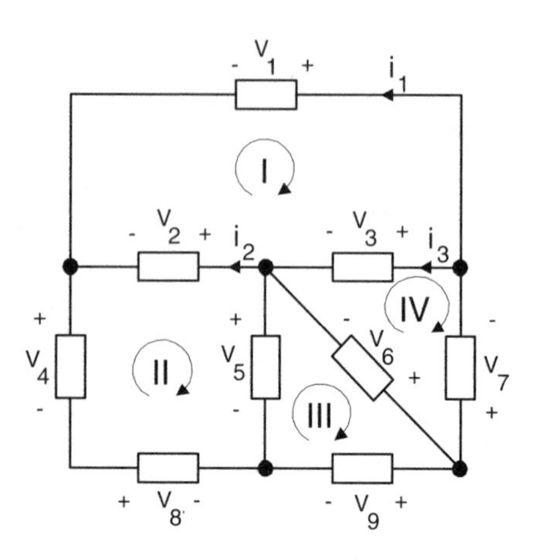

Figure 2.6: Example circuit with four loops shown as I, II, III, IV.

for this circuit). KVL for loop I in Figure 2.6 yields $-v_1 + v_3 + v_2 = 0$. Likewise, for loop III we will get $-v_5 - v_6 + v_9 = 0$ from KVL.

Since for passive elements, reference polarities for voltage and reference directions for current are coordinated, we can formulate the rule for determining the signs of branch voltages in a KVL equation in terms of the reference current directions. Namely, if the tracing direction coincides with the reference current direction, the branch voltage is taken with a positive sign in the KVL equation. Otherwise, the branch voltage requires a minus sign in the KVL equation. In loop I, the tracing direction coincides with reference directions of i_2 and i_3 but is opposite to i_1, so we would get $-v_1 + v_3 + v_2 = 0$ as we must. *Because the reference current directions and voltage polarities are not necessarily coordinated for sources, one should always use the previous rule for voltage polarities to evaluate the signs of the voltages across the sources in the KVL equations.*

EXAMPLE 2.2 Consider the circuit shown in Figure 2.7. Find all of the unknown voltages and currents.

Solution: KVL for loops (I)–(IV) yield, respectively:

$$-12 + v_1 + 2.5 = 0,$$

$$-2.5 + v_3 + v_2 = 0,$$

$$4 - 5 - v_3 = 0,$$

$$-v_4 + 2 - 4 = 0.$$

The first equation gives $v_1 = 9.5$ V; the third yields $v_3 = -1$ V; and the fourth requires $v_4 = -2$ V. Plugging v_3 into the second equation results in $v_2 = 3.5$ V.

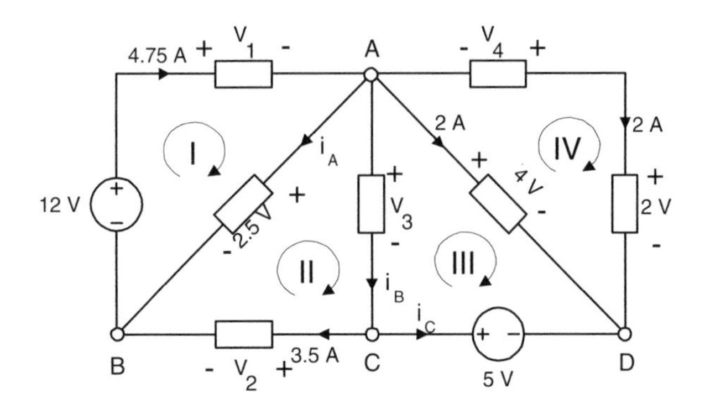

Figure 2.7: Example circuit with unknown currents and voltages.

KCL for nodes B, C, and D, respectively, yield:

$$i_A + 3.5 - 4.75 = 0,$$

$$i_B - 3.5 - i_C = 0,$$

$$2 + 2 + i_C = 0.$$

The first equation requires $i_A = 1.25$ A and the third implies that $i_C = -4$ A. Inserting this value into the second equation completes the solution of this problem by yielding $i_B = -1/2$ A. KCL at node A can be used to check for algebraic mistakes. The resulting equation is

$$4.75 - 1.25 \text{ A} - (-.5 \text{ A}) - 2 \text{ A} - 2 \text{ A} = 0,$$

which checks out correctly. ■

2.4 Linearly Independent Kirchhoff Equations

2.4.1 General Circuits

By writing equations according to Kirchhoff's laws, we end up with a system of linear algebraic equations. The task is to guarantee that the resulting Kirchhoff equations are *linearly independent* (i.e., solvable and without redundant information). We will define linearly independent equations as those in which each subsequent equation contains a variable (an unknown) which the previous equations in the system do not contain. Therefore, each equation cannot be obtained from the previous ones. In this sense, each equation contains essentially new information. The given definition of linearly independent equations is a very special (particular) case of the general definition of linearly independent equations used in linear algebra. However, the above definition will be quite sufficient in order to find a way to write linearly independent Kirchhoff equations.

Consider how linearly independent KCL equations can be written. To be specific, we will analyze the electric circuit shown in Figure 2.8. For nodes A, B, C, and D we, respectively, obtain:

$$i_1(t) + i_2(t) - i_3(t) = 0, \tag{2.3}$$

$$i_3(t) - i_4(t) - i_5(t) = 0, \tag{2.4}$$

$$i_4(t) + i_5(t) - i_6(t) = 0, \tag{2.5}$$

$$i_6(t) - i_1(t) - i_2(t) = 0. \tag{2.6}$$

The equations for nodes A, B, and C are linearly independent because each subsequent equation has a variable that the previous ones do not contain. However, the equation at node D is linearly dependent. Indeed, if we sum up the first three equations and multiply the resulting sum by -1, we can produce the equation for node D. Thus, this equation can be formed as a linear combination of the others, which means it is not linearly independent. From this example we can see that, in general, if we have n nodes we can write $n - 1$ linearly independent Kirchhoff current equations. This implies a general approach to writing linearly independent KCL equations: write one equation for each node except the last.

Consider the same electric circuit redrawn in Figure 2.9, and let us write the Kirchhoff voltage equations for the loops I, II, III, and IV, respectively:

$$v_1(t) - v_2(t) = 0, \tag{2.7}$$

$$v_2(t) + v_3(t) + v_4(t) + v_6(t) = 0, \tag{2.8}$$

$$v_5(t) - v_4(t) = 0, \tag{2.9}$$

$$v_1(t) + v_3(t) + v_5(t) + v_6(t) = 0. \tag{2.10}$$

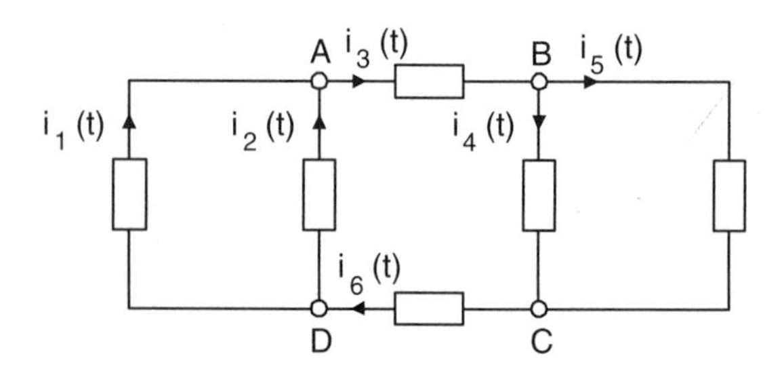

Figure 2.8: Sample circuit for KCL equations.

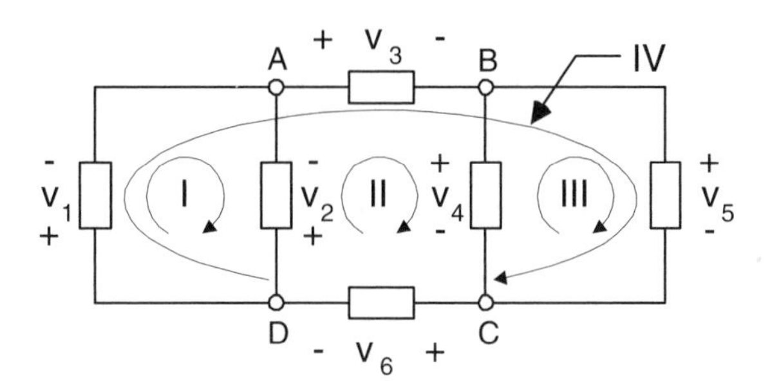

Figure 2.9: Sample circuit for KVL equations.

We see that the voltage equation for loop IV is not linearly independent because it can be obtained by summing up voltage equations for loops I, II, and III.

Since there are many possible loops for any given circuit, determining how many and which loops should be traced in order to find linearly independent KVL equations is not entirely obvious. One method which will always produce the correct number of linearly independent KVL equations is based on the use of a *graph tree*. Figure 2.10 shows a graph tree of the circuit shown in Figure 2.9. All four nodes A, B, C, D are connected together, and no loops are formed—exactly as the graph tree definition states. Now, as the figure shows, if a circuit has n nodes, then its graph tree will have $n - 1$ branches. This holds for all circuits and can be proved very simply. To make a graph tree of an arbitrary circuit, we first connect two nodes—in order to do this only one branch is necessary. There are now $n - 2$ nodes left to connect. To connect every other node to the tree only one branch has to be used, resulting in $n - 1$ branches for the graph tree:

$$1 + (n - 2) = n - 1. \tag{2.11}$$

Graph trees can be used to determine linearly independent KVL equations in a simple manner. Just add the missing branches and create loops (remember from the definition of a graph tree that there were no loops before). By adding a new branch we create a new loop and a new KVL equation, which will always contain a new

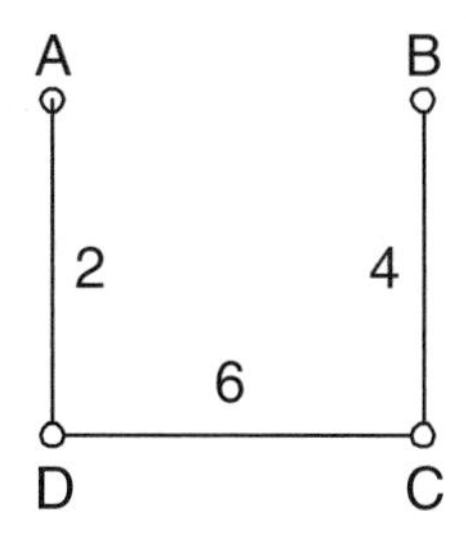

Figure 2.10: Graph tree of the previous circuit.

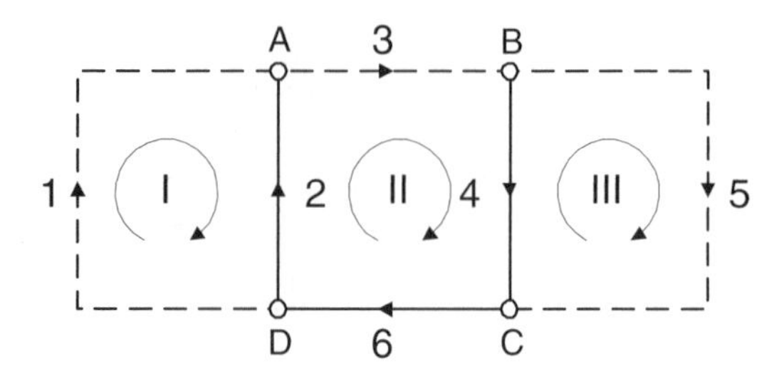

Figure 2.11: The same graph tree, with all linearly independent loops formed by using dashed branches.

variable in comparison with previously written KVL equations. This new variable is the voltage across the added branch. Consider Figure 2.11 as an example. By adding branch 1, we create loop I and the corresponding KVL equation (2.7) contains $v_1(t)$. By adding branch 3, we create loop II and the corresponding KVL equation (2.8) contains $v_3(t)$, which is not present in the previous KVL equation. Next, by adding branch 5, we create loop III, and the corresponding KVL equation (2.9) contains $v_5(t)$, which is not present in the previous equations. If the total number of branches is b, then the total number of linearly independent Kirchhoff voltage equations will be equal to the total number of added branches, which is $b - (n - 1)$.

Although by using graph trees we can produce linearly independent KVL equations for any circuit, this approach can be somewhat overcomplicated for most of the simple circuits encountered in practice. A much easier way to write the linearly independent KVL equations for planar circuits (circuits which can be drawn out without any lines crossing) is to write one KVL equation for each mesh. In the previously illustrated example, the circuit was planar, and the graph tree approach produced the loops which are meshes. In a way, the graph tree technique can be used to justify the mesh technique for writing KVL equations; however, we shall not delve further into this matter.

Thus, by using KVL and KCL, one arrives at b linearly independent equations: $(n - 1)$ linearly independent KCL equations and $b - (n - 1)$ linearly independent KVL equations. However, we have $2b$ unknowns, which are the branch currents and branch voltages. Thus, b additional equations are needed to solve the problem. Fortunately, these equations can be written by using the terminal relationships between voltages and currents which we derived in the previous chapter for the various circuit elements. These relationships can be written for resistors, inductors, and capacitors, respectively, as follows:

$$v_{k'}(t) = R_{k'} i_{k'}(t), \tag{2.12}$$

$$v_{k''}(t) = L_{k''} \frac{di_{k''}(t)}{dt}, \tag{2.13}$$

$$i_{k'''}(t) = C_{k'''} \frac{dv_{k'''}(t)}{dt}. \tag{2.14}$$

The variables on the right-hand side of the above equations can be used as *state variables* in the sense that we can write the KVL and KCL equations in terms of these variables. In order to achieve this, we substitute expressions (2.12) and (2.13) in the appropriate KVL equation (2.2) and expression (2.14) in the appropriate KCL equation (2.1). Once this is done, the result will be a system of linear differential equations which is written in terms of currents through the resistors and inductors and in terms of voltages across the capacitors. In this way, we reduce the total number of unknowns to b, which is the total number of equations. These differential equations should be supplemented by the initial conditions for $i_{k''}(t)$ and $v_{k'''}(t)$, respectively,

$$i_{k''}(0_+) = i_{k''}(0_-), \tag{2.15}$$

$$v_{k'''}(0_+) = v_{k'''}(0_-). \tag{2.16}$$

These initial conditions follow from the principles of continuity of electric currents through inductors and voltages across capacitors.

EXAMPLE 2.3 We shall illustrate the above discussion with an example of an electric circuit, shown in Figure 2.12. Note that the source voltage dependence $v_s(t)$ is assumed to be specified. We want to write b linearly independent equations with b unknowns. The first step is to specify the nodes and to assign reference directions. Next, we shall write Kirchhoff equations. By using KCL, we see that the current equations for nodes A, B, and C, respectively, are:

$$i_1(t) + i_2(t) - i_3(t) = 0, \tag{2.17}$$

$$i_3(t) - i_4(t) - i_5(t) = 0, \tag{2.18}$$

$$i_4(t) + i_5(t) - i_6(t) = 0. \tag{2.19}$$

Notice that we do not write an equation for node D because this equation would not be linearly independent. Notice also that there is a trivial node E that connects only the capacitor and the negative terminal of the voltage source. Rather than add another

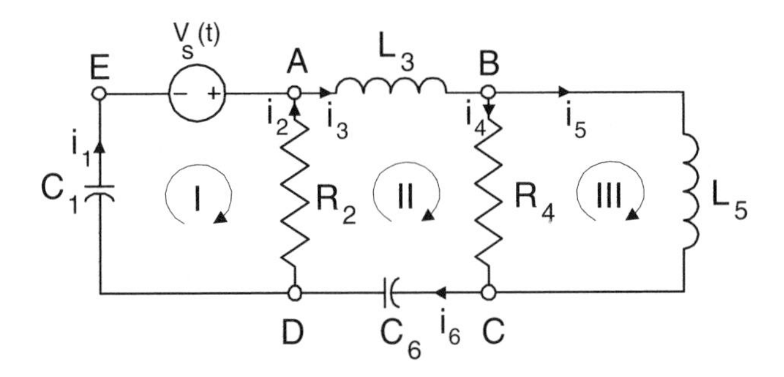

Figure 2.12: Example circuit, with nodes and reference directions shown.

equation and state variable (which for this example would be the current i_s through the source), we simply replace i_s by i_1 wherever it occurs.

We next write the voltage equations by using KVL for the three meshes in the circuit (since the circuit is planar, there is no need for a graph tree). The equations for meshes I, II, and III are

$$v_1(t) - v_s(t) - v_2(t) = 0, \tag{2.20}$$

$$v_2(t) + v_3(t) + v_4(t) + v_6(t) = 0, \tag{2.21}$$

$$-v_4(t) + v_5(t) = 0. \tag{2.22}$$

We chose the state variables to be $v_1(t)$, $i_2(t)$, $i_3(t)$, $i_4(t)$, $i_5(t)$, and $v_6(t)$. We can now supplement the previous Kirchhoff equations by the terminal relationships for the circuit elements:

$$i_1(t) = C_1 \frac{dv_1(t)}{dt}, \tag{2.23}$$

$$v_2(t) = R_2 i_2(t), \tag{2.24}$$

$$v_3(t) = L_3 \frac{di_3(t)}{dt}, \tag{2.25}$$

$$v_4(t) = R_4 i_4(t), \tag{2.26}$$

$$v_5(t) = L_5 \frac{di_5(t)}{dt}, \tag{2.27}$$

$$i_6(t) = C_6 \frac{dv_6(t)}{dt}. \tag{2.28}$$

By substituting (2.23) and (2.28) into the KCL equations, we find:

$$C_1 \frac{dv_1(t)}{dt} + i_2(t) - i_3(t) = 0, \tag{2.29}$$

$$i_3(t) - i_4(t) - i_5(t) = 0, \tag{2.30}$$

$$i_4(t) + i_5(t) - C_6 \frac{dv_6(t)}{dt} = 0. \tag{2.31}$$

By substituting (2.24)–(2.27) into the KVL equations, we obtain:

$$v_1(t) - R_2 i_2(t) = v_s(t), \tag{2.32}$$

$$R_2 i_2(t) + L_3 \frac{di_3(t)}{dt} + R_4 i_4(t) + v_6(t) = 0, \tag{2.33}$$

$$-R_4 i_4(t) + L_5 \frac{di_5(t)}{dt} = 0. \tag{2.34}$$

The equations (2.29)–(2.34) form a system of six linear differential and algebraic equations with six unknowns. These equations should be complemented with initial conditions which come from the continuity conditions for capacitors and inductors:

$$v_1(0_+) = v_1(0_-), \tag{2.35}$$

$$i_3(0_+) = i_3(0_-), \tag{2.36}$$

$$i_5(0_+) = i_5(0_-), \tag{2.37}$$

$$v_6(0_+) = v_6(0_-). \tag{2.38}$$

The circuit variables in the right-hand side of equations (2.35)–(2.38) are equal to zero if there was no electromagnetic process in the electric circuit prior to the time $t = 0$. Thus, in this case we will have zero initial conditions. If there was some electromagnetic process in the circuit before the time $t = 0$, then the values of the above circuit variables should be determined from the analysis of this process. ■

Differential equations (2.29)–(2.34) together with initial conditions (2.35)–(2.38) form the initial value problem. The essence of this problem is to find the solution to the above differential equations which satisfies the prescribed initial conditions. If this solution is somehow found, then by using expressions (2.23)–(2.28) we can also compute the currents through the capacitors and voltages across the inductors and resistors. In other words, we can find all voltages and currents in the electric circuit.

Initial value problems for ordinary differential equations are studied in mathematics and various techniques for their solutions have been developed. These techniques will be extensively used later in the text and they constitute an important component of the overall circuit analysis. However, it is not the goal of this discussion to present here the complete analysis of electric circuits. It is rather to demonstrate that Kirchhoff equations (2.1) and (2.2) together with terminal relationships (2.12)–(2.14) and initial conditions (2.15) and (2.16) allow one to reduce the analysis of electric circuits to well-studied problems in mathematics, i.e., initial value problems.

EXAMPLE 2.4 The derivation of the differential equations for the circuit shown in Figure 2.12 has been quite general in nature. In some cases this derivation can be simplified by exploiting a particular structure of an electric circuit. To demonstrate this, consider the electric circuit shown in Figure 2.13.

This circuit has four passive elements: two inductors and two resistors. For this reason, we may expect to end up with four coupled linear differential and algebraic equations with respect to the currents through the inductors and resistors. However, the analysis of this circuit can be significantly simplified by observing that nodes B and D are trivial (i.e., only two elements are connected by these nodes). By applying KCL to these nodes, we obtain:

$$i_1(t) = i_s(t), \tag{2.39}$$

$$i_3(t) = i_4(t). \tag{2.40}$$

Equation (2.39) tells us that the current $i_1(t)$ is essentially known because the source current $i_s(t)$ is given. Equation (2.40) further reduces the number of unknowns because $i_3(t)$ can be replaced by $i_4(t)$. Thus, we essentially have only two unknown currents:

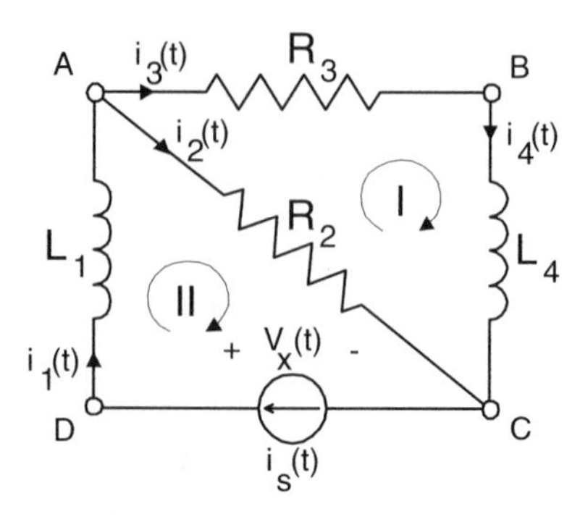

Figure 2.13: The electric circuit considered in Example 2.4.

$i_2(t)$ and $i_4(t)$. To derive the coupled differential and algebraic equations for these currents, we shall apply KCL to node A:

$$i_1(t) - i_2(t) - i_3(t) = 0, \tag{2.41}$$

and KVL to mesh I:

$$v_3(t) + v_4(t) - v_2(t) = 0. \tag{2.42}$$

Next, we shall use the terminal relationships:

$$v_2(t) = R_2 i_2(t), \tag{2.43}$$

$$v_3(t) = R_3 i_3(t), \tag{2.44}$$

$$v_4(t) = L_4 \frac{di_4(t)}{dt}. \tag{2.45}$$

By substituting (2.43)–(2.45) into (2.42) and by using (2.39) and (2.40) in (2.41) and (2.42) we end up with the following coupled equations for $i_2(t)$ and $i_4(t)$:

$$i_2(t) + i_4(t) = i_s(t), \tag{2.46}$$

$$R_3 i_4(t) + L_4 \frac{di_4(t)}{dt} - R_2 i_2(t) = 0. \tag{2.47}$$

Equations (2.46) and (2.47) should be complemented with the initial condition

$$i_4(0_+) = i_4(0_-) \tag{2.48}$$

which follows from the principle of continuity of electric current through an inductor.

Equations (2.46) and (2.47) together with initial condition (2.48) constitute the initial value problem. If this problem is solved, then by using (2.40) and (2.43)–(2.45) we can determine $i_3(t)$ and all voltages. After all branch voltages and currents in the circuit are determined, the voltage $v_x(t)$ across the current source can be found.

Indeed, by applying KVL to mesh II, we arrive at:

$$v_1(t) + v_2(t) - v_x(t) = 0, \tag{2.49}$$

which can be rewritten as:

$$v_x(t) = v_1(t) + v_2(t). \tag{2.50}$$

By using (2.39) and the terminal relationship for L_1, we obtain:

$$v_1(t) = L_1 \frac{di_s(t)}{dt}. \tag{2.51}$$

By substituting (2.43) and (2.51) into (2.50), we derive:

$$v_x(t) = L_1 \frac{di_s(t)}{dt} + R_2 i_2(t), \tag{2.52}$$

which is the expression for the voltage across the current source in terms of branch current i_2, which is presumed to be found. It is important to note that the reference polarity for $v_x(t)$ can be chosen arbitrarily. The actual polarity of $v_x(t)$ at any instant of time t can always be found from the sign of $v_x(t)$ at this instant of time and its reference polarity. The reference polarity of $v_x(t)$ should not necessarily be coordinated with the reference direction of the current source. However, the coordination of current reference directions and voltage reference polarities is very important for passive elements. This is because their terminal relationships (2.12), (2.13), and (2.14) have been derived for coordinated reference directions. For current and voltage sources, on the other hand, terminal relationships are specified by the expressions for $i_s(t)$ and $v_s(t)$, respectively. Thus, these terminal relationships do not depend on coordination of reference directions.

For this reason, one usually chooses a reference current for a voltage source to be in the expected direction of the actual current, which is typically out of the positive terminal. An analogous selection is made for a reference voltage polarity of a current source. ■

2.4.2 Resistive Circuits

Next, we shall consider purely resistive circuits. These circuits contain only sources and resistors. We start with KCL and KVL equations (2.1) and (2.2), respectively, and rewrite them in a slightly different form by moving all known driving (voltage and current) sources to the right-hand side of these equations:

$$\sum_k i_k(t) = -\sum_k i_{sk}(t), \tag{2.53}$$

$$\sum_k v_k(t) = -\sum_k v_{sk}(t). \tag{2.54}$$

We can always write $n-1$ linearly independent KCL equation (2.53) and $b-(n-1)$ linearly independent KVL equations (2.54). Thus, altogether we can write b linearly independent equations. However, we have $2b$ unknowns, which are currents through and voltages across the resistors. To circumvent this difficulty, we shall use the terminal relationships

$$v_k(t) = R_k i_k(t) \tag{2.55}$$

for the resistors and substitute them in the KVL equation (2.54). In this way, we arrive at the b linearly independent algebraic equations

$$\sum_k i_k(t) = -\sum_k i_{sk}(t), \tag{2.56}$$

$$\sum_k i_k(t) R_k = -\sum_k v_{sk}(t) \tag{2.57}$$

with respect to b unknown currents $i_k(t)$. Thus, we can see that the analysis of purely resistive circuits is reduced to the solution of simultaneous algebraic equations rather than to the solution of differential equations. This clearly suggests that calculus and differential equations enter circuit analysis when energy storage elements are present.

We shall illustrate the application of the equations (2.56) and (2.57) to the analysis of resistive circuits by the following two examples.

EXAMPLE 2.5 Consider the electric circuit shown in Figure 2.14. This circuit has three resistors and we want to write three linearly independent equations for the currents through these resistors. We first specify the nodes and introduce reference directions. Next, we shall write KCL and KVL equations. This circuit has only two nontrivial nodes, so only one KCL equation is linearly independent:

$$i_1(t) - i_2(t) - i_3(t) = 0. \tag{2.58}$$

This circuit is planar and has two meshes: I and II. KVL equations (2.57) for these meshes are:

$$R_1 i_1(t) + R_2 i_2(t) = v_{s1}(t), \tag{2.59}$$

$$-R_2 i_2(t) + R_3 i_3(t) = v_{s2}(t). \tag{2.60}$$

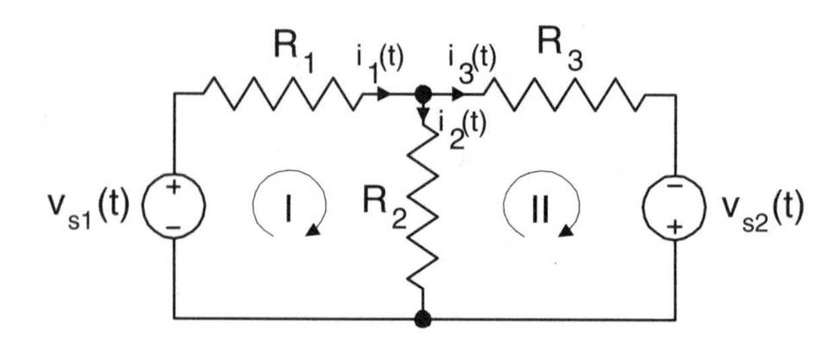

Figure 2.14: Resistive circuit with three resistors.

Thus, we end up with three linear algebraic equations, which is what is expected since there are only three resistors in the circuit. Systems of linear simultaneous equations can be solved in a variety of ways. For small systems of equations, the choice of solution methods is debatable. However, for large systems of equations the Gaussian elimination technique is a very attractive method of choice. This method is presented in Appendix B.

As far as equations (2.58)–(2.60) are concerned, the attractive method of solution is the method of substitutions. Indeed, from (2.59) and (2.60) we can find $i_1(t)$ and $i_3(t)$ in terms of $i_2(t)$:

$$i_1(t) = -i_2(t)\frac{R_2}{R_1} + \frac{v_{s1}(t)}{R_1}, \tag{2.61}$$

$$i_3(t) = i_2(t)\frac{R_2}{R_3} + \frac{v_{s2}(t)}{R_3}. \tag{2.62}$$

By substituting (2.61) and (2.62) into (2.58), we end up with the following equation for $i_2(t)$:

$$-i_2(t)\frac{R_2}{R_1} - i_2(t)\frac{R_2}{R_3} - i_2(t) = -\frac{v_{s1}(t)}{R_1} + \frac{v_{s2}(t)}{R_3}, \tag{2.63}$$

from which we find:

$$i_2(t) = \frac{-v_{s2}(t)R_1 + v_{s1}(t)R_3}{R_1R_2 + R_1R_3 + R_2R_3}. \tag{2.64}$$

After $i_2(t)$ is computed, expressions (2.61) and (2.62) can be used to calculate $i_1(t)$ and $i_3(t)$. This concludes the analysis of the above circuit. ■

EXAMPLE 2.6 Consider the resistive circuit shown in Figure 2.15. This circuit has four nodes (A, B, C, and D) and three meshes (I, II, and III). For this reason, one may expect to end up with the coupled linear algebraic equations for currents $i_1(t)$, $i_2(t)$, $i_3(t)$, $i_4(t)$, $i_x(t)$ and voltage $v_x(t)$. However, the analysis of this circuit can be simplified by observing that node C is a trivial one. By applying KCL to this node,

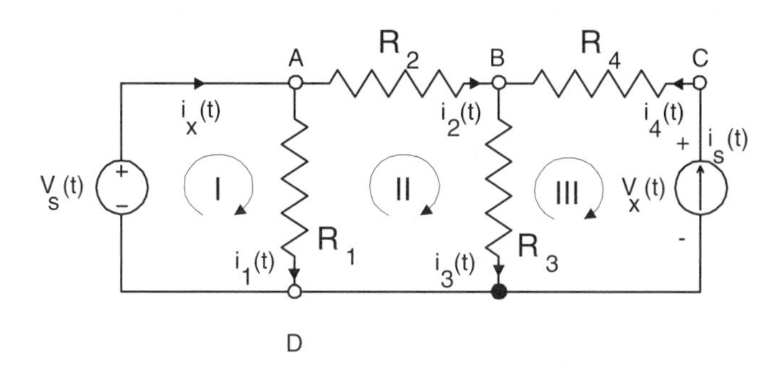

Figure 2.15: More complicated resistive circuit.

we obtain:

$$i_4(t) = i_s(t). \tag{2.65}$$

We can also observe that mesh I is a trivial mesh (which we define to be a mesh with only two branches). By applying KVL to this mesh, we find:

$$v_1(t) = v_s(t), \tag{2.66}$$

which can be rewritten as:

$$i_1(t)R_1 = v_s(t). \tag{2.67}$$

From (2.67), we obtain:

$$i_1(t) = \frac{v_s(t)}{R_1}. \tag{2.68}$$

Equations (2.65) and (2.68) tell us that the currents $i_4(t)$ and $i_1(t)$ are essentially known, which reduces the total number of unknown currents for which we have to solve. By writing KCL equations for nodes A and B and KVL equation for mesh II, we arrive at:

$$i_x(t) - i_1(t) - i_2(t) = 0, \tag{2.69}$$

$$i_2(t) - i_3(t) + i_4(t) = 0, \tag{2.70}$$

$$-i_1(t)R_1 + i_2(t)R_2 + i_3(t)R_3 = 0. \tag{2.71}$$

By using (2.65) in (2.70), (2.67) in (2.71), and (2.68) in (2.69), we obtain the following simultaneous linear algebraic equations for $i_2(t)$, $i_3(t)$, and $i_x(t)$:

$$i_x(t) - i_2(t) = \frac{v_s(t)}{R_1}, \tag{2.72}$$

$$i_2(t) - i_3(t) = -i_s(t), \tag{2.73}$$

$$i_2(t)R_2 + i_3(t)R_3 = v_s(t). \tag{2.74}$$

The above equations can be solved by the method of substitutions. Indeed, from (2.73) we find:

$$i_3(t) = i_2(t) + i_s(t). \tag{2.75}$$

By substituting (2.75) into (2.74), we arrive at the following equation for $i_2(t)$:

$$i_2(t)R_2 + i_2(t)R_3 + i_s(t)R_3 = v_s(t), \tag{2.76}$$

which can be further transformed as

$$i_2(t)[R_2 + R_3] = v_s(t) - i_s(t)R_3. \tag{2.77}$$

From the last equation we get:

$$i_2(t) = \frac{v_s(t) - i_s(t)R_3}{R_2 + R_3}. \tag{2.78}$$

By substituting (2.78) into (2.75), we find the expression for $i_3(t)$:

$$i_3(t) = \frac{v_s(t) + i_s(t)R_2}{R_2 + R_3}. \tag{2.79}$$

Finally, from (2.72) and (2.78), we derive:

$$i_x(t) = \frac{v_s(t)}{R_1} + \frac{v_s(t) - i_s(t)R_3}{R_2 + R_3}. \tag{2.80}$$

Now, we can also find the voltage $v_x(t)$ across the current source $i_s(t)$. By applying KVL to mesh III, we obtain:

$$v_x(t) - i_3(t)R_3 - i_4(t)R_4 = 0. \tag{2.81}$$

By using (2.65) and (2.79) in (2.81), we find:

$$v_x(t) = i_s(t)R_4 + \frac{v_s(t)R_3 + i_s(t)R_2R_3}{R_2 + R_3}. \tag{2.82}$$

This concludes the analysis of the above circuit. ■

2.5 Summary

In this chapter we have stressed the importance of the interconnections of circuit elements in determining the voltages and currents in electric circuits. Several key topological terms have been introduced to characterize the connectivity of electric circuits.

The two main relations for currents and voltages due to circuit topology have been presented. They are Kirchhoff's current and voltage laws. KCL states that *the algebraic sum of electric currents at any node of an electric circuit is equal to zero at every instant of time*. KVL tells us that *the algebraic sum of branch voltages around any loop of an electric circuit is equal to zero at every instant of time*. By using KCL we can write $n - 1$ linearly independent equations of the form:

$$\sum_k i_k(t) = 0, \tag{2.83}$$

where n is the total number of nodes of the electric circuit.

Similarly, by using KVL we can write $b-(n-1)$ linearly independent equations:

$$\sum_k v_k(t) = 0, \tag{2.84}$$

where b is the total number of branches of the electric circuit. In a planar circuit, the number of independent KVL equations corresponds exactly to the number of meshes.

Thus, by using KVL and KCL, one arrives at b linearly independent equations. However, these equations have $2b$ unknowns, which are the branch currents and branch voltages. To reduce the number of unknowns to b, the terminal relationships for resistors, inductors, and capacitors are used.

The chapter is closed by the discussion of resistive circuits which contain only sources and resistors. It is shown that the analysis of resistive circuits is reduced to the solution of simultaneous linear algebraic equations, and some examples of this analysis are given.

It should be stressed that the Kirchhoff equations (2.83) and (2.84), together with the terminal relationships (2.12)–(2.14) and the initial conditions (2.15) and (2.16), form the mathematical foundation of electric circuit theory. In the following chapters we will develop some powerful machinery that will help us to arrive at the solutions to circuit problems. This machinery will always be based on the foundation laid out in this chapter.

It is also clear from equations (2.29)–(2.38) that a circuit can be excited in various ways: by the initial conditions in the circuit, by sources, or by both. If a circuit is excited by initial conditions (such as some initial voltage across a capacitor which is going to be discharged), the response will gradually decay with time and the time evolution of circuit variables is called the *transient* response. If the circuit is driven by an ac source, the process in the circuit will go on indefinitely and with time the variation of circuit variables becomes repeatable and is called the *steady-state* response. In this text, the ac steady state is considered first in the next four chapters and the analysis of transient responses is delayed until Chapter 7. It will be shown in the next chapter that the basic circuit equations for the ac steady state are mathematically similar to the basic equations (2.56) and (2.57) for resistive electric circuits. This will allow us to treat the analysis of resistive circuits as a particular case of the ac steady-state analysis.

2.6 Problems

1. Draw the directed graphs which correspond to the circuits shown in Figure P2-1.

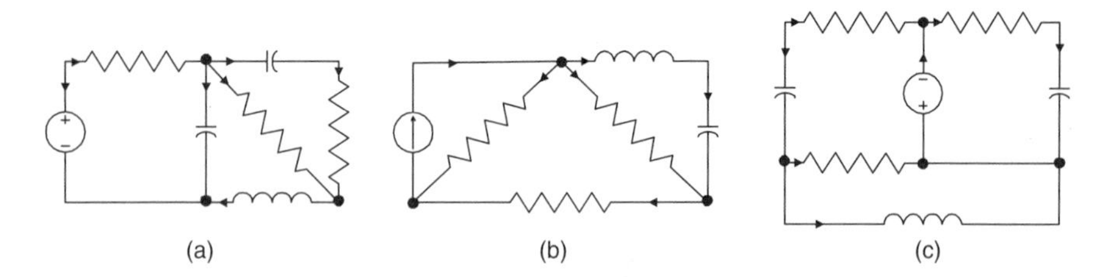

Figure P2-1

2. Use your imagination and the five basic circuit elements to create circuits which have the topology of the directed graphs shown in Figure P2-2.

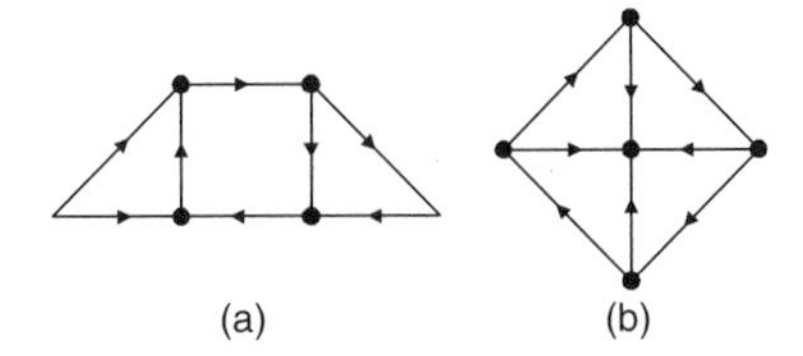

Figure P2-2

3. Clearly label the nodes in the circuit shown in Figure P2-3 and write KCL for each node.

4. Clearly label the nodes in the circuit shown in Figure P2-4 and write KCL for each node.

5. Choose three different cut sets of the elements in Figure P2-4. Write KCL for each cut set.

6. Choose three different cut sets of the elements in Figure P2-3. Write KCL for each cut set.

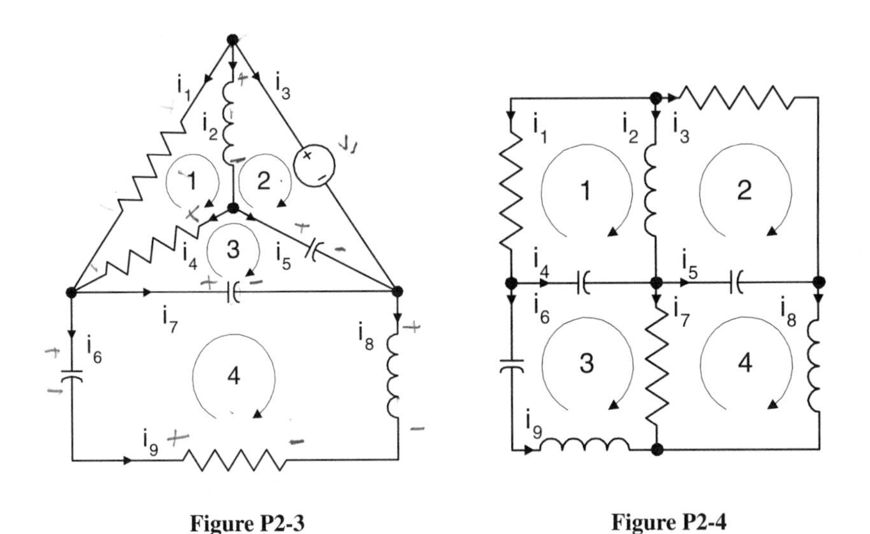

Figure P2-3 **Figure P2-4**

7. Consider the circuit shown in Figure P2-7. The currents listed in each individual row of the following table represent a different solution to the KCL equations. For each row of currents, fill in the missing currents.

i_1	i_2	i_3	i_4	i_5	i_6	i_7	i_8	i_9
1	?	?	2	?	1	?	?	3
?	0	1	2	?	?	1	?	?
?	?	−1	1	−1	1	?	?	?
4	?	7	?	?	3	?	?	9

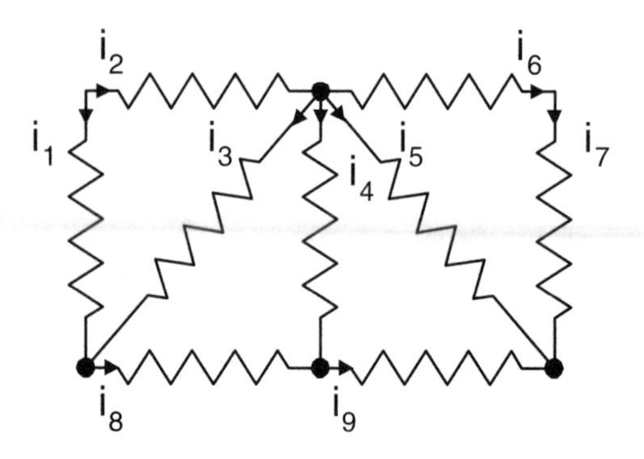

Figure P2-7

8. Determine the number of independent KCL equations for each circuit in Figure P2-1.

9. Write KVL for each loop indicated in the circuit shown in Figure P2-4.

10. Write KVL for each loop indicated in the circuit shown in Figure P2-3.

11. Determine the number of independent KVL equations for each circuit shown in Figure P2-1.

12. Consider the circuit shown in Figure P2-7. The voltages listed in each individual row of the following table represent a different solution to the KVL equations. For each row of voltages, fill in the missing voltages. (v_1 is the voltage across the component indicated by i_1, etc.)

v_1	v_2	v_3	v_4	v_5	v_6	v_7	v_8	v_9
1	1	?	−1	2	2	?	?	?
0	?	3	?	?	?	−2	4	2
?	1	−1	2	?	2	2	?	?
1	1	?	0	?	1	−1	?	?

13. Draw three different graph trees for the circuit shown in Figure P2-3.

14. Draw three different graph trees for the circuit shown in Figure P2-4.

In Problems 15–20, derive the complete system of first-order differential equations which describe the circuit shown in the figure indicated. Clearly identify the number of branches, label the nodes and meshes, and list the KCL equations, KVL equations, and terminal relationships used in the derivation. Use v_1 to denote the voltage drop across branch that has i_1 flowing through it, etc.

15. Figure P2-15.

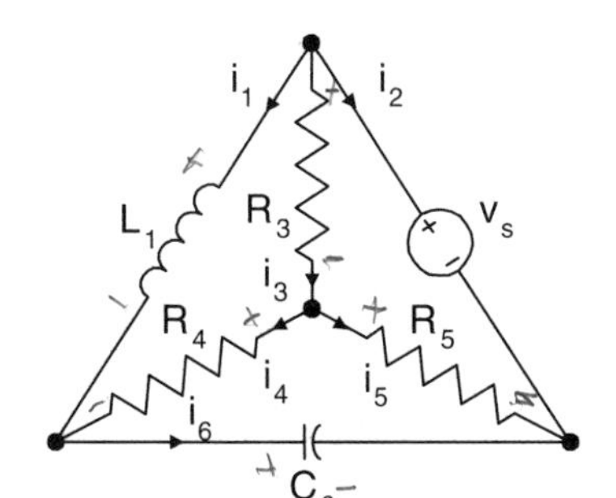

Figure P2-15

16. Figure P2-16.

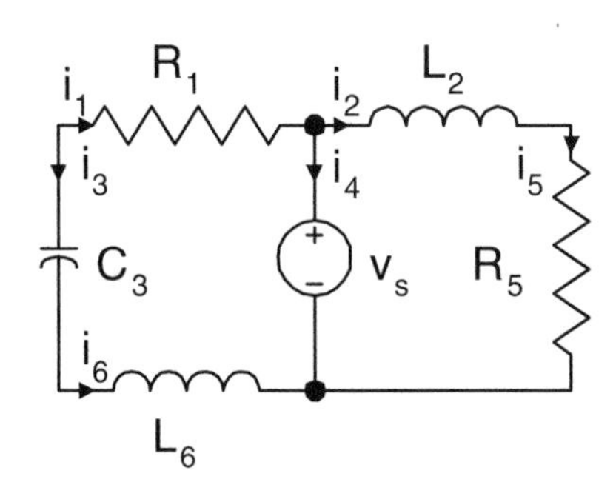

Figure P2-16

17. Figure P2-17.

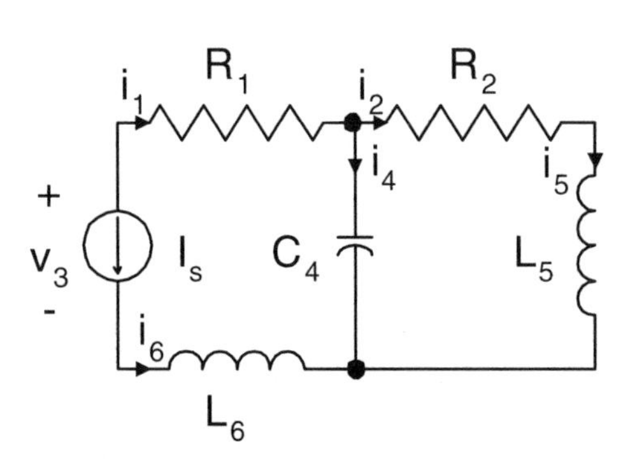

Figure P2-17

18. Figure P2-18.

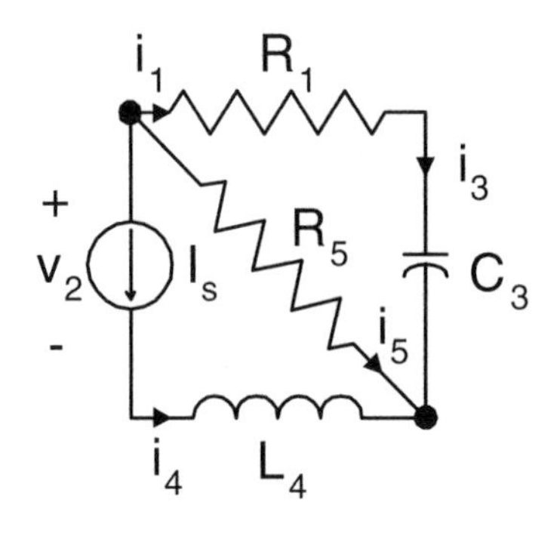

Figure P2-18

19. Figure P2-19.

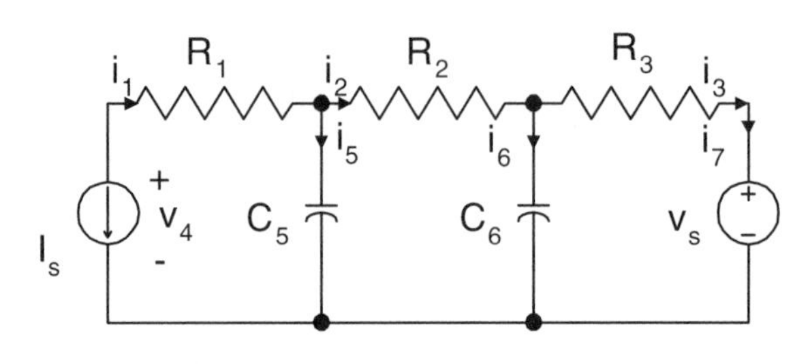

Figure P2-19

20. Figure 2.1(b) (page 34).

In Problems 21–26, find the currents through any voltage sources and the voltages across any current sources in the circuits shown in the figures indicated. Label nodes and meshes and define reference directions when needed. Assume the passive sign convention for all elements.

21. Figure P2-21.

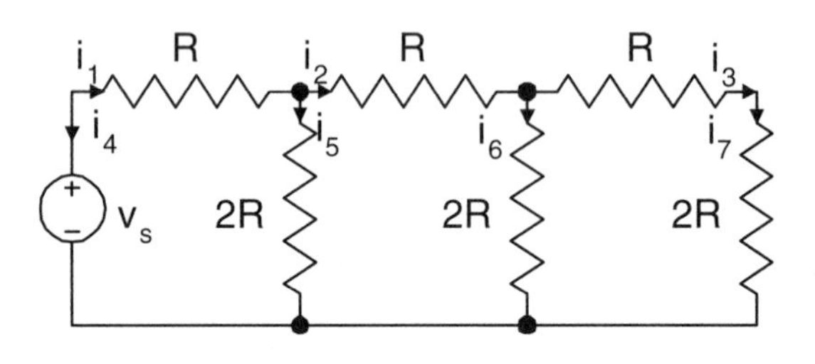

Figure P2-21

22. Figure P2-22.

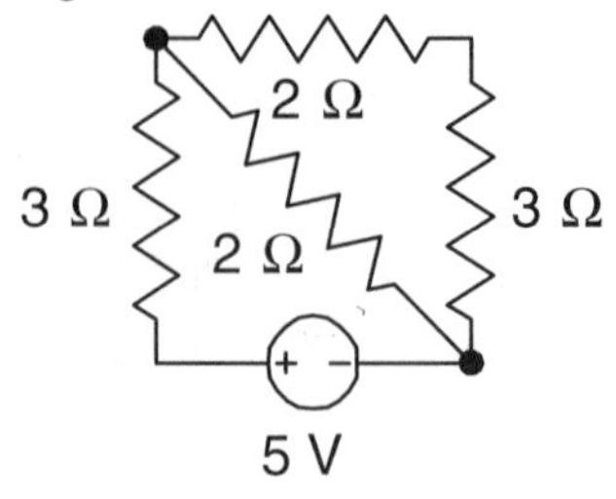

Figure P2-22

23. Figure P2-23.

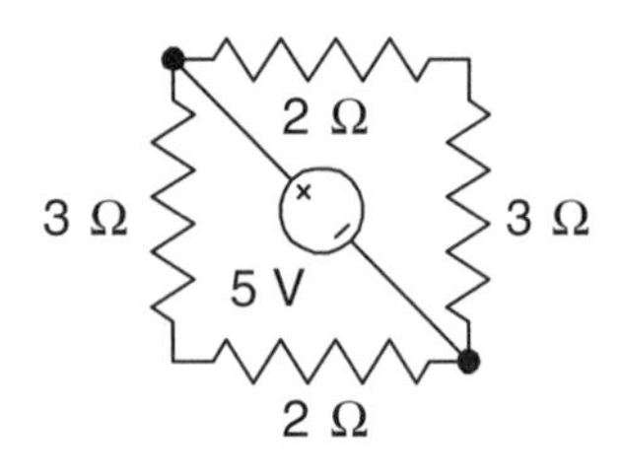

Figure P2-23

24. Figure P2-24.

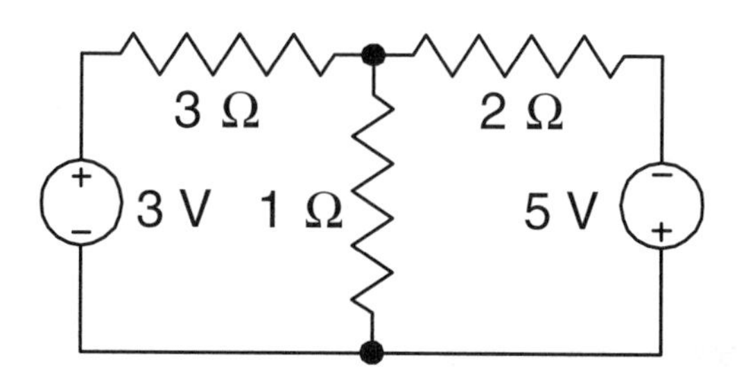

Figure P2-24

25. Figure P2-25.

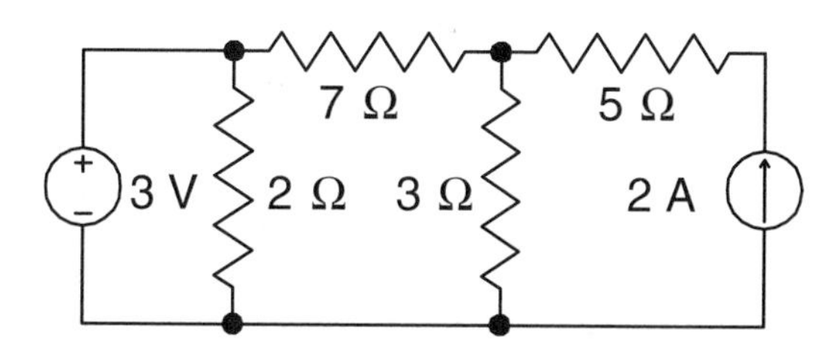

Figure P2-25

26. Figure P2-26.

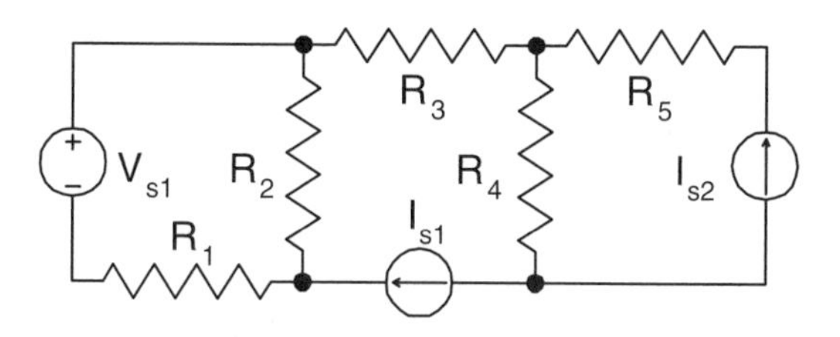

Figure P2-26

Chapter 3

AC Steady State

3.1 Introduction

In this chapter, we shall begin our study of electric circuits driven by ac sources. Under steady-state conditions, all voltages and currents in such circuits will be sinusoidal. A special and very powerful analysis technique exists which exploits this fact. It is known as the phasor technique and it uses complex numbers to represent sinusoidal time-varying quantities. The main power of the phasor technique is that it reduces the operations of calculus on sinusoidal quantities to simple algebraic operations on complex numbers (phasors). As a result, the basic differential equations of electric circuits discussed in the previous chapter are reduced to linear algebraic equations with respect to phasors. This significantly simplifies the analysis of electric circuits under ac steady-state conditions.

The phasor technique leads to the very important concept of impedance (and/or admittance). The notions of impedance as well as phasors themselves are ubiquitous in modern electrical engineering. For this reason, this text is structured in such a way that the phasors and impedances are introduced very early in the book. It is believed that this approach will allow students to become more familiar with these important notions and firmly absorb the related material.

This chapter includes a brief discussion of the important subject of ac power. The notions of root-mean-square (rms) values of voltage and current will be introduced along with the definition of power factor. We will demonstrate how phasors can be used to calculate average power and we will examine the conditions for maximum power transfer to a load. This chapter will close with a generalization of the phasor technique to complex frequency. This topic will be of special importance when we study transients.

3.2 AC Quantities

In this chapter we will consider the steady-state behavior of electric circuits driven by ac (sinusoidal) voltage or current sources.

Any ac voltage can be expressed mathematically as

$$v(t) = V_m \cos(\omega t + \phi_V). \tag{3.1}$$

In this equation V_m represents the peak or amplitude of the voltage, and ω represents the angular frequency. Angular frequency is related to frequency f by a factor of 2π:

$$\omega = 2\pi f. \tag{3.2}$$

Frequency is defined as the reciprocal of the period:

$$f = \frac{1}{T}. \tag{3.3}$$

The unit for frequency is s^{-1} and is called Hz (hertz). For example, in the United States, the frequency of the ac voltage in conventional power networks is 60 Hz, which corresponds to an angular frequency of about 377 s^{-1} (or rad/s).

The initial phase of the voltage is ϕ_V. When time is equal to zero, the expression for ac voltage becomes

$$v(0) = V_m \cos \phi_V. \tag{3.4}$$

This is the initial value of the voltage, which is determined by the amplitude and the initial phase. The graphical representation of ac voltage is shown in Figure 3.1.

The expression for ac current is very similar and is given by:

$$i(t) = I_m \cos(\omega t + \phi_I), \tag{3.5}$$

where I_m is the peak value of the current and ϕ_I is its initial phase. From equations (3.1) and (3.5) we see that if ω is zero then the voltage and current expressions are

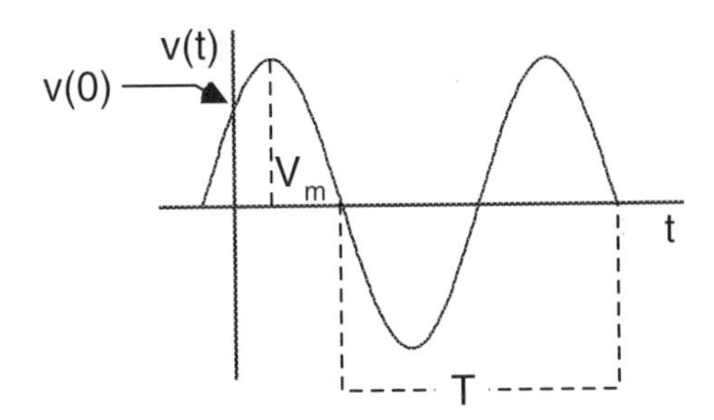

Figure 3.1: Graphical representation of ac voltage.

reduced to

$$v(t) = V_m \cos \phi_V = \text{constant}, \tag{3.6}$$

$$i(t) = I_m \cos \phi_I = \text{constant}. \tag{3.7}$$

These expressions yield constant values, and from this it is easy to deduce that dc (nonvarying) voltages and currents are simply particular cases of ac (sinusoidal) voltages and currents. As a result, dc analysis is a particular case of ac steady-state analysis, and it will be treated as such in this text.

Most often the frequency has some fixed predetermined value. Since the frequency is fixed, peak values and the initial phases completely determine the ac currents and voltages. For this reason, we would like to know the relationships between the peak values and initial phases of ac currents and voltages for every circuit element that we have studied.

3.3 Amplitude and Phase Relationships for Circuit Elements

We begin with a **resistor**. By using the expressions for ac voltage and ac current and substituting them into Ohm's law, we can derive the relationships between peak values and initial phases for a resistor:

$$v(t) = V_m \cos(\omega t + \phi_V), \tag{3.8}$$

$$i(t) = I_m \cos(\omega t + \phi_I), \tag{3.9}$$

$$V_m \cos(\omega t + \phi_V) = Ri(t) = RI_m \cos(\omega t + \phi_I). \tag{3.10}$$

From the last equation it follows that the peak and initial phase values of the current and voltage should be related as follows:

$$V_m = RI_m, \tag{3.11}$$

$$\phi_V = \phi_I. \tag{3.12}$$

Therefore, it is evident that the voltage and current through a resistor have the same initial phase angle; in other words, they are *in phase*. The graph of ac voltage and current in Figure 3.2 illustrates this fact.

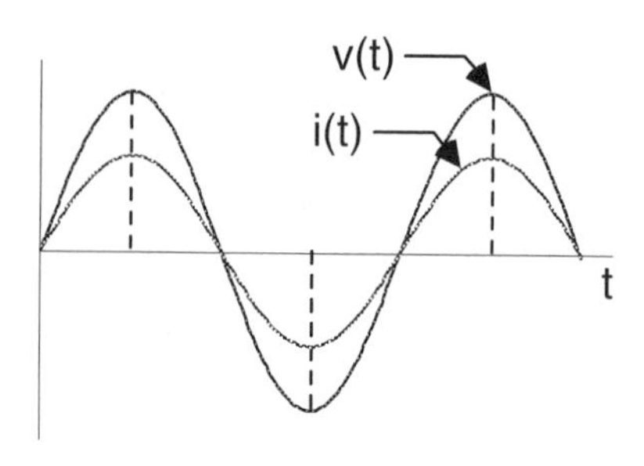

Figure 3.2: AC voltage and current in the case of a resistor.

Now, we consider the case of an **inductor**. Recall that the terminal relationship between the voltage and current in the case of an inductor is:

$$v(t) = L\frac{di(t)}{dt}. \tag{3.13}$$

By using this equation, we shall find the desired relationships. First, we take the time derivative of the current:

$$\frac{d}{dt}(I_m \cos(\omega t + \phi_I)) = -\omega I_m \sin(\omega t + \phi_I). \tag{3.14}$$

By substituting (3.1) and (3.14) into (3.13), we find:

$$V_m \cos(\omega t + \phi_V) = -L\omega I_m \sin(\omega t + \phi_I). \tag{3.15}$$

By using the trigonometric identity

$$\cos\left(\alpha + \frac{\pi}{2}\right) = -\sin\alpha, \tag{3.16}$$

the last equation can be rewritten as follows:

$$V_m \cos(\omega t + \phi_V) = L\omega I_m \cos\left(\omega t + \phi_I + \frac{\pi}{2}\right). \tag{3.17}$$

From this equation we find that:

$$V_m = \omega L I_m, \tag{3.18}$$

$$\phi_V = \phi_I + \frac{\pi}{2}. \tag{3.19}$$

Consequently, the ac voltage and current are not in phase, and the ac voltage across the inductor leads the ac current through the inductor by $\pi/2$. Another way to express the peak value and phase relationships in an inductor is

$$I_m = \frac{V_m}{\omega L}, \tag{3.20}$$

$$\phi_I = \phi_V - \frac{\pi}{2}. \tag{3.21}$$

The last expression suggests that the current lags behind the voltage by $\pi/2$. Figure 3.3 gives a graphical representation of ac voltage and current in the case of an inductor.

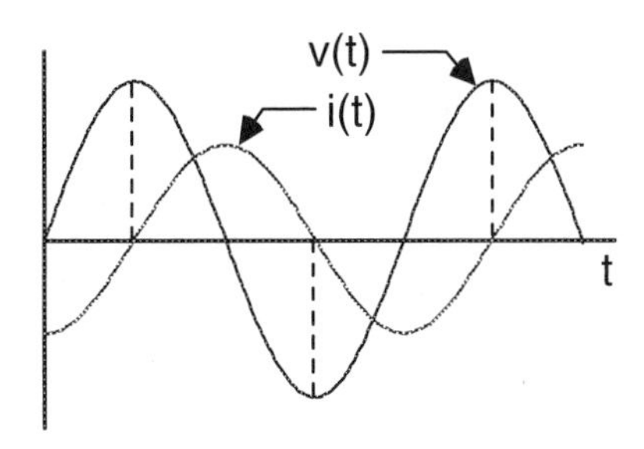

Figure 3.3: AC voltage and current in the case of an inductor.

Finally, we consider the case of a **capacitor**. Recall that the terminal relationship between the current and voltage in the case of a capacitor is:

$$i(t) = C\frac{dv(t)}{dt}. \tag{3.22}$$

We first take the time derivative of the voltage:

$$\frac{d}{dt}(V_m \cos(\omega t + \phi_V)) = -\omega V_m \sin(\omega t + \phi_V). \tag{3.23}$$

By substituting (3.23) and (3.5) into (3.22), we find:

$$I_m \cos(\omega t + \phi_I) = -C\omega V_m \sin(\omega t + \phi_V). \tag{3.24}$$

By making use of trigonometric identity (3.16), we transform (3.24) as follows:

$$I_m \cos(\omega t + \phi_I) = C\omega V_m \cos\left(\omega t + \phi_V + \frac{\pi}{2}\right). \tag{3.25}$$

From this expression we find that the peak values and phases are related by:

$$I_m = \omega C V_m, \tag{3.26}$$

$$\phi_I = \phi_V + \frac{\pi}{2}, \tag{3.27}$$

which can also be written as follows:

$$V_m = \frac{I_m}{\omega C}, \tag{3.28}$$

$$\phi_V = \phi_I - \frac{\pi}{2}. \tag{3.29}$$

Thus, in the case of capacitors, we see that the ac voltage lags behind current by $\pi/2$, or, in other words, the ac current leads the ac voltage by $\pi/2$. Figure 3.4 gives a graphical representation of ac voltage and current in the case of a capacitor.

EXAMPLE 3.1 Let's assume that we have a sinusoidal voltage source with a fixed amplitude of $V_m = 5$ V and a fixed frequency of $f = 1$ MHz. What would the

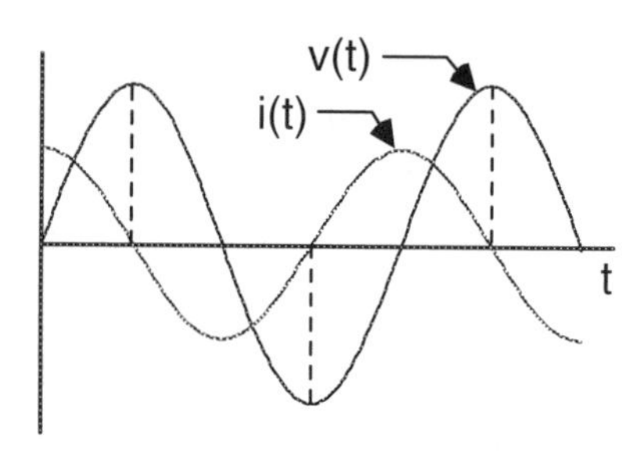

Figure 3.4: AC voltage and current in the case of a capacitor.

maximum current be through (a) a 5 Ω resistor, (b) a 5 μH inductor, and (c) a 5 μF capacitor?

Solution: (a) From (3.11) $I_m = V_m/R = 5\ \text{V}/5\ \Omega = 1$ A, (b) from (3.20) $I_m = V_m/\omega L = 5\ \text{V}/(2\pi \times 10^6\ \text{Hz} \times 5 \times 10^{-6}\ \text{H}) = 1/(2\pi)$ A $= 159$ mA, (c) from (3.26) $I_m = \omega C V_m = 2\pi \times 10^6\ \text{Hz} \times 5 \times 10^{-6}\ \text{F} \times 5\ \text{V} = 50\pi\ \text{A} = 157$ A.■

Based on the previous discussion, we can conclude that ac voltages across resistors, inductors, and capacitors result in ac currents, and vice versa. This fact follows from the linearity of the terminal relationships for resistors, capacitors, and inductors. Consequently, if all the voltage and current sources in a circuit are ac sources with the same frequency, then voltages and currents everywhere in the circuit are sinusoidal functions with that frequency. In the next few sections, we will present a circuit analysis method based on this fact.

3.4 Phasors

In the previous chapter, we discussed KCL and KVL for electric circuits and observed that they result in a system of b linear differential equations, where b is the number of branches in the circuit. By using these differential equations, we can solve for the currents and voltages in a circuit, and indeed, in some cases, it is the only method of choice. However, when we deal with sinusoidal voltages and currents (this is the case of ac steady state), there is a method which avoids solving a system of differential equations, a task which can be very difficult and time consuming. This method is called the *phasor technique* and its basic idea is to reduce differential equations for circuit variables to algebraic equations for phasors.

A *phasor* is a complex number[1] used for the representation of a sinusoidal quantity. To start the discussion, we consider an arbitrary sinusoidal quantity:

$$g(t) = G_m \cos(\omega t + \phi). \tag{3.30}$$

We wish to represent this sinusoidal quantity by a complex number. To do this, we make use of Euler's formula

$$e^{j\theta} = \cos\theta + j\sin\theta \tag{3.31}$$

where j is $\sqrt{-1}$ (electrical engineers often use j instead of the usual symbol i because the latter is used to represent currents). Notice that the right-hand side of Euler's formula is a complex number whose real part is $\cos\theta$ and whose imaginary part is $\sin\theta$. So it is clear that

$$\text{Re}\,(e^{j\theta}) = \cos\theta, \tag{3.32}$$

where Re means the real part of the complex number. From this equation, we can see how to rewrite the sinusoidal quantity $g(t)$ in complex form:

$$g(t) = G_m \cos(\omega t + \phi) = G_m\,\text{Re}(e^{j(\omega t + \phi)}). \tag{3.33}$$

[1]A brief review of complex numbers is included in Appendix A.

By using simple rules for exponents and complex numbers, we can rewrite the last equation as:

$$g(t) = G_m \operatorname{Re}(e^{j\phi} e^{j\omega t}), \tag{3.34}$$

$$g(t) = \operatorname{Re}(G_m e^{j\phi} e^{j\omega t}), \tag{3.35}$$

$$g(t) = \operatorname{Re}(\hat{G} e^{j\omega t}), \tag{3.36}$$

where

$$\hat{G} = G_m e^{j\phi}. \tag{3.37}$$

Equation (3.37) is the expression for the phasor of $g(t)$. (In this text, a phasor will be denoted by a capital letter with a "hat.") Phasors can be completely characterized by two quantities: the magnitude (absolute value) and the polar angle. According to equation (3.37) the magnitude is G_m and the polar angle is ϕ. The magnitude of the phasor is equal to the peak value of the sinusoidal quantity, while the polar angle of the phasor is equal to the initial phase of the sinusoidal quantity. It is important to note that the phasor $\hat{G}$ does not depend on the frequency ω of $g(t)$, and, therefore, we can conclude that phasors can be used only to describe sinusoidal quantities of the same fixed frequency.

We can freely convert from phasor representation to sinusoidal representation and vice versa using the following rules:

$$g(t) = G_m \cos(\omega t + \phi) \rightarrow \hat{G} = G_m e^{j\phi}, \tag{3.38}$$

$$\hat{G} = G_m e^{j\phi} \rightarrow g(t) = \operatorname{Re}(\hat{G} e^{j\omega t}) = G_m \cos(\omega t + \phi). \tag{3.39}$$

In ac steady-state analysis of electric circuits, the equations are written and solved in terms of phasors, and then the final answers are converted back into the time domain form by using (3.39).

Next, we consider some examples.

EXAMPLE 3.2 Given a sinusoidal voltage:

$$v(t) = 50 \cos\left(50t + \frac{\pi}{6}\right) \text{ V}, \tag{3.40}$$

find the expression for the phasor.

We know that the peak value of $v(t)$ is equal to the magnitude of the phasor and that the initial phase of $v(t)$ is equal to the polar angle of the phasor. So, the expression for the phasor is:

$$\hat{V} = 50e^{j\pi/6} = 50\left(\cos\frac{\pi}{6} + j\sin\frac{\pi}{6}\right) = 25\sqrt{3} + 25j \text{ V}, \tag{3.41}$$

where the phasor $\hat{V}$ has been written in algebraic form by using Euler's formula. ■

EXAMPLE 3.3 Given the phasor representation of the sinusoidal voltage:

$$\hat{V} = 30 + 30j \text{ V}, \tag{3.42}$$

find the expression for the corresponding ac voltage.

We start by finding the magnitude (or absolute value) of the phasor from (3.42):

$$|\hat{V}| = \sqrt{30^2 + 30^2} = 30\sqrt{2}\ \text{V}. \tag{3.43}$$

We then find the polar angle:

$$\tan \phi_V = \frac{30}{30} = 1, \tag{3.44}$$

$$\phi_V = \frac{\pi}{4}. \tag{3.45}$$

Now we can write the phasor expression as

$$\hat{V} = 30\sqrt{2}e^{j\pi/4}\ \text{V}. \tag{3.46}$$

Using equation (3.39) we can convert the phasor into a sinusoidal quantity:

$$v(t) = 30\sqrt{2}\cos\left(\omega t + \frac{\pi}{4}\right)\ \text{V}. \tag{3.47}$$

We choose the angular frequency arbitrarily to be ω because it will usually be specified in the problem and we do not have to worry about it. ■

Now consider another form of a sinusoidal voltage:

$$v(t) = A\cos\omega t - B\sin\omega t. \tag{3.48}$$

We want to find the phasor expression for $v(t)$. This problem appears somewhat different from the previous ones, and we do not yet have a formula which will immediately give the answer. We start by observing that $v(t)$ is written as a sum of sine and cosine terms of the same frequency. It is known that such a sum can always be reduced to the form:

$$v(t) = V_m\cos(\omega t + \phi_V) \tag{3.49}$$

in a simple manner. Equation (3.49) is in a form suitable to be converted into a phasor using one of the previously derived formulas. To convert (3.48) into form (3.49), we shall use the following transformation of (3.48):

$$v(t) = \sqrt{A^2 + B^2}\left(\frac{A}{\sqrt{A^2 + B^2}}\cos\omega t - \frac{B}{\sqrt{A^2 + B^2}}\sin\omega t\right). \tag{3.50}$$

Now if we introduce the notations:

$$V_m = \sqrt{A^2 + B^2}, \tag{3.51}$$

$$\cos\phi_V = \frac{A}{\sqrt{A^2 + B^2}}, \tag{3.52}$$

$$\sin\phi_V = \frac{B}{\sqrt{A^2 + B^2}}, \tag{3.53}$$

equation (3.50) will become:

$$v(t) = V_m(\cos\phi_V\cos\omega t - \sin\phi_V\sin\omega t). \tag{3.54}$$

cos

By using the trigonometric identity

$$\cos(\alpha + \beta) = \cos\alpha\cos\beta - \sin\alpha\sin\beta, \tag{3.55}$$

we can see that (3.50) is equivalent to

$$v(t) = V_m \cos(\omega t + \phi_V). \tag{3.56}$$

Once the sinusoidal quantity is in this form, it is trivial to convert it into a phasor.

By repeating the same line of reasoning as before we can show that a sinusoidal voltage

$$v(t) = A\cos\omega t + B\sin\omega t \tag{3.57}$$

can be reduced to

$$v(t) = V_m \cos(\omega t - \phi_V) \tag{3.58}$$

with the same expression as before for V_m and ϕ_V.

EXAMPLE 3.4 Consider the following problem. Express the voltage

$$v(t) = 40\cos\omega t - 40\sin\omega t \text{ V} \tag{3.59}$$

in phasor form.

Comparing equation (3.59) to equation (3.50) we identify $A = 40$ and $B = 40$. Next, we must find the magnitude of the phasor from equation (3.51):

$$|\hat{V}| = \sqrt{40^2 + 40^2} = 40\sqrt{2} \text{ V}. \tag{3.60}$$

Finally, we need the polar angle, which we can determine in many ways, one of which is to use equation (3.52)

$$\cos\phi_V = \frac{40}{40\sqrt{2}} = \frac{1}{\sqrt{2}}, \tag{3.61}$$

$$\phi_V = \frac{\pi}{4}. \tag{3.62}$$

Now we have all the information we need to express $v(t)$ as either a "single" sinusoidal quantity or a phasor:

$$v(t) = 40\sqrt{2}\cos\left(\omega t + \frac{\pi}{4}\right) \text{ V}, \tag{3.63}$$

$$\hat{V} = 40\sqrt{2}e^{j\pi/4} \text{ V}. \tag{3.64}$$

■

3.5 Impedance and Admittance

In Section 3.3 we discussed the relationships between ac voltages and currents for the basic two terminal elements. This time we shall express these relationships in phasor form.

To find the relationship between the phasors of ac current and voltage in the case of a **resistor**, we recall equation (3.10):

$$V_m \cos(\omega t + \phi_V) = RI_m \cos(\omega t + \phi_I). \tag{3.65}$$

Converting this equation into phasor form yields:

$$V_m e^{j\phi_V} = RI_m e^{j\phi_I}, \tag{3.66}$$

or written more succinctly:

$$\hat{V} = R\hat{I}. \tag{3.67}$$

We define the *impedance* of a circuit branch to be the ratio of the phasor voltage across the branch to the current through the branch. We denote impedances with an uppercase Z. The impedance for a resistor is simply equal to its resistance value:

$$Z = \hat{V}/\hat{I} = R. \tag{3.68}$$

Note that the units for impedance are ohms.

Now, consider the case of an **inductor**. We start with the sinusoidal relationship (see (3.17)):

$$V_m \cos(\omega t + \phi_V) = \omega L I_m \cos\left(\omega t + \phi_I + \frac{\pi}{2}\right). \tag{3.69}$$

Converting (3.69) into phasor form and manipulating algebraically yields:

$$V_m e^{j\phi_V} = \omega L I_m e^{j(\phi_I + \pi/2)} = \omega L I_m e^{j\phi_I} e^{j\pi/2} = j\omega L I_m e^{j\phi_I}, \tag{3.70}$$

where we made use of the fact that $e^{j\pi/2} = j$. The last expression is equivalent to:

$$\hat{V} = j\omega L\hat{I}, \tag{3.71}$$

which results in an impedance of

$$Z = \hat{V}/\hat{I} = j\omega L. \tag{3.72}$$

In the last derivation, the power of phasors is made evident. Recall from earlier sections that the voltage across an inductor is related to the current through an inductor by the equation

$$v(t) = L\frac{di(t)}{dt}. \tag{3.73}$$

But when using phasors we have

$$\hat{V} = j\omega L\hat{I}. \tag{3.74}$$

Therefore the operation of differentiation is reduced to multiplication by a factor of $j\omega$:

$$\frac{d}{dt} \rightarrow j\omega. \tag{3.75}$$

Thus, the differential operations on sinusoidal *functions* are replaced by algebraic operations on corresponding phasors, which are *complex numbers*.

To find the impedance of a **capacitor**, we begin with the sinusoidal relationship (see (3.25)):

$$V_m \cos\left(\omega t + \phi_V + \frac{\pi}{2}\right) = \frac{I_m}{\omega C}\cos(\omega t + \phi_I). \tag{3.76}$$

Converting (3.76) into phasor form and manipulating algebraically yields

$$V_m e^{j(\phi_V + \frac{\pi}{2})} = \frac{I_m}{\omega C} e^{j\phi_I}, \tag{3.77}$$

$$V_m e^{j\phi_V} e^{j\frac{\pi}{2}} = \frac{I_m}{\omega C} e^{j\phi_I}, \tag{3.78}$$

$$jV_m e^{j\phi_V} = \frac{I_m}{\omega C} e^{j\phi_I}, \tag{3.79}$$

$$\hat{V} = \frac{-j}{\omega C}\hat{I}, \tag{3.80}$$

$$Z = \hat{V}/\hat{I} = 1/j\omega C, \tag{3.81}$$

where again use was made of the identity $e^{j\pi/2} = j$.

Now, let us recall from earlier sections the expression relating the voltage across the capacitor to the current through the capacitor:

$$v(t) = \frac{1}{C}\int i(t)dt. \tag{3.82}$$

By comparing the previous equations we can see that the operation of integration of sinusoidal functions is reduced to multiplication of the corresponding phasors by a factor of $-j/\omega$ (which is also equivalent to division by $j\omega$):

$$\int \rightarrow \frac{-j}{\omega}. \tag{3.83}$$

Thus, the true power of the phasor technique is evident—it reduces the operations of calculus on sinusoidal quantities to those of simple algebra on phasors.

EXAMPLE 3.5 Calculate the impedances of a 50 Ω resistor, a 1 μH inductor, and a 100 μF capacitor at the following frequencies: $f = 0$ Hz, 1 kHz, 1 MHz, 1 GHz, ∞.

Solution: The results are given in Table 3.1 as a function of angular frequency and come from plugging the parameter values into equations (3.68), (3.72), and (3.81). We note that for dc signals ($\omega = 0$), inductors have zero impedance and can be replaced by short circuits (wires), while capacitors have infinite impedance and can be replaced by open circuits. As the frequency increases, inductor impedances increase while capacitor impedances decrease. At very high frequencies, inductors can be modeled by open circuits while capacitors can be approximated by short circuits.

Table 3.1: Sample impedances of passive components.

ω (rad/s)	$R = 50\ \Omega$	$L = 1\ \mu$H	$C = 100\ \mu$F
0	50 Ω	0	∞
$2\pi \times 10^3$	50 Ω	$j6.28$ mΩ	$-j1.59$ Ω
$2\pi \times 10^6$	50 Ω	$j6.28$ Ω	$-j1.59$ mΩ
$2\pi \times 10^9$	50 Ω	$j6.28$ kΩ	$-j1.59$ μΩ
∞	50 Ω	∞	0

■

The next discussion may seem trivial to some, but it is very important if we are to use phasors to represent sinusoidal quantities in the circuit equations. The question: Is the phasor of a sum of sinusoidal quantities equal to the sum of the phasors of the same sinusoidal quantities? The answer is affirmative.

Assume $g(t)$ is a sum of sinusoidal quantities:

$$g(t) = \sum_k G_{mk} \cos(\omega t + \phi_k), \tag{3.84}$$

then we want to prove that

$$\hat{G} = \sum_k \hat{G}_k, \tag{3.85}$$

where $\hat{G}$ is the phasor of $g(t)$, while $\hat{G}_k$ are the phasors of the corresponding sinusoidal terms of the sum in (3.84).

Proof:

$$\begin{aligned} g(t) &= \sum_k G_{mk} \operatorname{Re}\left(e^{j(\omega t + \phi_k)}\right) \\ &= \sum_k \operatorname{Re}\left(G_{mk} e^{j\omega t} e^{j\phi_k}\right) \\ &= \sum_k \operatorname{Re}\left(\hat{G}_k e^{j\omega t}\right) \\ &= \operatorname{Re}\left(\sum_k \hat{G}_k e^{j\omega t}\right) \\ &= \operatorname{Re}\left[\left(\sum_k \hat{G}_k\right) e^{j\omega t}\right] \\ &= \operatorname{Re}\left[\hat{G} e^{j\omega t}\right]. \end{aligned}$$

By looking at the last two lines of this expression, one can see that the phasor of the sum of sinusoidal quantities is indeed equal to the sum of the phasors of sinusoidal quantities. Now that this little proof is secure, we can proceed to the systematic use of phasors in ac steady-state analysis of electric circuits.

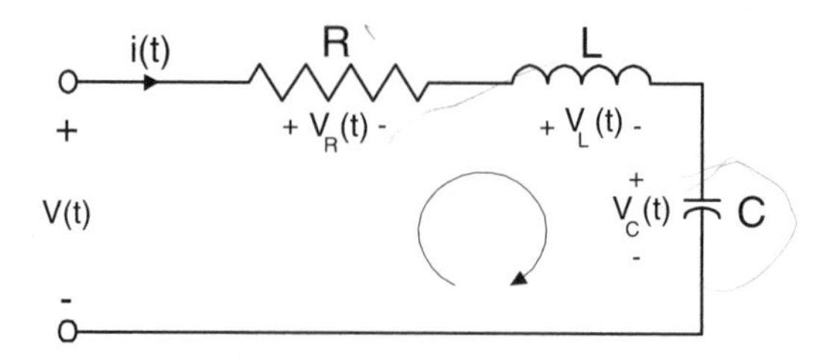

Figure 3.5: A general *RLC* branch.

To continue our study of ac circuits, we now consider a general branch. A diagram of this branch is shown in Figure 3.5. It is simply a resistor, a capacitor, and an inductor connected in series, which means that the same current flows through each of these elements. This general branch is called an *RLC* branch, and it is the basic branch of many ac circuits. We intend to show that each *RLC* branch can be characterized by an *impedance* just as we did with individual elements.

We wish to determine an expression for impedance in terms of resistance, inductance, and capacitance. To start, we apply KVL to the loop shown in Figure 3.5:

$$v(t) = v_R(t) + v_L(t) + v_C(t), \tag{3.86}$$

where $v(t)$ has been moved to the left-hand side of the equation. By assuming that all quantities in the branch are sinusoidal and by using (3.85), we now convert equation (3.86) into phasor form:

$$\hat{V} = \hat{V}_R + \hat{V}_L + \hat{V}_C. \tag{3.87}$$

We next use the terminal relationships for resistors, inductors and capacitors in the phasor form:

$$\hat{V}_R = R\hat{I}, \tag{3.88}$$

$$\hat{V}_L = j\omega L\hat{I}, \tag{3.89}$$

$$\hat{V}_C = \frac{-j}{\omega C}\hat{I}. \tag{3.90}$$

As a result, equation (3.87) becomes:

$$\hat{V} = R\hat{I} + j\omega L\hat{I} + \frac{-j}{\omega C}\hat{I} = \hat{I}\left[R + j\left(\omega L - \frac{1}{\omega C}\right)\right]. \tag{3.91}$$

Remember that the current $\hat{I}$ is the same for each element; that is why we factored it out in the last expression. Now, we define the impedance Z as:

$$Z = \left[R + j\left(\omega L - \frac{1}{\omega C}\right)\right], \tag{3.92}$$

and substitute (3.92) back into (3.91), which leads to:

$$\hat{V} = \hat{I}Z. \tag{3.93}$$

Thus, we can see that every *RLC* branch can be characterized by impedance, defined by equation (3.92), and the phasor of the branch voltage is related to the phasor of the branch current by expression (3.93), which is mathematically similar to Ohm's law.

As we can see from (3.92), impedance is a complex number whose real part is the *resistance* and whose imaginary part, called the *reactance*, is a combination of two terms which are due to inductance and capacitance, respectively. The reactance is due to the energy storage elements of the circuit and is denoted as X, leading to an alternative expression for impedance:

$$Z = R + jX, \tag{3.94}$$

where

$$X = \left(\omega L - \frac{1}{\omega C}\right). \tag{3.95}$$

Since impedance is a complex number, we can represent it in the polar form,

$$Z = |Z|e^{j\phi}, \tag{3.96}$$

where $|Z|$ is the absolute value and ϕ is the polar angle of the impedance. These quantities can be found by converting from Cartesian to polar form (see Appendix A). This leads to the following important formulas:

$$|Z| = \sqrt{R^2 + X^2} = \sqrt{R^2 + \left(\omega L - \frac{1}{\omega C}\right)^2}, \tag{3.97}$$

$$\tan\phi = \frac{X}{R}. \tag{3.98}$$

Rewriting equation (3.93) in polar form yields

$$V_m e^{j\phi_V} = I_m e^{j\phi_I}|Z|e^{j\phi}. \tag{3.99}$$

We can now relate the peak values and initial phases:

$$V_m = I_m|Z|, \tag{3.100}$$

$$\phi_V = \phi_I + \phi, \tag{3.101}$$

$$\phi_V - \phi_I = \phi. \tag{3.102}$$

From (3.102) we can see that the polar angle of the impedance is the phase difference between the ac branch voltage and ac branch current.

EXAMPLE 3.6 Now, we illustrate the previous discussion by the following example. Consider the circuit shown in Figure 3.5. Given $v(t) = V_m \cos(\omega t + \varphi_V)$ and the values of R, L, and C, we wish to find the expression for the current $i(t)$ by using the phasor method.

We begin with converting all known quantities into phasor-impedance form. First, we convert the given voltage $v(t)$ into a phasor form:

$$v(t) \rightarrow \hat{V} = V_m e^{j\varphi_V}. \tag{3.103}$$

Then, we introduce the impedance given by the expression:

$$Z = R + j\left(\omega L - \frac{1}{\omega C}\right) = |Z|e^{j\phi}. \tag{3.104}$$

We have represented the impedance in polar form to make the calculations easier. In general, when multiplying or dividing phasors, the polar form is more desirable, but when adding or subtracting phasors, the algebraic form is convenient. We can now use equation (3.93) to relate the phasors of branch voltage and current:

$$\hat{V} = \hat{I}Z, \tag{3.105}$$

$$\hat{I} = \frac{\hat{V}}{Z} = \frac{V_m e^{j\varphi_V}}{|Z|e^{j\phi}} = \frac{V_m}{|Z|}e^{j(\varphi_V - \phi)}. \tag{3.106}$$

We now have the phasor of the branch current, and we want to find the expression for the current in the time domain. Thus, we must convert the phasor of the current into the sinusoidal quantity which it represents:

$$\hat{I} = \frac{V_m}{|Z|}e^{j(\varphi_V - \phi)} \rightarrow i(t) = \frac{V_m}{|Z|}\cos(\omega t + \varphi_V - \phi). \tag{3.107}$$

Expression (3.107) is the solution to the problem. By using the phasor technique, we were able to obtain the expression for the current through the branch without resorting to any calculus. All ac steady-state problems will be solved in a similar fashion by using the basic phasor techniques outlined in this and subsequent sections. ■

If any one of the elements is removed from the general RLC branch, the new general impedance is found simply by deleting from (3.92) the term which corresponds to the deleted element. For example, the RC branch shown in Figure 3.6 has the impedance

$$Z = R - j\frac{1}{\omega C}. \tag{3.108}$$

Similarly, an RL branch would have an impedance of

$$Z = R + j\omega L \tag{3.109}$$

and an LC branch would have a purely imaginary impedance of

$$Z = j\left(\omega L - \frac{1}{\omega C}\right). \tag{3.110}$$

To calculate the impedance for a general branch, one must simply know the elements of the branch and take them into account according to equations (3.92), (3.108)–(3.110), or (3.68), (3.72), or (3.81) for single-element branches.

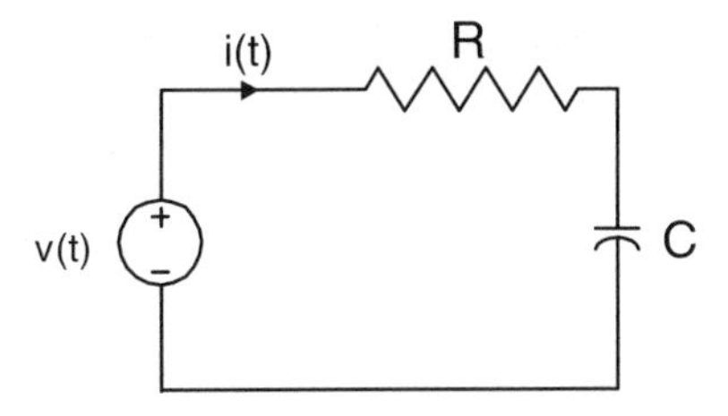

Figure 3.6: An RC branch driven by a voltage source.

EXAMPLE 3.7 An RC branch has a current $i(t) = 50\sin(377t)$ mA flowing through it when a voltage of $v(t) = 120\sqrt{2}\cos(377t - 150°)$ V is applied across it (see Figure 3.6). Find the resistance and the capacitance of the branch elements.

Solution: The impedance of the RC branch is given by (3.108):

$$Z = R - \frac{j}{\omega C} = \hat{V}/\hat{I} = \frac{120\sqrt{2}e^{-j150°}}{.05e^{-j90°}}\ \Omega$$

or

$$R - \frac{j}{\omega C} = 2{,}400\sqrt{2}e^{-j60°} = 1{,}697 - j2{,}939\ \Omega.$$

The real part of the previous equation yields the resistance:

$$R \approx 1.7\ \text{k}\Omega.$$

The imaginary part of the equation, coupled with the angular frequency $\omega = 377$ rad/s, yields:

$$C = \frac{1}{377(2939)} = 0.9\ \mu\text{F}.$$

■

In addition to impedance, there is another quantity which can be used to characterize any branch at ac steady state. This quantity is called *admittance* and is defined as the reciprocal of impedance:

$$Y = \frac{1}{Z}, \tag{3.111}$$

where Y is used to denote admittance. Admittance is analogous to the conductance when dealing with dc resistive circuits. However, the admittance is in general a complex number made up of both real and imaginary parts:

$$Y = G + jB, \tag{3.112}$$

where G is the *conductance* and B is called the *susceptance*. Like impedance, the admittance can also be used to relate the phasors of branch current and voltage:

$$\hat{I} = \hat{V}Y. \tag{3.113}$$

There is a simple relationship between conductance/susceptance and resistance/reactance. Note that:

$$R + jX = Z = \frac{1}{Y} = \frac{1}{G + jB} = \frac{1}{(G + jB)} \frac{G - jB}{(G - jB)} = \frac{G - jB}{G^2 + B^2}. \quad (3.114)$$

By equating real and imaginary parts we find that

$$R = \frac{G}{G^2 + B^2} \quad \text{and} \quad X = -\frac{B}{G^2 + B^2}. \quad (3.115)$$

Thus, the conductance is equal to the reciprocal of the resistance only if the susceptance is zero. Admittances will be used more frequently later in the text, when the method of node potentials is studied.

3.6 AC Steady-State Equations

We begin with the phasor versions of Kirchhoff's laws. Recall from earlier sections that the KCL and KVL equations are written in time domain form as

$$\sum_k i_k(t) = 0, \quad (3.116)$$

$$\sum_k v_k(t) = 0. \quad (3.117)$$

We now rewrite them in a slightly different form by moving all known driving (voltage and current) sources to the right-hand side of these equations:

$$\sum_k i_k(t) = -\sum_k i_{sk}(t), \quad (3.118)$$

$$\sum_k v_k(t) = -\sum_k v_{sk}(t). \quad (3.119)$$

In the case of ac steady state, all variables in equations (3.118) and (3.119) are sinusoidal. Consequently, by using the proven fact that the phasor of the sum of sinusoidal quantities is equal to the sum of phasors of the same sinusoidal quantities, the above equations can be converted into the phasor form:

$$\sum_k \hat{I}_k = -\sum_k \hat{I}_{sk}, \quad (3.120)$$

$$\sum_k \hat{V}_k = -\sum_k \hat{V}_{sk}. \quad (3.121)$$

For each branch, phasors of branch voltage and current are related to one another through the impedance of the branch:

$$\hat{V}_k = \hat{I}_k Z_k. \quad (3.122)$$

By substituting (3.122) into (3.121), we arrive at the basic ac steady-state equations:

$$\sum_k \hat{I}_k = -\sum_k \hat{I}_{sk}, \tag{3.123}$$

$$\sum_k \hat{I}_k Z_k = -\sum_k \hat{V}_{sk}. \tag{3.124}$$

These are simultaneous *linear* algebraic equations with respect to phasors of branch currents. The total number of these equations is equal to the total number of branches, which is b. By solving these equations, we can find $\hat{I}_k$ and then, by using (3.122), we can determine phasors of branch voltages $\hat{V}_k$. As soon as we know the phasors of branch voltages and currents, we can compute their instantaneous values using the rule (3.39) of conversion of phasors into sinusoidal functions. Equations (3.123) and (3.124) can be solved numerically by using the techniques developed in linear algebra, for instance, the Gaussian elimination technique.

Now, we summarize how the basic equations (3.123) and (3.124) can be used in circuit analysis. We assume that an electric circuit is given. This means resistances, capacitances, and inductances as well as the driving sources in the electric circuit are specified. Our goal is to solve for the branch voltages and currents.

We begin by introducing reference directions and reference polarities for all the branch currents and voltages. We then convert the driving sources into phasors. Next, we characterize each branch by its impedance. Finally, we write the basic equations (3.123) and (3.124) for all independent nodes and loops (meshes), respectively. We solve these equations and convert phasors into sinusoidal quantities. We illustrate this summary by the following examples.

EXAMPLE 3.8 Consider the circuit shown in Figure 3.7. We are given:

$$v_{s1}(t) = V_{ms1}\cos(\omega t + \phi_{s1}), \tag{3.125}$$

$$v_{s2}(t) = V_{ms2}\cos(\omega t + \phi_{s2}), \tag{3.126}$$

and L_1, R_1, C_2, R_3, L_3, and we want to find all currents and voltages.

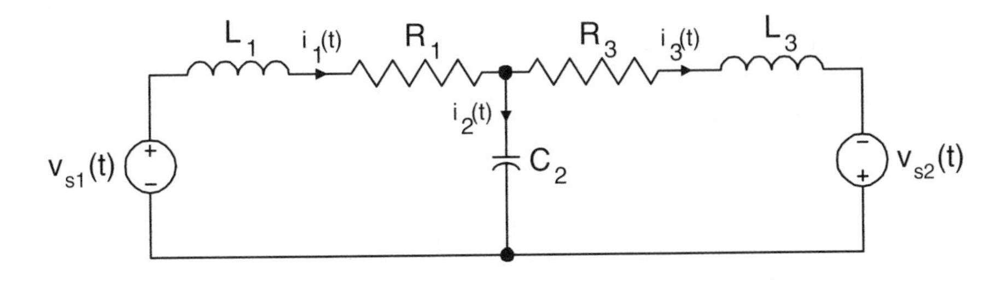

Figure 3.7: Example circuit.

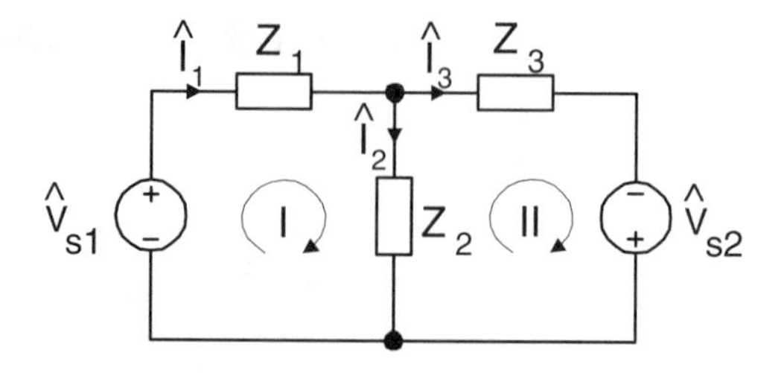

Figure 3.8: The same circuit in phasor notation.

First, we begin by introducing reference directions, already shown in Figure 3.7. Now, we convert the source voltages into phasors:

$$v_{s1}(t) \longrightarrow \hat{V}_{s1} = V_{ms1}e^{j\phi_{s1}}, \tag{3.127}$$

$$v_{s2}(t) \longrightarrow \hat{V}_{s2} = V_{ms2}e^{j\phi_{s2}}. \tag{3.128}$$

Then we characterize each branch by its impedance. It helps to redraw the circuit in phasor notation to easily discern all the branches. Figure 3.8 shows the same circuit, this time drawn in phasor notation with boxes indicating the three different branch impedances. The branch currents are also indicated in phasor form this time. We can easily calculate the three branch impedances:

$$Z_1 = R_1 + j\omega L_1, \tag{3.129}$$

$$Z_2 = -\frac{j}{\omega C_2}, \tag{3.130}$$

$$Z_3 = R_3 + j\omega L_3. \tag{3.131}$$

Once the impedances are found, the circuit equations can be set up. This circuit has only two nontrivial nodes, so only one KCL equation is linearly independent:

$$\hat{I}_1 - \hat{I}_3 - \hat{I}_2 = 0. \tag{3.132}$$

The circuit is also planar, resulting in two KVL equations—one for each mesh.
Mesh I:

$$\hat{I}_1 Z_1 + \hat{I}_2 Z_2 = \hat{V}_{s1} \tag{3.133}$$

Mesh II:

$$-\hat{I}_2 Z_2 + \hat{I}_3 Z_3 = \hat{V}_{s2} \tag{3.134}$$

That is all there is to it. We ended up with three equations, which is what was to be expected since there are only three branches.

We can quickly arrive at the solution of this problem by using the method of substitution. Indeed, from (3.133) and (3.134) we can find $\hat{I}_1$ and $\hat{I}_3$ in terms of $\hat{I}_2$:

$$\hat{I}_1 = -\hat{I}_2\frac{Z_2}{Z_1} + \frac{\hat{V}_{s1}}{Z_1}, \tag{3.135}$$

$$\hat{I}_3 = \hat{I}_2\frac{Z_2}{Z_3} + \frac{\hat{V}_{s2}}{Z_3}. \tag{3.136}$$

By substituting (3.135) and (3.136) into (3.132), we end up with the following equation for $\hat{I}_2$:

$$-\hat{I}_2\frac{Z_2}{Z_1} - \hat{I}_2\frac{Z_2}{Z_3} - \hat{I}_2 = -\frac{\hat{V}_{s1}}{Z_1} + \frac{\hat{V}_{s2}}{Z_3}, \tag{3.137}$$

from which we find:

$$\hat{I}_2 = \frac{-\hat{V}_{s2}Z_1 + \hat{V}_{s1}Z_3}{Z_1Z_2 + Z_1Z_3 + Z_2Z_3}. \tag{3.138}$$

After $\hat{I}_2$ is computed, expressions (3.135) and (3.136) can be used to calculate $\hat{I}_1$ and $\hat{I}_3$. ■

EXAMPLE 3.9 Consider the circuit shown in Figure 3.9. It is assumed that sources $v_s(t)$ and $i_s(t)$ are given:

$$v_s(t) = V_{ms}\cos(\omega t + \varphi_{V_s}), \tag{3.139}$$

$$i_s(t) = I_{ms}\cos(\omega t + \varphi_{I_s}), \tag{3.140}$$

as well as parameters $R_1, C_1, R_2, L_2, R_3, L_4$. We want to find all currents and voltages. First, we begin by introducing reference directions, already shown in Figure 3.9. Next, we convert the ac sources into phasors:

$$v_s(t) \longrightarrow \hat{V}_s = V_{ms}e^{j\varphi_{V_s}}, \tag{3.141}$$

$$i_s(t) \longrightarrow \hat{I}_s = I_{ms}e^{j\varphi_{I_s}}. \tag{3.142}$$

Then, we characterize each branch by its impedance:

$$Z_1 = R_1 - \frac{j}{\omega C_1}, \tag{3.143}$$

$$Z_2 = R_2 + j\omega L_2, \tag{3.144}$$

$$Z_3 = R_3, \tag{3.145}$$

$$Z_4 = R_4 + j\omega L_4. \tag{3.146}$$

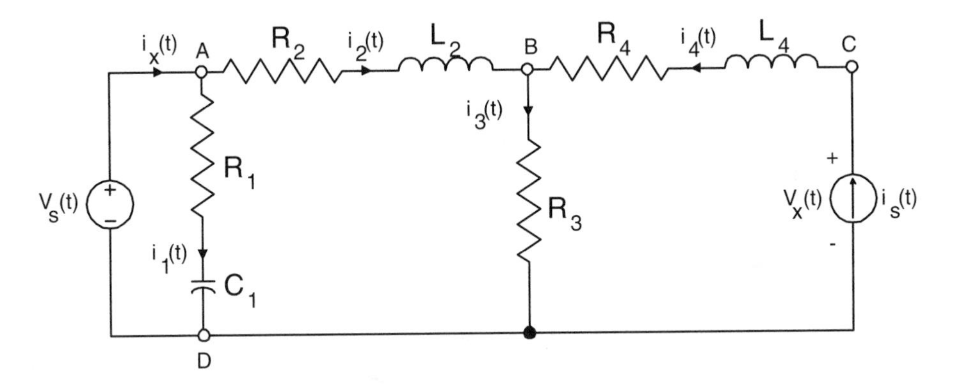

Figure 3.9: Example circuit.

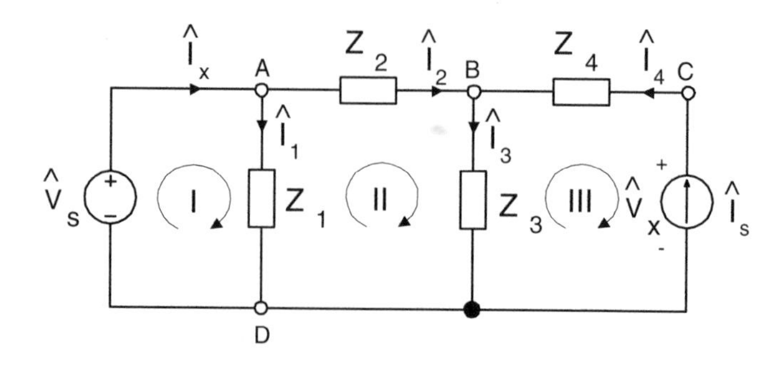

Figure 3.10: The previous circuit in phasor notation.

It is helpful to redraw the circuit in impedance-phasor notations to easily discern all the branches. The redrawn circuit is shown in Figure 3.10.

The analysis of this circuit is facilitated by the following observations. It is clear that node C is a trivial one. Consequently, by using KCL for this node, we obtain:

$$\hat{I}_4 = \hat{I}_s. \tag{3.147}$$

Next, it is also clear that mesh I is a trivial one. By using KVL for this mesh, we obtain:

$$\hat{I}_1 Z_1 = \hat{V}_s, \tag{3.148}$$

which leads to

$$\hat{I}_1 = \frac{\hat{V}_s}{Z_1}. \tag{3.149}$$

Now, we shall apply KCL to nodes A and B and KVL to mesh II. As a result, we arrive at the following equations:

$$\hat{I}_x - \hat{I}_1 - \hat{I}_2 = 0, \tag{3.150}$$

$$\hat{I}_2 - \hat{I}_3 + \hat{I}_4 = 0, \tag{3.151}$$

$$-\hat{I}_1 Z_1 + \hat{I}_2 Z_2 + \hat{I}_3 Z_3 = 0. \tag{3.152}$$

By using (3.147) in (3.151), (3.148) in (3.152), and (3.149) in (3.150), we obtain the following simultaneous linear equations:

$$\hat{I}_x - \hat{I}_2 = \frac{\hat{V}_s}{Z_1}, \tag{3.153}$$

$$\hat{I}_2 - \hat{I}_3 = -\hat{I}_s, \tag{3.154}$$

$$\hat{I}_2 Z_2 + \hat{I}_3 Z_3 = \hat{V}_s. \tag{3.155}$$

To solve these equations, we first use (3.154) in order to express $\hat{I}_3$ in terms of $\hat{I}_2$:

$$\hat{I}_3 = \hat{I}_2 + \hat{I}_s. \tag{3.156}$$

By substituting (3.156) into (3.155), we arrive at the following equation for $\hat{I}_2$:

$$\hat{I}_2 Z_2 + \hat{I}_2 Z_3 + \hat{I}_s Z_3 = \hat{V}_s, \tag{3.157}$$

which can be further transformed as follows:

$$\hat{I}_2(Z_2 + Z_3) = \hat{V}_s - \hat{I}_s Z_3. \tag{3.158}$$

The last equation yields the following expression for $\hat{I}_2$:

$$\hat{I}_2 = \frac{\hat{V}_s - \hat{I}_s Z_3}{Z_2 + Z_3}. \tag{3.159}$$

By substituting the last expression into (3.156), we find $\hat{I}_3$:

$$\hat{I}_3 = \frac{\hat{V}_s + \hat{I}_s Z_2}{Z_2 + Z_3}. \tag{3.160}$$

Finally, by using (3.159) in (3.153), we derive:

$$\hat{I}_x = \frac{\hat{V}_s}{Z_1} + \frac{\hat{V}_s - \hat{I}_s Z_3}{Z_2 + Z_3}. \tag{3.161}$$

To find the voltage $\hat{V}_x$ across the current source $\hat{I}_s$, we apply KVL to loop III:

$$\hat{V}_x - \hat{I}_3 Z_3 - \hat{I}_4 Z_4 = 0. \tag{3.162}$$

By using (3.147) and (3.160) in (3.162) we obtain:

$$\hat{V}_x = \hat{I}_s Z_4 + \frac{\hat{V}_s Z_3 + \hat{I}_s Z_2 Z_3}{Z_2 + Z_3}. \tag{3.163}$$

This concludes the analysis of the above circuit. ■

In conclusion, we would like to stress the mathematical similarity between the basic ac steady-state equations (3.123) and (3.124) and the equations (2.56) and (2.57) for resistive circuits. It is clear that these equations have identical mathematical forms. This explains why the analysis of electric circuits considered in Examples 3.8 and 3.9 closely parallels the analysis of resistive circuits considered in Examples 2.5 and 2.6. The only difference is that in the case of ac steady state we deal with current phasors and impedances, while in the case of resistive circuits we deal with instantaneous currents and resistances. This similarity suggests that the analysis of resistive circuits can be regarded as a particular case of ac steady-state analysis. Consequently, all the analysis techniques which are developed for ac steady state can be directly used for the analysis of resistive circuits. This is the approach which is adopted in this text. It allows one to avoid duplication in the exposition of circuit theory and, at the same time, to emphasize the generality of circuit analysis techniques.

3.7 AC Power

Consider a two-terminal electric device (Figure 3.11) whose terminal voltage $v(t)$ and current $i(t)$ are periodic functions of time with period T. The instantaneous electric power supplied to the device is given by:

$$p(t) = v(t)i(t). \tag{3.164}$$

This power varies with time. For this reason, it is customary to characterize power consumption by average power P_{av} which is defined as follows:

$$P_{av} = \frac{1}{T}\int_0^T p(t)dt. \tag{3.165}$$

First, let us consider a simple case in which the device can be electrically modeled as a pure resistor. In this case, we have the following terminal relationship:

$$v(t) = Ri(t). \tag{3.166}$$

By substituting expression (3.166) into formula (3.164) and then into (3.165), we end up with:

$$P_{av} = \frac{R}{T}\int_0^T i^2(t)dt. \tag{3.167}$$

Next, we introduce the important notion of **root-mean-square** (rms) value of electric current. This value is denoted as I_{rms} and it is defined as follows:

$$I_{rms} = \sqrt{\frac{1}{T}\int_0^T i^2(t)dt}. \tag{3.168}$$

By using (3.168) in (3.167), we obtain:

$$P_{av} = RI_{rms}^2. \tag{3.169}$$

Hence, the root-mean-square value of a periodic current is equal to the dc value of the current which delivers the same average power to a resistor.

By rewriting expression (3.166) as

$$i(t) = \frac{v(t)}{R} \tag{3.170}$$

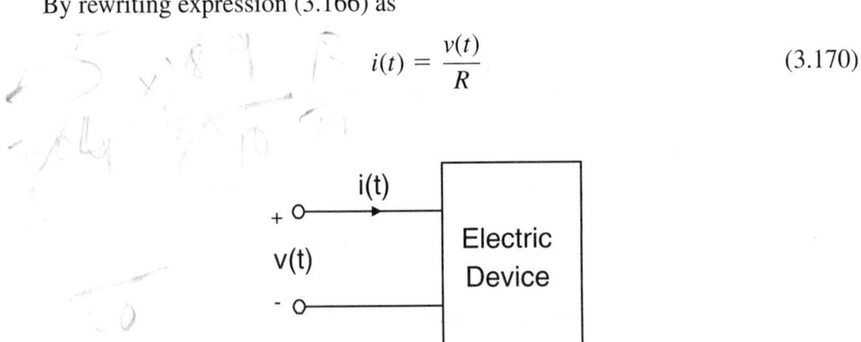

Figure 3.11: Electric device connected to a power network.

and by substituting (3.170) into (3.164) and then into (3.165), we obtain:

$$P_{av} = \frac{1}{R}\left(\frac{1}{T}\int_0^T v^2(t)dt\right). \tag{3.171}$$

This leads to the following definition of the rms values of a periodic voltage:

$$V_{rms} = \sqrt{\frac{1}{T}\int_0^T v^2(t)dt}. \tag{3.172}$$

By using (3.172) in (3.171), we arrive at:

$$P_{av} = \frac{V_{rms}^2}{R}. \tag{3.173}$$

Again, we see that the root-mean-square value of a periodic voltage is equal to the dc value of the voltage which delivers the same average power to a resistor. Thus, the idea behind the notion of root-mean-square value of a periodic time-varying quantity is to replace this time-varying quantity by a dc quantity which will guarantee the same average power.

For the sake of notational simplicity, in the subsequent discussion we shall omit the subscript "rms" and we shall use capital letters without any subscripts for the notations of root-mean-square values. In other words, we introduce the following convention:

$$I_{rms} = I, \qquad V_{rms} = V. \tag{3.174}$$

By using this convention, expressions (3.169) and (3.173) can be written as follows:

$$P_{av} = RI^2 = \frac{V^2}{R}. \tag{3.175}$$

It is important to stress here that root-mean-square values (3.168) and (3.172) can be introduced for **any** periodic current and voltage. However, in the case of a sinusoidal quantity, a very simple relation can be established between its rms and peak values. To see this, consider a sinusoidal current $i(t)$:

$$i(t) = I_m \cos(\omega t + \varphi_I). \tag{3.176}$$

Then:

$$i^2(t) = I_m^2 \cos^2(\omega t + \varphi_I) = \frac{I_m^2}{2}[1 + \cos(2\omega t + 2\varphi_I)]. \tag{3.177}$$

By substituting (3.177) into (3.168), we find

$$I = \sqrt{\frac{I_m^2}{2}\left[1 + \frac{1}{T}\int_0^T \cos(2\omega t + 2\varphi_I)dt\right]}. \tag{3.178}$$

It is clear that $\cos(2\omega t + 2\varphi_I)$ is a periodic function with period $T/2$. Consequently,

$$\int_0^T \cos(2\omega t + 2\varphi_I)dt = 0. \tag{3.179}$$

By substituting (3.179) into (3.178), we derive:

$$I = \frac{I_m}{\sqrt{2}}. \tag{3.180}$$

By using the same line of reasoning, we can show that in the case of a sinusoidal voltage we have:

$$V = \frac{V_m}{\sqrt{2}}. \tag{3.181}$$

Next, let us consider the general case of ac power supplied to an electric device shown in Figure 3.11. In the general case, this device cannot be modeled as a pure resistor. Hence, its terminal voltage and current are not in phase. Suppose that the voltage and current are given by the expressions:

$$v(t) = V_m \cos(\omega t + \varphi_V), \tag{3.182}$$

$$i(t) = I_m \cos(\omega t + \varphi_I). \tag{3.183}$$

By substituting (3.182) and (3.183) into (3.164), we obtain:

$$P(t) = V_m I_m \cos(\omega t + \varphi_V)\cos(\omega t + \varphi_I). \tag{3.184}$$

Now, we shall use the following trigonometric identity:

$$\cos\alpha \cos\beta = \frac{1}{2}[\cos(\alpha - \beta) + \cos(\alpha + \beta)]. \tag{3.185}$$

By assuming that $\alpha = \omega t + \varphi_V$ and $\beta = \omega t + \varphi_I$ and by using (3.185) in (3.184), we transform the latter as follows:

$$p(t) = \frac{V_m I_m}{2}[\cos(\varphi_V - \varphi_I) + \cos(2\omega t + \varphi_V + \varphi_I)]. \tag{3.186}$$

By substituting (3.186) into (3.165) and by taking into account that

$$\int_0^T \cos(2\omega t + \varphi_V + \varphi_I)dt = 0, \tag{3.187}$$

we obtain:

$$P_{av} = \frac{V_m I_m}{2} \cos(\varphi_V - \varphi_I). \tag{3.188}$$

Finally, by recalling (3.180) and (3.181) and by introducing the phase difference φ between terminal voltage and current

$$\varphi = \varphi_V - \varphi_I, \tag{3.189}$$

we arrive at the following very important formula:

$$P_{av} = VI \cos\varphi. \tag{3.190}$$

This formula can be used to compute an average ac power supplied to (or consumed by) an electric device when the rms values of terminal current and voltage as well as their phase difference are known. The same average power can also be computed by using the current and voltage phasors. To derive the appropriate expression, we introduce the phasor $\hat{V}$ of ac terminal voltage (3.182) and the complex conjugate $\hat{I}^*$ of the phasor of the ac terminal current (3.183):

$$\hat{V} = V_m e^{j\varphi_V}, \tag{3.191}$$

$$\hat{I}^* = I_m e^{-j\varphi_I}. \tag{3.192}$$

Now, we define the *complex power* $\hat{S}$ as follows:

$$\hat{S} = \frac{1}{2}\hat{V}\hat{I}^*. \tag{3.193}$$

From (3.191), (3.192), and (3.193) we find:

$$\hat{S} = \frac{1}{2}V_m I_m e^{j(\varphi_V - \varphi_I)}. \tag{3.194}$$

By recalling (3.180), (3.181), and (3.189), we transform the last expression as follows:

$$\hat{S} = VIe^{j\varphi}. \tag{3.195}$$

By comparing (3.190) with (3.195), we conclude that:

$$P_{av} = \text{Re}\,\{\hat{S}\}. \tag{3.196}$$

From (3.196) and (3.193), we obtain:

$$P_{av} = \text{Re}\left(\frac{1}{2}\hat{V}\hat{I}^*\right). \tag{3.197}$$

The last expression allows one to compute average power by using the phasors of the terminal ac voltage and current.

At this point, we shall return to expression (3.190) and shall discuss its practical implications. The factor $\cos\varphi$ in (3.190) is called the *power factor* and it has a significant economic impact as far as distribution and consumption of ac power are concerned. To understand this impact, we remark that ac power is usually supplied at specified constant rms voltage values. Actually, this is one of the main technical challenges for the electric power utility industry: to supply ac power at more or less constant rms voltage values in the face of permanently changing load (consumption). Since V in (3.190) is constant, then the same average power P_{av} can be delivered at: 1) larger rms current values I if the power factor $\cos\varphi$ is small and 2) smaller rms current values I if the power factor $\cos\varphi$ is close to one. Clearly, the first option is undesirable (and in many cases unacceptable) from the economic point of view. This is because the larger rms values I of the terminal current will result in larger ohmic losses [see expression (3.169)] in connecting distribution and transmission lines. Consequently, power utility companies have to generate more power to supply the same average power to a customer with a low power factor than would be required if the customer's power factor were high. This explains why the same average power delivered to a customer with a lower power factor should cost more. This is the reason why the power companies insist that their customers should maintain sufficiently high power factors, and usually adjust their rates to penalize consumers who do not meet their requirements. It is also important to keep in mind that line losses represent electric energy converted into heat and benefit no one. Actually, these losses have negative environmental effects directly and indirectly because they require more electric energy production.

The above discussion suggests that it is important to find an efficient way to raise the power factor of a load. It turns out that the power factor can be corrected by connecting a capacitor across the terminals of the device as shown in Figure 3.12a. This technique achieves its goal when the device has overall *inductive* reactance, that is, when the terminal current lags behind the terminal voltage. In this case, the device can be electrically modeled as a series connection of an inductor and a resistor (see Figure 3.12b). This situation is typical for electric motors, actuators, and various devices which contain multiturn coils. We shall next show how the capacitance C can be chosen to adjust the power factor to one.

For the terminal current $\hat{I}$ in Figure 3.12b we have:

$$\hat{I} = \hat{I}_c + \hat{I}_d, \tag{3.198}$$

where $\hat{I}_c$ and $\hat{I}_d$ are the phasors of electric currents through the capacitor and the device, respectively. Since the same voltage $\hat{V}$ is applied across the capacitor and RL branch, it is clear that the following expressions are valid for the phasors $\hat{I}_c$ and $\hat{I}_d$:

$$\hat{I}_c = j\omega C\hat{V}, \tag{3.199}$$

$$\hat{I}_d = \frac{\hat{V}}{R + j\omega L} = \left(\frac{R}{R^2 + \omega^2 L^2} - j\frac{\omega L}{R^2 + \omega^2 L^2}\right)\hat{V}. \tag{3.200}$$

By substituting (3.199) and (3.200) into (3.198), we derive:

$$\hat{I} = \frac{R}{R^2 + \omega^2 L^2}\hat{V} + j\left(\omega C - \frac{\omega L}{R^2 + \omega^2 L^2}\right)\hat{V}. \tag{3.201}$$

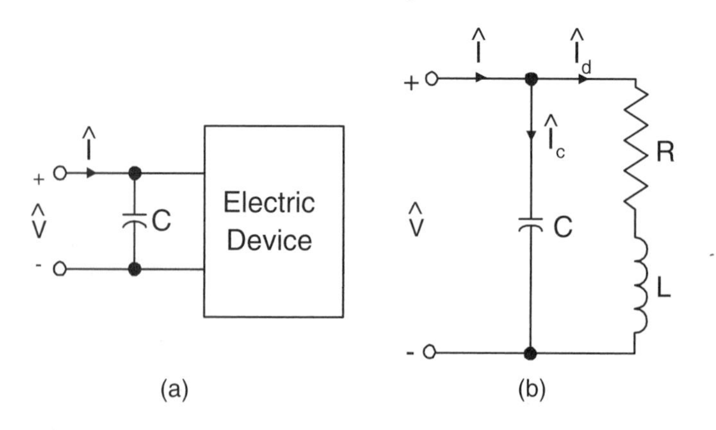

Figure 3.12: Correction of a power factor.

If the capacitance C is chosen to be

$$C = \frac{L}{R^2 + \omega^2 L^2}, \tag{3.202}$$

then from (3.201) we find:

$$\hat{I} = \frac{R}{R^2 + \omega^2 L^2} \hat{V}. \tag{3.203}$$

Since $R/(R^2 + \omega^2 L^2)$ is a real number, it is clear from expression (3.203) that the terminal current and voltage are in phase. Consequently,

$$\varphi = 0, \qquad \cos\varphi = 1. \tag{3.204}$$

It is also clear from (3.201) that the peak (and rms) value of the terminal current achieves its minimum value when the capacitance is chosen according to the expression (3.202).

Finally, the following two remarks are in order. First, by connecting a capacitor across the device terminals, we do not affect the overall average power consumed by the device. This is because a capacitor is an energy storage element and consumes no average power. This is also clear from the fact that for a capacitor $\varphi = \pi/2$, $\cos\varphi = 0$ and, according to (3.190), the average power is equal to zero. Second, by connecting a capacitor across the device, we do not affect the device terminal voltage. This would not be the case if the capacitor were connected in series with the device.

In some applications, it is of interest to find the maximum power that can be delivered to a load by a power network. To find this power, we will model the power network by a voltage source $\hat{V}_s$ and an impedance Z_s. Justification behind this model is the Thevenin theorem, which will be discussed in Chapter 5. By using the above model, we can determine the maximum power that the power network can supply and the manner in which to adjust the load to achieve maximum power transfer.

Consider the circuit shown in Figure 3.13. Here, the power network is represented by the voltage source $\hat{V}_s$ and the impedance Z_s, while the load is modeled by impedance Z_L. All the nodes in the circuit are trivial ones, so with a quick application of KVL around the one mesh we can find:

$$\hat{I}_L = \frac{\hat{V}_s}{Z_s + Z_L}, \tag{3.205}$$

$$\hat{V}_L = \hat{V}_s \frac{Z_L}{Z_s + Z_L}. \tag{3.206}$$

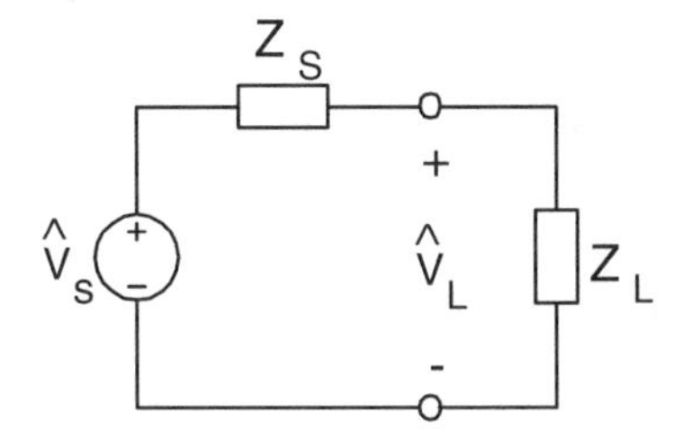

Figure 3.13: Representation of a power network by the voltage source and impedance.

From (3.205), we obtain the complex conjugate $\hat{I}_L^*$ of the load terminal current:

$$\hat{I}_L^* = \frac{\hat{V}_s^*}{(Z_s + Z_L)^*}. \tag{3.207}$$

Now, by using formula (3.197), from (3.205) and (3.207) we derive:

$$P_{av} = \frac{1}{2}|\hat{V}_s|^2 \frac{\mathrm{Re}\{Z_L\}}{|Z_s + Z_L|^2}. \tag{3.208}$$

The last expression can be transformed as follows:

$$P_{av} = \frac{1}{2}V_{ms}^2 \frac{R_L}{(R_s + R_L)^2 + (X_s + X_L)^2}. \tag{3.209}$$

Next, we would like to find such values of R_L and X_L for which P_{av} achieves its maximum. It is clear from (3.209) that for any fixed value of R_L average power achieves its maximum if

$$X_L = -X_s. \tag{3.210}$$

By using (3.210) in (3.209), we find

$$P_{av} = \frac{1}{2}V_{ms}^2 \frac{R_L}{(R_s + R_L)^2}. \tag{3.211}$$

Now, we shall find the value of R_L for which P_{av} given by (3.211) achieves its maximum. To this end, we differentiate (3.211) with respect to R_L:

$$\frac{dP_{av}}{dR_L} = \frac{1}{2}V_{ms}^2 \frac{(R_s + R_L)^2 - 2R_L(R_s + R_L)}{(R_s + R_L)^4} = \frac{1}{2}V_{ms}^2 \frac{R_s - R_L}{(R_s + R_L)^3}. \tag{3.212}$$

It is clear from (3.212) that for

$$R_L = R_s, \tag{3.213}$$

the above derivative is equal to zero and changes its sign from plus to minus (as R_L is increased above R_s). This means that under condition (3.213) the average power in (3.211) achieves its maximum value, which is equal to

$$\max\{P_{av}\} = \frac{V_{ms}^2}{8R_s}. \tag{3.214}$$

By combining (3.210) and (3.214), we find that the above maximum power transfer occurs when

$$Z_L = Z_s^*. \tag{3.215}$$

Expression (3.215) is typical of matching conditions which are encountered in similar forms in various areas of electrical engineering.

3.8 Complex Frequency

In our previous discussions of ac steady-state analysis of electric circuits, we have dealt with sinusoidal voltages and currents and we have used the phasor technique to calculate these voltages and currents. The main idea of the phasor technique is to reduce the operations of calculus on sinusoidal quantities to algebraic operations on their phasors. It turns out that this idea can be extended to a broader class of voltages and currents.

Consider the following voltage:

$$v(t) = V_m e^{\sigma t} \cos(\omega t + \varphi_V). \tag{3.216}$$

By using the Euler formula, we can transform this voltage as follows:

$$v(t) = V_m e^{\sigma t} \text{ Re } \left[e^{j(\omega t + \varphi_V)} \right] = \text{ Re } \left[V_m e^{j\varphi_V} e^{(\sigma + j\omega)t)} \right]. \tag{3.217}$$

Now, we introduce the voltage phasor $\hat{V}$:

$$\hat{V} = V_m e^{j\varphi_V}, \tag{3.218}$$

as well as the quantity s, which is called a *complex frequency*:

$$s = \sigma + j\omega. \tag{3.219}$$

The real part of this frequency characterizes the exponential decay (or growth) of the peak value of the voltage (3.216), while the imaginary part is the angular frequency. By using (3.218) and (3.219), we represent (3.217) in the form:

$$v(t) = \text{ Re } \left[\hat{V} e^{st} \right] \tag{3.220}$$

This formula extends the notion of phasors to voltages (3.216) which are characterized by complex frequency s. In a particular case when $\sigma = 0$, this formula is reduced to the familiar formula for sinusoidal voltages:

$$v(t) = V_m \cos(\omega t + \varphi_V) = \text{ Re } \left[\hat{V} e^{j\omega t} \right]. \tag{3.221}$$

If we have a current:

$$i(t) = I_m e^{\sigma t} \cos(\omega t + \varphi_I) = \text{ Re } \left[I_m e^{j\varphi_I} e^{st} \right], \tag{3.222}$$

then, by introducing phasor $\hat{I}$:

$$\hat{I} = I_m e^{j\varphi_I}, \tag{3.223}$$

we end up with a similar representation for $i(t)$:

$$i(t) = \text{ Re } \left[\hat{I} e^{st} \right]. \tag{3.224}$$

Next, we shall establish phasor terminal relationships for resistors, inductors, and capacitors for the case of voltages and currents of the same complex frequency s.

In the case of a resistor, we have:

$$v(t) = V_m e^{\sigma t} \cos(\omega t + \varphi_V) = Ri(t) = RI_m e^{\sigma t} \cos(\omega t + \varphi_I). \tag{3.225}$$

By expressing (3.225) in the phasor form, we find:

$$\text{Re } \left[\hat{V} e^{st} \right] = \text{ Re } \left[R\hat{I} e^{st} \right], \tag{3.226}$$

which suggests that:

$$\hat{V} = R\hat{I}. \tag{3.227}$$

Next, we consider an inductor and begin with its terminal relationship:

$$v(t) = L\frac{di(t)}{dt}. \tag{3.228}$$

By substituting (3.222) into (3.228) and by performing differentiation, we derive:

$$v(t) = \sigma L I_m e^{\sigma t}\cos(\omega t + \varphi_I) - \omega L I_m e^{\sigma t}\sin(\omega t + \varphi_I), \tag{3.229}$$

$$v(t) = \sigma L I_m e^{\sigma t}\cos(\omega t + \varphi_I) + \omega L I_m e^{\sigma t}\cos\left(\omega t + \varphi_I + \frac{\pi}{2}\right). \tag{3.230}$$

By using the Euler formula and the definition of current phasor, we transform (3.230) as follows:

$$v(t) = \text{Re}\left[\sigma L\hat{I}e^{st}\right] + \text{Re}\left[\omega L\hat{I}e^{j\pi/2}e^{st}\right], \tag{3.231}$$

$$v(t) = \text{Re}\left[(\sigma + j\omega)L\hat{I}e^{st}\right] = \text{Re}\left[sL\hat{I}e^{st}\right]. \tag{3.232}$$

By comparing (3.232) with (3.220), we derive:

$$\hat{V} = sL\hat{I}. \tag{3.233}$$

By using the same line of reasoning, we can establish the following relationship between the voltage and current phasors in the case of a capacitor:

$$\hat{I} = sC\hat{V}. \tag{3.234}$$

The last formula can also be written as:

$$\hat{V} = \frac{1}{sC}\hat{I}. \tag{3.235}$$

In the particular case of $\sigma = 0$, the last three formulas are reduced to the familiar expressions:

$$\hat{V} = j\omega L\hat{I}, \quad \hat{I} = j\omega C\hat{V}, \quad \hat{V} = \frac{1}{j\omega C}\hat{I}. \tag{3.236}$$

Now, we consider the circuit shown in Figure 3.14, which consists of an *RLC* branch and the voltage source:

$$v(t) = V_m e^{\sigma t}\cos(\omega t + \varphi_V). \tag{3.237}$$

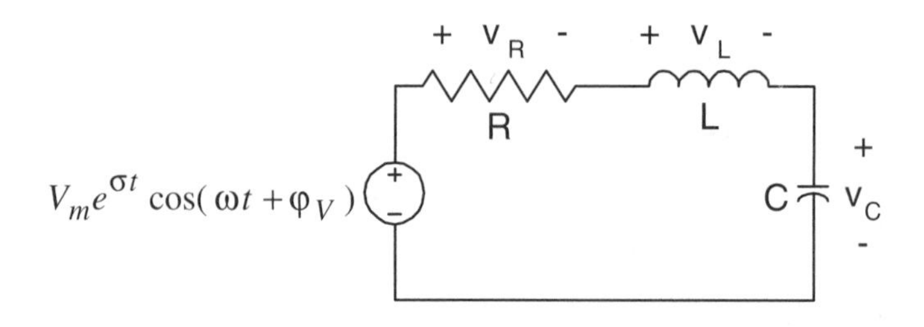

Figure 3.14: An *RLC* branch driven by a complex frequency source.

We look for the current in the form:

$$i(t) = I_m e^{\sigma t} \cos(\omega t + \varphi_I). \tag{3.238}$$

We shall use the phasor technique and formulas (3.227), (3.233), and (3.235) in order to find I_m and φ_I in (3.238). We begin by writing the KVL equation in phasor form:

$$\hat{V} = \hat{V}_R + \hat{V}_L + \hat{V}_C. \tag{3.239}$$

Voltage phasors in the right-hand side of (3.239) can be written in terms of the current phasor as follows:

$$\hat{V}_R = R\hat{I}, \qquad \hat{V}_L = sL\hat{I}, \qquad \hat{V}_C = \frac{1}{sC}\hat{I}. \tag{3.240}$$

By substituting (3.240) into (3.239), we derive:

$$\hat{V} = \left(R + sL + \frac{1}{sC}\right)\hat{I}. \tag{3.241}$$

We shall next introduce the impedance $Z(s)$ as a function of complex frequency s:

$$Z(s) = R + sL + \frac{1}{sC}. \tag{3.242}$$

In the particular case of $\sigma = 0$, the last expression is reduced to the familiar formula:

$$Z(j\omega) = R + j\omega L + \frac{1}{j\omega C}. \tag{3.243}$$

By using (3.242), we can rewrite (3.241) as follows:

$$\hat{V} = Z(s)\hat{I}, \tag{3.244}$$

which can now be solved for $\hat{I}$:

$$\hat{I} = \frac{\hat{V}}{Z(s)}. \tag{3.245}$$

Thus, in order to find the current phasor $\hat{I}$, we first find the voltage phasor $\hat{V} = V_m e^{j\varphi_V}$ from (3.237), then we evaluate the impedance $Z(s)$ for complex frequency $s = \sigma + j\omega$, and finally we use (3.245) to find $\hat{I}$. Knowing $\hat{I}$, we can easily find I_m and φ_I. It is clear that the described calculations are almost identical to the calculations which we have performed for ac steady-state analysis. **The only difference is that the impedance Z should be evaluated at the complex frequency $s = \sigma + j\omega$ rather than at $j\omega$.** This suggests that the analysis of electric circuits excited by voltage and/or current sources of complex frequency can be performed by using exactly the same techniques which we have developed for ac steady-state analysis.

In applications, it is quite rare that electric circuits are excited by **sources** of complex frequency. However, voltages and currents of complex frequency regularly appear in the case of transients in electric circuits. For this reason, the notion of complex frequency as well as the notion of impedance $Z(s)$ as a function of complex frequency s will be useful in the analysis of transients in electric circuits. That is why the notion of complex frequency is extensively used in Chapters 7, 8, and 9.

3.9 Summary

In this chapter, we introduced the concept of phasors, which were used to transform the complete set of differential equations for ac source-driven circuits into a linear system of algebraic equations.

The phasor representation of a current source, for example, is a complex number whose magnitude is equal to the peak value of the current source and whose polar angle is equal to the initial phase of the current source: $\hat{I}_s = I_{ms}e^{j\phi_{I_s}}$ if and only if $i_s(t) = I_{ms}\cos(\omega t + \phi_{I_s})$. A similar representation holds for voltage sources and for all sinusoidal voltages and currents in electric circuits under ac steady-state conditions.

The phasor method tacitly assumes that any and all voltages and currents are varying with the same angular frequency ω. Problems for which this assumption is not true must be deferred until the discussion of the superposition method in the next chapter.

We also introduced the concept of impedance, which is a far-reaching generalization of resistance in the case of the steady-state problems. For any branch of an electric circuit the impedance Z is defined as the ratio of the phasor of the branch voltage $\hat{V}$ to the phasor of the branch current $\hat{I}$:

$$Z = \hat{V}/\hat{I} = R + jX = 1/Y = \frac{1}{G + jB}. \tag{3.246}$$

It is clear that the resistance, R, and the reactance, X, are defined as the real and imaginary parts of the impedance, while the admittance Y is the reciprocal of the impedance. The conductance, G, and the susceptance, B, are the real and imaginary parts of the admittance. Table 3.2 summarizes the expressions for these quantities for various passive circuit elements.

Table 3.2: Expressions for impedances and admittances.

	Resistor	Inductor	Capacitor
Parameter	R (Ω)	L (H)	C (F)
Impedance	R	$j\omega L$	$1/j\omega C$
Resistance	R	0	0
Reactance	0	ωL	$-1/\omega C$
Admittance	$1/R$	$1/j\omega L$	$j\omega C$
Conductance	$1/R$	0	0
Susceptance	0	$-1/\omega L$	ωC

With the aid of the impedance concept, we transformed KCL and KVL equations into the standard ac steady-state equations:

$$\sum_k \hat{I}_k = -\sum_k \hat{I}_{s_k}, \tag{3.247}$$

$$\sum_k \hat{I}_k Z_k = -\sum_k \hat{V}_{s_k}. \tag{3.248}$$

These equations are mathematically similar to the basic equations (2.56) and (2.57) for resistive electric circuits. As a result, the analysis techniques developed for ac steady-state can be directly used for the analysis of resistive circuits. The only difference is that in the case of ac steady-state we deal with phasors and impedances (admittances) while in the case of resistive circuits we deal with instantaneous values and resistances (conductances).

It is important to stress here that the linear equations produced by the KCL and KVL equations are somewhat special. They are usually "sparse" because not every equation contains every variable. When standard matrix notations are used, this corresponds to the coefficient matrix containing many zeroes. Sparsity of Kirchhoff equations is closely related to the "connectivity" of electric circuits, that is, to the way circuit elements are interconnected. In circuit theory special analysis techniques have been developed to take advantage of the sparsity of the basic equations and reduce the number of needed computations. These techniques will be discussed in the following three chapters.

The discussion of ac power was presented. The notion of root-mean-square (rms) values was introduced and the expression for the average ac power was derived in terms of rms values of terminal voltage and current. This expression contains the power factor, which has a significant economic impact on distribution and consumption of ac power. The means of adjusting the power factor to unity were discussed along with the means of achieving the maximum power transfer from a power network to a load.

The chapter closed with a brief discussion of complex frequency. It was demonstrated that the phasor technique can be extended to the analysis of electric circuits excited by sources with complex frequency. It was remarked that voltages and currents of complex frequency regularly appear in the analysis of transients in electric circuits and it is there that the notion of complex frequency is most useful. That is why this notion will be extensively used in Chapters 7, 8, and 9.

3.10 Problems

1. The voltage across a 43 mH inductor is given by $v(t) = 50\cos(\omega t + 6)$ V. Find the peak current at a frequency $f =$ (a) 1/50 Hz, (b) 1 Hz, (c) 30 Hz, and (d) 1 MHz.

2. The current through a 10 μF capacitor is given by $i(t) = 0.1\sin(2\omega t)$ A. Find the peak voltage at a frequency $f =$ (a) 1/20 Hz, (b) 2 Hz, (c) 1 kHz, and (d) 30 MHz.

3. Find the phasors which represent the following sinusoidal currents: $i(t) =$ (a) $6\cos(\omega t + 1/2)$ A, (b) $4\sin(\omega t + \pi/3)$ mA, (c) $3\cos(\omega t - 3\pi/8)$ A, and (d) $2\cos(\omega t) - 4\sin(\omega t)$ A.

4. Find the phasors which represent the following sinusoidal voltages: $v(t)$ = (a) $1/2\cos(\omega t - \pi)$ V, (b) $7\cos(2\omega t + \pi/2)$ mV, (c) $2\sin(\omega t - \pi/7)$ kV, and (d) $3\sin(\omega t) - \cos(\omega t)$ V.

5. Transform the following voltage phasors to the standard form $v(t) = V_m \cos(\omega t + \phi_v)$ V: $\hat{V}$ = (a) $20e^{j\pi/6}$, (b) $5e^{-j\pi/2}$, (c) $3(1/2 - j\sqrt{3}/2)$, and (d) $5 + j7$ (all in volts).

6. Transform the following current phasors to the standard form $i(t) = I_m \cos(\omega t + \phi_I)$ A: $\hat{I}$ = (a) $0.01e^{-2j}$, (b) $3e^{j\pi/4}$, (c) $1/2 - j3$, and (d) $2.7\,(1/3 + j2\sqrt{2}/3)$ (all in amps).

7. Find the impedance of a 4.7 pF capacitor at the following frequencies: (a) 1 Hz, (b) 5 kHz, (c) 1 MHz, and (d) 2 GHz.

8. Find the admittance of a 1.1 mH inductor at the following frequencies: (a) 10 mHz, (b) 3.3 Hz, (c) 27 kHz, and (d) 5 MHz.

9. A voltage $v(t) = 10\cos(10t + \pi/3)$ V is applied across a two-terminal device with an impedance $Z = 8 - j6\ \Omega$. Find the expression for the time variation of the ac current flowing through the device.

10. A current $i(t) = 0.02\sin(1000t + \pi/4)$ A flows through a two-terminal device with an impedance $Z = 50 - j40\ \Omega$. Find the expression for the time variation of the ac voltage across the device.

11. The impedance of a branch is given by the expression $Z = 50 + j40\ \Omega$. Find the: (a) reactance, (b) admittance, (c) conductance, and (d) susceptance of the branch.

12. The admittance of a branch is given by $Y = 12 - j5$ mΩ. Find the: (a) susceptance, (b) impedance, and (c) reactance of the element.

13. A voltage $v(t) = 120\sqrt{2}\cos(377t + 120°)$ V is applied across an RL branch with $R = 10\ \Omega$ and $L = 10$ mH. Find an expression for the time variation of the current through the branch.

14. A current of $i(t) = 3\cos(t) - 4\sin(t)$ A is made to flow through an RLC branch with $R = 2\ \Omega$, $L = 2$ H, and $C = 1$ F. Find an expression for the time variation of the voltage across the branch.

15. A voltage $v(t) = 100\sin(50t)$ mV is applied across an RC branch with $R = 5$ kΩ and $C = 10\ \mu$F. Find an expression for the time variation of the current through the branch.

16. Find the time dependence of the voltage source that must be applied to an RLC branch ($R = 10\ \Omega$, $L = 1$ H, and $C = 3/2$ F) in order to achieve a current of $i(t) = \cos(10t)$ mA through the branch.

17. A voltage $v(t) = 120\sqrt{2}\cos(2400t)$ V is applied across an LC branch with $L = 1$ mH and $C = 1\ \mu$F. Find an expression for the time variation of the current through the branch.

18. When a voltage of $v(t) = 20\cos(100t + 30°)$ V is applied across an RL branch, the resulting current through the branch is $i(t) = 5\cos(100t - 45°)$ A. Find the values of the resistance and inductance for this branch.

19. Consider the circuit shown in Figure P3-19a. Figure P3-19b represents a concise version of the circuit where some of the elements have been combined into single branches. By

comparing the two figures, write the impedance for each branch of the circuit in Figure P3-19b. Let Z_1 be the impedance for the branch with i_1 flowing through it, etc.

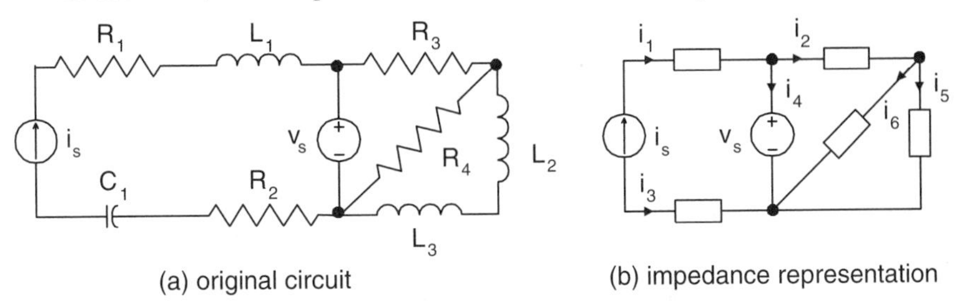

Figure P3-19

20. Consider the circuit shown in Figure P3-20a. Figure P3-20b represents a concise version of the circuit where some of the elements have been combined into single branches. By comparing the two figures, write the impedance for each branch of the circuit in Figure P3-20b. Let Z_1 be the impedance for the branch with i_1 flowing through it, etc.

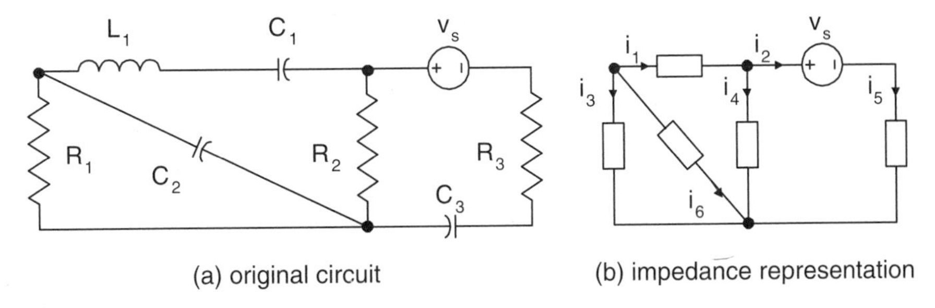

Figure P3-20

Problems 21–25. For the circuits shown in the figures indicated, label the nodes and meshes and write the ac steady-state KCL and KVL equations. Write all equations in terms of the phasor currents and the source phasors.

21. Figure P3-21.

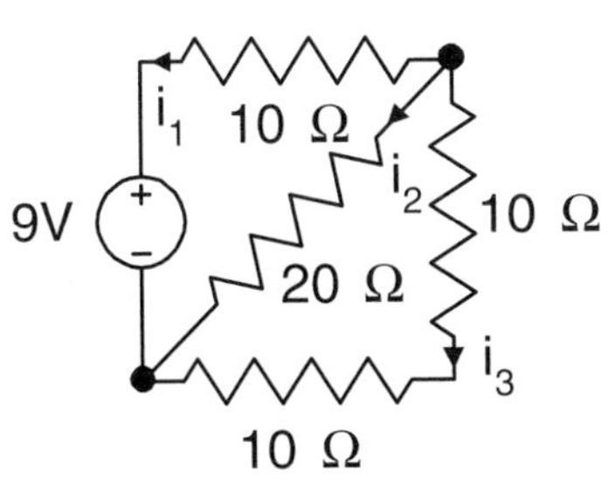

Figure P3-21

22. Figure P3-22.

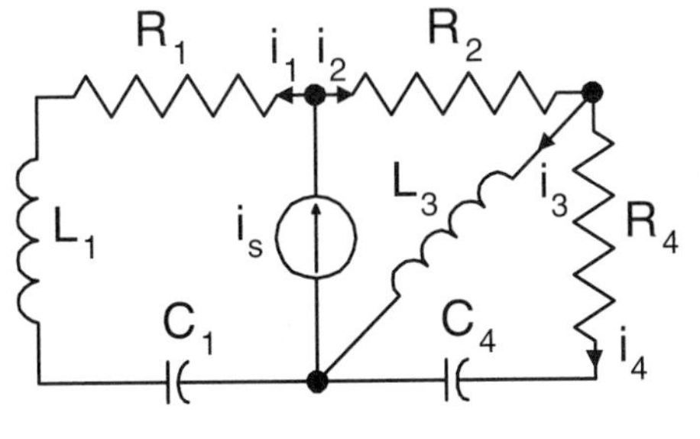

Figure P3-22

23. Figure P3-23.

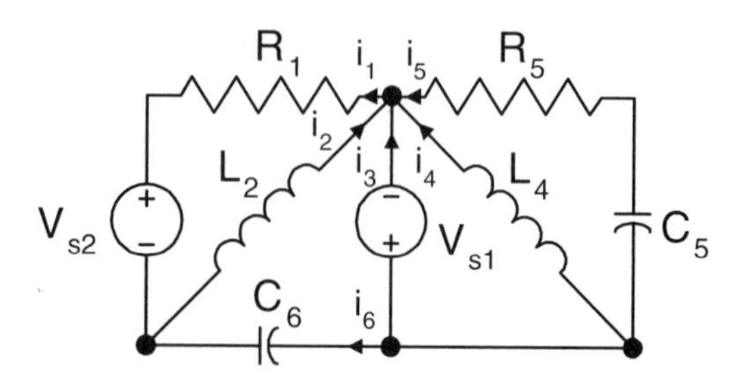

Figure P3-23

24. Figure P3-24.

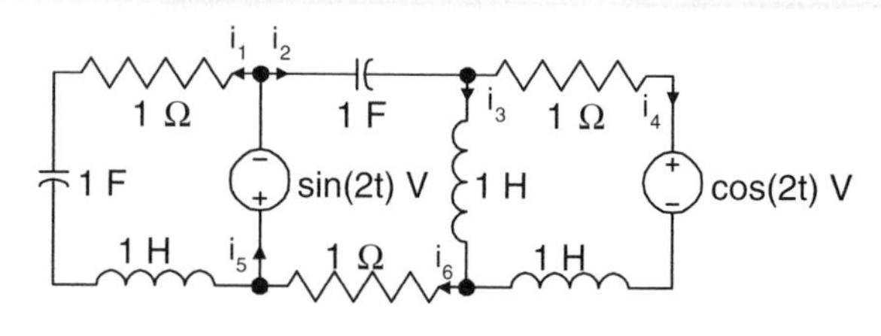

Figure P3-24

25. Figure P3-25.

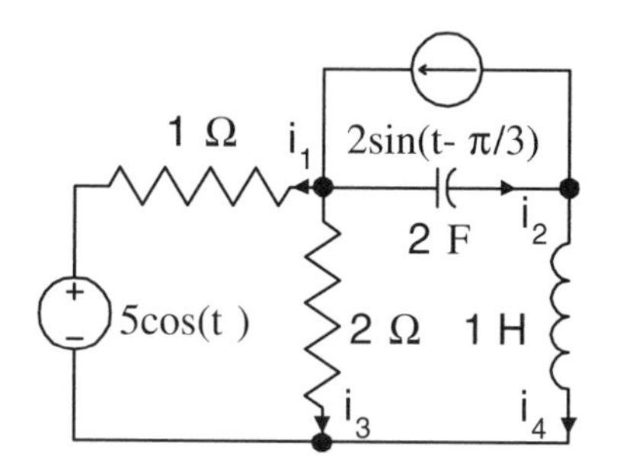

Figure P3-25

Problems 26–30. An alternate form for the basic ac steady-state equations arises from substituting into KCL the expression $\hat{I}_k = Y_k \hat{V}_k$ for each branch:

$$\sum_k Y_k \hat{V}_k = -\sum_k \hat{I}_{s_k}; \quad \sum_k \hat{V}_k = -\sum_k \hat{V}_{s_k}.$$

For the circuits shown in the figures indicated, label the nodes and meshes and write the ac steady-state KCL and KVL equations. Write all equations in terms of the voltage phasors and the source phasors.

26. Figure P3-25.

27. Figure P3-27.

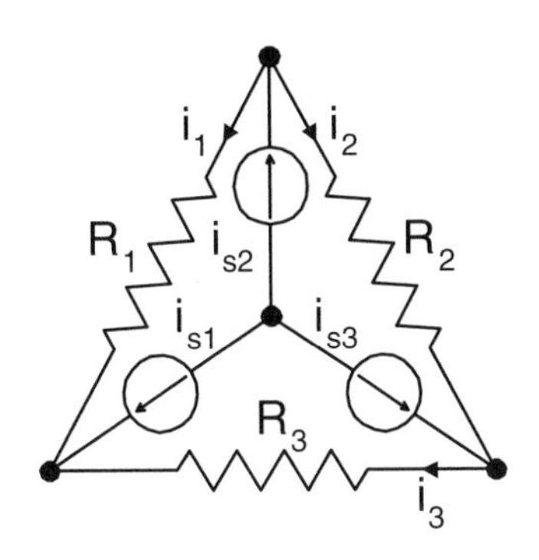

Figure P3-27

28. Figure P3-28.

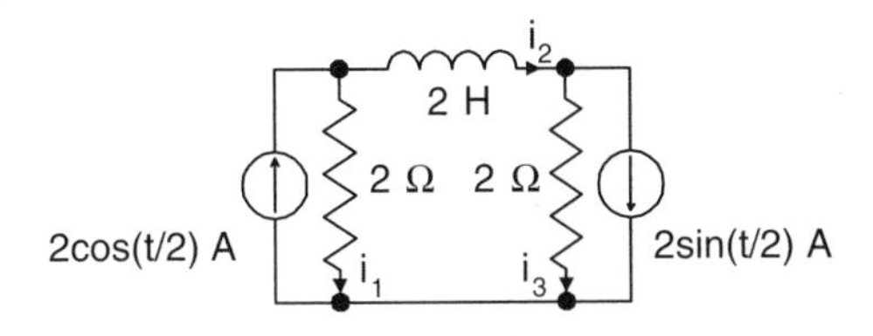

Figure P3-28

29. Figure P3-29.

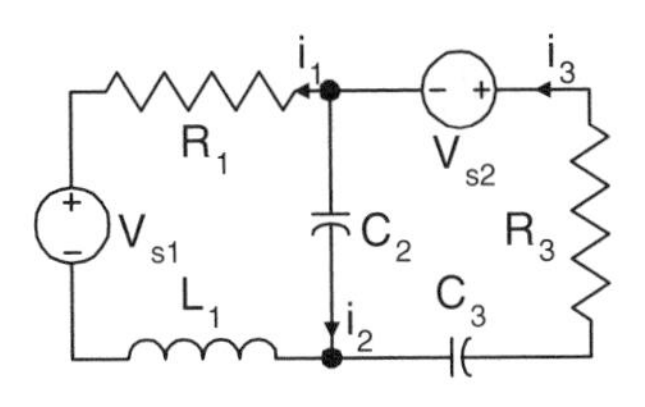

Figure P3-29

30. Figure P3-30.

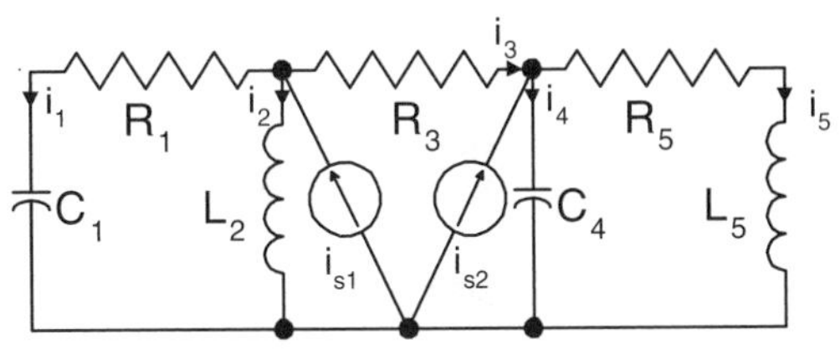

Figure P3-30

31. Find the rms value of the periodic ramp voltage source shown in Figure 1.20c. The expression for the voltage when $|t| > T/2$, where T is the period, is given by $v(t) = 2V_m t/T$.

32. The periodic pulse voltage source, shown in Figure 1.20d, is given in the interval $0 < t < T$ by

$$v(t) = \begin{cases} V_m & 0 < t < \tau, \\ 0 & \tau < t < T \end{cases}$$

(where T is the period). Find an expression for the rms voltage as a function of τ/T and show that it approaches a reasonable value as $\tau/T \rightarrow 1$.

33. Consider the circuit shown in Figure P3-33. Find the rms current and power factor of the load for the parameters listed in the following table.

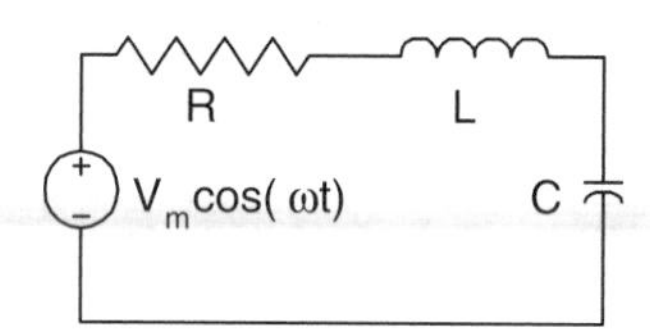

Figure P3-33

Case	V_m (V)	ω (rad/s)	R (Ω)	L (H)	C (F)
a	169.7	377	20	0.01	0.001
b	169.7	377	1 k	0.3	∞
c	100	2513	5	0	100 μ
d	100	10	100	2	0.005

34. Consider the circuit shown in Figure P3–34. Find the rms voltage and power factor of the load for the parameters listed in the following table.

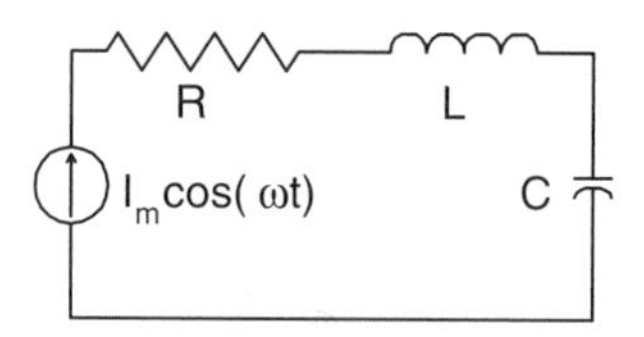

Figure P3-34

Case	I_m (A)	ω (rad/s)	R (Ω)	L (H)	C (F)
a	1	377	5	0	470 μ
b	10	377	50	0.1	∞
c	1 m	1 M	5 k	0.2 m	100 p
d	25 m	10	50	5	0.002

35. Consider the circuit shown in Figure P3-35. Find the impedance of the *RLC* branch and the branch current for the cases indicated in the table below.

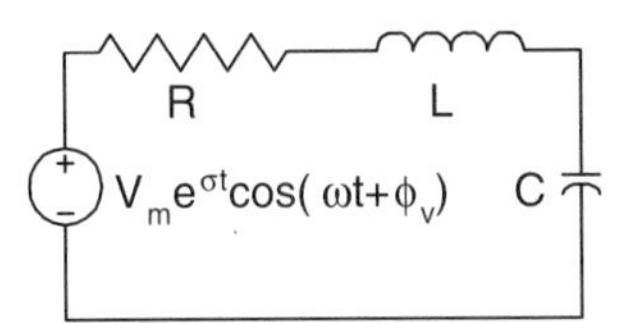

Figure P3-35

Case	V_m (V)	σ (s^{-1})	ω (rad/s)	ϕ_v (deg)	R (Ω)	L (H)	C (F)
a	10	−1	1	30	10	1	∞
b	50	−50	1 k	45	50	0	12 m
c	100	−0.2 M	1 M	75	300	0.01 m	47 μ

36. Consider the circuit shown in Figure P3-35. Find the current through the *RLC* branch for the cases indicated in the table below.

Case	V_m (V)	σ (s^{-1})	ω (rad/s)	ϕ_v (deg)	R (Ω)	L (H)	C (F)
a	170	−2	377	−90	50	0.01	470 μ
b	5	−2	4π	−45	10	0	0.01
c	12	−2 M	10 M	30	300	0.1 m	10 μ

37. A voltage source $v(t) = 10e^{-t}\sin(7t)$ V generates a current of $i(t) = 0.2e^{-t}\cos(7t-30°)$ A when connected to an *RC* branch. Find the values of resistance and capacitance of the branch.

38. When a current source $i(t) = e^{-20t}\cos(80t + 45°)$ A is connected to an *RL* branch, the resulting voltage across the source is measured to be $v(t) = -80\sqrt{2}e^{-20t}\sin(80t)$ V. Find the values of resistance and inductance of the branch.

Chapter 4

Equivalent Transformations of Electric Circuits

4.1 Introduction

This chapter deals with equivalent transformations of electric circuits. Through the use of *equivalent transformations*, complex circuits requiring numerous computations for their analysis can be reduced to simpler *equivalent circuits*, sometimes dramatically reducing the computational effort required to analyze them.

We first consider passive circuits and derive general formulas for transforming an arbitrary number of series and parallel elements. We next introduce rules which are known as the current divider and the voltage divider rules. We then consider the concept of input impedance and give several examples of equivalent transformations of electric circuits which utilize alternating applications of series and parallel transformations. We end the study of transformations of passive circuits with a discussion of symmetry.

For active circuits, we introduce the superposition principle and demonstrate via several examples how it can be used to simplify circuit analysis. We also detail the solution technique for circuits with ac sources which operate at different frequencies.

The chapter is closed with an introduction to the MicroSim PSpice circuit simulation package. This introduction walks the reader through Microsoft® Windows™–based PSpice simulations of three examples from this chapter.

4.2 Series and Parallel Connections

Consider the branch shown in Figure 4.1, which is connected to the rest of the electric circuit. The impedances shown in this branch are said to be connected in *series*. This means that they all have the same identical current flowing through them. This fact

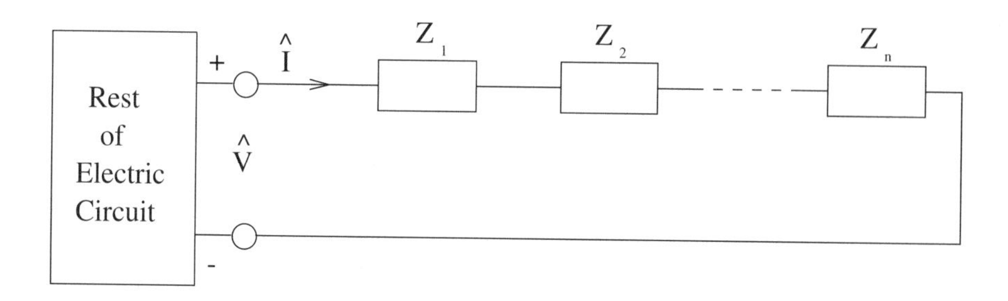

Figure 4.1: Branch with impedances connected in series.

can be easily concluded from KCL and happens because each node that connects two elements together is a trivial one.

We wish to simplify this branch by representing all the impedances with a single equivalent impedance, shown in Figure 4.2. But how can we guarantee that the replacement by one equivalent impedance will not affect the rest of the circuit? The answer is that the electric circuit in the box will not be affected by this equivalent transformation if this transformation *preserves the terminal relationship between the phasors of branch voltage* $\hat{V}$ *and branch current* $\hat{I}$. Indeed, if the above terminal relationship is preserved, then the circuits shown in Figures 4.1 and 4.2 will be described by the *identical* sets of linear algebraic equations. To prove this, we remark that the electric circuits in the boxes are the same and, consequently, they are described by identical sets of basic ac steady-state equations. These basic equations are complemented by the same terminal ($\hat{V}$ versus $\hat{I}$) relationship for the branches outside the box. This makes the overall sets of ac steady-state equations for the above circuits *identical*. Consequently, by solving these equations, we find the same branch currents for both circuits. This proves that the rest of the circuit will not be affected by the equivalent transformation.

In a way, we have already formulated and proved a very general principle of equivalent transformation. According to this principle, we can divide an electric circuit into two parts (I and II) which are connected through terminals A and B

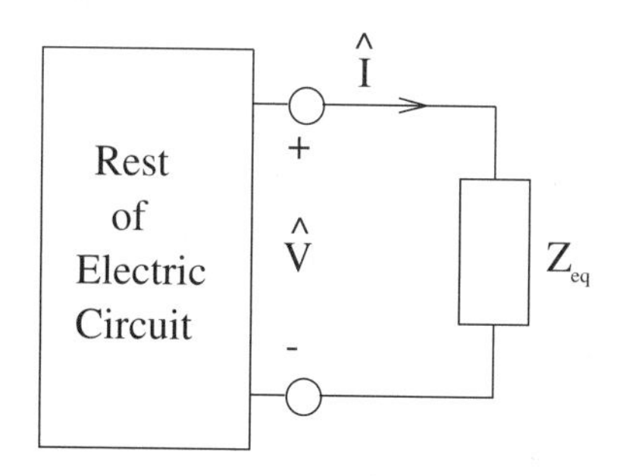

Figure 4.2: The equivalent impedance.

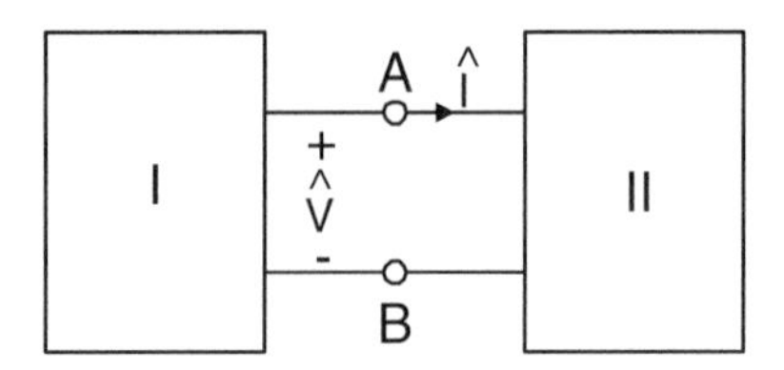

Figure 4.3: An electric circuit divided into two parts.

(see Figure 4.3). Then, any transformation of part II, which *preserves the terminal $\hat{V}$–$\hat{I}$ relationship*, will not affect the currents and voltages in part I of the circuit. In this sense, this transformation is an *equivalent* one. This principle of equivalent transformations based on preserving voltage-current terminal relationships will be frequently used throughout this text. We shall first apply this principle to the circuits shown in Figures 4.1 and 4.2.

We write KVL for the loop in Figure 4.1:

$$\hat{V} = \hat{V}_1 + \hat{V}_2 + \cdots + \hat{V}_n. \tag{4.1}$$

Now, by using the formula $\hat{V}_k = \hat{I}Z_k$, we find:

$$\hat{V} = \hat{I}Z_1 + \hat{I}Z_2 + \cdots + \hat{I}Z_n, \tag{4.2}$$

$$\hat{V} = \hat{I}(Z_1 + Z_2 + \cdots + Z_n). \tag{4.3}$$

From Figure 4.2 we obtain:

$$\hat{V} = \hat{I}Z_{eq}. \tag{4.4}$$

Clearly, the $\hat{V}$–$\hat{I}$ terminal relationships (4.3) and (4.4) will be the same if:

$$Z_{eq} = Z_1 + Z_2 + \cdots + Z_n = \sum_k Z_k. \tag{4.5}$$

Therefore, the equivalent impedance of a series connection of impedances is merely the sum of those impedances. From equation (4.5) and the definition of impedance we can see that the equivalent resistance of a branch composed of resistors connected in series is:

$$R_{eq} = \sum_k R_k. \tag{4.6}$$

For a branch composed of inductors connected in series, we have:

$$j\omega L_{eq} = \sum_k j\omega L_k, \tag{4.7}$$

which leads to:

$$L_{eq} = \sum_k L_k. \tag{4.8}$$

Similarly, for a branch composed of capacitors connected in series, we find:

$$-\frac{j}{\omega C_{eq}} = \sum_k \frac{-j}{\omega C_k}, \tag{4.9}$$

which results in:

$$\frac{1}{C_{eq}} = \sum_k \frac{1}{C_k}, \tag{4.10}$$

or:

$$C_{eq} = \frac{1}{\sum_k 1/C_k}. \tag{4.11}$$

In the particular case of a series connection of identical impedances Z, from (4.5) we find:

$$Z_{eq} = nZ, \tag{4.12}$$

where n is the total number of impedances.

Now, consider the circuit shown in Figure 4.4. The impedances in this circuit are said to be connected in *parallel*, which means that the same voltage is applied across each of them. This happens when corresponding terminals of the circuit elements are connected together to a pair of nodes by wires.

Again, we wish to derive an equivalent transformation which allows one to replace the circuit with just one equivalent impedance (Figure 4.2). We begin by writing the KCL equation for the circuit shown in Figure 4.4:

$$\hat{I} = \hat{I}_1 + \hat{I}_2 + \cdots + \hat{I}_n. \tag{4.13}$$

Note that there are only two nodes in this circuit, one on the top and one on the bottom where all the branches are connected together. This is because the lines connecting circuit elements have zero impedance and therefore can all be consolidated into one node on the top and bottom, respectively.

Applying the relation $\hat{I}_k = \hat{V}/Z_k$, we derive:

$$\hat{I} = \frac{\hat{V}}{Z_1} + \frac{\hat{V}}{Z_2} + \cdots + \frac{\hat{V}}{Z_n}, \tag{4.14}$$

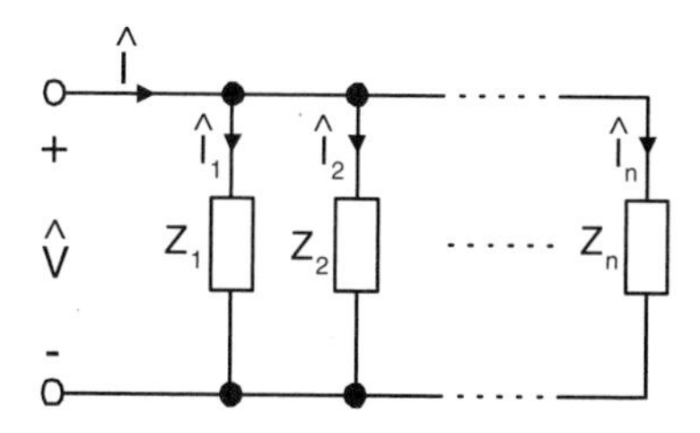

Figure 4.4: Circuit with impedances connected in parallel.

$$\hat{I} = \hat{V}\left(\frac{1}{Z_1} + \frac{1}{Z_2} + \cdots + \frac{1}{Z_n}\right), \tag{4.15}$$

$$\hat{V} = \hat{I}\left(\frac{1}{\frac{1}{Z_1} + \frac{1}{Z_2} + \cdots + \frac{1}{Z_n}}\right). \tag{4.16}$$

Now we can see that the $\hat{V}$–$\hat{I}$ terminal relationships (4.16) and (4.4) will be the same, if:

$$Z_{eq} = \frac{1}{\frac{1}{Z_1} + \frac{1}{Z_2} + \cdots + \frac{1}{Z_n}} = \frac{1}{\sum_k \frac{1}{Z_k}}. \tag{4.17}$$

In a particular case of equal impedances connected in parallel, from (4.17) we derive:

$$Z_{eq} = \frac{Z}{n}, \tag{4.18}$$

where n is the total number of impedances.

From equation (4.17) and the definition of impedance we find that if the branches in parallel are all composed of resistors, then:

$$\frac{1}{R_{eq}} = \frac{1}{R_1} + \frac{1}{R_2} + \cdots + \frac{1}{R_n}. \tag{4.19}$$

If they contain only inductors, then:

$$\frac{1}{L_{eq}} = \frac{1}{L_1} + \frac{1}{L_2} + \cdots \frac{1}{L_n}. \tag{4.20}$$

And in the case of capacitors, we have:

$$C_{eq} = C_1 + C_2 + \cdots + C_n. \tag{4.21}$$

It is sometimes more convenient to describe the relationship for parallel connections in terms of admittance. Equation (4.17) can be rewritten as $1/Z_{eq} = \sum_k 1/Z_k$, and since we defined the admittance as the reciprocal of impedance, we get $Y_{eq} = 1/Z_{eq} = \sum_k Y_k$. In other words, the equivalent admittance of a parallel connection is simply the sum of the individual admittances.

EXAMPLE 4.1 Consider a 10 Ω, a 22 Ω, and a 47 Ω resistor. What is the equivalent resistance if they are all connected in series? What is the equivalent resistance if they are connected in parallel?

For a series connection we have $R_{eq} = 10\ \Omega + 22\ \Omega + 47\ \Omega = 79\ \Omega$. For a parallel connection $R_{eq} = 1/(1/10 + 1/22 + 1/47) \approx 6\ \Omega$. ■

EXAMPLE 4.2 Consider a 10 Ω resistor, a 470 μF capacitor, and a 1.1 mH inductor. What is the equivalent impedance if these elements are connected in series and driven with a 60 Hz source? What is the equivalent impedance if these elements are connected in parallel and driven by a 400 Hz source (a frequency often used in aviation)?

The impedance of the capacitor at 60 Hz is $Z_c = -j/\omega C = -j/(2\pi \times 60 \times 470 \times 10^{-6}) = -j5.644\ \Omega$ and the impedance of the inductor is $Z_L =$

$j\omega L = j2\pi \times 60 \times 1.1 \times 10^{-3} = j0.415\ \Omega$. The equivalent series impedance is $Z_{eq} = 10 - j5.644 + j0.415 = 10 - j5.229\ \Omega$.

The admittance of the capacitor at 400 Hz is $Y_c = j\omega C = 2\pi \times 400 \times 470 \times 10^{-6} = j1.181\ \mho$ and the admittance of the inductor is $Y_L = -j/\omega L = -j/(2\pi \times 400 \times 1.1 \times 10^{-3}) = -j0.362\ \mho$. The equivalent parallel admittance is $Y_{eq} = 0.10 + j1.181 - j0.362 = 0.10 + j0.819\ \mho$. The equivalent impedance is $Z_{eq} = 1/Y_{eq} = (0.10 - j0.819)/0.676 = 0.148 - j1.212\ \Omega$. ■

4.3 Voltage and Current Divider Rules

Sometimes in circuit analysis, only the voltage across one specific element in the circuit is desired. To find the voltage across one impedance which is in series with many other impedances, we can utilize the *voltage divider rule*. The voltage divider rule is based on the fact that the total voltage is proportionately divided between impedances in series. Consider the circuit in Figure 4.5. Assume we wish to find $\hat{V}_1$. We know that the voltage across Z_1 is simply the product of the current and the impedance:

$$\hat{V}_1 = \hat{I}Z_1. \tag{4.22}$$

We also know that (see (4.3)):

$$\hat{V}_s = \hat{I}(Z_1 + Z_2 + \cdots + Z_n). \tag{4.23}$$

We now look at the ratio of $\hat{V}_1$ to $\hat{V}_s$:

$$\frac{\hat{V}_1}{\hat{V}_s} = \frac{\hat{I}Z_1}{\hat{I}(Z_1 + Z_2 + \cdots + Z_n)}, \tag{4.24}$$

which leads to:

$$\hat{V}_1 = \hat{V}_s \frac{Z_1}{Z_1 + Z_2 + \cdots + Z_n}. \tag{4.25}$$

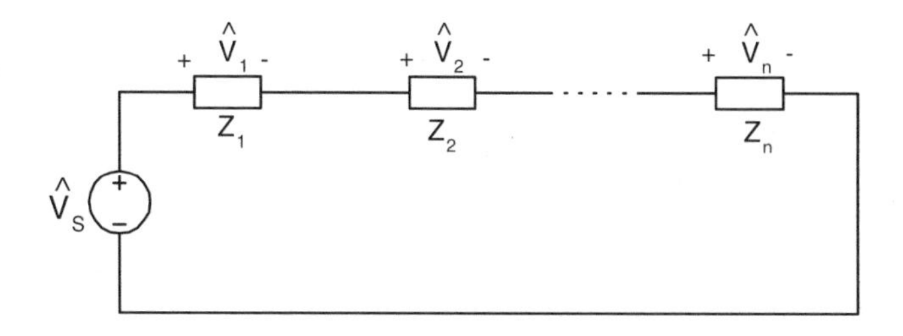

Figure 4.5: Circuit with impedances in series.

Similarly, it can be derived that the voltage across the kth element is:

$$\hat{V}_k = \hat{V}_s \frac{Z_k}{Z_1 + Z_2 + \cdots + Z_n}. \tag{4.26}$$

The *current divider rule* is useful for finding the currents through impedances that are connected in parallel. If the total current and all the impedances are known, then the current divider rule can be used to determine the individual currents through each impedance. It is derived in a fashion similar to the voltage divider rule. Consider the circuit shown in Figure 4.6. Assume we want to find $\hat{I}_1$. We know that the current through Z_1 is the voltage across it divided by the impedance:

$$\hat{I}_1 = \frac{\hat{V}}{Z_1}. \tag{4.27}$$

We also know that (see (4.15)):

$$\hat{I}_s = \hat{V}\left(\frac{1}{Z_1} + \frac{1}{Z_2} + \cdots + \frac{1}{Z_n}\right). \tag{4.28}$$

We now find the ratio of $\hat{I}_1$ to $\hat{I}_s$:

$$\frac{\hat{I}_1}{\hat{I}_s} = \frac{\frac{\hat{V}}{Z_1}}{\hat{V}\left(\frac{1}{Z_1} + \frac{1}{Z_2} + \cdots + \frac{1}{Z_n}\right)}, \tag{4.29}$$

which leads to:

$$\hat{I}_1 = \frac{\hat{I}_s\left(\frac{1}{Z_1}\right)}{\frac{1}{Z_1} + \frac{1}{Z_2} + \cdots + \frac{1}{Z_n}}. \tag{4.30}$$

By using the relationship between impedance and admittance, from (4.30) we find:

$$\hat{I}_1 = \hat{I}_s \frac{Y_1}{Y_1 + Y_2 + \cdots + Y_n}. \tag{4.31}$$

Similarly, it can be derived that the current through the kth element is:

$$\hat{I}_k = \hat{I}_s \frac{Y_k}{Y_1 + Y_2 + \cdots + Y_n}. \tag{4.32}$$

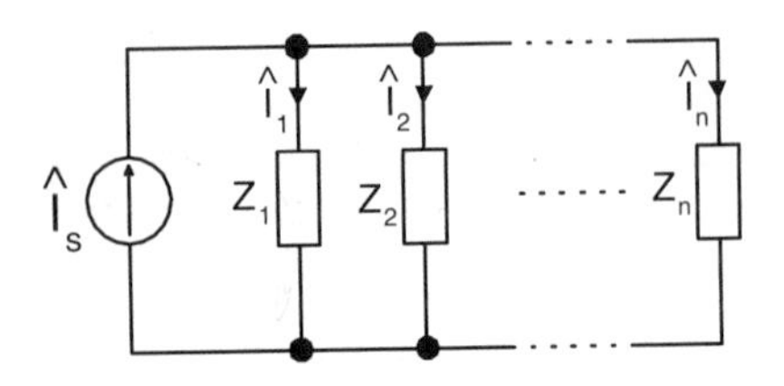

Figure 4.6: Circuit with impedances in parallel.

The current divider rule is quite often used in the case of two impedances connected in parallel. In this case the rule is very easy to remember. Consider the simple circuit in Figure 4.7. From (4.30) we can find the expression for $\hat{I}_1$:

$$\hat{I}_1 = \frac{\hat{I}_s\left(\frac{1}{Z_1}\right)}{\frac{1}{Z_1} + \frac{1}{Z_2}}, \tag{4.33}$$

which can be simplified by multiplying the numerator and denominator by $Z_1 Z_2$:

$$\hat{I}_1 = \frac{\hat{I}_s Z_2}{Z_1 + Z_2}. \tag{4.34}$$

Therefore, whenever two impedances are connected in parallel, the current through one is equal to the total current times the adjacent impedance divided by the sum of the two impedances connected in parallel.

EXAMPLE 4.3 Consider a voltage divider consisting of only two resistors: $R_1 = 10\ \Omega$ and $R_2 = 5\ \Omega$. Find the ratio of the voltage across R_2 to the input voltage $\hat{V}_s$. If instead you configure the two resistors as a current divider, what is the ratio of the current through R_2 to the input current $\hat{I}_s$?

From equation (4.26), we find $\hat{V}_2/\hat{V}_s = Z_2/(Z_1 + Z_2) = 5/(5 + 10) = 1/3$. On the other hand, the current divider rule for two elements (4.34) yields $\hat{I}_2/\hat{I}_s = Z_1/(Z_1 + Z_2) = 10/(5 + 10) = 2/3$. ■

EXAMPLE 4.4 Design a capacitive voltage divider to measure a 50,000 *V* signal with a 10 V probe. The largest capacitor you have available to use has a capacitance of 1 μF.

For this problem it is most convenient to write the voltage divider rule for two elements in terms of admittances. By using (4.26), it is easy to show that $\hat{V}_2/\hat{V}_s = Y_1/(Y_1 + Y_2) = j\omega C_1/(j\omega C_1 + j\omega C_2) = C_1/(C_1 + C_2)$. For the design problem, we need to find values for C_1 and C_2. We need to have $\hat{V}_2/\hat{V}_s = 10/50{,}000 = 0.0002$. This will require that C_2 be much larger than C_1. Let $C_2 = 1\ \mu$F. Then $0.0002 = C_1/(C_1 + 1\mu\text{F})$, which can be solved to find $C_1 \approx 200$ pF. Similar capacitive voltage dividers are quite useful in high-voltage, pulsed power systems. Note that C_1 must physically be very large to sustain (without breakdown) over 50,000 volts. It is often made by immersing two large electrodes in a dielectric oil. C_2 can be a standard off-the-shelf capacitor. ■

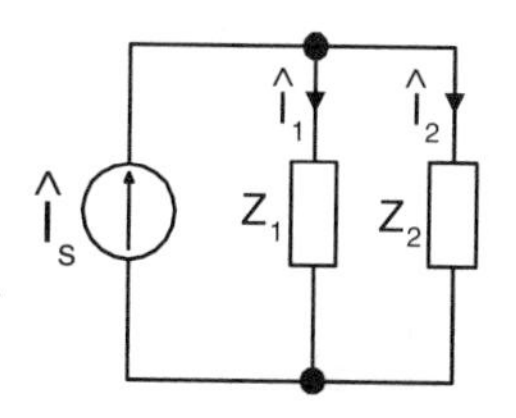

Figure 4.7: Simple circuit with two parallel impedances.

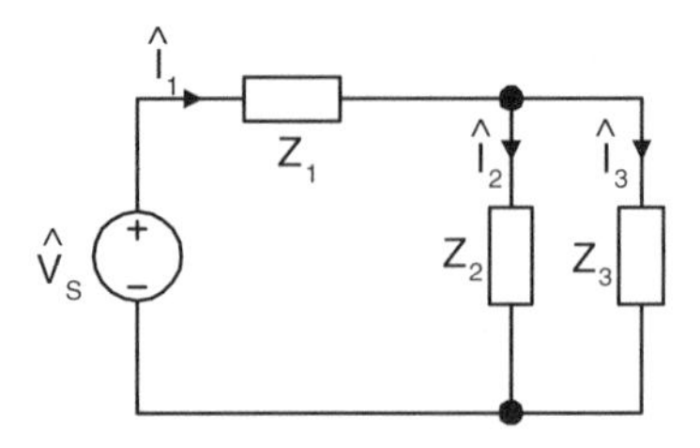

Figure 4.8: Example circuit.

EXAMPLE 4.5 Consider a two-branch current divider consisting of a capacitor C and a resistor R. Derive the expression for the magnitude of the ratio of the output current through the resistor to the input current as a function of angular frequency ω. What is the divider ratio at the limit of very low and very high frequencies?

Equation (4.32) for this example becomes $\hat{I}_2/\hat{I}_s = Y_2/(Y_1+Y_2) = (1/R)/(1/R+j\omega C) = 1/(1+j\omega RC)$. The magnitude of the ratio is just $|\hat{I}_2/\hat{I}_s| = 1/\sqrt{1+(\omega RC)^2}$. As $\omega \to 0$, the divider ratio approaches 1 and the output signal $\hat{I}_2$ equals the input signal. As $\omega \to \infty$, the output signal goes to zero. This device is called a passive low-pass filter and will be discussed in Chapter 9. ■

We now demonstrate how combinations of equivalent transformations can be used to simplify problems in circuit analysis.

Consider the circuit shown in Figure 4.8. It is assumed that $\hat{V}_s$, Z_1, Z_2, and Z_3 are given and we want to find all the currents $\hat{I}_1$, $\hat{I}_2$, and $\hat{I}_3$. First, we find $\hat{I}_1$. To do this, we notice that Z_2 and Z_3 are connected in parallel, and we use the equivalent transformation for parallel connections to combine them into one equivalent impedance Z_{23} (see Figure 4.9):

$$Z_{23} = \frac{Z_2 Z_3}{Z_2 + Z_3}. \tag{4.35}$$

To further simplify the circuit, we use the equivalent transformation for series connections to combine Z_1 and Z_{23} into one equivalent impedance Z_{eq}. Thus, the total

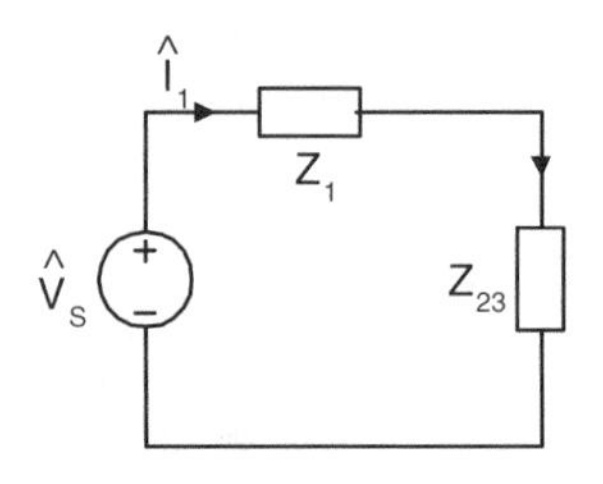

Figure 4.9: Z_2 and Z_3 combined into one impedance.

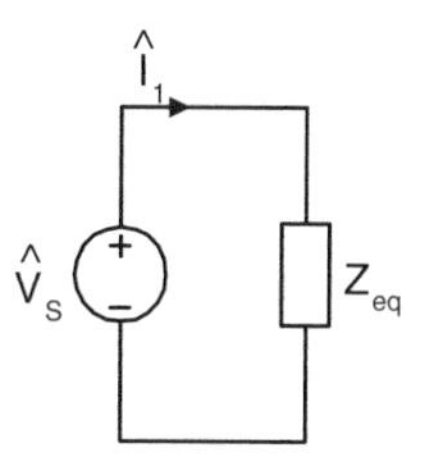

Figure 4.10: All impedances combined into Z_{eq}.

impedance of the circuit (Figure 4.10) is:

$$Z_{eq} = Z_1 + Z_{23} = Z_1 + \frac{Z_2 Z_3}{Z_2 + Z_3}. \tag{4.36}$$

We can now find $\hat{I}_1$:

$$\hat{I}_1 = \frac{\hat{V}_s}{Z_{eq}}. \tag{4.37}$$

Now, to find $\hat{I}_2$ and $\hat{I}_3$ we use the current divider rule, because from Figure 4.8 it is evident that Z_2 and Z_3 are connected in parallel and we already know I_1 (the total current). So, we have

$$\hat{I}_2 = \frac{\hat{I}_1 Z_3}{Z_2 + Z_3}, \tag{4.38}$$

$$\hat{I}_3 = \frac{\hat{I}_1 Z_2}{Z_2 + Z_3}. \tag{4.39}$$

The circuit is now completely analyzed, and no system of equations has been set up or solved. This is the advantage of using equivalent transformations for circuit analysis. These transformations explicitly exploit the connectivity of electric circuits.

4.4 Input Impedance

The *input impedance* is defined as the equivalent impedance with respect to the input terminals of the source. It can sometimes be found through the successive use of the equivalent transformations of parallel and series connections, although the circuits may originally be drawn in such a way as to obfuscate the connections between the elements. Consider the following examples.

EXAMPLE 4.6 Consider the circuit shown in Figure 4.11. This circuit may seem to be complex with impedance Z_4 connected diagonally between the other four impedances, but in actuality this circuit is no more difficult to analyze than any of the others we have already considered. As a proof, we redraw the above circuit (see

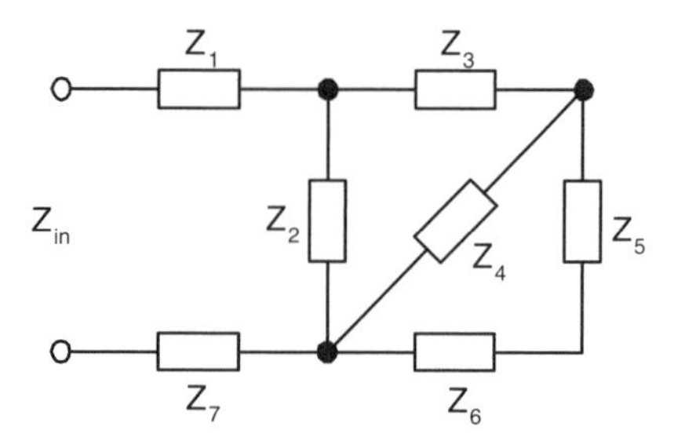

Figure 4.11: Example circuit.

Figure 4.12). It is the same circuit as before, just represented in a more familiar manner which reveals the parallel and series connections. This circuit is easy to transform by using the methods already discussed. Indeed, it is now easy to see that impedances Z_5 and Z_6 are in series and that they together are in parallel with Z_4. The three together are in series with Z_3, and that whole group is then in parallel with Z_2. Finally, it is obvious that the group of five impedances mentioned above is in series with Z_1 and Z_7. Therefore, the whole circuit can be reduced to one equivalent impedance:

$$Z_{in} = \frac{\left(\frac{Z_4(Z_5+Z_6)}{Z_4+Z_5+Z_6} + Z_3\right) Z_2}{\frac{Z_4(Z_5+Z_6)}{Z_4+Z_5+Z_6} + Z_3 + Z_2} + Z_1 + Z_7. \tag{4.40}$$

Although writing a formula like this one may seem complicated, one should try to become fluent at reading circuits and recognizing the connections. Figuring out input impedances should become natural with practice. ■

EXAMPLE 4.7 Consider the circuit shown in Figure 4.13. This is another circuit which might be confusing at first glance. This circuit also features diagonally arranged impedances, which give some appearance to this circuit of being more complex than it really is. Again, we can redraw this circuit in a more "enlightening" manner which

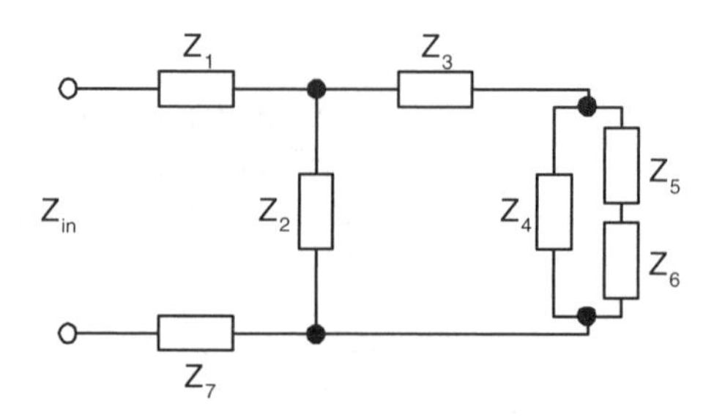

Figure 4.12: The example circuit redrawn more clearly.

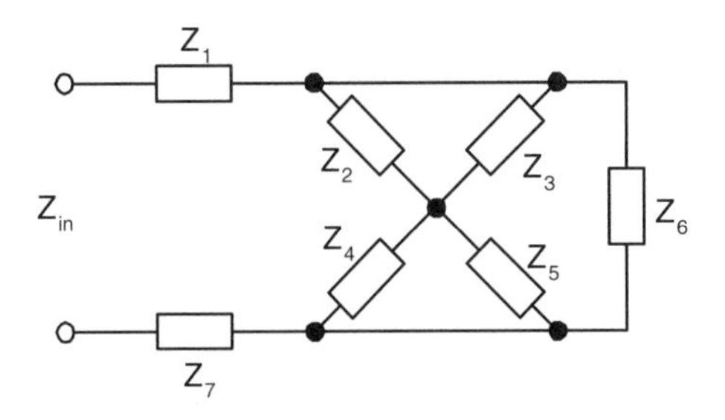

Figure 4.13: Another example circuit.

reveals its hidden simplicity (Figure 4.14). By joining the nodes connected only by wires with no impedances, we can now clearly see that Z_2 and Z_3 are in parallel, as well as Z_4 and Z_5. Those two groups are in series with each other and together are in parallel with Z_6. Impedances Z_1 and Z_7 are then connected in series with the rest of the circuit. Therefore, the whole circuit can be reduced to the equivalent impedance:

$$Z_{in} = \frac{Z_6\left(\frac{Z_2Z_3}{Z_2+Z_3} + \frac{Z_4Z_5}{Z_4+Z_5}\right)}{Z_6 + \frac{Z_2Z_3}{Z_2+Z_3} + \frac{Z_4Z_5}{Z_4+Z_5}} + Z_1 + Z_7. \tag{4.41}$$

■

EXAMPLE 4.8 The circuit for this example is known as the $R-2R$ ladder circuit and is depicted in Figure 4.15a for a three-stage configuration. What is the equivalent input resistance of this system?

The two rightmost resistors have resistance R and are connected in series to yield an equivalent resistance of $2R$. This equivalent resistor is in parallel with a $2R$ resistor as indicated in Figure 4.15b and the equivalent resistance of this combination is $2R \times 2R/(2R + 2R) = R$. The cycle repeats itself as this equivalent R resistor is in series with another R resistor and that combination is in parallel with a $2R$ resistor (Figure 4.15c) to produce an R equivalent resistance which is in series with the final

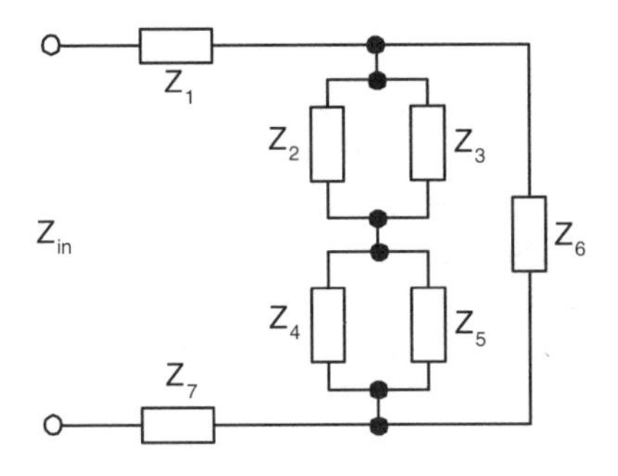

Figure 4.14: The previous circuit redrawn.

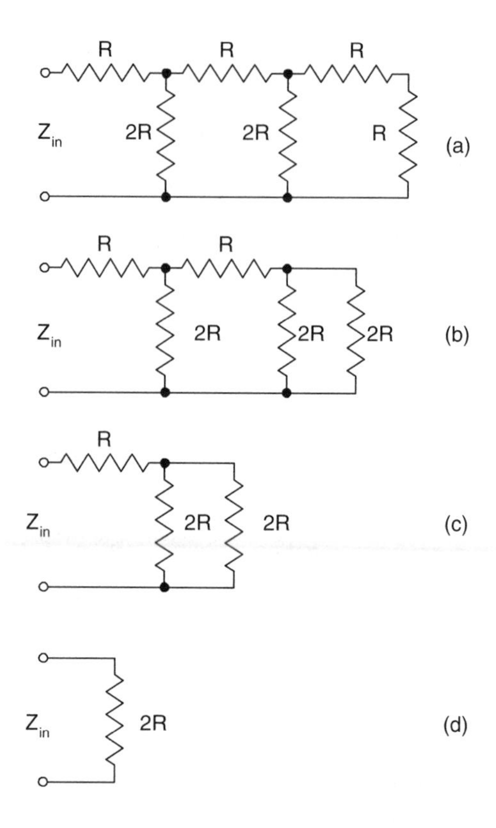

Figure 4.15: An $R-2R$ ladder.

R resistor. Thus the equivalent input resistance is $2R$ as shown in Figure 4.15d. It is left as an exercise to show that the equivalent resistance of an *infinite* $R-2R$ ladder is also equal to $2R$. ■

4.5 Symmetry

Sometimes no matter how a circuit is redrawn, it is just not possible to simplify it any further by using only the tools we have already developed. For some of these circuits, *symmetry* can be successfully used for analysis. Consider the circuit shown in Figure 4.16a. The input impedance for this circuit is not immediately apparent. The impedance connected between the nodes A and B prevents the use of the equivalent transformation of parallel connections, so something else must be done. In this example, we are aided by the observation that all the impedances in the vertical branches are of the same value, Z. For this reason, the two vertical paths that the current could possibly take are identical (they are indistinguishable, which is a reflection of their symmetry). As a result, the current will be evenly divided between

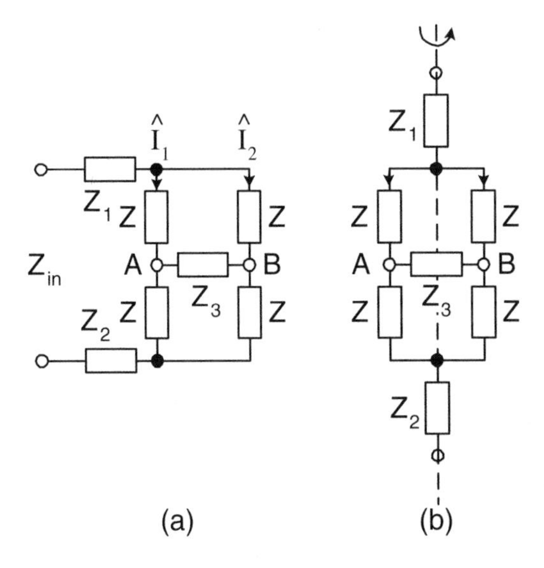

Figure 4.16: A circuit solvable by symmetry: (a) original circuit and (b) circuit redrawn to show rotational symmetry.

them. Another way to look at the symmetry of the circuit is to redraw it as shown in Figure 4.16b. The circuit looks exactly the same if we rotate it around the indicated axis through 180°, even though this interchanges the currents paths for $\hat{I}_1$ and $\hat{I}_2$. Thus, these paths are indistinguishable and the currents must be the same. Therefore, the voltage drops across the two top impedances will be the same. This means that the potentials of nodes A and B are equal. In other words, the potential difference between these points is zero. If the potential difference between A and B is zero then no current flows between these nodes, and the impedance Z_3 between them can be omitted. Thus, we can just redraw the circuit as shown in Figure 4.17 and analyze it easily.

The key to the above argument is the observation that the currents flowing through each vertical branch will be the same, since there is nothing to distinguish

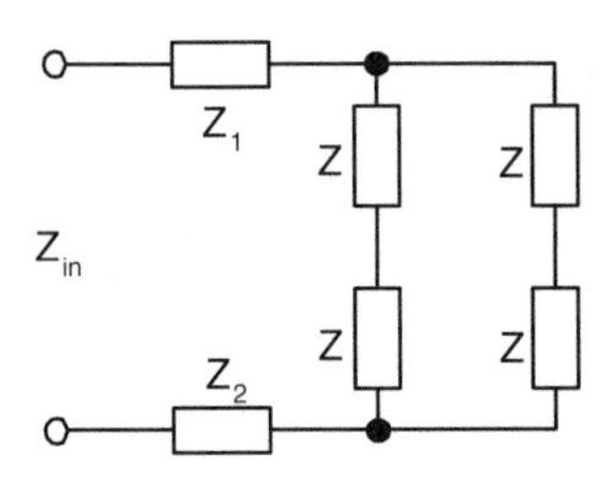

Figure 4.17: Circuit with middle branches removed.

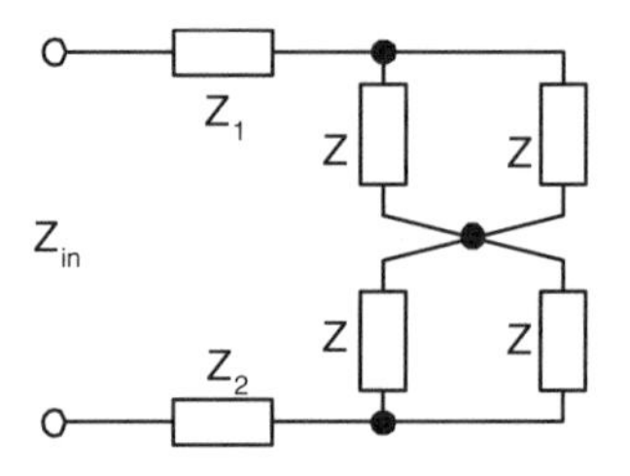

Figure 4.18: The circuit with equipotential points connected.

one branch from another. This is an example of a symmetry argument. Most symmetry arguments use similar approaches to eliminate unnecessary branches or to connect different nodes together into one. For example, the circuit in Figure 4.16 can also be redrawn as shown in Figure 4.18. This is because the equipotential nodes can be connected together into one node. Of course, analyzing the circuit in this manner will yield the same input impedance. Indeed, from Figure 4.17 we find:

$$Z_{in} = \frac{2Z}{2} + Z_1 + Z_2 = Z + Z_1 + Z_2. \tag{4.42}$$

From Figure 4.18 we derive:

$$Z_{in} = \frac{Z}{2} + \frac{Z}{2} + Z_1 + Z_2 = Z + Z_1 + Z_2. \tag{4.43}$$

Thus, we arrive at identical expressions for Z_{in}, which supports our previous assertion.

Sometimes a circuit is drawn in a manner which hides the symmetry it contains. As an example, consider the circuits shown in Figure 4.19. The left circuit contains the same type of symmetry present in the Figure 4.16, although it may not be immediately visible. But when redrawn as the right circuit, we can clearly see it. By the same argument as in the first example, we can conclude that there will be no current flow through the horizontal branch, and, thus, we can remove it without altering the input impedance. The problem is now reduced to a simple parallel transformation, yielding

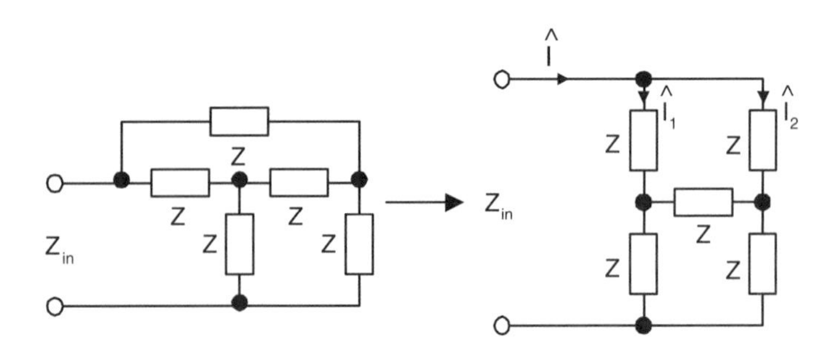

Figure 4.19: A circuit with hidden symmetry revealed.

the input impedance:

$$Z_{in} = Z. \tag{4.44}$$

We can alter the previous circuit in a small way to demonstrate another aspect of symmetry. Consider the same circuit as before, but this time with impedances that are not all equal (refer to the circuit on the left in Figure 4.20). The symmetry argument can still be applied to find the input impedance, although the impedances are no longer all equal. First we observe that there is symmetry between the upper and lower parts of the circuit, that is, the symmetry with respect to impedance Z_2. In other words, the upper and lower parts of the circuit are identical. Because of that, the voltage drops over the upper and lower parts are the same. Thus, the potentials at the nodes A and B are equal, meaning that no current will flow between them and that we can remove the impedance in between. This results in the circuit shown on the right in Figure 4.20. Now we can find the input impedance using series and parallel transformations:

$$Z_{in} = \frac{(2Z)(2Z_1)}{2Z + 2Z_1} = \frac{2ZZ_1}{Z + Z_1}. \tag{4.45}$$

EXAMPLE 4.9 This is a somewhat more challenging problem with symmetry. Figure 4.21 depicts a three-dimensional cube with each branch containing an impedance Z. We wish to find the input impedance between points 1 and 7. This circuit cannot be simplified by using standard transformations, so we shall try to use a symmetry argument. We start by considering the current entering at node 1. By symmetry, we can see that the three paths the current can take are equivalent, so the current will be evenly divided. (This is because a rotation of the circuit through 120° or 240° around an axis which goes through nodes 1 and 7 leaves the circuit unchanged.) Consequently, the potentials at nodes 2, 4, and 5 will be equal, and we can connect these three nodes together. A similar argument can be used for nodes 3, 6, and 8 by considering the incoming current at node 7, which is also evenly divided. This means that these three nodes can be connected together as well. This results in the circuit

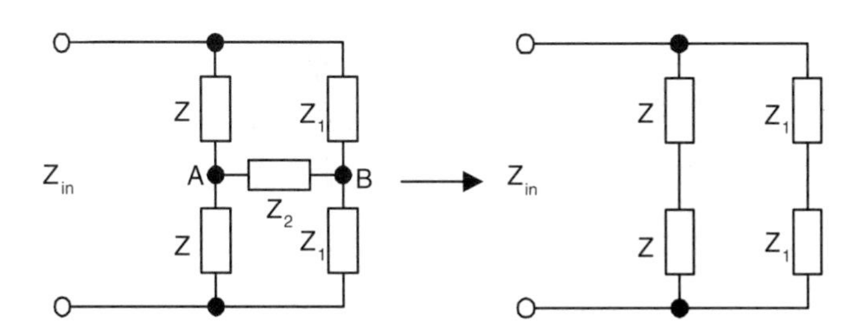

Figure 4.20: Symmetrical circuit with different impedances.

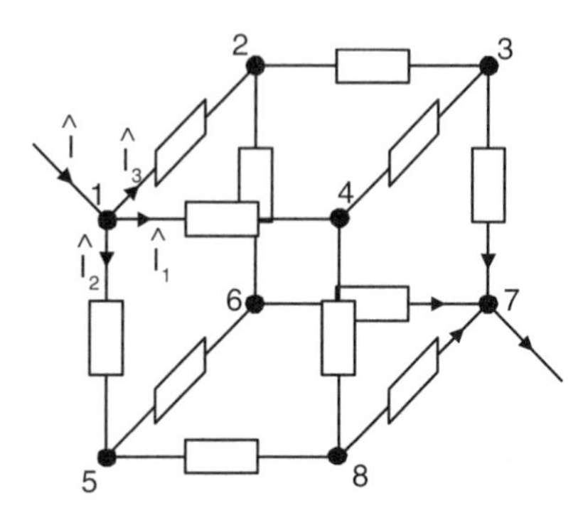

Figure 4.21: A cube of impedances.

shown in Figure 4.22. Now, we can see that the input impedance is:

$$Z_{in} = \frac{Z}{3} + \frac{Z}{6} + \frac{Z}{3} = \frac{5Z}{6}. \tag{4.46}$$

■

EXAMPLE 4.10 Consider the problem of measuring the resistance of resistor R connected to some (unknown) circuit by using an ohmmeter and only *one* measurement. By using the conventional technique and connecting the ohmmeter to the terminals 1 and 2 of the resistor, we will produce a reading which corresponds to the equivalent resistance of R in parallel with the input resistance of the circuit. Thus, the conventional technique does not work. It may first seem that the posed problem cannot be solved without disconnecting the resistor. However, the solution of the problem can be accomplished by using the measuring scheme shown in Figure 4.23. Here, the ohmmeter has one of its terminals connected to nodes 1 and 2, while the other termi-

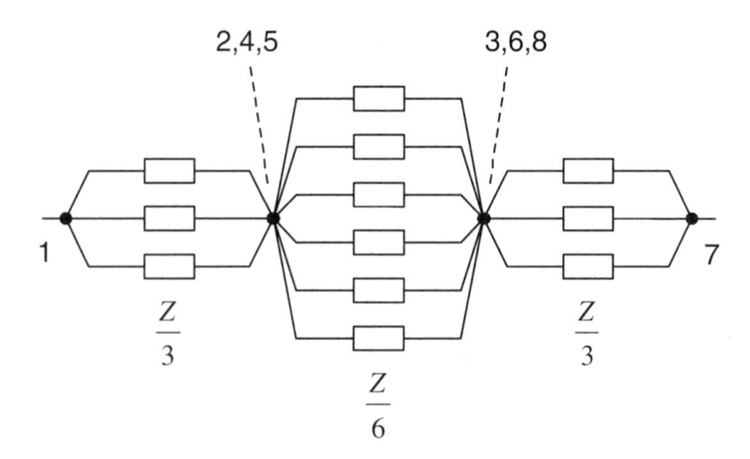

Figure 4.22: The simplified impedance cube.

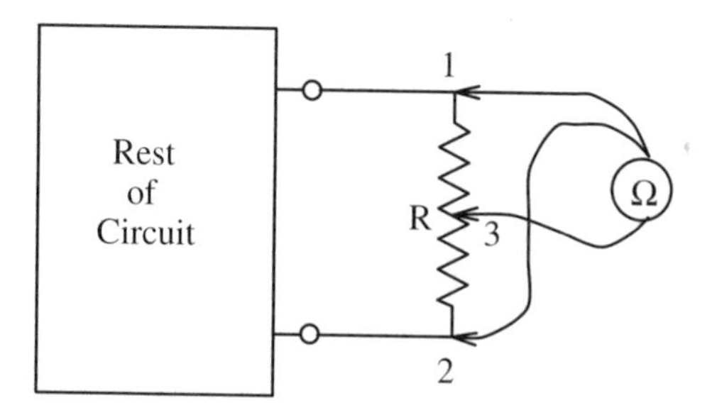

Figure 4.23: Ohmmeter connected to a resistor.

nal is connected to the middle of the resistor, at point 3. With the ohmmeter "hooked" in this manner, the resistor R is "electrically disconnected" from the rest of the circuit and its resistance can be easily measured.[1] To see this, consider the potentials of nodes 1 and 2. Since they are both connected to the same terminal of the ohmmeter, they are at the same potential. Thus, no current will flow between them through the connected circuit in the box. The current will flow in symmetrical fashion from node 3 to nodes 1 and 2. Therefore, we can redraw the diagram as shown in Figure 4.24. From Figure 4.24 it is clear that the ohmmeter will record a measurement of $R/4$ since the two halves of the resistor are in parallel. In this manner, we can measure the resistance of the resistor R without having it physically disconnected from the circuit.

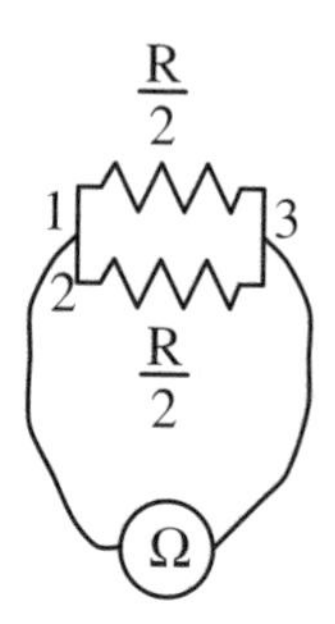

Figure 4.24: Equivalent circuit for the measuring technique. ■

4.6 The Superposition Principle

In the previous sections, we analyzed electric circuits containing only one source. We used equivalent transformations to reduce these circuits to forms which yielded

[1](Of course, not all resistors are exposed in the middle.)

easily obtainable solutions. This approach can be generalized to electric circuits with several sources. This can be done by using the principle of superposition.

The superposition principle is a very important concept in the circuit theory. This principle can be formulated as follows. Consider an electric circuit which contains many sources. We can subdivide these sources into several (nonintersecting) groups and consider different regimes of the original circuit when only sources of one group are *active* while all other sources *are set to zero*. It is clear that the number of different regimes is equal to the number of the groups into which all sources are divided. The superposition principle states that the actual branch currents in the original circuit are equal to the superposition (the algebraic sum) of the corresponding branch currents for all different regimes of the circuit. This principle is based on the *linearity* of the basic circuit equations and can be easily understood (or proved) from the following mathematical fact. If the right-hand side of the basic equations (3.123) and (3.124) is represented as an algebraic sum of several right-hand sides, then the solution of the basic equations is equal to the algebraic sum of solutions of these equations obtained for each of the right-hand sides. Since the right-hand side of the basic equations (3.123) and (3.124) contains only sources, the subdivision of their right-hand side into several right-hand sides is equivalent to the subdivision of all circuit sources into several groups. This remark clearly suggests that the above-mentioned mathematical fact is equivalent to the superposition principle.

There are several facts which should be clearly understood while using the superposition principle. First, the superposition principle is valid for *any* subdivision of sources into several groups. A choice of subdivision is usually suggested by the structure (connectivity) of the electric circuit being analyzed. There is nothing which prohibits having each group consist of only one source. In this case, we shall have as many regimes as the total number of sources in the electric circuit. Second, when we set all sources (except those which belong to an active group) to zero, it means that we set the corresponding voltages and currents of those sources to zero. If a voltage source is set to zero, it means that it is replaced by a short-circuit branch (see Figure 4.25), because the voltage across this branch is equal to zero. If a current source is set to zero, it means that it is replaced by an open branch (see Figure 4.26), because the current through an open branch is equal to zero. These replacements of sources by short-circuited and open branches often lead to significant simplifications of electric circuits, simplifications which allow one to exploit a particular connectivity of the electric circuit being analyzed.

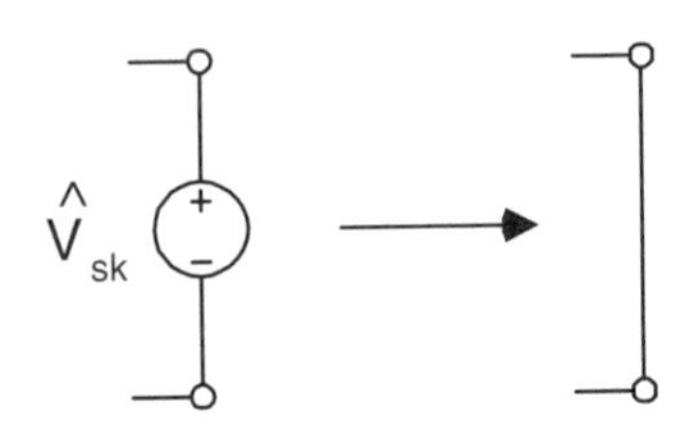

Figure 4.25: Voltage source set to zero.

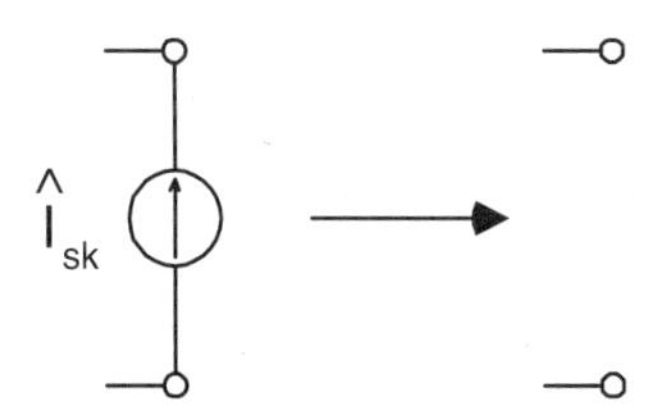

Figure 4.26: Current source set to zero.

EXAMPLE 4.11 Consider the circuit shown in Figure 4.27. It is not much different from the circuit shown in Figure 4.8, except that it is driven by two sources. We want to find all the branch currents.

We begin, as before, by representing the circuit in phasor-impedance notation (Figure 4.28) and by introducing reference directions (these directions are especially important in the superposition principle). We calculate all the impedances and convert the sources into phasor form:

$$Z_1 = R_1 + j\omega L_1 = 1 + j(2)\left(\frac{1}{2}\right) = 1 + j\,\Omega, \tag{4.47}$$

$$Z_2 = \frac{-j}{\omega C_2} = \frac{-j}{(2)(\frac{1}{4})} = -2j\,\Omega, \tag{4.48}$$

$$Z_3 = R_3 + j\omega L_3 = 1 + j(2)\left(\frac{3}{2}\right) = 1 + 3j\,\Omega, \tag{4.49}$$

$$v_1(t) = \cos 2t \rightarrow \hat{V}_1 = 1\text{ V}, \tag{4.50}$$

$$v_2(t) = \sqrt{2}\cos(2t - 45^o) \rightarrow \hat{V}_2 = \sqrt{2}e^{-j\frac{\pi}{4}} = 1 - j\text{ V}. \tag{4.51}$$

Now we apply the superposition principle. We will consider two separate regimes, each with one source. Because both sources in the above circuit are voltage sources, we set them to zero by replacing the corresponding sources with short-circuit branches. We can now consider two separate circuits, corresponding to the two separate regimes. These circuits are denoted as a and b, respectively, in Figure 4.29. The first circuit contains only the left voltage source, while the second circuit has only the right one.

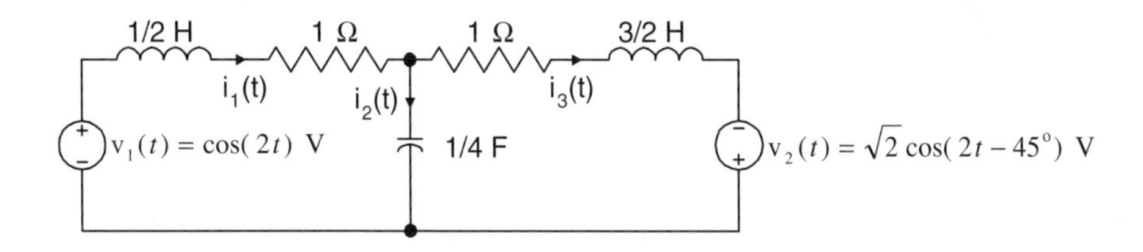

Figure 4.27: A circuit excited by two sources.

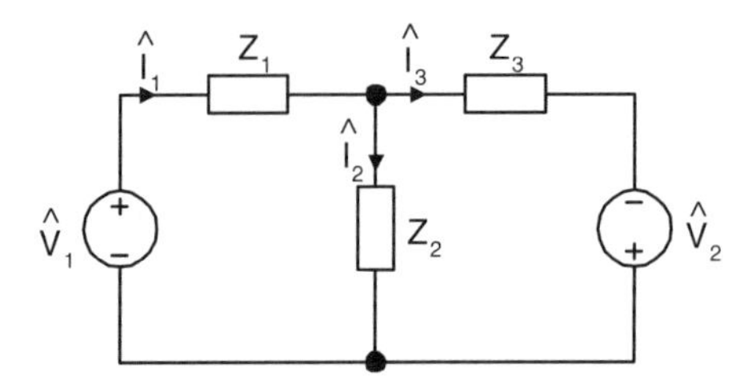

Figure 4.28: The same circuit represented in phasor-impedance notation.

It is important to introduce reference directions for these circuits, but they do not have to be the same as those introduced for the original circuit. For this problem, it is convenient to reverse the direction of the current through Z_2 in circuit b to conform to the model of the current divider. Note that the branch currents shown in the new circuits are not the same as the original branch currents. They represent two different, separate components of actual branch currents which must be added together to find the actual currents.

Now circuits a and b can be analyzed by using the techniques of equivalent transformations. We omit the details since the procedure is identical to the one being used for the analysis of the circuit shown in Figure 4.8. For circuit a, we first find $\hat{I}_1^a$ by dividing the voltage of the source by the equivalent impedance with respect to the terminals of the first source:

$$\hat{I}_1^a = \frac{\hat{V}_1}{Z_1 + \frac{Z_3 Z_2}{Z_3 + Z_2}}. \tag{4.52}$$

Now $\hat{I}_2^a$ and $\hat{I}_3^a$ can be found by applying the current divider rule to the parallel branches:

$$\hat{I}_2^a = \frac{\hat{I}_1^a (Z_3)}{Z_3 + Z_2}, \tag{4.53}$$

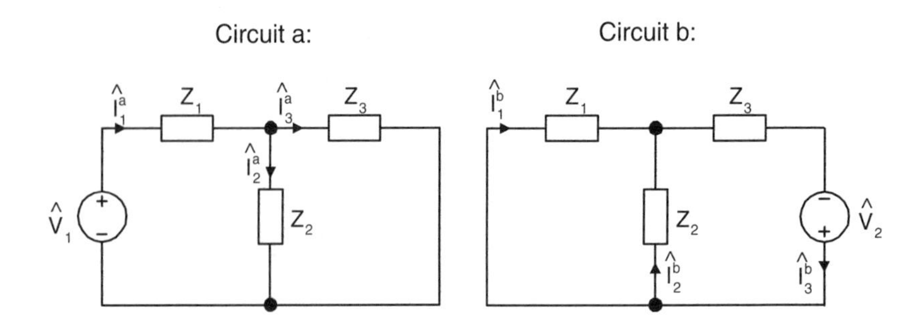

Figure 4.29: Separate circuits a and b.

$$\hat{I}_3^a = \frac{\hat{I}_1^a(Z_2)}{Z_3 + Z_2}. \tag{4.54}$$

The currents in circuit b can be found in a similar manner. Dividing the voltage by the equivalent impedance of the circuit with respect to the terminals of the second source, we find $\hat{I}_3^b$:

$$\hat{I}_3^b = \frac{\hat{V}_2}{Z_3 + \frac{Z_2 Z_1}{Z_1 + Z_2}}. \tag{4.55}$$

Next, we use the current divider rule:

$$\hat{I}_1^b = \frac{\hat{I}_3^b(Z_2)}{Z_1 + Z_2}, \tag{4.56}$$

$$\hat{I}_2^b = \frac{\hat{I}_3^b(Z_1)}{Z_1 + Z_2}. \tag{4.57}$$

Now that the branch currents for each of the two separate regimes have been found, we must add them with the proper signs to find the actual currents. When adding the regime currents, the proper signs are determined by comparing the initial reference directions with the reference directions for each regime. If the reference directions of the currents in circuits a and b coincide with the reference directions in the original circuit, then the currents are taken with a plus sign; otherwise they are taken with a minus sign. By examining the reference directions in circuit a, we can see that all of them coincide with the reference directions in the original circuit. However, for circuit b, the reference direction for $\hat{I}_2^b$ is opposite to the reference direction of $\hat{I}_2$, so it will be taken with a minus sign. Consequently, the expressions for the branch currents are

$$\hat{I}_1 = \hat{I}_1^a + \hat{I}_1^b, \tag{4.58}$$

$$\hat{I}_2 = \hat{I}_2^a - \hat{I}_2^b, \tag{4.59}$$

$$\hat{I}_3 = \hat{I}_3^a + \hat{I}_3^b. \tag{4.60}$$

The numerical values obtained according to equations (4.52)–(4.60) are listed (in amperes) below:

$$\hat{I}_1^a = \frac{1+j}{6}, \qquad \hat{I}_1^b = \frac{-(1+j)}{3}, \qquad \hat{I}_1 = \frac{-(1+j)}{6},$$

$$\hat{I}_2^a = \frac{1+3j}{6}, \qquad \hat{I}_2^b = \frac{1}{3}, \qquad \hat{I}_2 = \frac{-1+3j}{6},$$

$$\hat{I}_3^a = -j/3, \qquad \hat{I}_3^b = -j/3, \qquad \hat{I}_3 = -4j/6.$$

Converting back to the time domain we arrive at:

$$i_1(t) = \frac{\sqrt{2}}{6}\cos(2t - 135^\circ)\ \text{A},$$

$$i_2(t) = \frac{\sqrt{10}}{6}\cos(2t + 108.4^\circ)\ \text{A},$$

$$i_3(t) = \frac{2}{3}\cos(2t - 90^\circ)\ \text{A}.$$

This concludes the analysis of the circuit. ■

As is clear from the above example, the superposition principle presents a way to analyze electric circuits with multiple current and voltage sources. This method will also work for circuits with multiple sources of different frequencies. In this case, we first use the superposition principle and then convert a circuit into phasor-impedance form for each regime (frequency) separately. Next, we analyze the separate regimes as before (using different frequencies to calculate the impedances) and convert the phasors back into time domain form *before* summing for the actual currents. This procedure will preserve the proper frequency dependence of the various sources and their respective responses.

EXAMPLE 4.12 Consider the same circuit in Figure 4.27 but this time with the source:

$$v_2(t) = \sqrt{2}\cos(4t - 45^\circ)\ \text{V}. \tag{4.61}$$

Regime (a) of the problem remains exactly the same as before. We convert the currents for this regime to the time domain separately to get:

$$i_1^a(t) = \frac{\sqrt{2}}{6}\cos(2t + 45^\circ)\ \text{A}, \tag{4.62}$$

$$i_2^a(t) = \frac{\sqrt{10}}{6}\cos(2t + 71.6^\circ)\ \text{A}, \tag{4.63}$$

$$i_3^a(t) = \frac{1}{3}\cos(2t - 90^\circ)\ \text{A}. \tag{4.64}$$

For regime (b), the phasor $\hat{V}_2$ is still equal to $1 - j$. However, we get new values for the impedances when we insert $\omega = 4$ into equations (4.47)–(4.49):

$$Z_1 = R_1 + j\omega L_1 = 1 + 2j\ \Omega, \tag{4.65}$$

$$Z_2 = -j/\omega C_2 = -j\ \Omega, \tag{4.66}$$

$$Z_3 = R_3 + j\omega C_3 = 1 + 6j\ \Omega. \tag{4.67}$$

Inserting these values into equations (4.55)–(4.57) results in:

$$\hat{I}_1^b = (3 - j)/15\ \text{A}, \tag{4.68}$$

$$\hat{I}_2^b = (-1 + j)/3\ \text{A}, \tag{4.69}$$

$$\hat{I}_3^b = -2(1 - 2j)/15\ \text{A}. \tag{4.70}$$

When we convert these phasors to the time domain and add to the results of the first regime, we arrive at the final answer:

$$i_1(t) = \frac{\sqrt{2}}{6}\cos(2t + 45°) + \frac{\sqrt{10}}{15}\cos(4t - 18.4°)\text{ A}, \tag{4.71}$$

$$i_2(t) = \frac{\sqrt{10}}{6}\cos(2t + 71.6°) - \frac{\sqrt{2}}{3}\cos(4t + 135°)\text{ A}, \tag{4.72}$$

$$i_3(t) = \frac{1}{3}\cos(2t - 90°) + \frac{2\sqrt{5}}{15}\cos(4t + 116.6°)\text{ A}. \tag{4.73}$$

■

EXAMPLE 4.13 We now consider a rectangular grid of impedances which extends to infinity (practically this can be realized by using a large grid of impedances, while only considering measurements away from the edges).

We want to find the input impedance between nodes A and B, shown in Figure 4.30. To solve this problem, we must make use of both symmetry and the superposition principle. We begin by writing the expression for Z_{in} in terms of the voltage applied across A and B and the current flowing through the grid:

$$Z_{in}^{(A,B)} = \frac{\hat{V}_{AB}}{\hat{I}}. \tag{4.74}$$

This is actually the definition of input impedance.

The approach we take is to find $\hat{V}_{AB}$ in terms of the current $\hat{I}$ by using the superposition principle. We consider two separate regimes, illustrated in Figure 4.31. In the first regime the current $\hat{I}$ is introduced into the grid at the node A (and flows out of the circuit infinitely far away), while in the second regime the current $\hat{I}$ is

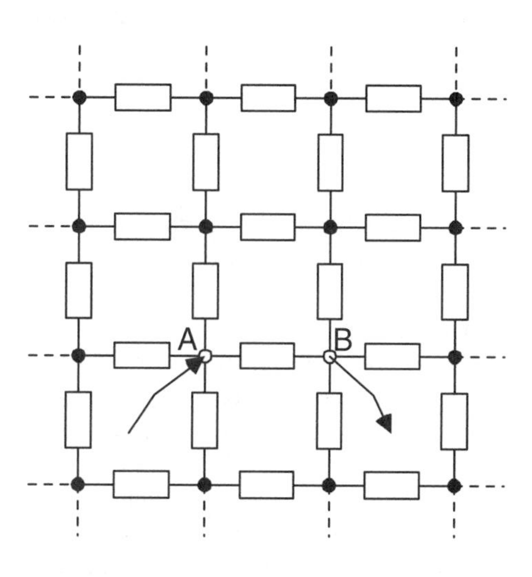

Figure 4.30: An infinite grid of impedances.

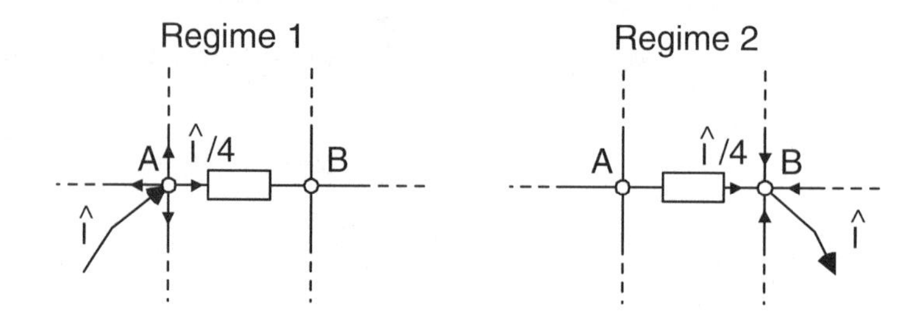

Figure 4.31: Two regimes for infinite grid problem.

taken from the grid at node B. In the first regime, we conclude from symmetry that the current will be evenly divided into four equal parts since no path from A can be distinguished from another. Thus, the current flowing through the impedance between points A and B is $\hat{I}/4$ and the voltage is then:

$$\hat{V}_{AB}^{(1)} = \frac{\hat{I}}{4}Z. \tag{4.75}$$

For the second regime, we can make the same argument to show that:

$$\hat{V}_{AB}^{(2)} = \frac{\hat{I}}{4}Z. \tag{4.76}$$

Thus, by using the superposition principle we can find that the total voltage is the sum of these two voltages:

$$\hat{V}_{AB} = \hat{V}_{AB}^{(1)} + \hat{V}_{AB}^{(2)} = \frac{\hat{I}}{4}Z + \frac{\hat{I}}{4}Z = \frac{\hat{I}}{2}Z. \tag{4.77}$$

If we substitute (4.77) into (4.74), we will find the desired input impedance:

$$Z_{in}^{(A,B)} = \frac{\frac{\hat{I}}{2}Z}{\hat{I}} = \frac{Z}{2}. \tag{4.78}$$

■

EXAMPLE 4.14 As a final example of the superposition principle we consider the three-stage digital-to-analog (D/A) signal converter shown in Figure 4.32a. We want to convert a binary number into an output voltage (or current). We will use b_0 to denote the least significant digit, b_1 to represent the intermediate digit, and b_2 to indicate the most significant digit. We can express the binary number as $b_2b_1b_0$. Thus, the voltage sources b_0, b_1, and b_2 will either be set to zero volts or set to one volt to reflect their digital nature. For three stages we can go from 0 (000) up to 7 (111). The output voltage V_0 is the voltage across the output $2R$ resistor as indicated in the figure. We want to show that this voltage is proportional to the number represented by $b_2b_1b_0$.

There are three independent voltage sources and so we will consider three separate regimes. First, let us set b_1 and b_0 to zero and find the dependence of V_0 on b_2. The resulting circuit is shown in Figure 4.32b. The part of the circuit enclosed

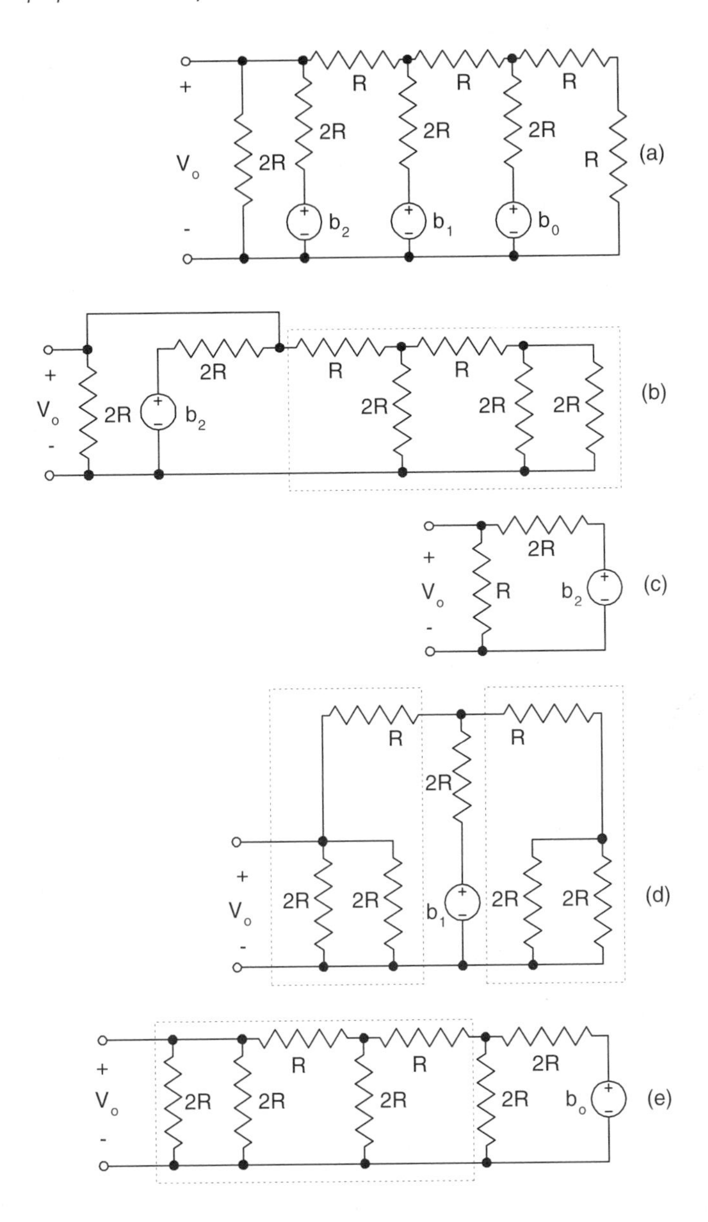

Figure 4.32: A three-stage D/A converter.

in the box is just the $R-2R$ ladder that was discussed in Section 4.4. Its equivalent resistance of $2R$ is in parallel with the output resistance. The simplified equivalent circuit, shown in Figure 4.32c, is just a voltage divider. Using equation (4.26), we see that the sought relationship is given by $V_0 = b_2 \times R/(R + 2R) = b_2/3$.

The circuit that results when we set b_2 and b_0 to zero is shown in Figure 4.32d. The equivalent resistance of the resistors in both boxes is $2R$. We must be careful when we make this simplification because we lose direct information about V_0. However, by the voltage divider rule, we know that V_0 will be one-half the voltage across the equivalent $2R$ resistance. The simplified equivalent circuit for this case looks just like that in Figure 4.32c if we replace b_2 by b_1 and V_0 by $2V_0$. Therefore, another application of the voltage divider formula results in $V_0 = b_1/6$. The circuit that results when we set b_2 and b_1 to zero is shown in Figure 4.32e. The part of the circuit enclosed in the box is another $R-2R$ ladder. Again we lose direct information about V_0 when simplifying, but careful analysis will yield the expression $V_0 = b_0/12$.

The final result comes from summing up the three individual answers, $V_0 = (4b_2+2b_1+b_0)/12$, and gives us the proper conversion formula (with a proportionality constant). ■

4.7 An Introduction to Electric Circuit Simulation with MicroSim PSpice

The use of computer simulation tools for the analysis of electric and electronic circuits pervades the modern workplace. These codes are typically employed when the circuits contain a sufficient number of elements to render impractical the computation of circuit voltages and currents by hand. This is the case for many circuits of interest. While most of these circuits can be broken down into subcircuits that can be analyzed by hand, analysis of the performance of the complete circuit is often of paramount importance and can be done only via computer. These simulation tools are also useful for investigating circuits of any size which have nonlinear elements. Furthermore, they can also be used to evaluate the dependence of output quantities on the parameter tolerances or the nonideal properties of various circuit elements.

In this book we will make use of numerical simulations only to examine the relatively small, linear, electric circuits that we can also analyze by hand. The purpose of this is threefold. First, because of the prevalent use of simulators in industry, you should have a working knowledge of at least one circuit simulation package and have a general understanding of their capabilities. Second, once you have the ability to simulate circuits, you can use the codes to check your hand calculations and perhaps even to investigate deviations from ideal circuit performance. Finally, if you utilize a code which contains a good graphics package (like the one described here), you can readily obtain visual information about any circuit variable in either the time or the frequency domain.

It cannot be overemphasized that **a circuit simulator is not a substitute for the knowledge of basic circuit theory principles.** It will not design circuits from scratch for you. It will not troubleshoot a circuit for you if it is not performing according to expectations. Perhaps most important, the code will likely, without hesitation or complaint, give you incorrect or unreasonable answers if you give it incorrect or unreasonable input data. The only way you can trust the simulation results is to have a solid understanding of Kirchhoff's laws, terminal relations, and the rest of the electric circuit theory concepts.

In this section we will give a brief introduction, via several examples, to the operation of the MicroSim PSpice circuit simulator and auxiliary routines. At the end of Chapters 6 through 10 you will find examples and homework problems which are based on this simulator. PSpice was chosen for two reasons. First, it is based on the SPICE2 simulator developed at University of California Berkeley in the 1970s, which is the basic engine (albeit often modified) for a number of simulators that are widely used in industry. Second, there is available an evaluation copy of MicroSim's family of products, which is comprised of a schematic generator, the PSpice simulator for analog and digital components, and the PROBE graphics package. This software is available free to professors and can be copied for students with the encouragement of the manufacturer, MicroSim Corporation.[2] This discussion is based on the IBM-PC compatible version 6.2 for Microsoft Windows. Furthermore, there are often site licenses for the regular SPICE simulator on the various workstations that are typically available to students. If one is familiar with the operating system and can locate the SPICE code, this introduction should be sufficient to enable one to simulate circuits in that environment as well.

This introduction is by no means a substitute for an extensive operating manual for PSpice. It contains only a bare-bones description designed to give the user the ability to perform the basic operations of circuit simulation. There are a number of excellent texts dedicated to PSpice simulation. Several of them are listed in Appendix C. The Reference guide from MicroSim Corporation is also indicated there. However, after mastering the basics, one can also learn more about the code's capabilities by roaming through the various drop-down menus and experimenting with the various options.

Before proceeding, the evaluation copy must be loaded onto an IBM-PC. Version 6.2 is usually supplied on five 3 1/2 inch 1.44 MB floppy disks or a CD-ROM. Installation follows the usual procedure for any Windows 3.1 software.[3] Disk 1 is loaded into the floppy drive and the setup.exe program on that disk is executed either by double-clicking on the "setup.exe" listing in the File Manager or by using the "Run" command on the "File" drop-down menu in the Program Manager. (Windows 95 users can use the Windows Explorer or the control panel to load the program.) Follow the directions that appear on the screen to load the software, generate a Program Group entitled "MicroSim Eval 6.2," and make any necessary changes to your autoexec.bat file. The files will nominally be saved in a directory called "MSIMPR62." The pro-

---

[2]Submit a request on educational letterhead to: Product Marketing Department, MicroSim Corporation, 20 Fairbanks, Irvine, CA 92718. Their phone number is (714) 770-3022.

[3]Users of older versions of Windows may have to also install "Win32s" in order to run PSpice. This code is also supplied on the PSpice CD-ROM and can be transferred to two floppy disks for distribution. Alternatively, one can use an older version of MicroSim's products. Version 6.0, for example, has most of the Version 6.2 features and runs without "Win32s."

gram group should contain at least five items: 1) MicroSim Schematics, 2) MicroSim PSpice, 3) MicroSim Probe, 4) MicroSim Stimulus Editor, and 5) MicroSim Parts. There will also be a "README" file which has some information about the latest versions of the codes. We will always begin our examples by double-clicking on the Schematics icon; all of the other programs can be invoked at the proper time from the Schematics window. (Unless otherwise stated, "clicking" always refers to pressing the left button on a two-button mouse.)

The first circuit we will analyze is the current divider of Example 4.5. The first step will be to draw the schematic of the circuit. Then we must select the types of analyses that we would like PSpice to perform. In this example we claim that this circuit functions as a low-pass (frequency) circuit, so we will plot the current through the resistor as a function of frequency. Third, we must run the PSpice simulator. Finally, we will use Probe to draw the resistor current over the specified frequency range.

We begin by opening the "MicroSim Eval" Program Group in Windows and double-clicking on the Schematics icon. (Windows 95 (or later) users must use the start button and move through the various menus to arrive at the "schematics" label.) From the drop-down menus at the top of the window, click on "Draw" and then on "Get New Part... " A window entitled "Add Part" will appear and in the dialog box you should enter "C" and then press the "OK" button (by clicking on it). The box will disappear and the cursor arrow will have a capacitor attached to it. Move the capacitor to a desirable location and place it by clicking the left mouse button. You can continue to add capacitors by repeating the last two operations. Click the right mouse button to stop adding capacitors. If you have placed any parts unintentionally, you can delete them by clicking on them (they change color to show that they have been selected) and then hitting the delete key. Next click the "Edit" drop-down menu and select "Rotate" to get the capacitor vertical on the page.

Default values for the capacitor's name and capacitance value are indicated in the schematic. The capacitor's name is typically Cn, where n is an ascending integer. The name can be changed by double-clicking on it so that the "Edit Reference Designator" dialog box appears. Type in a new name and hit "OK." Double-click on the capacitance (whose default is typically 1 nF) to invoke the "Set Attribute Value" dialog box. For this example set the capacitance to 1 F by entering a "1" (farad is assumed — if you type 1 F it will assume that you meant one femtofarad). The multiplying factors that can be entered to scale the units are essentially the same as those listed in Table 1.1. Two exceptions are that "u" is used instead of "μ" to mean 10^{-6} and "MEG" must be used to represent 10^6 because "m" and "M" are both interpreted by PSpice to mean 10^{-3}.

Add a resistor to the circuit by repeating the steps above for the capacitor, except that in the "Add Part" dialog box you should enter "R" for a resistor. Change the resistance to "1" (ohms are assumed) and move the part parallel to the capacitor by "dragging" it with the mouse (holding the left mouse button down while moving the mouse). Connect the two parts in parallel using the "Draw" drop-down menu and selecting "Wire." Click once after moving the pointer to the top of the capacitor and once again at the top of the resistor. Note that the wires are only drawn orthogonally and that they "snap" to the grid points. You can change this in the drop-down

"Options" menu by selecting "Display Options," but the settings are adequate for our purposes. You can select "Repeat" from the "Draw" menu to place the wire between the two lower ends of the elements.

It is important to note that PSpice will automatically number the nodes in a circuit, but that the user must supply the ground node in order for the simulation to run. You do this by selecting "AGND" in the "Add Part" dialog box. Connect this ground to the base of the capacitor. The final element of this circuit is the current source. The notation for this element in the "Add Part" dialog box is "ISRC." Rotate it twice to get the direction arrow pointing up and then add two wires to place it in parallel with the other elements. To set the parameters of the current source, double-click on the arrow to bring up the window which is titled "I1 Part name: ISRC." There are many options that can be adjusted. Because we want to perform an ac analysis, click on the line that says "AC=." There are two boxes at the top of the window. The left one is entitled "Name" and "AC" should appear in that box. Click somewhere in the box on the right, enter a "1" for one ampere, and press the "Save Attr" button. Finally, press the "OK" button. The circuit has now been completely drawn and should look like the circuit in Figure 4.33.[4] Note that you can use the "Text" option of the "Draw" drop-down menu if you want to add comments to your circuit.

You are now ready to save this schematic on disk. This is accomplished by using the "File" drop-down menu and clicking on the "Save" option. The usual file extension for Schematics drawings is *.SCH; we will call our schematic example1.sch. The schematic can be printed (if a printer is connected to your computer) by selecting "File" and "Print." After selecting any options you want in the window that appears, press the "OK" button to initiate the hardcopy generation.

Once the file is saved, go to the "Analysis" drop-down menu and click on "Create Netlist." This will convert the schematic to the tabular form that is required by PSpice. Also in the "Analysis" menu you can click on "Examine Netlist" to view the data in this form. This will launch the Windows Notepad editor to allow you to examine

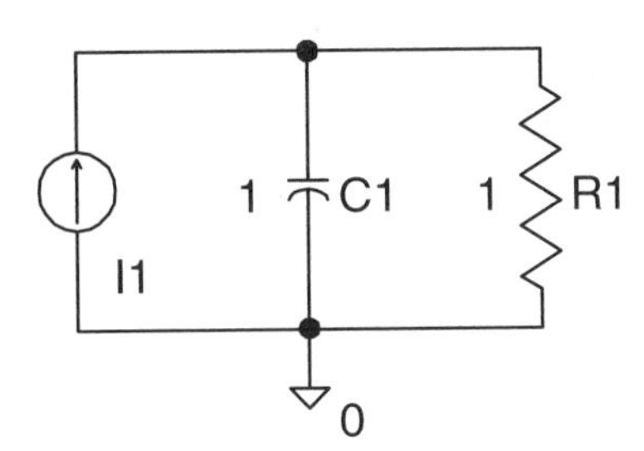

Figure 4.33: The PSpice schematic for Example 4.5.

[4]In this edition, it was not possible to incorporate the PSpice screen images directly into the text. Instead, graphics packages were used to imitate the PSpice output. Consequently, there will be slight differences occasionally between the figures and your computer screen images.

Table 4.1: The circuit Netlist for Example 4.5.

Schematics Netlist			
C_C1	0	N_0001	1
R_R1	0	N_0001	1
I_I	0	$N_0001 AC	1

the Netlist, which has been saved on disk as example1.net. The result is shown in Table 4.1. For the capacitor and the resistor (the second and third rows in the Netlist, respectfully), the first column is the name of the element, the next two entries are the nodes to which the elements are connected, and the fourth and final entry is the value (capacitance and resistance, respectfully). The first node entered is always considered the "plus" node in terms of the assumed reference voltage polarity. *Note that PSpice uses the passive sign convention for all elements, including sources!* The plus node in this example is always the ground node and is indicated by a zero. The only other node in this example is given the name $N_0001. The final row in the Netlist represents the current source. The columns are similar to those of the passive elements except that an "AC" precedes the current value. We could have also assigned currents for dc analysis, transient analysis, etc. in the appropriate dialog box and they would have been indicated on this line as well.

Before PSpice can be executed we must indicate the type of analyses to be run. Go to the "Analysis" drop-down menu and select "Setup." In the "Analysis Setup" dialog box that appears, normally only "dc Bias Point Detail" is selected, as indicated by the "x" in the box to the left of that button. Calculation of the dc bias point is always the first step of a PSpice run. Enable the ac analysis by clicking on the box to the left of "AC Sweep" so that an "x" appears. Next, push the ac sweep button. A window will appear entitled "AC Sweep and Noise Analysis" that will allow us to modify the parameters of the frequency sweep. First, change the "AC Sweep Type" to "Decade" by clicking in the appropriate circle. Next, in the "Sweep Parameter Box" set the "Pts/Decade" to 21, the "Start Freq." to 0.01 (Hz), and the "End Freq." to 100. Press the "OK" button at the bottom of that window. Finally, press the "Close" button in the "Analysis Setup" window to return to the schematic. We are now ready to run PSpice.

From the "Analysis" drop-down menu select "Simulate." First a small window entitled "Schematics" will appear announcing that the Netlist is being generated (unless this has already been done). Next, a "PSpice" window will appear that will update you on the status of the simulation. First it will say that the program is reading and checking the circuit. If any errors are encountered, an error message will be printed in the window and the simulation will stop. The user must then determine the error and correct it before attempting another simulation. If there are no errors, PSpice will proceed with the bias point calculation and then move on to the AC analysis.

After the successful completion of the simulation, the "PROBE" program will be initiated automatically. There are two points to note. First, even though "PROBE" usually comes up "full screen," the schematics generator, PSpice, and PROBE are all running at the same time in different windows and one can move back and forth between them as necessary. Second, PROBE need not be launched automatically after PSpice is run; this option is selected in the "Analysis" drop-down menu under "Probe Setup. . . ." When the PROBE window appears, the frequency axis is drawn but no curves are drawn. Notice that the independent axis has a log scale as requested in an earlier dialog box. Select the "Trace" drop-down menu and click on "Add." In the window that appears, double-click on "I(R1)." The dependence of the resistor current on frequency is then displayed, along with a dependent axis scale and a legend. The current is nearly equal to the source current (of 1 A) at low frequency and goes to zero as the frequency is increased. This is the principal characteristic of the low-pass filter. Note that we can delete any trace if necessary by clicking on the appropriate name in the legend (its color will then change) and hitting the delete key. There are many modifications that we could make to this plot, but we will just modify the labels and leave other changes for later examples. Click on the "Edit" drop-down menu and select "Modify Title. . . ." In the resulting dialog box enter "Example 1" and press "OK." From the "Plot" drop-down menu select "Y Axis Settings," type "Resistor Current" in the "Axis Title" box, and then hit "OK." Selecting "File" and then "Print" will bring up a window that can be used to generate a hard copy. This graph is shown in Figure 4.34.

We conclude this rather long example with a brief discussion of the files that are generated during a typical circuit simulation. The files example1.sch and exam-

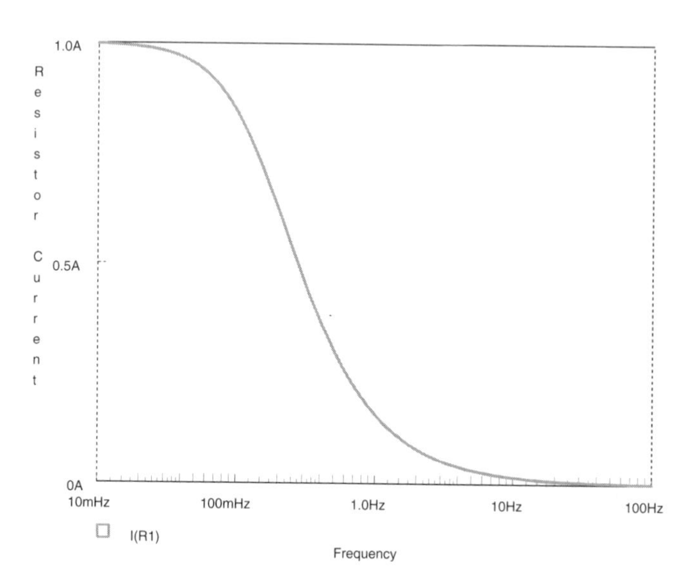

Figure 4.34: The probe output for Example 4.5.

Table 4.2: The PSpice input file for Example 4.5.

```
* D:\ CIRCUITS\ EXAMPLE1.SCH
* Schematics Version 6.2 - April 1995
* Sun Jun 25 15:31:14 1995
** Analysis setup **
.ac DEC 21 .01 100
.OP
* From [SCHEMATICS NETLIST] section of msim.ini:
.lib nom.lib
.INC "EXAMPLE1.net"
.INC "EXAMPLE1.als"
.probe
.END
```

ple1.net have already been described. The input to PSpice is actually read from the file example1.cir, a listing of which is given in Table 4.2. The first line is automatically a comment line and gives the name of the schematic file. Additional comment lines are generated by placing an asterisk in the first position. Comments can also be placed at the end of input lines by separating them from the data by a semicolon. The first two noncomment lines tell PSpice what types of analyses to run. The first line initiates the ac analysis and the trailing parameters indicate the method of determining the frequency points. They come directly from our earlier dialog box entries. The next line tells PSpice to find the dc operating point. The line which begins with ".lib" indicates the name of a library that will be used to specify the parameters of various components. This is discussed in detail at the end of Chapter 7 after diodes are introduced. The two lines beginning with ".inc" tell PSpice to read in the data from the two files indicated. We have already seen the Netlist file. The other file contains aliases, i.e., a list of alternate ways to identify nodes. We will examine this file in the next example. The second to last line invokes the PROBE program at the end of the simulation and the final line identifies the end of the PSpice data.

The final file that we will point out is the example.out file. This file contains the output from the PSpice run and can be loaded into Notepad from the schematics window via the drop-down "Analysis" window by clicking on "Examine Output." It can also be examined from the PSpice window from the "File" drop-down menu by clicking on "Examine Output." Scrolling through it, one can find a listing of the input information as well as some details of the simulation. Error messages that are generated if the input file is not correct will be given in this file. One curious point is that next to total power dissipation the output shows 0.00 W. This is because PSpice calculates only the power dissipated by voltage sources. Another thing that you may notice is that nowhere are the frequency dependences of the circuit voltages and currents to be found in this output file. These numbers are stored in another file that is typically readable by PROBE, but they are not in a simple text format. If we want to generate a table of the resistor current as a function of frequency in the output file, for example, we must insert the line: ".PRINT AC I(R_R1)" into the example.cir file somewhere before the ".END" statement and rerun PSpice. Notepad, or any other editor that is at your disposal, can be readily used to accomplish the required change.

Next, we will use PSpice to analyze the circuit from Example 4.11. We will perform a transient analysis on this circuit and compare the simulations with the analytic results. The circuit redrawn by the schematic editor is shown in Figure 4.35. The resistors, the capacitor, and ground are added and their values are modified as described in the previous example. The inductors are added by selecting "L" in the "Add Part" dialog box. Their values (in henries) are modified by double-clicking on the default values in the schematic.

The name for the voltage sources in the "Add Part" dialog box is "VSIN." By double-clicking on the left voltage source one brings up a window entitled "V1 Part Name: VSIN." Click on the "Voff =" line and then move to the upper right box and type a "0" and push the "Save Attr" button. This sets the source's dc offset to zero. Next, click on the "Vampl =" line, type a "1" in the box, and hit "Save Attr" to set the sine wave amplitude to 1 V. Repeat the steps outlined above to set the frequency to Freq = 0.31831 (2 rad/s) and the phase to 90°. Note that zero degrees gives a sine function for the source, so 90 is required to achieve a cosine dependence. When adding the right voltage source, be sure to rotate it twice to get the positive terminal facing downward. For this element we need to enter Voff = 0, Vampl = 1.414213 ($\sqrt{2}$), Freq = 0.31831, and Phase = 45°. The Netlist for this problem is shown in Table 4.3. There is a total of six nodes (including ground) because PSpice must label even the trivial ones.

To select the analysis type we click on "Setup" in the "Analysis" drop-down menu. We enable the Transient analysis by clicking on the box to the left of the "Transient..." button. Then, by depressing this button a window appears that allows us to define the transient analysis. Click on "Skip initial transient solution." Because the period is about 3.14 seconds, set the "Final Time" to 10 (seconds). That will allow over three full periods to be simulated. For a total of about 200 printed points, select the "Print Step" to be 50m. Then click the "OK" button in that window and the "Close" button in the "Analysis Setup" window. Save this circuit as example2.sch. We are now ready to run the second simulation.

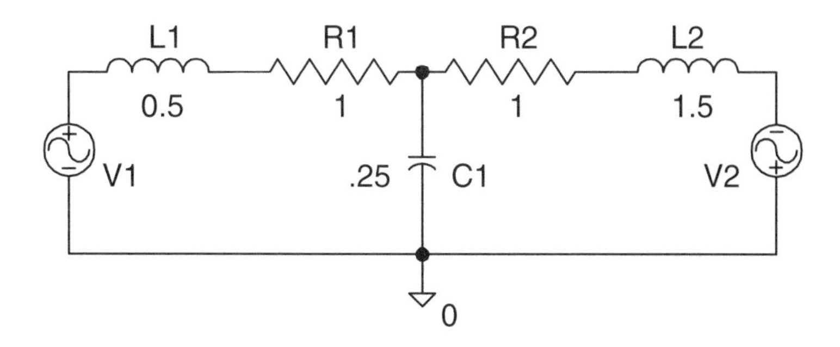

Figure 4.35: The PSpice schematic for Example 4.11.

Table 4.3: The circuit Netlist for Example 4.11.

R_R1	N_0002	N_0001	1			
R_R2	N_0001	N_0003	1			
L_L1	N_0004	N_0002	0.5			
L_L2	N_0003	N_0005	1.5			
C_C1	0	N_0001	0.25			
V_V1	N_0004	0				
+SIN	0	1	0.31831	0	0	90
V_V2	0	N_0005				
+SIN	0	1.414213	0.31831	0	0	45

From the "Analysis" drop-down menu select "Simulate." The usual sequence of windows will come and go and, if there are no errors, the "PROBE" program will be initiated automatically. The input file for PSpice is displayed in Table 4.4, and the alias list is given in Table 4.5. The only new input in the example2.cir file is the directive for the transient analysis, which includes the print step, the final time, and an instruction to skip the initial conditions. The alias file just gives a list of alternate expressions for the nodes in terms of the circuit elements. For example, the first line after the .ALIASES line means that R1:1 should be interpreted as $N_0002 and R1:2 is the same as $N_0001. In the next line and in the seventh line, R2:1 and C1:2 are also made equivalent to $N_0001, respectively, since they are also connected to that node.

When the PROBE window appears, the time axis is drawn but no curves are present. Select the "Trace" drop-down menu and click on "Add." In the window that appears, one can select time, five voltages that correspond to each of the nonzero node potentials via aliases, a current through any of the passive or active elements, or the ground potential. One can also select combinations of the variables. For example, the voltage across the resistor R1 can be plotted as follows. Click on the "Trace Command" line and then enter "V(R1:1)-V(R1:2)" and press "OK." We can

Table 4.4: The PSpice input file for Example 4.11.

```
* D:\ CIRCUITS\EXAMPLE2.SCH
* Schematics Version 6.2 - April 1995
* Mon Jun 26 02:29:11 1995
** Analysis setup **
.tran 50m 10 SKIPBP
.OP
* From [SCHEMATICS NETLIST] section of msim.ini:
.lib nom.lib
.INC "EXAMPLE2.net"
.INC "EXAMPLE2.als"
.probe
.END
```

Table 4.5: The PSpice alias file for Example 4.11.

```
* Schematics Aliases *
.ALIASES
R_R1          R1(1=$N_0002 2=$N_0001 )
R_R2          R2(1=$N_0001 2=$N_0003 )
L_L1          L1(1=$N_0004 2=$N_0002 )
L_L2          L2(1=$N_0003 2=$N_0005 )
C_C1          C1(1=0 2=$N_0001 )
V_V1          V1(+=$N_0004 -=0 )
V_V2          V2(+=0 -=$N_0005 )
.ENDALIASES
```

plot the analytic result for comparison as follows. Select the "Trace" drop-down menu again and click on "Add." Click on the "Trace Command" line and then enter "0.2357 $*$ cos(2 $*$ time $-$ 2.35619)" and then press "OK." [The analytic result is $v_{R_1}(t) = (\sqrt{2}/6)\cos(2t - 135°)$.]

Probe can be used to make multiple plots. Click on "Add Plot" in the "Plot" drop-down menu. Repeat the above sequences, but enter "V(R2:1)-V(R2:2)" to plot the voltage across R2. The analytic result for R2 is $v_{R_2}(t) = (2/3)\sin(2t)$. The resulting plots are shown in Figure 4.36. Note that the curves do not agree well at early times but are nearly indistinguishable at later times. The difference between the curves is called the transient response and is the subject of Chapter 7.

Our final example will be that of the three-stage digital-to-analog converter that was analyzed in Example 4.14. We will perform only a detailed bias point analysis, but we will use the "Parametric" analysis to change the voltage output of some of the sources. The circuit as it is redrawn in PSpice is shown in Figure 4.37. The resistors and the ground are drawn and the resistances are modified as in the previous examples. The name for the voltage sources in the "Add Part" dialog box is "VSRC." By double-clicking on the left voltage source one brings up a window entitled "V1 Part Name: VSRC." Click on the "DC =" line then move to the upper right box and type "{Vt}" and push the "Save Attr" button and then "OK." Vt will be the name of the parameter that we will assign two values to: 0 V for "off" and 5 V for "on." Symbolic values will not work properly without the braces {}. Repeat the above procedure for the middle voltage source. Set the "DC =" voltage on the rightmost voltage source to 5 V. Don't use braces around the 5, of course, since it is a numeric value.

From the "Analysis" drop-down menu select "Setup." Click the box to the left of the "Parametric" button to place an "x" and activate this sweep. Press the "Parametric" button and a window will appear that will allow us to define Vt. Set the "Swept Var. Type" to "Global Parameter" and set the "Sweep Type" to "Value List" by clicking on the circle to the left of each name (or on the name itself). Click on the "Name" box in the upper right corner and enter "Vt." Click on the "Value" box in the lower right corner and enter "0, 5." Finally, hit the "OK" and "Close" buttons and save the

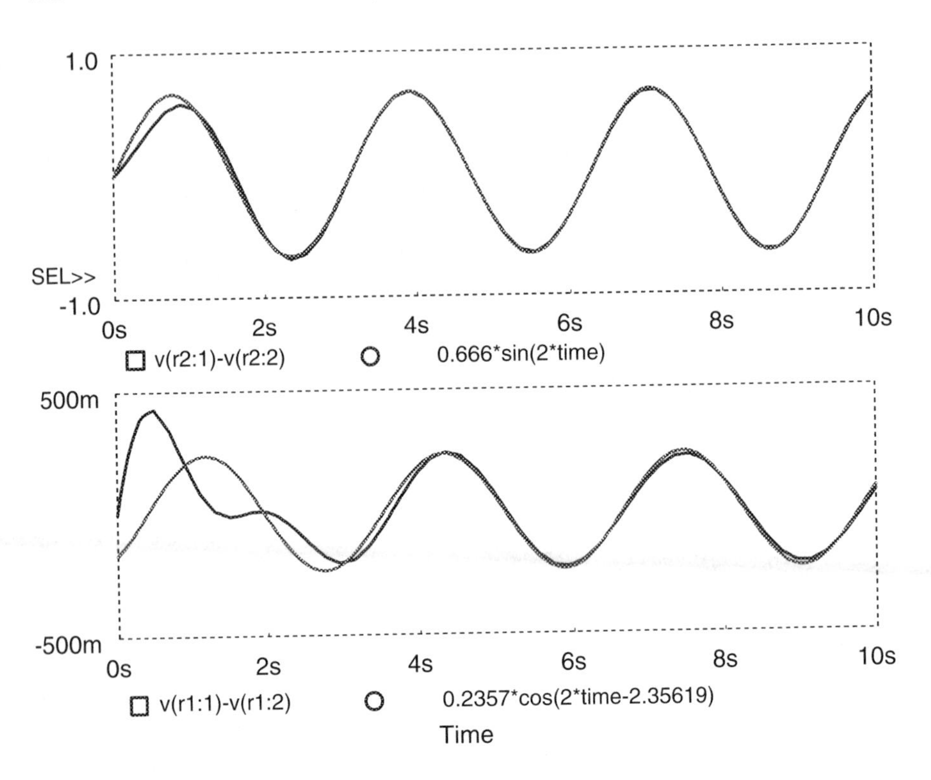

Figure 4.36: The PROBE output for Example 4.11.

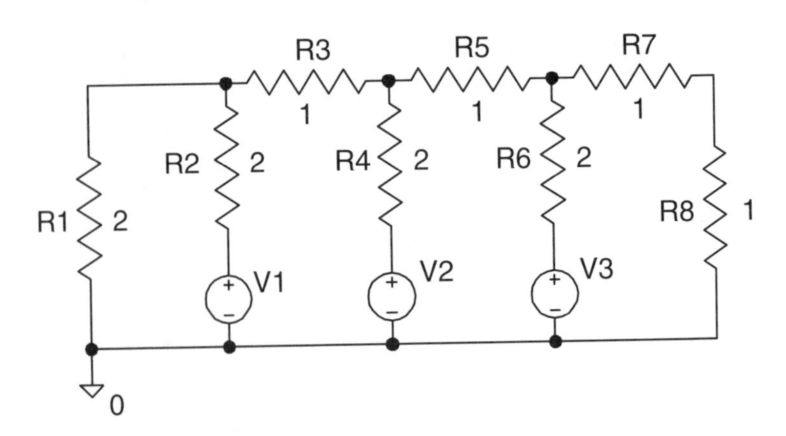

Figure 4.37: The PSpice schematic for Example 4.14.

Table 4.6: The analysis and Netlist sections of the output file for the PSpice Example 4.14.

```
** Analysis setup **
.STEP PARAM Vt LIST
+ 0,5
.OP
* Schematics Netlist *
R_R1          0 $N_0001           2
R_R2          $N_0002 $N_0001     2
R_R3          $N_0001 $N_0003     1
R_R4          $N_0004 $N_0003     2
R_R5          $N_0003 $N_0005     1
R_R6          $N_0006 $N_0005     2
R_R7          $N_0005 $N_0007     1
R_R8          0 $N_0007           1
V_V1          $N_0002             0   DC
+{Vt}
V_V2          $N_0004             0   DC
+{Vt}
V_V3          $N_0006             0   DC   5
```

schematic as example3.net. (Note that you cannot run PSpice until the schematic has been saved.) We are now ready to perform the simulation.

Select "Simulate" from the "Analysis" drop-down menu to execute PSpice. When the analysis is complete, select "Examine Output" from the "File" drop-down menu in the PSpice window. The example3.out file will be loaded into the Notepad program for you to view. Selected lines from that file are listed in Table 4.6, and Table 4.7. The ".STEP" line in the Analysis setup section in Table 4.6 indicates the values of Vt at which the circuit variables will be computed. The lines for V_V1 and V_V2 in the Schematics Netlist section show the parameterized dc voltage.

Table 4.7 contains the results of the two DC bias point analyses for the different values of Vt. Note that the first case corresponds to the digital number 1 because V3 is the only non-zero voltage source. The output resistor is R1 and the expected output voltage for this case according to Example 4.14 is one-twelfth of the "on" voltage. We do in fact see $5/12 = 0.4167$ V for $N_0001 in the node voltage table. Note that the current flowing through V3 is given as -1.667 A. This source is supplying 8.33 W so clearly current is flowing out of the plus terminal of the source. The negative sign is a direct consequence of the passive sign convention. Current is flowing into the other two sources, but no power is consumed since their voltages are zero.

All the voltage sources are "on" in the second case. This corresponds to a digital 7 and the output voltage at node $N_0001 is seven times that of the first case. Note that all voltage source currents are now negative as they are all supplying power to the circuit.

Table 4.7: The DC bias result sections of the output file for the PSpice Example 4.14.

```
**** 06/28/95 03:51:28 ******* Win32s Evaluation PSpice (April 1995) *********
* D:\CIRCUITS\EXAMPLE3.SCH
*** SMALL SIGNAL BIAS SOLUTION TEMPERATURE = 27.000 DEG C
*** CURRENT STEP PARAM VT = 0
************************************************************************
NODE VOLTAGE NODE VOLTAGE
($N_0001) .4167 ($N_0002) 0.0000
($N_0003) .8333 ($N_0004) 0.0000
($N_0005) 1.6667 ($N_0006) 5.0000
($N_0007) .8333
VOLTAGE SOURCE CURRENTS
NAME CURRENT
V_V1 2.083E-01
V_V2 4.167E-01
V_V3 -1.667E+00
TOTAL POWER DISSIPATION 8.33E+00 WATTS
*** 06/28/95 03:51:28 ******* Win32s Evaluation PSpice (April 1995) *********
D:\CIRCUITS\EXAMPLE3.SCH
*** SMALL SIGNAL BIAS SOLUTION TEMPERATURE = 27.000 DEG C
*** CURRENT STEP PARAM VT = 5
************************************************************************
NODE VOLTAGE NODE VOLTAGE
($N_0001) 2.9167 ($N_0002) 5.0000
($N_0003) 3.3333 ($N_0004) 5.0000
($N_0005) 2.9167 ($N_0006) 5.0000
($N_0007) 1.4583
VOLTAGE SOURCE CURRENTS
NAME CURRENT
V_V1 -1.042E+00
V_V2 -8.333E-01
V_V3 -1.042E+00
TOTAL POWER DISSIPATION 1.46E+01 WATTS
```

4.8 Summary

In this chapter we introduced a number of important concepts and rules that can be used to simplify the analysis of electric circuits. Along with these concepts came many definitions which we will continue to use throughout the text. The main concepts and definitions are summarized below.

- Equivalent impedance is the impedance of a single element that can be used to replace a collection of elements at a pair of nodes without affecting any of the voltages and currents throughout the rest of the circuit.

- The principle of equivalent transformations is to preserve the terminal $\hat{V}$–$\hat{I}$ relationships under these transformations.
- Series connection—two (or more) elements are said to be connected in series if they always have identical currents flowing through them.
- The equivalent impedance of a series connection is equal to the sum of the individual impedances.
- Parallel connection—two (or more) elements are said to be connected in parallel if they always have the same voltage across them.
- The equivalent admittance of a parallel connection is equal to the sum of the individual admittances.
- The voltage divider formula gives the voltage across an individual element (say the jth one) which is in series with n elements in terms of the applied voltage and the impedances: $\hat{V}_j = \hat{V}_s Z_j / Z_{eq}$.
- The current divider formula gives the current through an individual element (say the jth one) which is in parallel with n elements in terms of the total current and the admittances: $\hat{I}_j = \hat{I}_s Y_j / Y_{eq}$.
- The input impedance of a circuit is the equivalent impedance with respect to the input terminals.
- The symmetry of some electric circuits can be exploited to simplify their analysis. The main idea in exploiting symmetry of electric circuits is to identify equipotential nodes in symmetric electric circuits and connect them together into one node.
- The superposition principle states that the voltages and currents in a complex circuit with multiple sources can be found by summing the voltages and currents found for several regimes of the original circuit when only some subsets of original sources are active while other sources are set to zero. These subsets of sources are formed as a result of subdivision of all original sources into several nonintersecting groups.
- When voltage sources are set to zero, they are replaced by short-circuit branches.
- When current sources are set to zero, they are replaced by open-circuit branches.

4.9 Problems

Problems 1–6. Find the equivalent impedances of the circuits shown in the figures indicated.

1. Figure P4-1.

2. Figure P4-2.

3. Figure P4-3. Assume $f = 2$ kHz.

4. Figure P4-4. Assume $f = 10$ kHz.

5. Figure P4-5.

6. Figure P4-6.

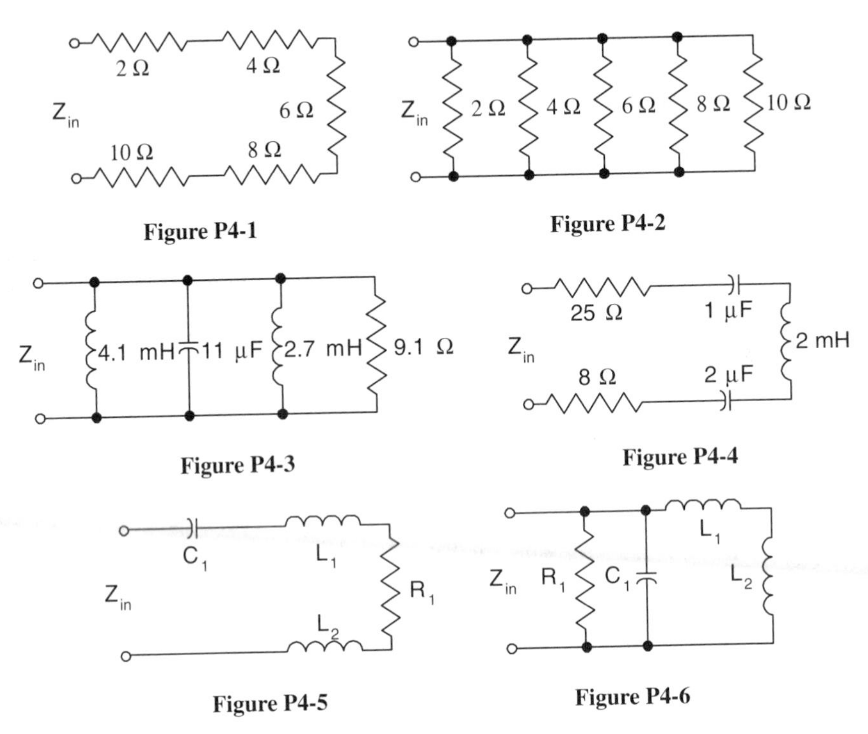

Figure P4-1

Figure P4-2

Figure P4-3

Figure P4-4

Figure P4-5

Figure P4-6

The divider ratios in Problems 7–10, denoted $a : b$, indicate the relative proportion of the source quantity to the output quantity. For example, $\hat{V}_o/\hat{V}_s = b/a$ for a voltage divider.

7. Design a resistive voltage divider with the input resistance and divider ratio given in the following table. (Find R_1 and R_2.)

Divider	Input resistance (Ω)	Divider ratio
a	50	2:1
b	1 k	10:1
c	2 M	10,000:1

8. Design a resistive current divider with the input resistance and divider ratio given in the following table. (Find R_1 and R_2.)

Divider	Input resistance (Ω)	Divider ratio
a	50	5:2
b	300	50:1
c	10 k	300:1

9. Design an inductive current divider with the input impedance magnitude and divider ratio given in the following table. (Find L_1 and L_2.)

Divider	Input impedance (Ω)	Frequency (rad/s)	Divider ratio
a	50	10 k	3:1
b	2 k	1 M	20:1
c	180 k	22 M	100:1

10. Design a capacitive voltage divider with the input impedance magnitude and divider ratio given in the following table. (Find C_1 and C_2.)

Divider	Input impedance (Ω)	Frequency (rad/s)	Divider ratio
a	50	7	7:3
b	20 k	1 k	40:1
c	47 k	1 M	50,000:1

11. Find the ac current $i(t)$ in the circuit shown in Figure P4-11.

12. Find the ac voltage $v(t)$ in the circuit shown in Figure P4-12.

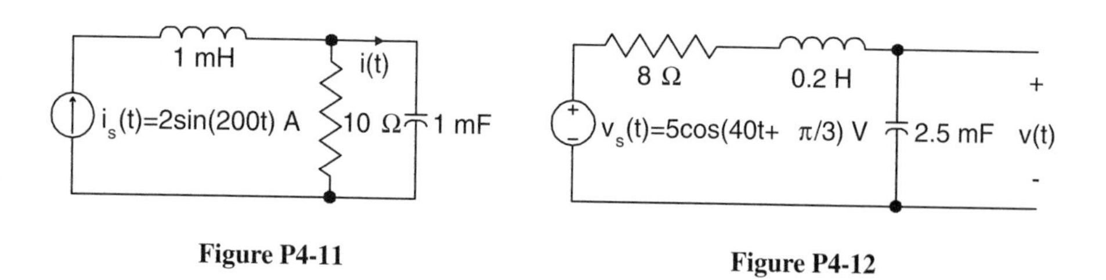

Figure P4-11

Figure P4-12

Problems 13–18. Find the equivalent admittances of the circuits shown in the figures indicated.

13. Figure P4-13. **15.** Figure P4-15. **17.** Figure P4-17 (use symmetry arguments).

14. Figure P4-14. **16.** Figure P4-16. **18.** Figure P4-18 (use symmetry arguments).

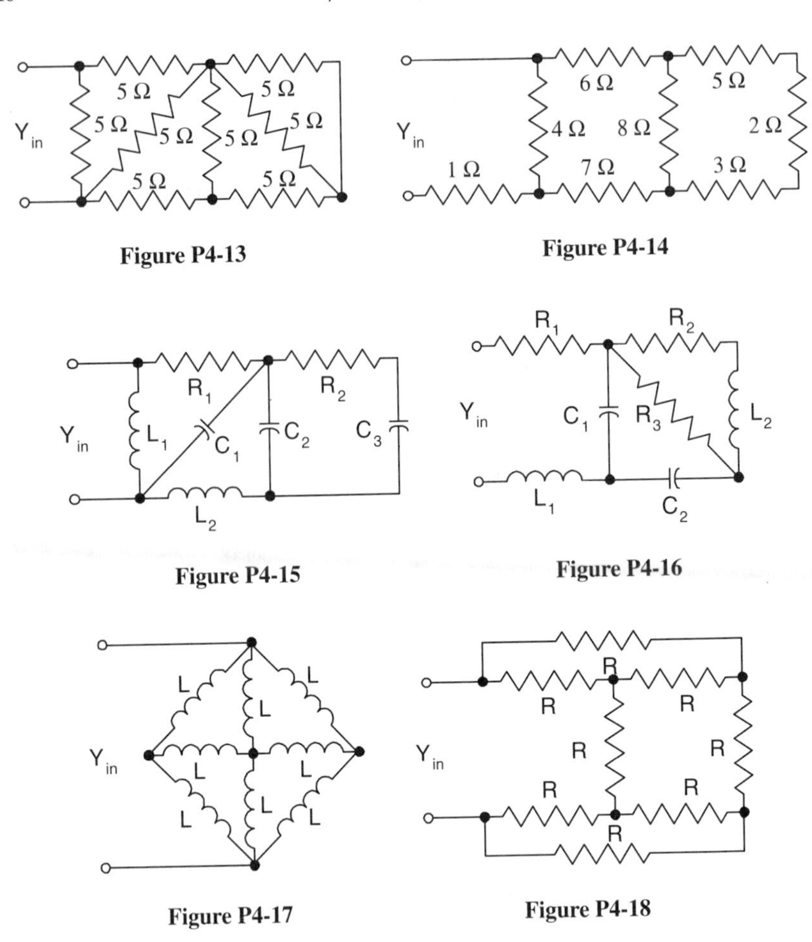

Figure P4-13

Figure P4-14

Figure P4-15

Figure P4-16

Figure P4-17

Figure P4-18

In Problems 19 and 20, assume that each branch has an impedence of Z.

19. Find the input admittance of the symmetric cube across a side as shown in Figure P4-19.

20. Find the input impedance of the symmetric cube across a face as shown in Figure P4-20.

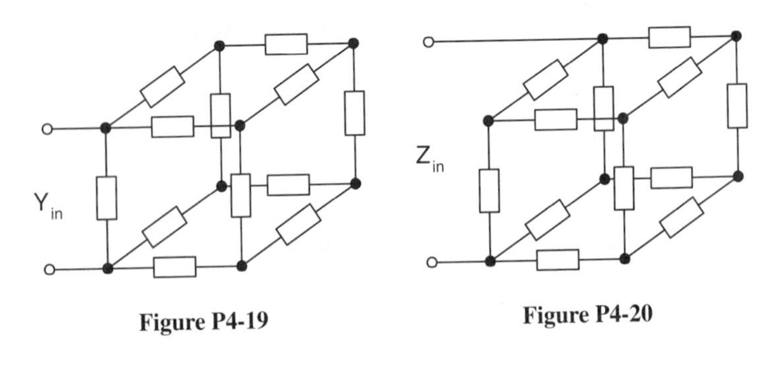

Figure P4-19

Figure P4-20

21. Prove that the input impedance of an R–$2R$ ladder with an infinite number of stages is $2R$.

22. Find the input impedance of an R–$3R$ ladder with an infinite number of stages.

23. Design a five-stage D/A converter that produces 1 A through the output $2R$ resistor when the input voltages (either 0 V or 1 V) correspond to the number 30 (decimal). That is, draw the circuit and indicate the values of all resistances.

In Problems 24–29, use the superposition principle to find the required circuit variable.

24. Find the expression for the current indicated in the circuit shown in Figure P4-24.

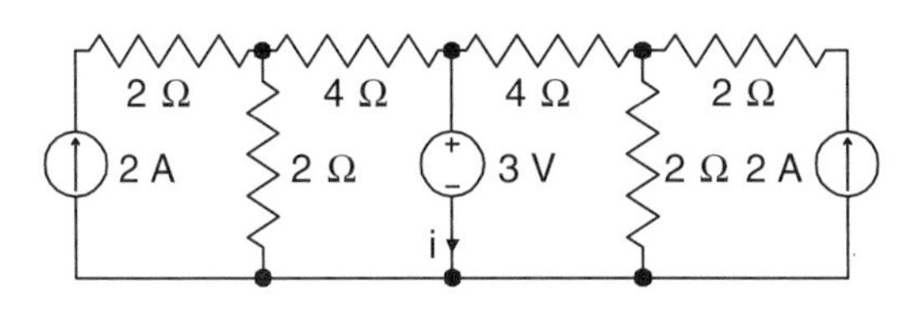

Figure P4-24

25. Find the expression for the time dependence of the voltage indicated in the circuit shown in Figure P4-25.

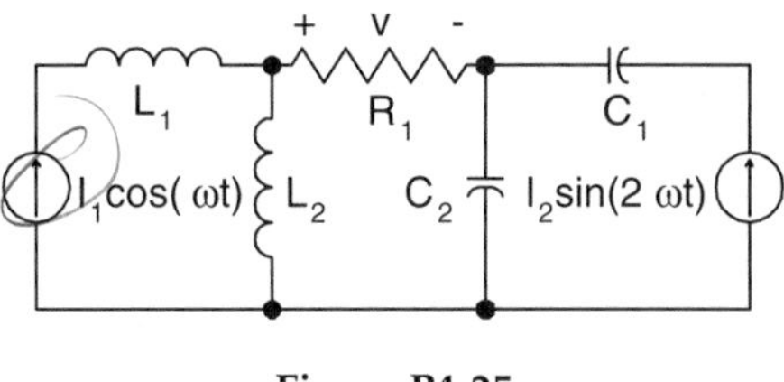

Figure P4-25

26. Find the phasor current $\hat{I}$ indicated in the circuit shown in Figure P4-26.

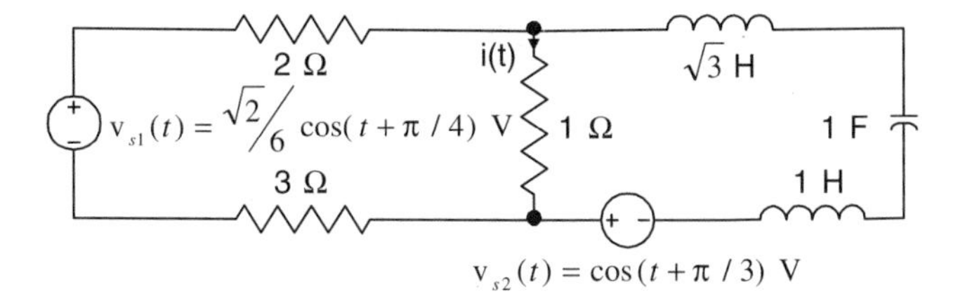

Figure P4-26

27. Find the voltage V_o indicated in the circuit shown in Figure P4-27.

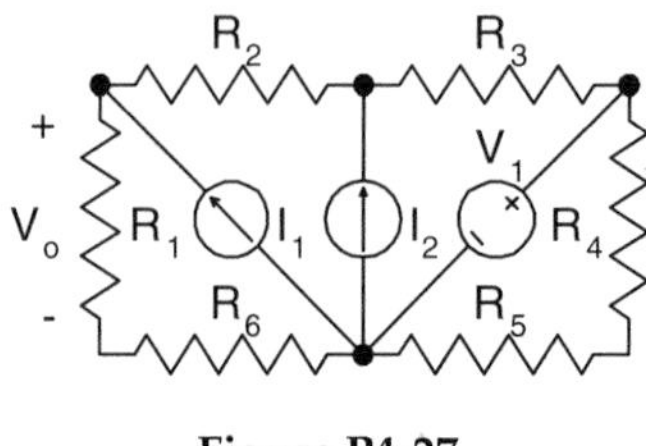

Figure P4-27

28. Find the time-dependent current i_o indicated in the circuit shown in Figure P4-28.

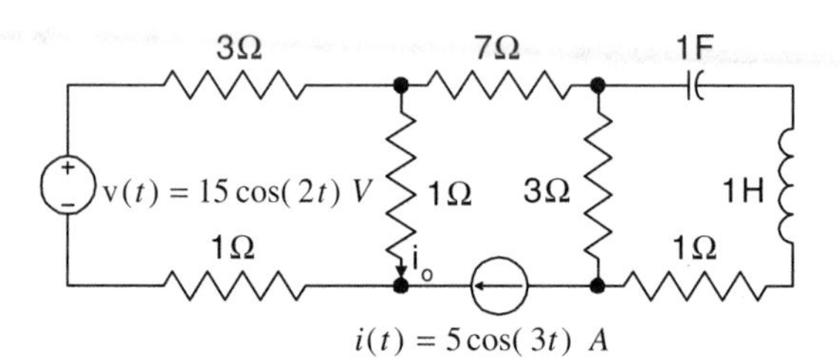

Figure P4-28

29. Find the voltage v indicated in the circuit shown in Figure P4-29.

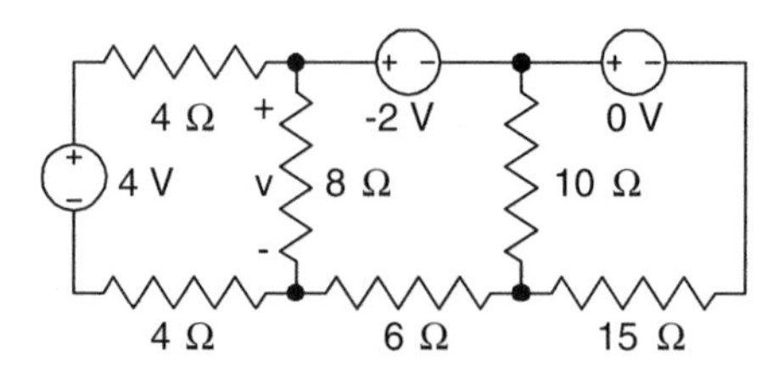

Figure P4-29

Chapter 5

Thevenin's Theorem and Related Topics

5.1 Introduction

In this chapter we first discuss the properties of "real-world" circuit components and define nonideal circuit elements (active and passive). Next, we demonstrate the equivalent nature of nonideal voltage and current sources.

We then present and prove two most central theorems in circuit theory, which are attributed to Thevenin and Norton. The essence of these theorems is that any linear circuit with numerous sources can be replaced at a given pair of nodes by a *nonideal* independent source. After proving these theorems, we present several examples of their applications.

The chapter is closed by a discussion of resistive circuits with single nonlinear resistors and the Thevenin theorem for these circuits is then developed.

5.2 Nonideal Two-Terminal Circuit Elements

All the circuit elements so far studied in this text have been *ideal* elements. However, real circuit elements are always *nonideal*, which means that they possess certain imperfections which make their observed behavior deviate slightly from the mathematical definitions we assigned to ideal elements. It turns out that we can model some of the deviations from ideal behavior by using impedances and resistors. In this way, we arrive at nonideal elements. Although nonideal circuit elements are important in modeling real-life circuit elements, their usefulness is not limited only to this purpose. It turns out that nonideal elements play an important role in the theory of equivalent transformations of electric circuits.

We shall first describe the model of a *nonideal voltage source*. The circuit notation for a nonideal voltage source is shown in Figure 5.1a. A nonideal voltage

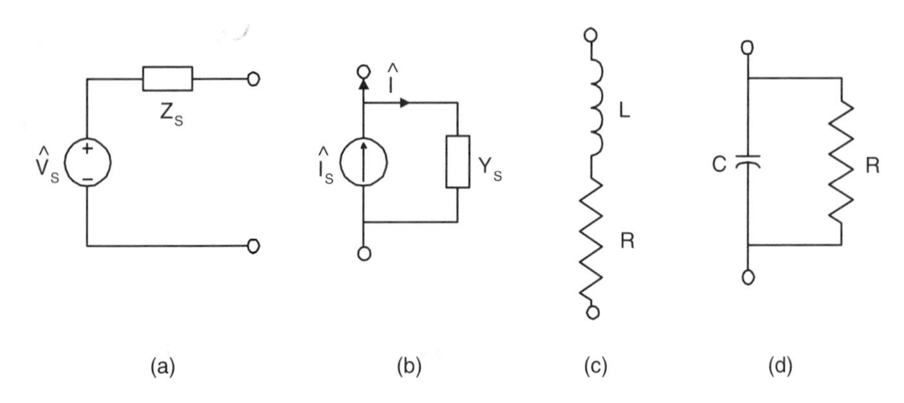

Figure 5.1: Nonideal voltage source (a), nonideal current source (b), nonideal inductor (c), and nonideal capacitor (d).

source consists of an ideal voltage source connected in series with a source impedance. This impedance allows for a slight variation in voltage across the source output due to a variation in the current through the source. When used as a model for a real-life voltage source, this source impedance is usually very small.

Next, we consider a *nonideal current source*. The circuit notation for nonideal current source is shown in Figure 5.1b. A nonideal current source consists of an ideal current source connected in parallel with a source admittance. This admittance allows for a slight variation in source output current due to a variation in the voltage across the source. The small current which flows through the admittance's branch is called the *leakage current*. This leakage current is a function of the voltage across the source. In this way the dependence of the total (external) current on the voltage is introduced. When used as a model for a real-life current source, the source admittance is usually small.

Now, we discuss a *nonideal inductor*. The circuit notation for the low-frequency model of a nonideal inductor is shown in Figure 5.1c. This model consists of an ideal inductor connected in series with a resistor. This resistor represents the small resistance of the wire from which the coil of the inductor is made. Since the energy losses of the inductor are proportional to the square of the current, the resistor is connected in series. A figure of merit of the quality of an inductor is given by the ratio L/R (which is related to a decay time as we will see in Chapter 7).

Finally, we consider a *nonideal capacitor*. The circuit notation for the low-frequency model of a nonideal capacitor is shown in Figure 5.1d. This model consists of an ideal capacitor connected in parallel with a resistor. In real capacitors there are dielectric losses which are taken into account by the resistor. Because these losses are proportional to the square of the voltage, the resistor is connected in parallel. At high frequencies, one must consider the inductance of the leads as well. A figure of merit for the quality of a capacitor is given by the ratio $C/G = RC$. Usually, real capacitors are closer to ideal than are real inductors, i.e., $C/G >> L/R$. Because of this, the shock hazard from capacitors should never be underestimated. For this reason, high-voltage capacitors should always be stored with their terminals shorted.

In conclusion, we remark that resistors are also not ideal. Resistors that are constructed by winding a fine wire in a helical coil, for example, often have appreciable inductances and must be modeled with the circuit in Figure 5.1c. They are usually not suitable for high-frequency applications.

5.3 Equivalent Transformations of Nonideal Voltage and Current Sources

Consider the equivalent transformation of a nonideal voltage source into a nonideal current source, and vice versa. Such a transformation may first seem superfluous. However, it has proved to be very useful in circuit analysis, especially in the node and mesh analysis techniques, which will be discussed in the next chapter.

According to the previously discussed general principle of equivalent transformations, the transformation of a nonideal voltage source into a nonideal current source will be equivalent if it preserves the terminal $\hat{V}$ versus $\hat{I}$ relationship. Being guided by this principle, we first find the terminal relationship for the left circuit in Figure 5.2. By using KVL, we derive:

$$-\hat{V}_s + \hat{I}Z_s + \hat{V} = 0. \tag{5.1}$$

Thus,

$$\hat{V} = \hat{V}_s - \hat{I}Z_s. \tag{5.2}$$

In order to find the terminal relationship for the right circuit in Figure 5.2, we use KCL at node A:

$$\hat{I} = \hat{I}_s - \hat{I}'. \tag{5.3}$$

Consequently,

$$\hat{I} = \hat{I}_s - \hat{V}Y_s \tag{5.4}$$

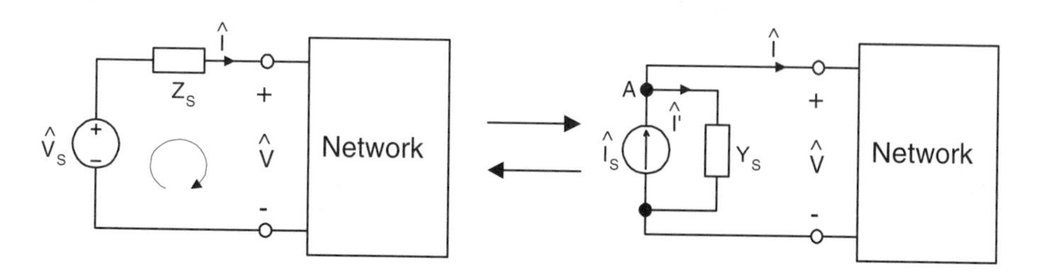

Figure 5.2: Nonideal voltage source – nonideal current source transformation.

where the expression $\hat{I}' = \hat{V}Y_s$ was used in (5.4). Equation (5.4) can be solved for voltage, which yields:

$$\hat{V} = \frac{\hat{I}_s}{Y_s} - \hat{I}\frac{1}{Y_s}. \tag{5.5}$$

Comparing terminal relationships (5.2) and (5.5) we see that they will be identical when:

$$Z_S = \frac{1}{Y_s}, \tag{5.6}$$

$$\hat{V}_S = \frac{\hat{I}_S}{Y_s} = \hat{I}_s Z_S. \tag{5.7}$$

Formulas (5.6) and (5.7) constitute the rules for the equivalent transformation of nonideal sources. The polarities of the equivalent nonideal sources are also critical and are indicated in Figure 5.2. For example, when replacing a nonideal voltage source with a nonideal current source, the "tail" of the current arrow is connected to the node that was originally connected to the negative terminal of the voltage source.

5.4 Thevenin's Theorem

Thevenin's theorem is the central result of basic circuit theory. For this reason, it is essential to understand this theorem, know how to prove it and how to use it in solving problems.

Consider a branch with impedance Z connected to an arbitrary active circuit (network). Thevenin's theorem basically states that as far as the current through this impedance is concerned, the active linear network can be replaced by an equivalent nonideal voltage source. This theorem is illustrated in Figure 5.3. In the following section, the proof of Thevenin's theorem is given.

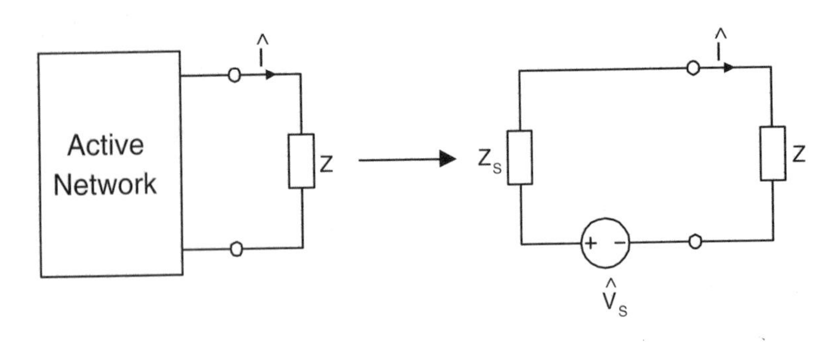

Figure 5.3: Graphical illustration of Thevenin's theorem.

5.4.1 Proof of Thevenin's Theorem

The proof consists of the following three steps:

Step I:

First, we introduce two voltage sources into the branch with impedance Z (see Figure 5.4). Both have the same peak value of voltage but opposite polarities. We can see that doing this will not affect the current through Z at all, because the two voltage sources will cancel out in any KVL equations due to their opposite polarities.

Step II:

Next, we shall use the superposition principle. To do this, we divide all the sources in the network into two groups:

1. The first group consists of all the sources in the active network and the left voltage source introduced into the branch with impedance Z.
2. The second group consists of only the right voltage source in the same branch.

Now, we can consider two separate regimes, shown in Figure 5.5. The first regime contains the sources of the first group, and the second regime contains the only source of the second group. Note that, in the second regime, the active network has been replaced by a passive network. This passive network is formed when the sources in the active network are set to zero. This is accomplished by replacing voltage sources with short-circuit branches and current sources with open branches.

By using the superposition principle we find that the total current through Z is equal to the sum of the branch currents in the above two regimes:

$$\hat{I} = \hat{I}^{(1)} + \hat{I}^{(2)}. \tag{5.8}$$

Step III:

Consider the first regime. We choose $\hat{V}_s$ in a way that causes $\hat{I}^{(1)}$ to be zero. This is equivalent to having the branch with Z open. Consequently, the voltage across the

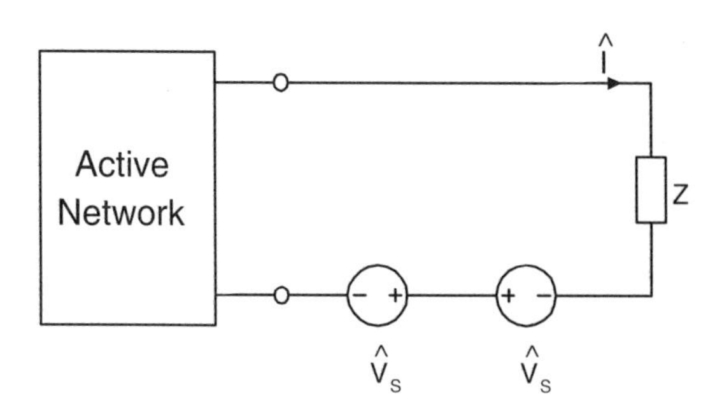

Figure 5.4: Introduce two voltage sources.

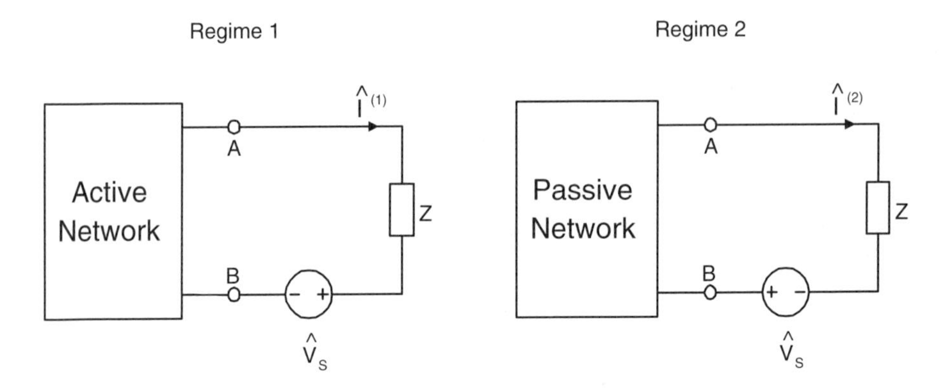

Figure 5.5: The two regimes.

terminals A and B will be equal to the open-circuit voltage $\hat{V}_{OC}$. This situation is shown in Figure 5.6. To find such a $\hat{V}_s$ which guarantees that $\hat{I}^{(1)}$ will be equal to zero, we write the KVL for the loop shown in Figure 5.6:

$$\hat{V}_s - \hat{V}_{OC} = 0, \tag{5.9}$$

$$\hat{V}_s = \hat{V}_{OC}. \tag{5.10}$$

Therefore, $\hat{V}_s$ must be equal to the open-circuit voltage in order to force $\hat{I}^{(1)} = 0$. According to equation (5.8), we find that the total current is determined by the second regime:

$$\hat{I} = \hat{I}^{(2)}. \tag{5.11}$$

Next, consider the second regime. The passive network contains no sources and can be replaced by the equivalent input impedance with respect to the terminals A and B. This replacement is shown in the rightmost circuit in Figure 5.7. This series combination of Z_s and $\hat{V}_s$ constitutes a nonideal voltage source which (according to

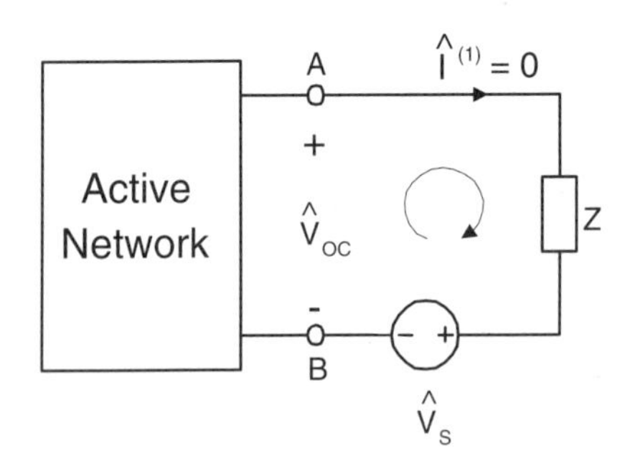

Figure 5.6: Regime 1.

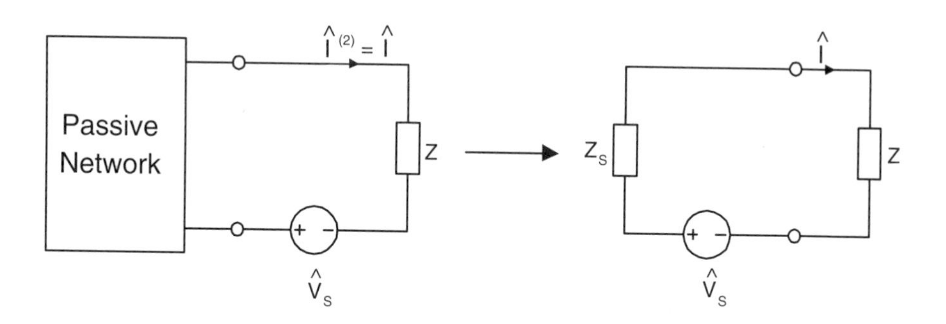

Figure 5.7: Regime 2.

(5.11)) generates the same current through the load impedance Z that the original network does. This concludes the proof of the theorem.

As a useful by-product, we have also found the following expressions for the source voltage and source impedance:

$$\hat{V}_s = \hat{V}_{oc}, \tag{5.12}$$

$$Z_s = Z_{in}. \tag{5.13}$$

By using these expressions and the Thevenin equivalent circuit, we derive the following formula:

$$\hat{I} = \frac{\hat{V}_s}{Z_s + Z} = \frac{\hat{V}_{oc}}{Z_{in} + Z} \tag{5.14}$$

which relates the branch current to the open-circuit voltage and the input impedance. This equation is essential in the application of Thevenin's theorem to circuit analysis.

5.4.2 Using Thevenin's Theorem in Analysis

Thevenin's theorem and equation (5.14) can be successfully used in circuit analysis to solve for currents through specific branches. All we need to do is to find two quantities, the open-circuit voltage and the input impedance, and use them in equation (5.14) to arrive at the solution. Finding these quantities is fairly straightforward, and the entire method can be reduced to the following three steps.

Step I:

To calculate the current through any particular branch, the open-circuit voltage must be found. Consequently, the first step is to remove the branch through which the current is desired and determine the voltage with respect to the open terminals (see Figure 5.8). It is this first step of removing the branch which usually simplifies the circuit.

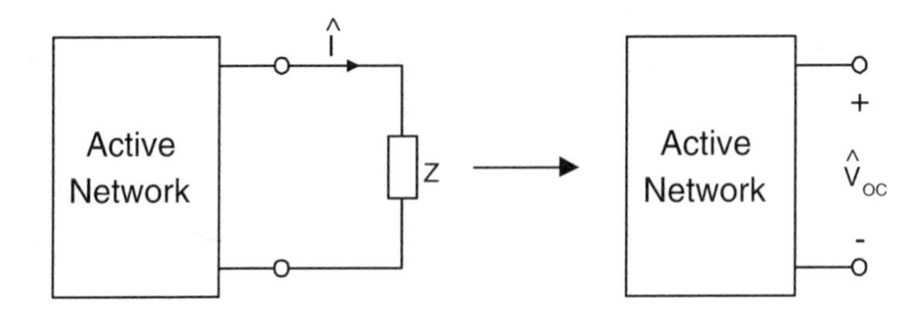

Figure 5.8: Step I: Open branch, find open-circuit voltage.

Step II:

The second step is to find the input impedance with respect to the open terminals of the removed branch. The rest of the circuit must be transformed to a passive network by setting all sources to zero (see Figure 5.9). Recall that a voltage source that is set to zero is replaced by a short-circuit branch, while a current source set to zero is replaced by an open branch.

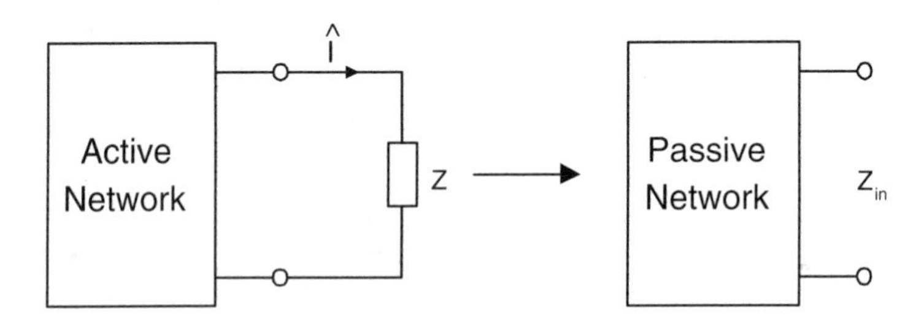

Figure 5.9: Step II: Find input impedance of the passive network.

Step III:

Once the open-circuit voltage and the input impedance are known, finding the current through the branch with equation (5.14) is simple. The justification behind using this equation is Thevenin's theorem itself.

EXAMPLE 5.1 As an example of using Thevenin's theorem in the analysis of electric circuits, we consider a "bridge" circuit, pictured on the left in Figure 5.10. It is assumed that Z_1, Z_2, Z_3, Z_4, Z_5, $\hat{V}_s$ are given. We want to find $\hat{I}_5$.

We start the solution of this problem with step one, namely, we remove the branch in question and find the open-circuit voltage. Once we remove the branch, the circuit becomes the one shown on the right in Figure 5.10, which is now much simplified. To find the open-circuit voltage, we write KVL for the loop shown:

$$\hat{V}_{oc} - \hat{I}_3 Z_3 + \hat{I}_1 Z_1 = 0. \tag{5.15}$$

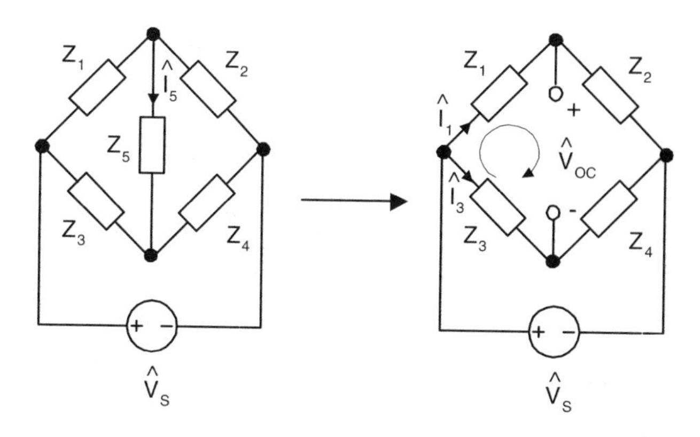

Figure 5.10: A bridge circuit, with center branch removed.

We want to express the open-circuit voltage in terms of the given source voltage, so we must find $\hat{I}_1$ and $\hat{I}_3$ in terms of this source voltage. Notice that the circuit is now quite simple; it contains two branches in parallel with the same voltage $\hat{V}_s$ across them. (One branch consists of the series combination of Z_1 and Z_2 and the other is the series combination of Z_3 and Z_4.) Hence, we can calculate the currents through these branches by dividing $\hat{V}_s$ by the impedances of the branches:

$$\hat{I}_1 = \frac{\hat{V}_s}{Z_1 + Z_2}, \tag{5.16}$$

$$\hat{I}_3 = \frac{\hat{V}_s}{Z_3 + Z_4}. \tag{5.17}$$

Substituting (5.16) and (5.17) into (5.15), we find the open-circuit voltage to be:

$$\hat{V}_{oc} = \hat{V}_s \left(\frac{Z_3}{Z_3 + Z_4} - \frac{Z_1}{Z_1 + Z_2} \right). \tag{5.18}$$

Now, we move on to step two, finding the input impedance. The circuit must be replaced by a passive network—in this case the voltage source must be replaced by a short-circuit branch. The resulting passive network (see Figure 5.11) may still seem complicated; however, redrawing the circuit reveals its simplicity. The essence of redrawing is to merge into one node the two nodes which are connected by the short-circuit branch. In Figure 5.11 the passive network is redrawn to emphasize its simplicity. In this case, the input impedance is easy to calculate. It is merely two parallel connections in series with one another:

$$Z_{in} = \frac{Z_1 Z_2}{Z_1 + Z_2} + \frac{Z_3 Z_4}{Z_3 + Z_4}. \tag{5.19}$$

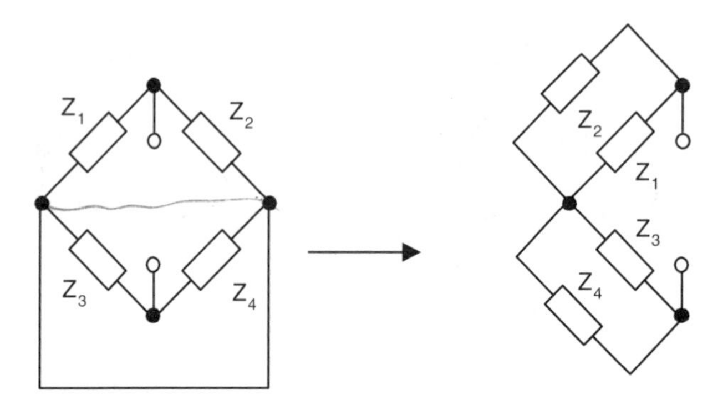

Figure 5.11: The passive network.

Now that the open-circuit voltage and the input impedance have been found, equation (5.14) can be used to find the current $\hat{I}_5$. This constitutes the final step:

$$\hat{I}_5 = \frac{\hat{V}_s \left(\frac{Z_3}{Z_3+Z_4} - \frac{Z_1}{Z_1+Z_2} \right)}{\frac{Z_3 Z_4}{Z_3+Z_4} + \frac{Z_1 Z_2}{Z_1+Z_2} + Z_5}. \tag{5.20}$$

■

EXAMPLE 5.2 As a second example of using Thevenin's theorem in the analysis of electric circuits, consider the same bridge circuit as in the previous example. This time we want to find the current through one of the side branches, say $\hat{I}_1$ (Figure 5.12).

Again, we follow the same three steps as before. First, we open the branch with Z_1 and find the open-circuit voltage. Writing KVL for the circuit shown on the right hand in Figure 5.12 yields

$$\hat{V}_{oc} = \hat{I}_5 Z_5 + \hat{I}_3 Z_3. \tag{5.21}$$

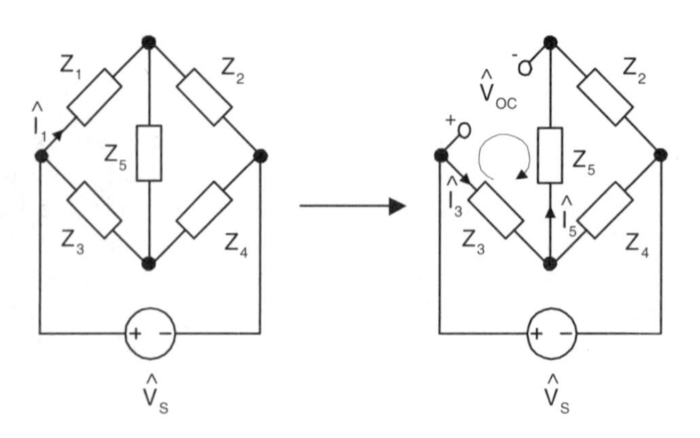

Figure 5.12: Circuit for Example 5.2.

Now, the currents $\hat{I}_5$ and $\hat{I}_3$ must be found in terms of the given source voltage. From Figure 5.12 it is apparent that $\hat{I}_3$ is equal to the source voltage divided by the total circuit impedance with respect to the terminals of the given voltage source:

$$\hat{I}_3 = \frac{\hat{V}_s}{Z_3 + \frac{(Z_2+Z_5)Z_4}{Z_2+Z_4+Z_5}}, \tag{5.22}$$

or

$$\hat{I}_3 = \frac{\hat{V}_s(Z_2 + Z_4 + Z_5)}{Z_3(Z_2 + Z_4 + Z_5) + (Z_2 + Z_5)Z_4}. \tag{5.23}$$

To facilitate the understanding of how the total impedance has been evaluated, we remark that according to the right-hand circuit in Figure 5.12 impedances Z_5 and Z_2 are connected in series and together are in parallel with Z_4, and then they are all together in series with Z_3.

Once $\hat{I}_3$ is found, it can be used to calculate $\hat{I}_5$ by using the current divider rule:

$$\hat{I}_5 = \hat{I}_3 \frac{Z_4}{Z_2 + Z_4 + Z_5} = \hat{V}_s \frac{Z_4}{Z_3(Z_2 + Z_4 + Z_5) + (Z_2 + Z_5)Z_4}. \tag{5.24}$$

Now, the open-circuit voltage can be found in terms of the known source voltage by substituting (5.23) and (5.24) into (5.21), which yields:

$$\hat{V}_{oc} = \hat{V}_s \frac{Z_3(Z_2 + Z_4 + Z_5) + Z_4 Z_5}{Z_3(Z_2 + Z_4 + Z_5) + (Z_2 + Z_5)Z_4}. \tag{5.25}$$

Now, according to step two, the input impedance must be calculated by setting the source to zero and forming a passive network. In this case, redrawing the circuit is also helpful in determining the proper input impedance (Figure 5.13). From the redrawn circuit, it is clear that Z_4 and Z_3 are connected in parallel with each other and that combination is in series with Z_5. This group of three impedances is then

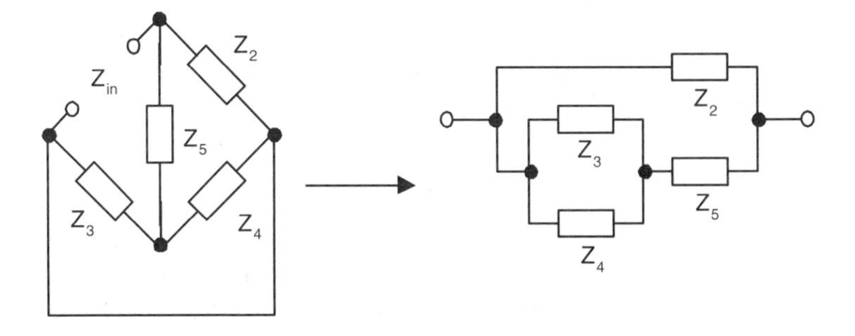

Figure 5.13: The passive network.

connected in parallel with Z_2. So, the resulting input impedance is:

$$Z_{in} = \frac{\left(\frac{Z_3 Z_4}{Z_3+Z_4} + Z_5\right) Z_2}{\frac{Z_3 Z_4}{Z_3+Z_4} + Z_5 + Z_2}, \tag{5.26}$$

which after a simple transformation yields:

$$Z_{in} = \frac{(Z_3 Z_4 + Z_5 Z_3 + Z_5 Z_4) Z_2}{Z_3 Z_4 + Z_5 Z_3 + Z_3 Z_2 + Z_4 Z_5 + Z_4 Z_2}. \tag{5.27}$$

In the third step, we use (5.25) and (5.27) in equation (5.14), and after simple transformations we find the desired current $\hat{I}_1$:

$$\hat{I}_1 = \frac{\hat{V}_{oc}}{Z_{in} + Z_1},$$

$$\hat{I}_1 = \hat{V}_s \frac{Z_3(Z_2 + Z_4 + Z_5) + Z_4 Z_5}{(Z_3 Z_4 + Z_3 Z_5 + Z_4 Z_5)(Z_2 + Z_1) + Z_1(Z_3 Z_2 + Z_4 Z_2)}. \tag{5.28}$$

■

We have seen from the preceding examples how Thevenin's theorem is useful from an analytical standpoint. Now, we discuss how this theorem can be useful from an experimental point of view. Recall that to find the current through a branch with impedance Z by using equation (5.14), the open-circuit voltage must be determined. This can be accomplished by simply measuring this voltage with a voltmeter. But how could the input impedance be found experimentally? The answer is to short-circuit the branch with impedance Z and measure the short-circuit current, $\hat{I}_{sc}$. According to Thevenin's theorem and (5.14), the open-circuit voltage and the short-circuit current are related by the following expression:

$$\hat{I}_{sc} = \frac{\hat{V}_{oc}}{Z_{in}} \rightarrow Z_{in} = \frac{\hat{V}_{oc}}{\hat{I}_{sc}}. \tag{5.29}$$

Thus, if the open-circuit voltage and short circuit current are both measured, the input impedance can be found. This technique enables the experimental use of Thevenin's theorem in some practical applications. Measuring the open-circuit voltage and the short-circuit current is accomplished by using two "extreme" tests shown in Figures 5.14 and 5.15, respectively. By using the result of these two tests, the current $\hat{I}$ through the branch with any impedance Z can be predicted:

$$\hat{I} = \frac{\hat{V}_{oc}}{\frac{\hat{V}_{oc}}{\hat{I}_{sc}} + Z}. \tag{5.30}$$

It is important to note that the described experimental approach does not require any *a priori* knowledge concerning the active network. For this reason, the open-circuit and short-circuit tests are widely used for characterizations of many devices such as transformers, induction motors, and synchronous generators.

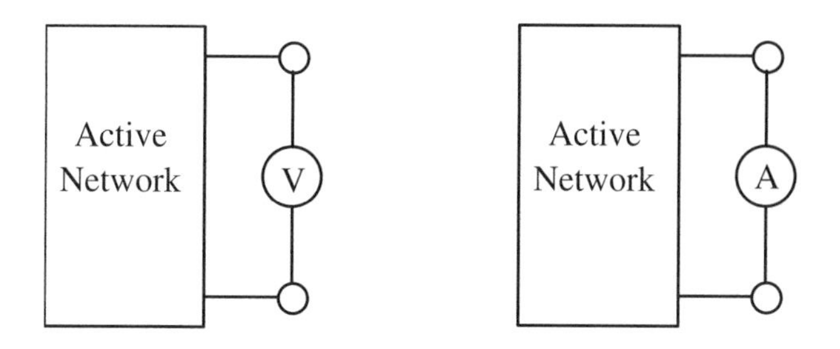

Figure 5.14: Open-circuit test. **Figure 5.15:** Short-circuit test.

5.5 Norton's Theorem

Norton's theorem is very similar to Thevenin's theorem. Basically it states that as far as the current through an impedance Z is concerned any active network can be replaced by a non-ideal current source. It is graphically represented in Figure 5.16. This theorem is trivial to prove by using Thevenin's theorem.

Proof: We know that according to Thevenin's theorem the active network in the left circuit of Figure 5.16 can be equivalently replaced by a nonideal voltage source. Since any nonideal voltage source can be equivalently transformed to a nonideal current source (see Figure 5.17), Norton's theorem is proven.

We know the equivalence between nonideal voltage and current sources means that

$$Y_s = \frac{1}{Z_s}, \tag{5.31}$$

$$\hat{I}_s = \frac{\hat{V}_s}{Z_s} = \frac{\hat{V}_{OC}}{\frac{\hat{V}_{OC}}{\hat{I}_{sc}}} = \hat{I}_{SC}. \tag{5.32}$$

Therefore, the current source in Norton's theorem is equal to the short-circuit current, and the admittance is equal to the input admittance of the corresponding passive

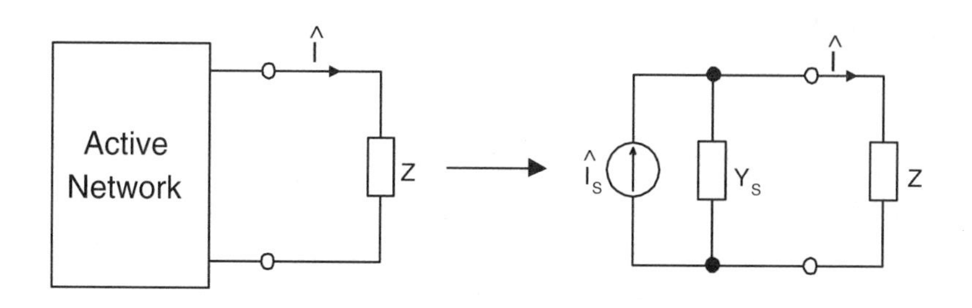

Figure 5.16: Norton's theorem.

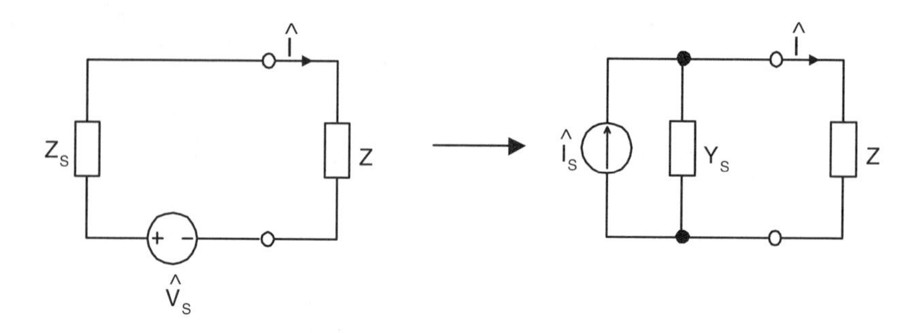

Figure 5.17: "Proof" of Norton's theorem.

network:

$$\hat{I}_s = \hat{I}_{sc}, \tag{5.33}$$

$$Y_s = \frac{1}{Z_{in}} = Y_{in}. \tag{5.34}$$

Norton's theorem can be used in a manner similar to Thevenin's theorem.

EXAMPLE 5.3 We want to find the Norton equivalent nonideal current source for the circuit shown in Figure 5.18.

First, we can find the equivalent impedance by setting the ideal sources to zero. For this example, that entails shorting the voltage source $\hat{V}_s$ and opening the current source $\hat{I}_s$. The resulting passive circuit is given in Figure 5.19. Because Z_1 is disconnected on the left side, it cannot have any current flow through it and thus has no effect on the equivalent impedance. Thus, the elements Z_2 and Z_3 are effectively in series and that combination is in parallel with Z_4. The resulting Norton equivalent admittance is:

$$Y_s = Y_{NO} = \frac{1}{Z_2 + Z_3} + \frac{1}{Z_4} = \frac{Z_2 + Z_3 + Z_4}{(Z_2 + Z_3)Z_4}. \tag{5.35}$$

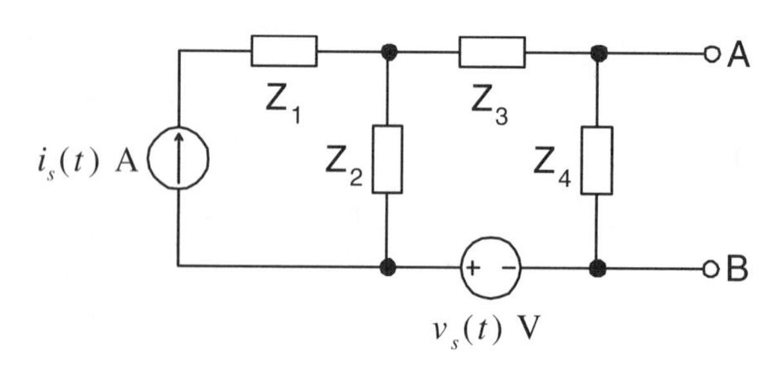

Figure 5.18: Circuit for the first Norton example.

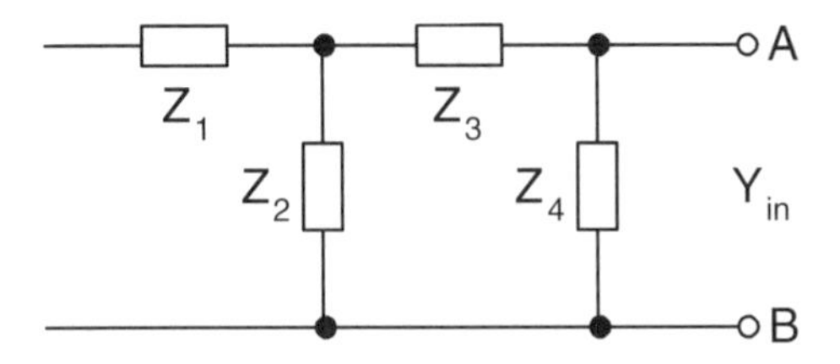

Figure 5.19: The passive circuit used to find the input admittance.

Next, we will use the principle of superposition to find the Norton current source. The two circuits that we need to analyze are given in Figures 5.20a and b. In both cases, we short-circuit the output terminals. This results in zero voltage across Z_4 and subsequently zero current through that impedance. Therefore, it has no effect on the circuit and we don't need to consider it further. We must find the short-circuit current for each of the two problems and add the results to get the current of the equivalent non-ideal current source. Problem (a) resembles a current divider and the current we need is also the current through Z_3. The appropriate formula for current dividers yields: $\hat{I}^a_{sc} = \hat{I}_s Z_2/(Z_2 + Z_3)$. Problem (b) is best solved with KVL by taking a path through the voltage source, Z_2, Z_3, and the short-circuit branch. The resulting current is found to be: $\hat{I}^b_{sc} = \hat{V}_s/(Z_2 + Z_3)$. Therefore, the Norton equivalent current is:

$$\hat{I}_{sc} = \hat{I}_{NO} = (\hat{V}_s + \hat{I}_s Z_2)/(Z_2 + Z_3). \tag{5.36}$$

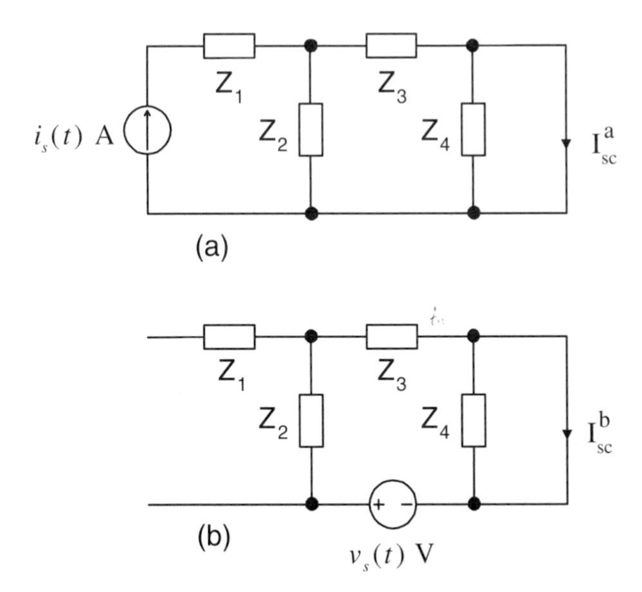

Figure 5.20: The two problems that arise from the superposition principle.

Notice that Z_1 does not enter into the equivalent circuit in any way. Can you explain why this is so? ■

EXAMPLE 5.4 Consider a nonideal voltage source with $v_s = 120\sqrt{2}\cos(377t + 120°)$ V and an internal resistance of 0.1 Ω, connected in parallel with a 1 μF capacitor and having a lead inductance of 0.1 mH. The resulting circuit is given in Figure 5.21. Again, we want to find the Norton equivalent circuit.

This time, we shall find the Thevenin equivalent circuit and arrive at the final result via the rules for the transformation of nonideal sources. When we convert the circuit to phasor form, we find that $\hat{V}_s = 120\sqrt{2}e^{j2\pi/3}$ V, $R_s = 0.1\Omega$, $Z_L = j0.0377$ Ω, and $Z_C = -j2.653$ kΩ. To find the Thevenin impedance, we short the voltage source and notice that the equivalent impedance is equal to the series combination of Z_L with the parallel combination of R_s and Z_C:

$$Z_{in} = Z_{TH} = Z_L + R_s Z_C/(R_s + Z_C) \approx 0.1 + 0.0377j\ \Omega \tag{5.37}$$

To find the Thevenin voltage, we notice that for the open-circuit configuration, there is no current through the inductor and therefore no voltage drop across it. That means that the open-circuit voltage is equal to the voltage across the capacitor. We can find this voltage by the voltage divider rule:

$$\hat{V}_{oc} = \hat{V}_{TH} = \hat{V}_C = \hat{V}_s Z_C/(R_s + Z_C) \approx \hat{V}_s(1 - j/26,525) \approx 120\sqrt{2}e^{j2\pi/3}\ \text{V}. \tag{5.38}$$

The Norton current just comes from the relation:

$$\hat{I}_{NO} = \hat{V}_{TH}/Z_{TH} = \frac{12 - \sqrt{2}e^{j2\pi/3}}{0.1 + 0.0377j} \approx 1.588e^{j1.734}\ \text{kA}. \tag{5.39}$$

Notice that in this example the capacitance is sufficiently small that it doesn't significantly affect the equivalent circuit. ■

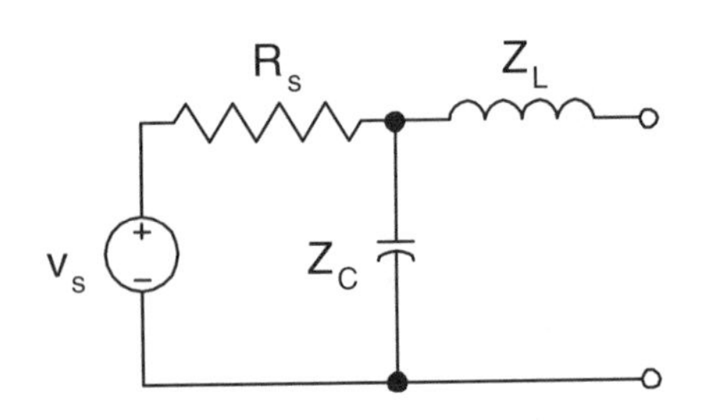

Figure 5.21: Circuit for the second Norton example.

5.6 Nonlinear Resistive Circuits

In the previous sections, we have analyzed electric circuits containing only **linear** circuit elements such as resistors, inductors, and capacitors. The term "linear" means that the values of resistances, inductances, and capacitances do not depend on voltages across the corresponding circuit elements or currents through these elements. For these reasons, the terminal relationships for such elements are linear and the circuits with these elements are described by linear equations such as equations (2.29)–(2.34). It has also been assumed in our previous discussions that the values of resistances, inductances, and capacitances do not change with time. Such circuit elements are called time-invariant and circuits with linear and time-invariant elements are described by linear equations with **constant** coefficients. Equations (2.29)–(2.34) can be again used as an example. Linear and time-invariant (LTI) circuits are important representatives of LTI systems and comprehensive mathematical study of these systems belongs to a course on linear systems and signals. However, it is useful to stress here the two most fundamental properties of such circuits and systems. The first property is the **invariance with respect to time shifting**, while the second property is the **superposition principle**. To express these properties mathematically, consider an LTI electric circuit with two terminals shown in Figure 5.22. Suppose that terminal voltage $v(t)$ results in terminal current response $i(t)$. Then, the invariance with respect to time shifting means that terminal voltage $v(t - \tau)$ will result in terminal current response $i(t - \tau)$, where τ is an arbitrary shift in time. This property is transparent from the physical point of view and simply states that for linear **time-invariant** circuits there is nothing which singles out the origin of time. This origin can be chosen arbitrarily without affecting the mathematical description of LTI electric circuits, and the change of time origin is equivalent to time shifting.

To demonstrate the superposition principle, let us suppose that two terminal voltages $v_1(t)$ and $v_2(t)$ result in two terminal current responses $i_1(t)$ and $i_2(t)$, respectively. Then, according to the superposition principle, the terminal voltage

$$v(t) = c_1 v_1(t) + c_2 v_2(t) \tag{5.40}$$

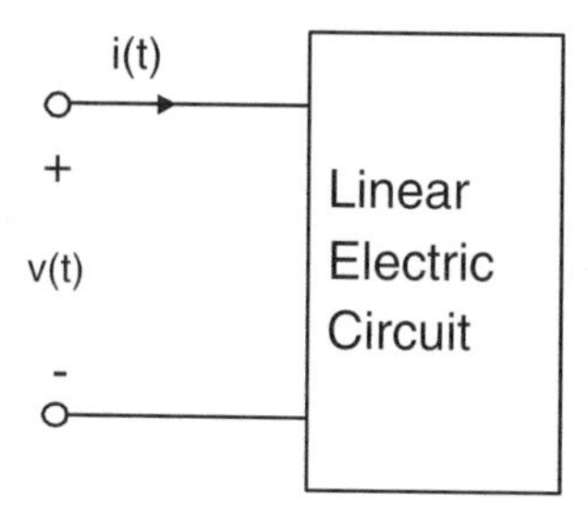

Figure 5.22: Input voltage and terminal current response for an LTI electric circuit.

will result in the terminal current response

$$i(t) = c_1 i_1(t) + c_2 i_2(t), \tag{5.41}$$

where c_1 and c_2 are arbitrary numbers.

It has been stressed before that the superposition principle is a mathematical consequence of **linearity** of circuit equations.

The time shifting invariance and the superposition principle can be combined into the following statement: if $i_1(t)$ and $i_2(t)$ are terminal current responses to terminal voltages $v_1(t)$ and $v_2(t)$, respectively, then

$$i(t) = c_1 i_1(t - \tau_1) + c_2 i_2(t - \tau_2) \tag{5.42}$$

will be the current response to the terminal voltage

$$v(t) = c_1 v_1(t - \tau_1) + c_2 v_2(t - \tau_2), \tag{5.43}$$

where τ_1 and τ_2 are arbitrary time shifts. The last statement can be considered as another definition of LTI circuits, the definition which is extensively used in the linear system theory.

In many applications, the values of resistances, inductances, and capacitances may depend on voltages across or currents through the circuit elements. Such circuit elements are nonlinear, and electric circuits with nonlinear elements are described by nonlinear differential and algebraic equations for which the superposition principle is not valid. For this reason, the theory of nonlinear electric circuits is much more complex. In this section, we consider only some simple facts and results related to nonlinear *resistive* circuits. Such circuits are described by nonlinear algebraic equations.

First, we shall recall the definition of linear resistors. These resistors are characterized by the following terminal relationship:

$$v = Ri, \tag{5.44}$$

where v and i are the instantaneous voltage across and the current through a resistor, and R is the resistance, which does not depend on the values of v and i.

Expression (5.44) can be graphically illustrated by a straight line shown in Figure 5.23. The slope of this straight line is equal to the value of the resistance:

$$R = \frac{v}{i} = \tan \alpha. \tag{5.45}$$

The circuit notation for a nonlinear resistor which will be used throughout this section is shown in Figure 5.24. The nonlinear resistor cannot be characterized completely by one value of the resistance. Usually, the nonlinear resistor is characterized by a "$v(i)$" curve like the one shown in Figure 5.25. It is clear from this figure that if we define the resistance as the ratio

$$R(i) = \frac{v(i)}{i}, \tag{5.46}$$

the value of this resistance will vary with the current through the resistor. This fact is emphasized by the notation $R(i)$. Indeed, the values of the resistance for two different

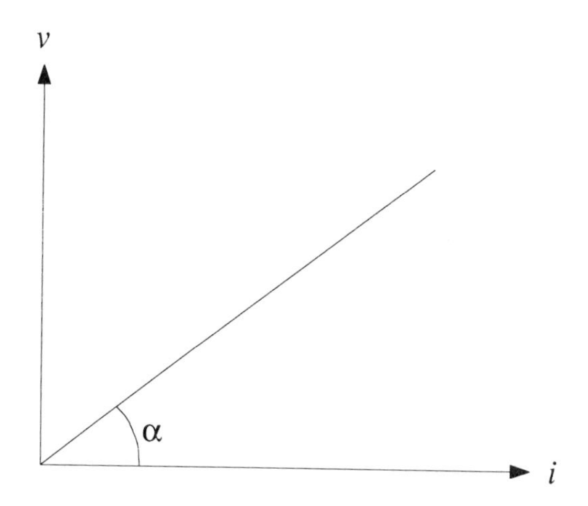

Figure 5.23: v versus i representation of a linear resistor.

values i_1 and i_2 of the current through the resistor will be equal to two different slopes:

$$R(i_1) = \frac{v_1}{i_1} = \tan \alpha_1, \tag{5.47}$$

$$R(i_2) = \frac{v_2}{i_2} = \tan \alpha_2. \tag{5.48}$$

This explains why $v(i)$ curves are used for characterization of nonlinear resistors.

There are many physical mechanisms which lead to a nonlinear dependence of resistance on electric current through a resistor. One simple mechanism, for example, is based on the fact that the resistance generally increases with temperature. As the current through a resistor is increased, the amount of power dissipated (as heat) increases, and its temperature goes up. This results in the increase of resistance.

Analysis of electric circuits with nonlinear resistors requires the solution of nonlinear algebraic equations. For this reason, this analysis encounters some difficulties which can be somewhat circumvented by using graphical techniques. We shall illustrate this by an example of a simple nonlinear circuit shown in Figure 5.26.

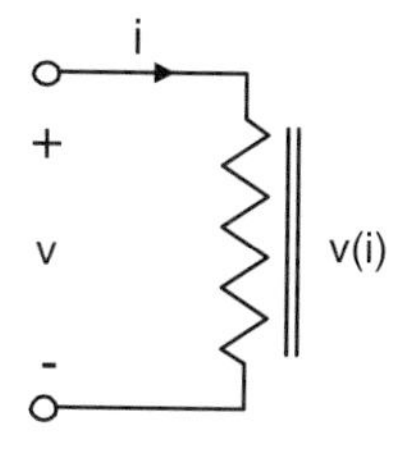

Figure 5.24: Circuit notation for a nonlinear resistor.

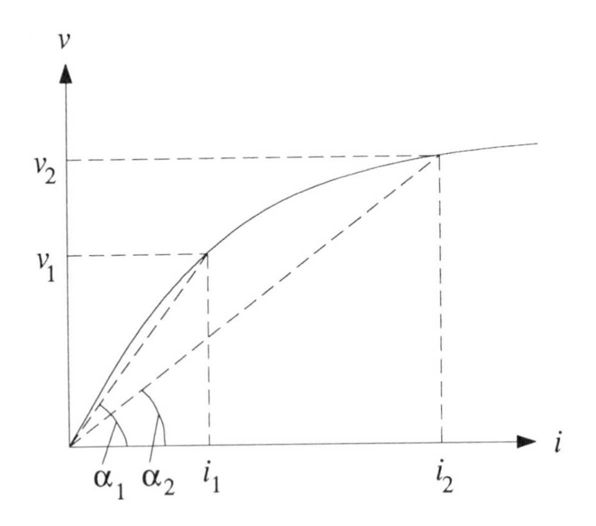

Figure 5.25: $v(i)$ curve for a nonlinear resistor.

By applying KVL to this circuit, we end up with the equation:

$$v_s = R_s i + v_1(i). \tag{5.49}$$

This is a nonlinear algebraic equation. The nonlinearity of this equation stems from the last term $v_1(i)$, which is the "voltage versus current" description of the nonlinear resistor. We shall solve this nonlinear equation graphically. To this end, we transform it as follows:

$$v_s - R_s i = v_1(i). \tag{5.50}$$

Next, we shall plot the left-hand side and right-hand side of this equation as shown in Figure 5.27. The straight line which corresponds to the left-hand side of (5.50) is usually called the "load line." The solution of equation (5.50) is the value i of the current for which both sides in (5.50) are equal. It is clear from Figure 5.27 that

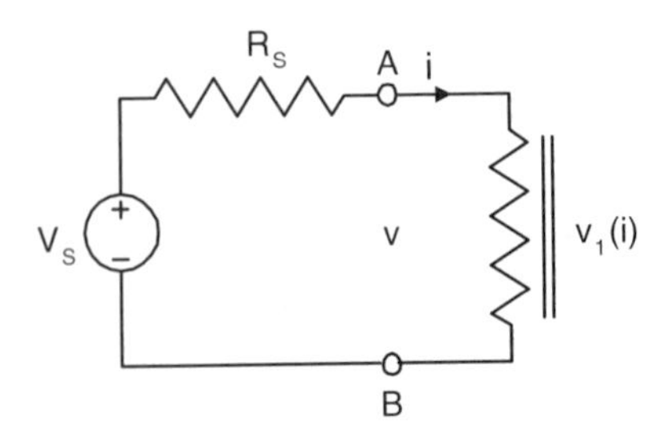

Figure 5.26: Electric circuit with a nonlinear resistor.

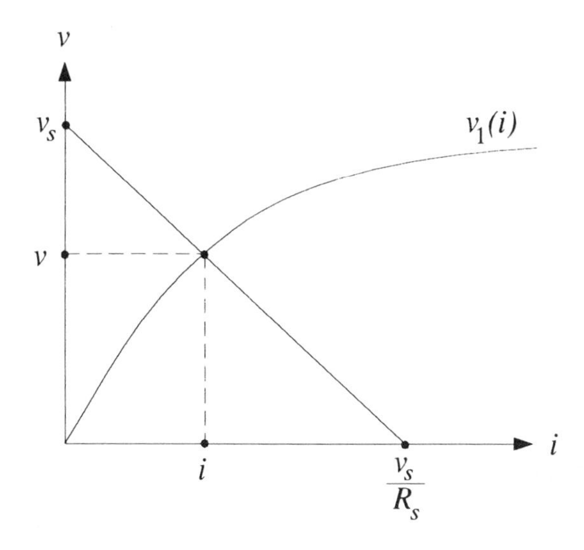

Figure 5.27: Graphical technique for the analysis of a nonlinear electric circuit.

both sides in (5.50) will be equal to one another for the value of the current which corresponds to the intersection of the plots of these sides. Thus, we can graphically find the current i through the circuit and the voltage v across the nonlinear resistor (see Figure 5.27).

The discussed problem is a very simple one. However, arbitrary resistive electric circuits with **single** nonlinear resistors can be reduced to the circuit shown in Figure 5.26. This statement is the Thevenin theorem for resistive circuits with single nonlinear resistors.

To prove this theorem, consider the circuit shown in Figure 5.28. Suppose i is the current through the nonlinear resistor. The voltage v across the terminals "A-B" **will remain the same** if we replace this resistor with a current source $i_s = i$ (see Figure 5.29). This is true because this replacement preserves the current through the terminals A-B of the active linear resistive circuit. In other words, this linear resistive

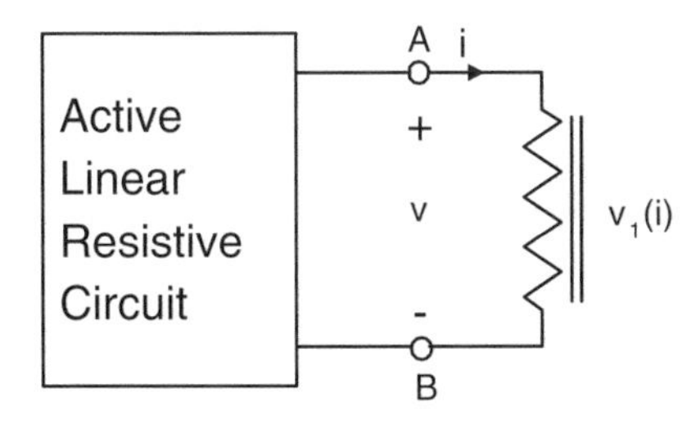

Figure 5.28: Resistive circuit with a single nonlinear resistor.

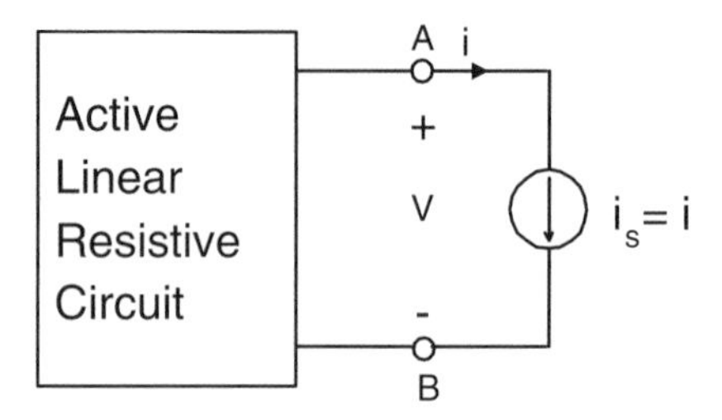

Figure 5.29: Replacement of a nonlinear resistor with a current source.

circuit cannot distinguish between the nonlinear resistor and the current source if each of them results in the same terminal current. Consequently, in both cases the voltage across the terminals A-B will be the same. From a purely mathematical point of view, this is true because the linear resistive circuits in Figure 5.28 and Figure 5.29 are described by **identical** linear algebraic equations. Consequently, by solving these equations, we find the same voltage across the terminals A-B.

Next, we shall apply the superposition principle to the circuit in Figure 5.29. To do this, we divide all the sources in this circuit into two groups: 1) all the sources in the active linear resistive circuit, 2) a single current source $i_s = i$. Now, we consider two separate regimes shown in Figure 5.30. According to the superposition principle, the voltage v across the terminals A-B is equal to the sum of voltages $v^{(1)}$ and $v^{(2)}$ across the same terminals for the first and second regimes, respectively:

$$v = v^{(1)} + v^{(2)}. \tag{5.51}$$

It is clear that $v^{(1)}$ is equal to the open-circuit voltage

$$v^{(1)} = v_{oc}, \tag{5.52}$$

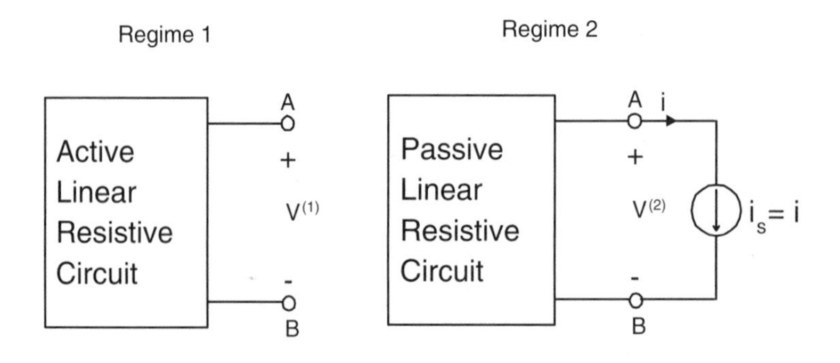

Figure 5.30: Two regimes.

while $v^{(2)}$ can be expressed in terms of the equivalent input resistance R_{in} of the passive linear resistive circuit and the current i as follows:

$$v^{(2)} = -R_{in}i. \tag{5.53}$$

Please note that the minus sign in (5.53) has appeared because the reference directions of the current i through R_{in} and voltage $v^{(2)}$ across the same resistance are not coordinated (are opposite).

By substituting (5.52) and (5.53) into (5.51), we obtain:

$$v = v_{oc} - R_{in}i. \tag{5.54}$$

If we consider the Thevenin equivalent circuit shown in Figure 5.26, we find that:

$$v = v_s - R_s i. \tag{5.55}$$

Formula (5.54) is the expression for the voltage versus current at the terminals A-B of the original circuit shown in Figure 5.28, while formula (5.55) is the expression for the voltage versus current at the terminals A-B of the circuit shown in Figure 5.26. These two expressions will be identical if:

$$v_s = v_{oc}, \tag{5.56}$$

$$R_s = R_{in}. \tag{5.57}$$

Under conditions (5.56) and (5.57), the voltage versus current relationship at the terminals A-B in the circuits shown in Figures 5.28 and 5.26 will be the same. This guarantees the same currents through the nonlinear resistor for both circuits. Thus, the proof of the Thevenin theorem for resistive circuits with single nonlinear resistors is concluded.

The given proof of the Thevenin theorem is more subtle than the one presented in Section 5.4. The main distinction is that this proof uses the superposition principle only for the active (linear) part of the circuit rather than for the entire circuit.

We have also established already familiar expressions (5.56) and (5.57) for the Thevenin voltage v_s and resistance R_s.

By using the Thevenin theorem, we can formulate the following two-step algorithm for the analysis of the generic circuit shown in Figure 5.28.

Step I:

Find the open circuit voltage across the terminals A-B for the active linear resistive circuit and the equivalent input resistance across the same terminals for the corresponding passive linear resistive circuit.

Step II:

Replace the active linear resistive circuit by the Thevenin equivalent circuit (see Figure 5.26) and use the graphical technique shown in Figure 5.27 for the analysis of the transformed circuit.

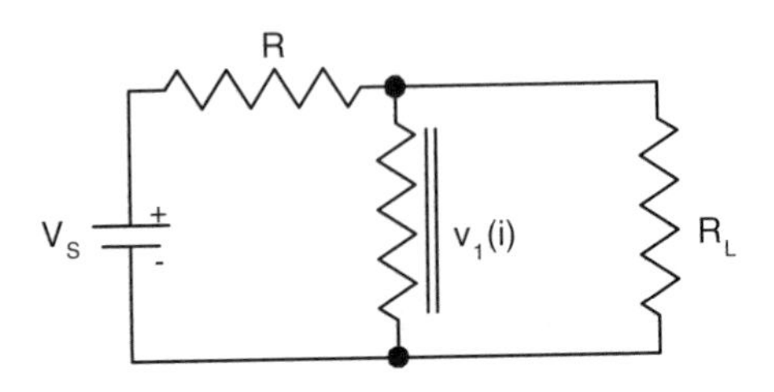

Figure 5.31: Voltage regulator circuit.

We shall illustrate the above algorithm by the following two examples.

EXAMPLE 5.5 A Voltage Regulator Circuit. Consider the circuit shown in Figure 5.31, where a load resistor R_L is connected in parallel with a nonlinear resistor characterized by the $v_1(i)$ curve shown in Figure 5.32. This curve exhibits "voltage saturation." In other words, it has an almost horizontal (flat) portion which starts from small current values. We would like to find all currents and voltages in this circuit.

First, we redraw this circuit to represent it in the form similar to the generic circuit in Figure 5.28. The redrawn circuit is shown in Figure 5.33. Next, we shall find the Thevenin voltage v_s which according to (5.56) is equal to the open circuit voltage, that is, the voltage across the load resistor R_L when the nonlinear resistor is removed. By using the voltage divider rule, it is easy to find that:

$$v_s = v_{oc} = V_s \frac{R_L}{R + R_L}. \tag{5.58}$$

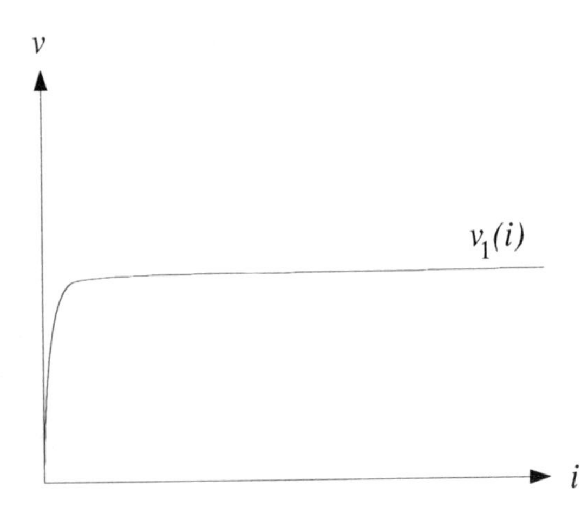

Figure 5.32: Voltage-current relationship for a nonlinear resistor used in a voltage regulator circuit.

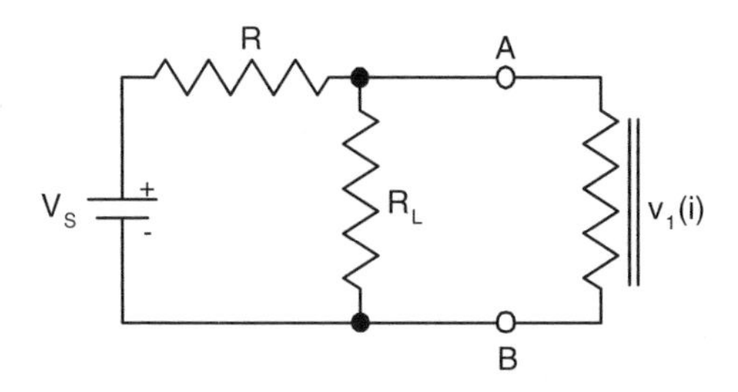

Figure 5.33: Redrawn voltage regulator circuit.

Now, we shall find the Thevenin resistance R_s, which is equal to the resistance across the terminals A-B when the nonlinear resistor is removed and dc voltage source V_s is short-circuited. It is easy to see that this resistance is equal to the equivalent resistance of R and R_L connected in parallel:

$$R_s = R_{in} = \frac{RR_L}{R + R_L}. \tag{5.59}$$

Thus, we have found the parameters of the equivalent circuit shown in Figure 5.26 and now we can use the graphical technique demonstrated in Figure 5.27. For our case, this technique is illustrated in Figure 5.34. The current i through the nonlinear resistor corresponds to the point of intersection of the straight load line and $v_1(i)$

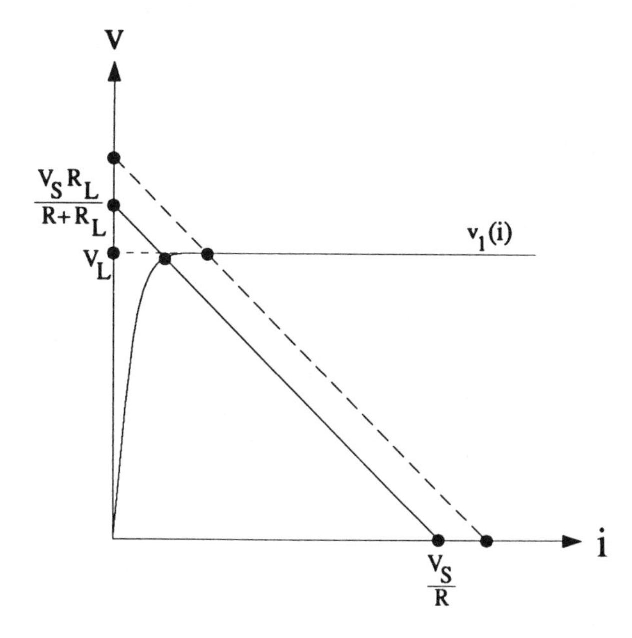

Figure 5.34: Graphical analysis of a voltage regulator circuit.

curve. The voltage v_L which corresponds to this point is the voltage across the load resistance. Thus, by dividing this voltage by R_L, we find the current i_L through the load resistance. By summing up this current with i, we find the current through the resistor R, and this concludes the analysis of the above circuit.

Now, we shall demonstrate a peculiar feature of this circuit. Suppose that the source voltage V_s does not remain constant and changes somewhat with time. Then, the load line will change its position but will remain parallel to the original load line (see Figure 5.34). New load voltage v_L will correspond to a new intersection point. However, since the $v_1(i)$ curve exhibits voltage saturation, this new load voltage will be practically the same as before. This explains why this circuit is called a voltage regulator circuit. To build circuits like this, nonlinear resistors with the property of voltage saturation are needed. It turns out that this property is exhibited by Zener diodes, which can be used for the design of the voltage regulator circuits. Zener diodes are usually studied in some detail in courses on electronic circuits.

The regulator circuit of the type shown in Figure 5.33 is often regarded as inefficient. The reason is the following. If V_s tends to increase, this results in a substantial increase in the current i through the nonlinear resistor (Zener diode). Since the load voltage and, consequently, load current remain the same, this leads to an increase in the current through the resistor R. This causes an increase in the voltage drop across this resistor which offsets the increase in V_s. Thus, we can see that the voltage regulation is achieved at the expense of increases in currents through the nonlinear resistor and resistor R and, consequently, at the expense of increase in power dissipation in these resistors. For this reason, these voltage regulator circuits are mostly used in low-power applications. For high-power applications, more efficient voltage regulator circuits have been designed. ■

EXAMPLE 5.6 A Bridge Circuit. This circuit is shown in Figure 5.35. It would be very difficult to analyze this circuit without the use of the Thevenin theorem. The utilization of this theorem simplifies the analysis appreciably.

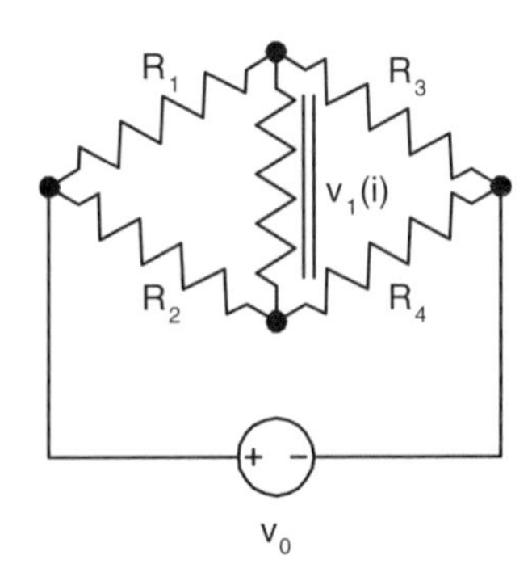

Figure 5.35: "Bridge" circuit with a nonlinear resistor.

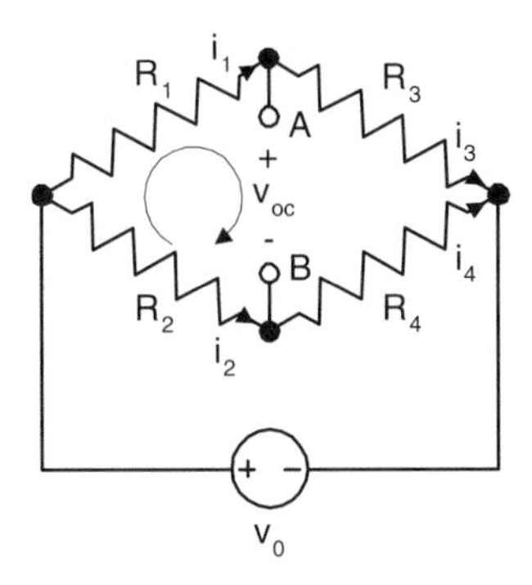

Figure 5.36: Electric circuit for the calculation of v_{oc}.

First, we shall find the Thevenin voltage v_s which is equal to the open-circuit voltage, that is, the voltage across the terminals A-B when the nonlinear resistor is removed (see Figure 5.36). It is easy to see that

$$i_1 = i_3 = \frac{v_0}{R_1 + R_3}, \tag{5.60}$$

$$i_2 = i_4 = \frac{v_0}{R_2 + R_4}. \tag{5.61}$$

By using (5.60), (5.61), and KVL for the loop consisting of R_1, the open terminals, and R_2, we derive:

$$v_s = v_{oc} = R_2 i_2 - R_1 i_1 = v_0 \left(\frac{R_2}{R_2 + R_4} - \frac{R_1}{R_1 + R_3} \right). \tag{5.62}$$

Now, we shall find the Thevenin resistance R_s which is equal to the input resistance across the terminals A-B when the voltage source v_0 is short-circuited (see Figure 5.37). It is easy to see that this resistance is given by:

$$R_s = R_{in} = \frac{R_1 R_3}{R_1 + R_3} + \frac{R_2 R_4}{R_2 + R_4}. \tag{5.63}$$

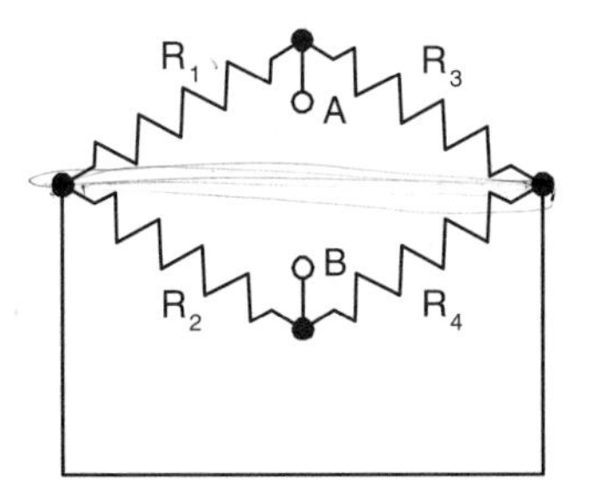

Figure 5.37: Electric circuit for the calculation of R_{in}.

Finally, by using the graphical technique demonstrated in Figure 5.27, we can find the current through and the voltage across the nonlinear resistor. Then it is easy to find all other voltages and currents in the circuit. ■

5.7 Summary

In this chapter we introduced a number of important concepts and rules that can be used to simplify the analysis of electric circuits. Along with these concepts came some definitions which we will continue to use throughout the text. The main concepts and definitions are summarized below.

- A nonideal voltage source is modeled by an ideal independent voltage source in series with a (normally small) impedance.
- A nonideal current source is modeled by an ideal independent current source in parallel with a (normally small) admittance.
- Nonideal voltage and current sources can be converted equivalently into one another via the formulas: $Z_s = 1/Y_s$ and $\hat{V}_s = \hat{I}_s Z_s$.
- Thevenin's theorem states that any active network can be replaced at a given pair of nodes by a nonideal voltage source. The equivalent voltage source is equal to the open-circuit voltage $\hat{V}_{oc}$, while the source impedance is equal to the input impedance Z_{in}.
- Norton's theorem states that any active network can be replaced at a given pair of nodes by a nonideal current source. The equivalent current source is the short-circuit current $\hat{I}_{sc}$, while the source admittance is equal to the input admittance Y_{in}.
- The input impedance is related to the open-circuit voltage and short-circuit current by the equation: $Z_{in} = \hat{V}_{oc}/\hat{I}_{sc}$.
- Thevenin's theorem can be extended to nonlinear circuits with single nonlinear resistors.

5.8 Problems

1. A nonideal inductor is connected in series with a nonideal capacitor. Find the input impedance if the circuit parameters are as indicated in the table.

Circuit	Frequency	Inductor		Capacitor	
	ω (rad/s)	L (mH)	R_L(mΩ)	C (μF)	G_C (m℧)
a	377	.001	100.	15	.01
b	2 k	2.2	10.	.001	1
c	3.1 M	470	1.	47	100

2. A nonideal inductor is connected in parallel with a nonideal capacitor. Find the input admittance if the circuit parameters are as indicated in the table.

Circuit	Frequency	Inductor		Capacitor	
	ω (rad/s)	L (mH)	R_L (mΩ)	C (μF)	G_C (m℧)
a	1.1 M	.003	1000.	.0047	.002
b	5 k	91	10.	1.0	5
c	377	20	1000.	56	20

3. A nonideal current source has a phasor current $\hat{I}_s = 20\sqrt{3}e^{j\pi/3}$. Find the equivalent nonideal voltage source if Y_s = (a) 1 k℧, (b) 2.1 M℧, (c) $-j0.07$ M℧, and (d) $15 + 3j$ k℧.

4. A nonideal voltage source has a phasor voltage $\hat{V}_s = 12\sqrt{2}e^{j\pi/6}$. Find the equivalent nonideal current source if Z_s = (a) 3 kΩ, (b) 20 $\mu\Omega$, (c) $j0.07$ Ω, and (d) $33 - 10j$ mΩ.

In Problems 5–11, find the Thevenin equivalent circuits with respect to the terminals indicated in the following figures.

5. Figure P5-5. **6.** Figure P5-6. **7.** Figure P5-7. **8.** Figure P5-8.

9. Figure P5-9. **10.** Figure P5-10. **11.** Figure P5-11.

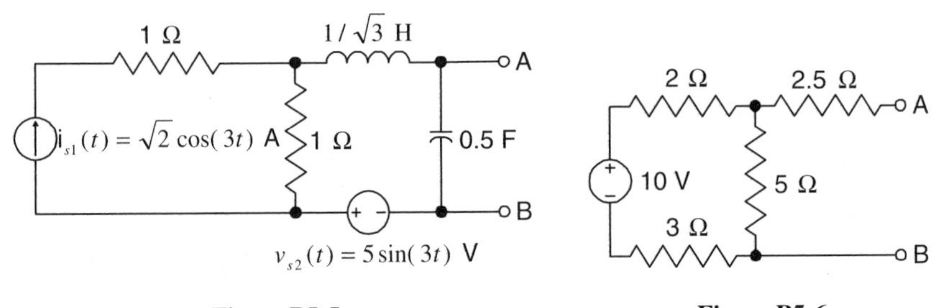

Figure P5-5 **Figure P5-6**

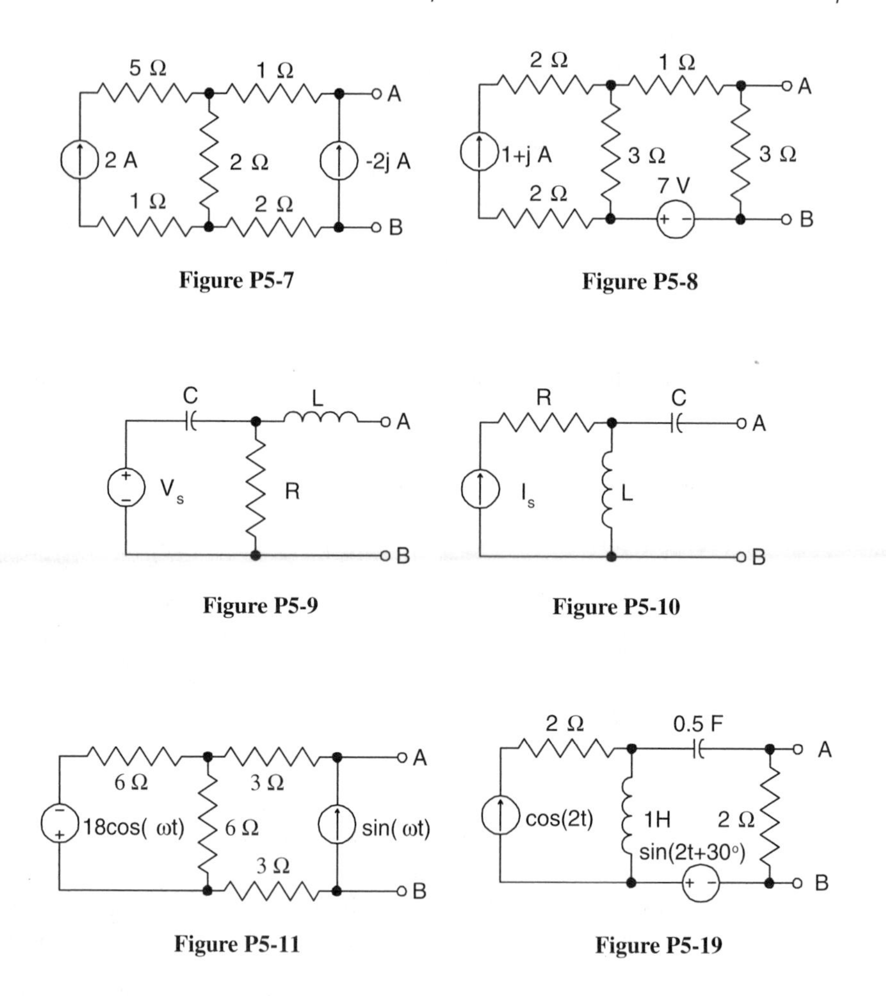

Figure P5-7

Figure P5-8

Figure P5-9

Figure P5-10

Figure P5-11

Figure P5-19

12. Prove Norton's theorem from basic principles (without invoking Thevenin's theorem).

In Problems 13–19, find the Norton equivalent circuits with respect to the terminals indicated in the following figures.

13. Figure P5-6. **14.** Figure P5-7. **15.** Figure P5-8. **16.** Figure P5-9.

17. Figure P5-10. **18.** Figure P5-11. **19.** Figure P5-19.

20. Find the value of resistance R in the circuit shown in Figure P5-20 that results in a Thevenin equivalent voltage of 5 V at the terminals indicated. What is the corresponding equivalent resistance?

21. An RL load is connected to a voltage source as shown in the circuit in Figure P5-21. For the parameters listed in the following table, find the capacitance which, when added in parallel to the load, adjusts the power factor to unity. Find the average power delivered by the source after the capacitor is connected.

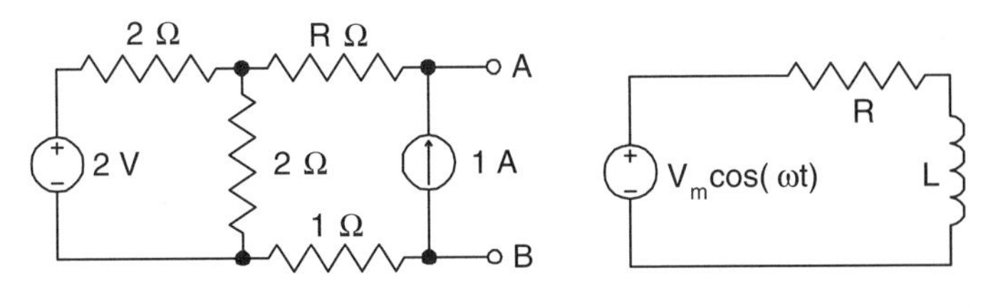

Figure P5-20 **Figure P5-21**

Case	V_m (V)	ω (rad/s)	R (Ω)	L (mH)
a	5	10	1	10
b	12	2513	10	1
c	170	377	100	10
d	294	377	1 k	500

22. Consider the circuit shown in Figure P5-22. Find the value of Z_L which maximizes the power transferred to this element. What is the average power dissipated in the load?

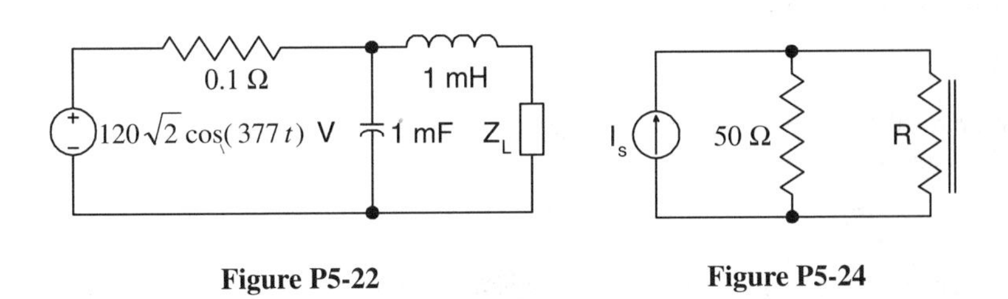

Figure P5-22 **Figure P5-24**

23. The resistance of a nonlinear resistor can be described by $R = 10 + .5i\ \Omega$, where i is the current through the resistor in amperes. Use the graphical technique to find the voltage across the load when it is connected to a nonideal voltage source with $V_s = 169.7$ V and $R_s = 10\ \Omega$.

24. A varistor is an element which is designed to protect sensitive elements from large current spikes. It has high resistance at low current and low resistance at high current. If a varistor's resistance is given by $R = 1000e^{-i/2}\ \Omega$, where i is the current in amperes, find the current through the 50 Ω resistor in Figure P5-24 when (a) $I_s = 1$ A, (b) $I_s = 10$ A, and (c) $I_s = 17.5$ A. Use the graphical technique to solve the problem.

Chapter 6

Nodal and Mesh Analysis

6.1 Introduction

It has been shown in earlier chapters that the basic method for ac steady-state analysis involves setting up and solving a system of linear equations derived from Kirchhoff's current and voltage laws. Here, we recall that each branch in a circuit can be characterized by either impedance or admittance (the latter is merely the reciprocal of the former). Given these impedances or admittances, the KCL and KVL equations can be set up for the nodes and loops of the circuit in the following manner.

First, we use KCL:

$$\sum_k \hat{I}_k = -\sum_k \hat{I}_{sk}, \tag{6.1}$$

which results in $n - 1$ linearly independent equations, where n is the number of nodes.

Then, we use KVL:

$$\sum_k \hat{I}_k Z_k = -\sum_k \hat{V}_{sk}, \tag{6.2}$$

which results in $b - (n - 1)$ linearly independent equations, where b is the number of branches. In total, b equations will be produced, which could be a very large system of equations for complex circuits. Fortunately, these equations are generally *sparse*, a property which gives rise to special techniques for circuit analysis. Two of these techniques are *nodal analysis* and *mesh current analysis*. They are presented in this chapter.

In the final section, the PSpice circuit simulation package is used to analyze several of the examples from this chapter.

6.2 Nodal Analysis

The main idea of nodal analysis is to represent KCL equations in terms of node potentials. To describe the method of nodal analysis (also called the method of node potentials), we consider an example circuit shown in Figure 6.1. Note that the

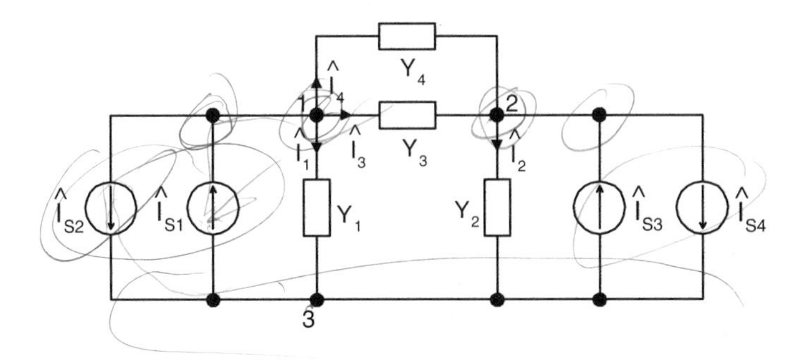

Figure 6.1: Sample circuit for method of node potentials.

branches are characterized by admittances—this is not accidental. When using nodal analysis, it is important to convert impedances into admittances before beginning circuit analysis. This circuit also contains only current sources that are in parallel with other elements. This is to simplify our initial discussion, and later this restriction on sources will be removed.

To start the analysis, we introduce three new variables, $\hat{v}_1$, $\hat{v}_2$, and $\hat{v}_3$. These variables are the node potentials; that is, they are the potentials of the corresponding nodes. Now, we recall from Chapter 1 that voltages can be expressed as potential differences. Thus, we can write the voltage across any branch in the circuit in terms of these node potentials. For instance, the voltage across the branch Y_3 (between nodes 1 and 2) is just the difference in the potentials at nodes 1 and 2. The values of the node potentials must be taken with respect to some reference. Because we are only interested in the difference in potentials and not their actual values, we are free to choose whatever reference we wish. To simplify things, we set the potential of some chosen node equal to zero. This is called the *reference node*, and it is generally chosen to be the node with the most branches connected to it. In the example circuit, the node with the greatest connectivity is node 3, so its potential is set to zero:

$$\hat{v}_3 = 0. \tag{6.3}$$

Now, we write the KCL equations for nodes 1 and 2:

$$\hat{I}_1 + \hat{I}_3 + \hat{I}_4 = \hat{I}_{s1} - \hat{I}_{s2}, \tag{6.4}$$

$$\hat{I}_2 - \hat{I}_3 - \hat{I}_4 = \hat{I}_{s3} - \hat{I}_{s4}. \tag{6.5}$$

Recalling the relationship between current, voltage and admittance,

$$\hat{I}_k = Y_k \hat{V}_k, \tag{6.6}$$

we can write the branch currents in terms of node potentials and admittances:

$$\hat{I}_1 = Y_1(\hat{v}_1 - \hat{v}_3) = Y_1\hat{v}_1, \tag{6.7}$$

$$\hat{I}_2 = Y_2(\hat{v}_2 - \hat{v}_3) = Y_2\hat{v}_2, \tag{6.8}$$

$$\hat{I}_3 = Y_3(\hat{v}_1 - \hat{v}_2), \tag{6.9}$$

$$\hat{I}_4 = Y_4(\hat{v}_1 - \hat{v}_2). \tag{6.10}$$

Using the above relations, we can now express the KCL equations (6.4) and (6.5) in terms of node potentials and admittances,

$$Y_1\hat{v}_1 + Y_3(\hat{v}_1 - \hat{v}_2) + Y_4(\hat{v}_1 - \hat{v}_2) = \hat{I}_{s1} - \hat{I}_{s2}, \tag{6.11}$$

$$Y_2\hat{v}_2 - Y_3(\hat{v}_1 - \hat{v}_2) - Y_4(\hat{v}_1 - \hat{v}_2) = \hat{I}_{s3} - \hat{I}_{s4}. \tag{6.12}$$

After combining similar terms in the above equations, we have:

$$(Y_1 + Y_3 + Y_4)\hat{v}_1 - (Y_3 + Y_4)\hat{v}_2 = \hat{I}_{s1} - \hat{I}_{s2}, \tag{6.13}$$

$$-(Y_3 + Y_4)\hat{v}_1 + (Y_2 + Y_3 + Y_4)\hat{v}_2 = \hat{I}_{s3} - \hat{I}_{s4}. \tag{6.14}$$

The last two equations represent the essence of the method of node potentials. These equations can be solved for $\hat{v}_1$ and $\hat{v}_2$, and, once these node potentials are known, all the currents in the circuit can be found by using expressions (6.7)–(6.10).

We rewrite the obtained equations in matrix form in order to make the benefits of nodal analysis even more apparent. The matrix form is as follows:

$$\tilde{Y}\,\vec{v} = \vec{I}_s, \tag{6.15}$$

where $\tilde{Y}$ is an admittance matrix, and $\vec{v}$ and $\vec{I}_s$ are the vector of unknown node potentials and the "source" vector, respectively:

$$\tilde{Y} = \begin{bmatrix} Y_{11} & Y_{12} \\ Y_{21} & Y_{22} \end{bmatrix}, \tag{6.16}$$

$$\vec{v} = \begin{bmatrix} \hat{v}_1 \\ \hat{v}_2 \end{bmatrix}, \tag{6.17}$$

$$\vec{I}_s = \begin{bmatrix} \hat{I}_s^{(1)} \\ \hat{I}_s^{(2)} \end{bmatrix}. \tag{6.18}$$

By using the introduced matrix notations, equations (6.13) and (6.14) can be written as follows:

$$\begin{bmatrix} Y_1 + Y_3 + Y_4 & -Y_4 - Y_3 \\ -Y_4 - Y_3 & Y_2 + Y_3 + Y_4 \end{bmatrix} \begin{bmatrix} \hat{v}_1 \\ \hat{v}_2 \end{bmatrix} = \begin{bmatrix} \hat{I}_{s1} - \hat{I}_{s2} \\ \hat{I}_{s3} - \hat{I}_{s4} \end{bmatrix}. \tag{6.19}$$

Thus, we can see that:

$$Y_{11} = Y_1 + Y_3 + Y_4, \tag{6.20}$$

$$Y_{12} = Y_{21} = -(Y_3 + Y_4), \tag{6.21}$$

$$Y_{22} = Y_2 + Y_3 + Y_4, \tag{6.22}$$

$$\hat{I}_s^{(1)} = \hat{I}_{s1} - \hat{I}_{s2}, \tag{6.23}$$

$$\hat{I}_s^{(2)} = \hat{I}_{s3} - \hat{I}_{s4}. \tag{6.24}$$

Writing the equations in matrix form is useful because it reveals the *pattern* that the matrices and source vectors of nodal analysis follow. For the admittance matrix in the example, we can see that Y_{11} is equal to the sum of all branch admittances connected to node 1. Likewise, Y_{22} is equal to the sum of all branch admittances connected to node 2. The other two admittances, Y_{12} and Y_{21}, are identical and equal to the negative of the sum of the branch admittances connected between nodes 1 and 2. In the case of the source vector, $\hat{I}_s^{(1)}$ is equal to the algebraic sum of all current sources connected to node 1, while $\hat{I}_s^{(2)}$ is equal to the algebraic sum of all current sources connected to node 2. In this case, "algebraic sum" means that currents entering the node are taken with positive signs while those leaving the node are taken with negative signs.

This pattern for forming the coefficient matrices and the source vectors of the nodal equations can be generalized to an arbitrary circuit with any number of nodes. To write the matrix equations for any circuit, three simple rules must be followed. It is this fast and easy way to produce the matrix equations as well as the relatively small number of these equations that makes the method of node potentials so attractive.

Consider the general case of a circuit with $n + 1$ nodes. This circuit can be characterized by $n + 1$ node potentials, which give a total of n equations (one node is taken to be the reference node; its potential is set to zero and does not appear in the equations). Then, a general form of the matrix equations for a circuit with $n + 1$ nodes is

$$\begin{bmatrix} Y_{11} & Y_{12} & \cdots & Y_{1n} \\ Y_{21} & Y_{22} & \cdots & Y_{2n} \\ \vdots & \vdots & \ddots & \vdots \\ Y_{n1} & Y_{n2} & \cdots & Y_{nn} \end{bmatrix} \begin{bmatrix} \hat{v}_1 \\ \hat{v}_2 \\ \vdots \\ \hat{v}_n \end{bmatrix} = \begin{bmatrix} \hat{I}_s^{(1)} \\ \hat{I}_s^{(2)} \\ \vdots \\ \hat{I}_s^{(n)} \end{bmatrix}. \tag{6.25}$$

The three rules for determining the elements of the admittance matrix and the source vector are as follows.

For diagonal matrix elements we have:

$$Y_{ii} = \sum_k^{\prime} Y_k, \tag{6.26}$$

which means that a diagonal element Y_{ii} of the coefficient matrix is the sum of all branch admittances connected to node i, and superscript "$\prime$" in (6.26) suggests that the summation is performed over all these branches.

For off-diagonal matrix elements we have:

$$Y_{ij} = Y_{ji} = -\sum_k^{\prime\prime} Y_k, \tag{6.27}$$

which means that an element which is on an intersection of the ith row and jth column of the matrix is the negative of the sum of all branch admittances directly connected between nodes i and j, and superscript "$\prime\prime$" in (6.27) indicates that the summation is performed over all branches between those nodes.

For the source vector elements we have:

$$\hat{I}_s^{(i)} = \sum_k^{\prime} \hat{I}_{sk}, \tag{6.28}$$

which means that the ith element of the source vector is the algebraic sum of all current sources connected to node i.

Only these three simple rules are required to form the matrix equations for the method of node potentials. It is clear that even for very complex circuits, these equations can easily be formed by inspecting an electric circuit. We shall illustrate the previous discussion by the following example.

EXAMPLE 6.1 Consider the circuit shown in Figure 6.2. We would like to write the node potential equations in order to find all branch currents.

This circuit is already given in phasor-admittance form, but, if this had not been the case, the first step in the analysis would have been to convert it to this form. Recall that for the method of node potentials, the branches must be characterized by admittances. These can be found by taking the reciprocal of the impedances.

We now determine how many nodes the circuit has. In this case, there are four nodes, which means there will be three equations or, equivalently, a 3×3 admittance matrix.

Once the nodes are identified and labeled (done already in Figure 6.2), the node potentials are introduced, and the reference node is chosen. For our example, we choose node 4 to be the reference node *since it has the greatest connectivity*. Therefore, we set $\hat{v}_4$ equal to zero and proceed to set up the matrix equations to solve for the remaining three node potentials. To set up the matrix, we simply follow the three rules outlined above. In this way, we end up with:

$$\begin{bmatrix} Y_1 + Y_4 + Y_6 & -Y_4 & -Y_6 \\ -Y_4 & Y_2 + Y_4 + Y_5 & -Y_5 \\ -Y_6 & -Y_5 & Y_6 + Y_5 + Y_3 \end{bmatrix} \begin{bmatrix} \hat{v}_1 \\ \hat{v}_2 \\ \hat{v}_3 \end{bmatrix} = \begin{bmatrix} \hat{I}_{s1} - \hat{I}_{s2} \\ 0 \\ \hat{I}_{s3} \end{bmatrix}. \tag{6.29}$$

These matrix equations can be solved by using a variety of methods. In most cases, the recommended method of solving linear systems of equations is the Gaussian elimination technique (see Appendix B). This method involves triangulation of the coefficient matrix and back-substitution of the variables.

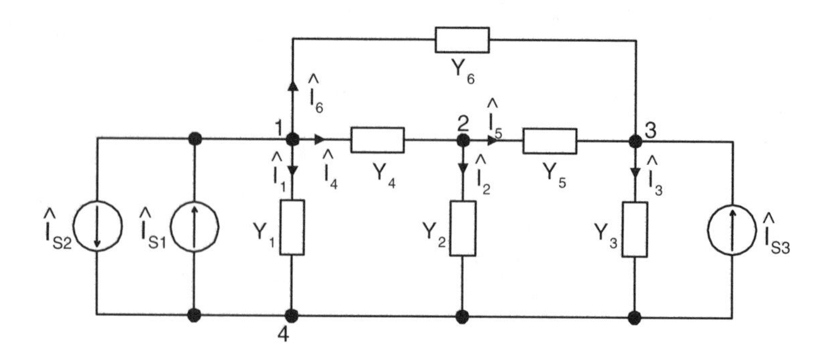

Figure 6.2: Example circuit for nodal analysis.

Once the node potentials are found, the branch currents can be determined by using the relationship between currents, admittances, and node potentials:

$$\hat{I}_1 = Y_1\hat{v}_1, \tag{6.30}$$

$$\hat{I}_2 = Y_2\hat{v}_2, \tag{6.31}$$

$$\hat{I}_3 = Y_3\hat{v}_3, \tag{6.32}$$

$$\hat{I}_4 = Y_4(\hat{v}_1 - \hat{v}_2), \tag{6.33}$$

$$\hat{I}_5 = Y_5(\hat{v}_2 - \hat{v}_3), \tag{6.34}$$

$$\hat{I}_6 = Y_6(\hat{v}_1 - \hat{v}_3). \tag{6.35}$$

■

EXAMPLE 6.2 Consider the circuit shown in Figure 6.3. We want to use nodal analysis to find the current through the $1/2\ \Omega$ resistor.

Comparing Figure 6.3 with Figure 6.2, we see that the circuits are identical if we let $Y_6 = 2\ \mho$, $Y_2 = 3\ \mho$, $Y_1 = Y_3 = Y_4 = Y_5 = 1\ \mho$, $I_{s1} = 2$ A, $I_{s2} = 7$ A, and $I_{s3} = 4$ A. By plugging these values into the matrix equation (6.29), we obtain:

$$\begin{bmatrix} 4 & -1 & -2 \\ -1 & 5 & -1 \\ -2 & -1 & 4 \end{bmatrix} \begin{bmatrix} v_1 \\ v_2 \\ v_3 \end{bmatrix} = \begin{bmatrix} -5 \\ 0 \\ 4 \end{bmatrix}. \tag{6.36}$$

To solve the above equations, we can use the Gaussian elimination technique that is described in Appendix B. Following the procedure in the appendix, we can arrive at the following upper-triangular matrix:

$$\begin{bmatrix} 4 & -1 & -2 \\ 0 & 19/4 & -3/2 \\ 0 & 0 & 48/19 \end{bmatrix} \cdot \begin{bmatrix} v_1 \\ v_2 \\ v_3 \end{bmatrix} = \begin{bmatrix} -5 \\ -5/4 \\ 21/19 \end{bmatrix}. \tag{6.37}$$

We then use back-substitution to find $v_3 = 0.4375$ V, $v_2 = -0.125$ V, and $v_1 = -1.0625$ V.

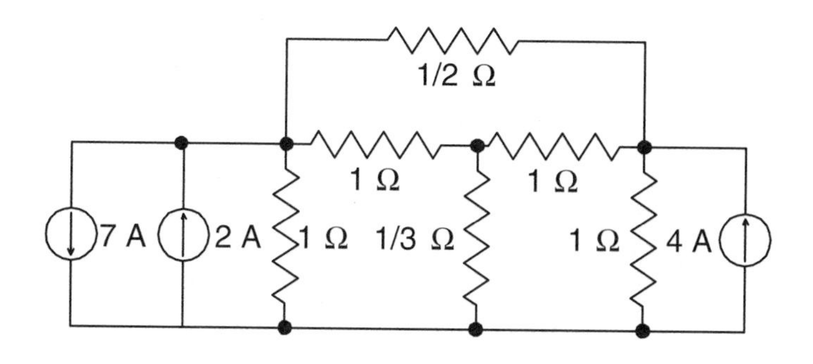

Figure 6.3: A numerical example of nodal analysis.

Finally, by using equation (6.35) we find the desired current: $i_6 = 2(-1.0625 - 0.4375) = -3$ A.

This concludes the analysis of the problem. ■

In the above treatment of nodal analysis, only circuits containing current sources connected in parallel with the admittances have been considered. We now consider the three other possibilities: (1) branches with current sources in series with admittances, (2) branches with voltage sources in parallel with admittances, and (3) branches with voltage sources in series with admittances. The development of techniques to deal with these cases will make the method of node potentials applicable to any general circuit.

The case of branches with current sources in series with admittances is the simplest of the three—in fact, it is trivial. This is because admittances in series with current sources will have no effect on the branch currents and, consequently, on the KCL equations. Thus, they will not affect the matrix equations of nodal analysis. This means that, for the purpose of finding *node potentials*, we can treat the series admittance as though it wasn't there (i.e., replace it with a short). However, this does not mean that these admittances have no effect on the circuit at all. Actually, these admittances affect how much power is absorbed or supplied by current sources. Indeed, the voltage drops across the current sources must be known to calculate power. The currents are already known (they are the source currents) and the voltages can be found as follows. First, we find the potential differences between the nodes to which the above branches are connected; then the voltage drops across the admittances in these branches are determined. Subtracting these two quantities will yield the voltages across the current sources, which can then be used to calculate power.

The second case of branches with only voltage sources is somewhat more complicated. This case is shown in Figure 6.4, which shows the same circuit as in Example 6.1 except that it has one added voltage source between nodes 2 and 3. For this case, we recall that voltage sources define the potential difference between the nodes

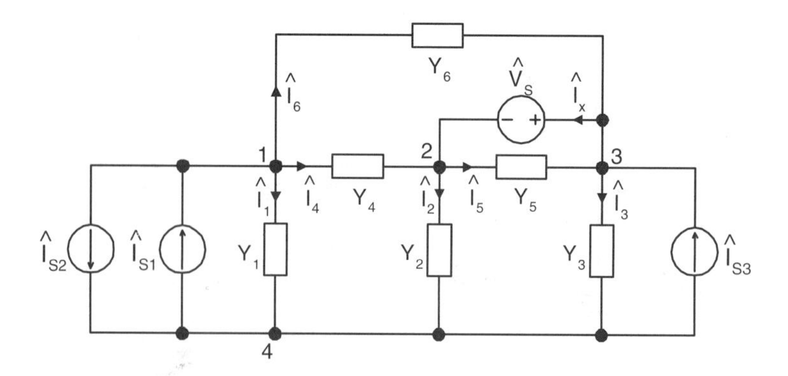

Figure 6.4: A voltage source connected in parallel.

to which they are connected. Since the potentials at nodes 2 and 3 are $\hat{v}_2$ and $\hat{v}_3$, respectively, we can conclude that

$$\hat{v}_3 - \hat{v}_2 = \hat{V}_s, \tag{6.38}$$

$$\hat{v}_3 = \hat{v}_2 + \hat{V}_s. \tag{6.39}$$

Thus, because of the voltage source, $\hat{v}_2$ and $\hat{v}_3$ are no longer independent variables, and we can redefine $\hat{v}_3$ in terms of $\hat{v}_2$ and $\hat{V}_s$ as shown in (6.39). This effectively reduces the number of unknown variables in the equations by one. However, there is a complication: the current $\hat{I}_x$ through voltage source $\hat{V}_s$ is not known and it cannot be expressed through known quantities. To circumvent this difficulty, we shall temporarily treat $\hat{I}_x$ as a known quantity and write the nodal equations, which are similar to (6.29):

$$\begin{bmatrix} Y_1 + Y_4 + Y_6 & -Y_4 & -Y_6 \\ -Y_4 & Y_2 + Y_4 + Y_5 & -Y_5 \\ -Y_6 & -Y_5 & Y_6 + Y_5 + Y_3 \end{bmatrix} \begin{bmatrix} \hat{v}_1 \\ \hat{v}_2 \\ \hat{v}_3 \end{bmatrix} = \begin{bmatrix} \hat{I}_{s1} - \hat{I}_{s2} \\ \hat{I}_x \\ \hat{I}_{s3} - \hat{I}_x \end{bmatrix}. \tag{6.40}$$

We next eliminate $\hat{I}_x$ by summing up the second and third equations in (6.40). This results in the following set of two equations with three unknowns:

$$\begin{bmatrix} Y_1 + Y_4 + Y_6 & -Y_4 & -Y_6 \\ -Y_4 - Y_6 & Y_2 + Y_4 & Y_3 + Y_6 \end{bmatrix} \begin{bmatrix} \hat{v}_1 \\ \hat{v}_2 \\ \hat{v}_3 \end{bmatrix} = \begin{bmatrix} \hat{I}_{s1} - \hat{I}_{s2} \\ \hat{I}_{s3} \end{bmatrix}. \tag{6.41}$$

Finally, we reduce the number of unknowns by replacing $\hat{v}_3$ by $\hat{v}_2 + \hat{V}_s$ (see (6.39)). After simple transformations, we end up with the following two equations with two unknowns:

$$\begin{bmatrix} Y_1 + Y_4 + Y_6 & -Y_4 - Y_6 \\ -Y_4 - Y_6 & Y_2 + Y_3 + Y_4 + Y_6 \end{bmatrix} \begin{bmatrix} \hat{v}_1 \\ \hat{v}_2 \end{bmatrix} = \begin{bmatrix} \hat{I}_{s1} - \hat{I}_{s2} + Y_6 \hat{V}_s \\ \hat{I}_{s3} - (Y_3 + Y_6)\hat{V}_s \end{bmatrix}. \tag{6.42}$$

By solving these equations, we can find all nodal potentials and then compute all currents.

The third case of voltage sources connected in series with admittances is somewhat easier than the previous one. As an example, consider the circuit shown on the left in Figure 6.5. Note that the voltage sources in this circuit can actually be considered as nonideal voltage sources since they are connected in series with impedances. For this reason, we simply use the equivalent transformation of nonideal voltage sources into nonideal current sources. After performing the transformation, the circuit will look like the one shown on the right in Figure 6.5. This circuit has current sources in parallel with admittances. The values for the new current sources are

$$\hat{I}_{s1} = \hat{V}_{s1} Y_1, \tag{6.43}$$

$$\hat{I}_{s2} = \hat{V}_{s2} Y_2. \tag{6.44}$$

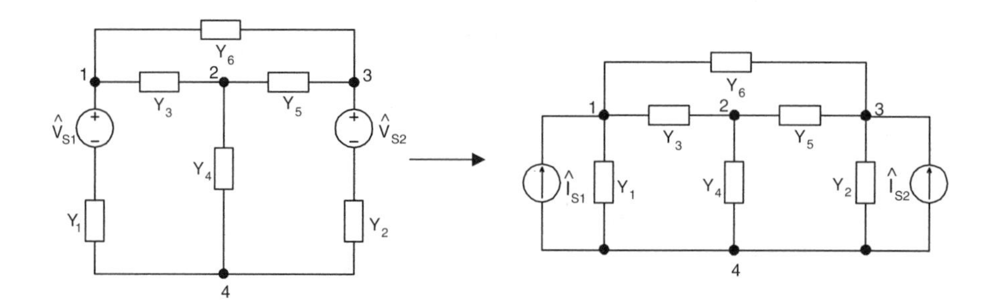

Figure 6.5: Example of voltage sources in series with admittances.

The direction of the current sources can be determined from the polarity of the voltage sources as described at the end of Section 5.3. In our case, this means that the new current sources will be entering nodes 1 and 3.

Once the voltage sources have been transformed, we end up with precisely the same case which has been considered in the original derivation of nodal equations. Therefore, we already know how to write these nodal equations. They are as follows:

$$\begin{bmatrix} Y_1 + Y_3 + Y_6 & -Y_3 & -Y_6 \\ -Y_3 & Y_3 + Y_4 + Y_5 & -Y_5 \\ -Y_6 & -Y_5 & Y_2 + Y_5 + Y_6 \end{bmatrix} \begin{bmatrix} \hat{v}_1 \\ \hat{v}_2 \\ \hat{v}_3 \end{bmatrix} = \begin{bmatrix} \hat{I}_{s1} \\ 0 \\ \hat{I}_{s2} \end{bmatrix}. \tag{6.45}$$

By substituting (6.43) and (6.44) into (6.45), we end up with:

$$\begin{bmatrix} Y_1 + Y_3 + Y_6 & -Y_3 & -Y_6 \\ -Y_3 & Y_3 + Y_4 + Y_5 & -Y_5 \\ -Y_6 & -Y_5 & Y_2 + Y_5 + Y_6 \end{bmatrix} \begin{bmatrix} \hat{v}_1 \\ \hat{v}_2 \\ \hat{v}_3 \end{bmatrix} = \begin{bmatrix} Y_1\hat{V}_{s1} \\ 0 \\ Y_2\hat{V}_{s2} \end{bmatrix}. \tag{6.46}$$

By comparing equations (6.46) with the original (not transformed) circuit shown in Figure 6.5, we conclude that we can write these matrix equations just by inspection of the original circuit. The first two rules concerning the formation of the admittance matrix remain the same. However, the third rule, which is used for the formation of the source vector, should be somewhat modified. According to the modified rule, the elements of the source vector are given by the expression:

$$\hat{I}_s^{(i)} = \sum_k^{l} \hat{I}_{sk} + \sum_k^{l} Y_k \hat{V}_{sk}. \tag{6.47}$$

Here, the first term in the right-hand side of (6.47) has the same meaning as in (6.28) and it accounts for all current sources connected to node i. The second term in (6.47) is the algebraic sum of products $Y_k \hat{V}_{sk}$ and summation is taken over all voltage sources connected to node i. The word "algebraic" means that voltage sources are taken with positive signs if their "positive" terminals are connected to node i, and voltage sources are taken with negative signs if their "negative" terminals are connected to node i.

By using rules (6.26), (6.27), and (6.47) we can immediately write the nodal equations in matrix form just *by inspecting* an electric circuit. This is true for all circuits except those which contain branches with voltage sources alone. These "exceptional" circuits require special treatment which has already been discussed.

EXAMPLE 6.3 As an application of nodal analysis, we consider a three-phase circuit, pictured in Figure 6.6. This three-phase circuit consists of three voltage sources and three impedances arranged into "star" connections. Each of the voltage sources has the same peak value and frequency, but they are out of phase with each other by 120°. There is a neutral wire connecting nodes 1 and 2 which keeps them at almost the same potential if the impedance Z_n of the neutral wire is very small. This wire is present so that variations of any (load) impedance in the circuit will not affect significantly voltages across other load impedances. In other words, the neutral wire is used to maintain almost constant voltages across load impedances. Such three-phase circuits are used by utility companies to supply electric power at almost constant voltages in the face of permanently and drastically changing loads. Ideally, the impedance of the neutral wire should be zero, but realistically there is always some small impedance present. We would like to analyze this circuit by using nodal techniques in order to understand the influence of Z_n on the variation of load voltages.

We begin by introducing node potentials $\hat{v}_2$ and $\hat{v}_1$ for each of the two nodes in the circuit. Because there are only two nodes, we will end up with one nodal equation. We choose $\hat{v}_2$ to be the reference node, so

$$\hat{v}_2 = 0. \tag{6.48}$$

Now, we must convert the impedances into admittances, because nodal analysis is carried out for circuits characterized by admittances. Therefore, we have:

$$Y_1 = \frac{1}{Z_1}, \quad Y_2 = \frac{1}{Z_2}, \quad Y_3 = \frac{1}{Z_3}, \quad Y_n = \frac{1}{Z_n}. \tag{6.49}$$

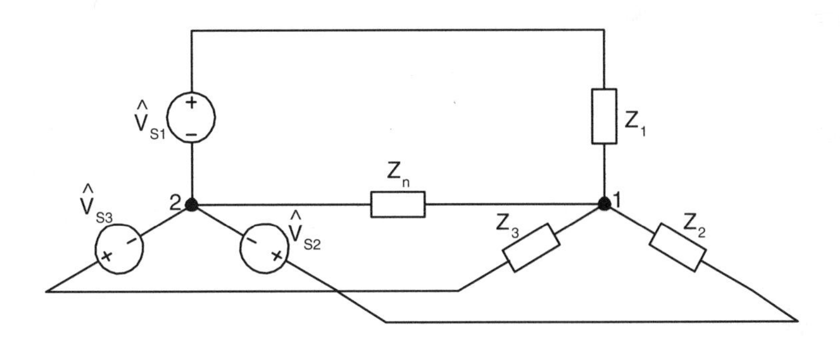

Figure 6.6: A three-phase network.

We now form the node equation. In our case, the admittance matrix consists of only one element, which is equal to the sum of all admittances connected to node 1. We also note that this circuit contains voltage sources in series with admittances, so we can use the modified rule (6.47). This leads to:

$$(Y_1 + Y_2 + Y_3 + Y_n)\hat{v}_1 = Y_1\hat{V}_{s1} + Y_2\hat{V}_{s2} + Y_3\hat{V}_{s3}, \tag{6.50}$$

which results in:

$$\hat{v}_1 = \frac{Y_1\hat{V}_{s1} + Y_2\hat{V}_{s2} + Y_3\hat{V}_{s3}}{Y_1 + Y_2 + Y_3 + Y_n}. \tag{6.51}$$

Equation (6.51) is the central formula of three-phase circuit theory. By using this formula, we can make a few general observations.

First, when the load is balanced, that is, when all the admittances are equal, then the neutral wire is not needed. To demonstrate this, we take advantage of the fact that in the case of a balanced load:

$$Y_1 = Y_2 = Y_3 = Y. \tag{6.52}$$

Then equation (6.51) is reduced to

$$\hat{v}_1 = \frac{Y(\hat{V}_{s1} + \hat{V}_{s2} + \hat{V}_{s3})}{3Y + Y_n}. \tag{6.53}$$

Although it is not immediately obvious, this equation means that when the load is balanced, $\hat{v}_1$ will be equal to zero and, consequently, it will be equal to $\hat{v}_2$ for any value of Y_n. To prove this, recall that the three voltage sources have the same peak values and are out of phase by 120°. Thus, we can write them in phasor form as follows:

$$\hat{V}_{s1} = V_m, \tag{6.54}$$

$$\hat{V}_{s2} = V_m e^{j2\pi/3}, \tag{6.55}$$

$$\hat{V}_{s3} = V_m e^{j4\pi/3}. \tag{6.56}$$

Through a little algebraic manipulation, we can show that the sum of these three expressions is zero, making $\hat{v}_1$ in (6.53) also equal to zero. Indeed,

$$V_m + V_m e^{j2\pi/3} + V_m e^{j4\pi/3} = V_m\left(1 + \cos\frac{2\pi}{3} + j\sin\frac{2\pi}{3} + \cos\frac{4\pi}{3} + j\sin\frac{4\pi}{3}\right), \tag{6.57}$$

$$\hat{V}_{s_1} + \hat{V}_{s_2} + \hat{V}_{s_3} = V_m\left(1 - .5 + \frac{\sqrt{3}}{2}j - .5 - \frac{\sqrt{3}}{2}j\right) = 0. \tag{6.58}$$

Therefore, the value of Y_n does not matter when the load is balanced, and the neutral wire can be omitted.

Next, consider the case of an unbalanced load (i.e., when (6.52) is not valid). In this case, if $|Z_n|$ is very small, $|Y_n|$ is very large, and, in the limit of $|Z_n|$ approaching zero, from (6.51) we find:

$$\hat{v}_1 = 0. \tag{6.59}$$

Thus, in the case of an unbalanced load but an ideally conducting neutral wire we still have

$$\hat{v}_1 = \hat{v}_2, \tag{6.60}$$

which should be expected from the configuration of the three-phase circuit. In this ideal case, the voltages across load impedances Z_k are equal to $\hat{V}_{sk}$, respectively, and they do not depend on load impedances.

If Y_n is finite, then from (6.51) we can find $\hat{v}_1$ and voltages across loads Z_k:

$$\hat{V}_k = \hat{V}_{sk} - \hat{v}_1. \tag{6.61}$$

Expressions (6.51) and (6.61) allow one to estimate variations of load voltages due to an unbalanced load.

Why do we need three-phase power? One reason is that, in order to generate electricity or to convert it into mechanical power, synchronous generators and induction motors are utilized. The principle of operation of these electric machines is based on *uniformly rotating* magnetic fields. It is easy to create these rotating magnetic fields by using three-phase circuits. Another reason is that the total power consumed by the loads in a balanced three-phase system is constant in time. This can be seen by summing the expressions for instantaneous power (3.184) for each of the three loads and performing a few simple algebraic manipulations. This property is desirable for power plants, which prefer to generate energy at a fairly constant rate. ■

EXAMPLE 6.4 As an example of nodal analysis of general circuits, let us find the current through the 1 Ω resistor in the circuit shown in Figure 6.7a.

We see that the 4 V voltage source has a series resistor and consequently, it can be converted into a nonideal current source. The 3 Ω resistor that is connected in series with the 3 A current source can be ignored for the purpose of calculating node potentials. The 3 V voltage source cannot be converted, so we assume that some unknown current I_x is flowing through it. We consider the node at the bottom of the circuit to be the reference node. The modified circuit is shown in Figure 6.7b and the corresponding matrix equation is given below:

$$\begin{bmatrix} 5/6 & -1/2 & 0 \\ -1/2 & 5/6 & -1/3 \\ 0 & -1/3 & 4/3 \end{bmatrix} \begin{bmatrix} v_1 \\ v_2 \\ v_3 \end{bmatrix} = \begin{bmatrix} 2 - I_x \\ 1 \\ I_x \end{bmatrix}. \tag{6.62}$$

Because the 3 V voltage source is connected between node 1 and node 3, $v_1 = 3 + v_3$. Now, we can reduce the number of equations to two by adding the first and third rows and by inserting the above expression for v_1 into the unknown node potential vector

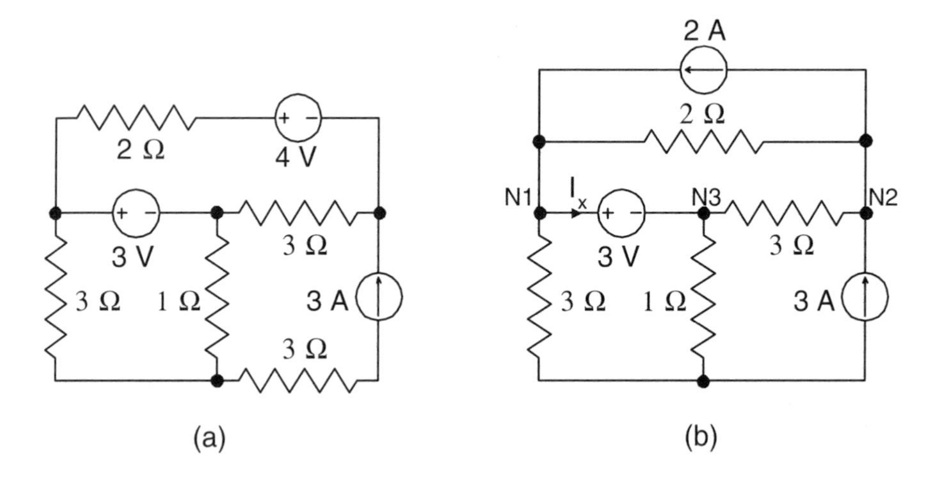

Figure 6.7: A general circuit for nodal analysis: (a) the original circuit and (b) an equivalent circuit.

and shifting the constants to the source column:

$$\begin{bmatrix} 5/6 & -5/6 \\ -5/6 & 13/6 \end{bmatrix} \begin{bmatrix} v_2 \\ v_3 \end{bmatrix} = \begin{bmatrix} 5/2 \\ -1/2 \end{bmatrix}. \tag{6.63}$$

By using Gaussian elimination, we can find the desired current to be $i_{1\Omega} = 1.5$ A. ■

6.3 Mesh Current Analysis

Mesh current analysis is a technique which is dual in form to the method of node potentials. *The basic idea of the mesh current technique is to represent KVL equations in terms of mesh currents.*

To arrive at the method of mesh currents, we consider the circuit shown in Figure 6.8.

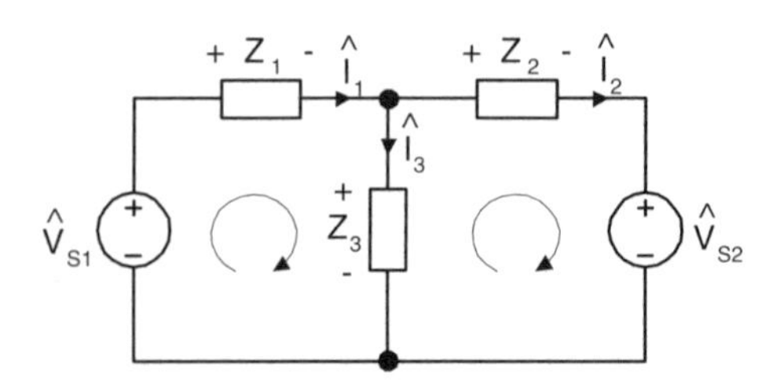

Figure 6.8: Example circuit for mesh current analysis.

To simplify our discussion, we have deliberately constructed a circuit that contains only voltage sources in series with impedances. Later, we will extend the mesh current technique to include other source configurations as well. We begin with KCL. There are two non-trivial nodes, so there will be only one KCL equation:

$$\hat{I}_1 - \hat{I}_2 - \hat{I}_3 = 0, \tag{6.64}$$

or, equivalently,

$$\hat{I}_3 = \hat{I}_1 - \hat{I}_2. \tag{6.65}$$

Now, we write KVL for the two meshes in the circuit:

$$\hat{I}_1 Z_1 + \hat{I}_3 Z_3 - \hat{V}_{s1} = 0, \tag{6.66}$$

$$-\hat{I}_3 Z_3 + \hat{I}_2 Z_2 + \hat{V}_{s2} = 0. \tag{6.67}$$

If we substitute (6.65) into the KVL equations, we will end up with two equations in terms of only two unknowns: $\hat{I}_1$ and $\hat{I}_2$.

$$\hat{I}_1 Z_1 + (\hat{I}_1 - \hat{I}_2) Z_3 = \hat{V}_{s1}, \tag{6.68}$$

$$-(\hat{I}_1 - \hat{I}_2) Z_3 + \hat{I}_2 Z_2 = -\hat{V}_{s2}. \tag{6.69}$$

Rearranging the terms, we obtain:

$$\hat{I}_1 (Z_1 + Z_3) - \hat{I}_2 Z_3 = \hat{V}_{s1}, \tag{6.70}$$

$$-\hat{I}_1 Z_3 + \hat{I}_2 (Z_2 + Z_3) = -\hat{V}_{s2}. \tag{6.71}$$

These two equations are the same equations that will be produced below by using mesh current analysis. However, the mesh current analysis provides a much easier way to do this by just looking at the circuit.

Mesh current analysis uses new mathematical quantities called *mesh currents*. These are fictitious, virtual circulating currents which are introduced to simplify calculations. They have no physical meaning but are very helpful in writing equations for circuits which contain many meshes. We introduce one mesh current for each mesh of the circuit, shown in Figure 6.9. Note that reference directions for all mesh currents are coordinated. We shall also choose a tracing direction for each loop

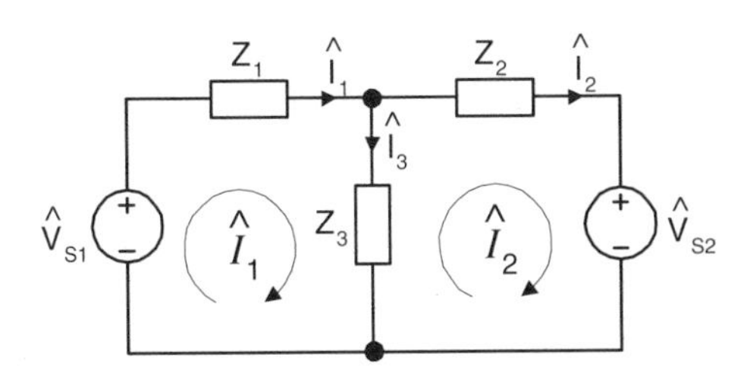

Figure 6.9: Circuit with mesh currents.

(mesh) coinciding with the reference direction of the mesh current through this loop. *The following rules are introduced* in order to write KVL equations in terms of mesh currents. When mesh current $\hat{\mathcal{I}}_k$ flows through impedance Z_k, this results in the voltage drop $\hat{\mathcal{I}}_k Z_k$. This voltage drop is taken with positive sign if the reference direction of the mesh current coincides with the tracing direction of a loop, otherwise it is taken with negative sign. For instance, if we consider the first loop in Figure 6.9, then the voltage drop in this loop due to mesh current $\hat{\mathcal{I}}_1$ is $\hat{\mathcal{I}}_1(Z_1 + Z_3)$, while the voltage drop due to mesh current $\hat{\mathcal{I}}_2$ is $-\hat{\mathcal{I}}_2 Z_3$. By using the above rules, we obtain the following KVL equations:

$$\hat{\mathcal{I}}_1(Z_1 + Z_3) - \hat{\mathcal{I}}_2 Z_3 = \hat{V}_{s1}, \tag{6.72}$$

$$-\hat{\mathcal{I}}_1 Z_3 + \hat{\mathcal{I}}_2(Z_3 + Z_2) = -\hat{V}_{s2}. \tag{6.73}$$

It is clear that equations (6.72) and (6.73) are mathematically identical to equations (6.70) and (6.71). Thus, we can conclude that:

$$\hat{I}_1 = \hat{\mathcal{I}}_1, \tag{6.74}$$

$$\hat{I}_2 = \hat{\mathcal{I}}_2, \tag{6.75}$$

$$\hat{I}_3 = \hat{\mathcal{I}}_1 - \hat{\mathcal{I}}_2. \tag{6.76}$$

The above three equations relate the purely mathematical mesh currents to the actual branch currents in the circuit. From this example, we can formulate two general rules of relating mesh currents to actual branch currents.

1. If a branch belongs to only one mesh, then the branch current is equal to the mesh current if their reference directions coincide, otherwise it is the negative of the mesh current.
2. If a branch is shared between two meshes, then the branch current is the algebraic sum of the two mesh currents. The word "algebraic" means that a mesh current is taken with a positive sign if its reference direction coincides with the reference direction of the branch current; otherwise the mesh current is taken with a minus sign.

Like the method of node potentials, the mesh current analysis generates a system of linear equations which can be represented in the matrix form:

$$\tilde{Z}\vec{\mathcal{I}} = \vec{V}_s. \tag{6.77}$$

For a two-mesh circuit, the above form of the matrix equations is:

$$\begin{bmatrix} Z_{11} & Z_{12} \\ Z_{21} & Z_{22} \end{bmatrix} \begin{bmatrix} \hat{\mathcal{I}}_1 \\ \hat{\mathcal{I}}_2 \end{bmatrix} = \begin{bmatrix} \hat{V}_s^{(1)} \\ \hat{V}_s^{(2)} \end{bmatrix}. \tag{6.78}$$

In our example, these equations can be specified as follows:

$$\begin{bmatrix} Z_1 + Z_3 & -Z_3 \\ -Z_3 & Z_3 + Z_2 \end{bmatrix} \begin{bmatrix} \hat{\mathcal{I}}_1 \\ \hat{\mathcal{I}}_2 \end{bmatrix} = \begin{bmatrix} \hat{V}_{s1} \\ -\hat{V}_{s2} \end{bmatrix}. \tag{6.79}$$

Comparing the last matrix equation with equation (6.78) we find that:

$$Z_{11} = Z_1 + Z_3, \tag{6.80}$$

$$Z_{22} = Z_3 + Z_2, \tag{6.81}$$

$$Z_{12} = Z_{21} = -Z_3, \tag{6.82}$$

$$\hat{V}_s^{(1)} = \hat{V}_{s1}, \tag{6.83}$$

$$\hat{V}_s^{(2)} = -\hat{V}_{s2}. \tag{6.84}$$

As in nodal analysis, a specific *pattern* emerges when the mesh current equations are written in matrix form. Indeed, Z_{11} is equal to the sum of all impedances in the first mesh, while Z_{22} is equal to the sum of all impedances in the second mesh. Impedances Z_{12} and Z_{21} are the same and equal to the negative of the sum of impedances shared between the two meshes. Voltages $\hat{V}_s^{(1)}$ and $\hat{V}_s^{(2)}$ are equal to the algebraic sums of voltage sources around meshes 1 and 2, respectively.

Of course, these patterns can be generalized for any planar circuit with any number of meshes. As with nodal analysis, there are three simple rules to follow.

For a general circuit with m meshes, the matrix equation has the form:

$$\begin{bmatrix} Z_{11} & Z_{12} & \cdots & Z_{1m} \\ Z_{21} & Z_{22} & \cdots & Z_{2m} \\ \vdots & \vdots & \ddots & \vdots \\ Z_{m1} & Z_{m2} & \cdots & Z_{mm} \end{bmatrix} \begin{bmatrix} \hat{I}_1 \\ \hat{I}_2 \\ \vdots \\ \hat{I}_m \end{bmatrix} = \begin{bmatrix} \hat{V}_s^{(1)} \\ \hat{V}_s^{(2)} \\ \vdots \\ \hat{V}_s^{(m)} \end{bmatrix}. \tag{6.85}$$

The three rules for writing these matrix equations are the following.

For diagonal elements of the impedance matrix we have:

$$Z_{ii} = \sum_k^{I} Z_k, \tag{6.86}$$

which means that a diagonal element Z_{ii} of the coefficient matrix is the sum of all branch impedances in mesh i.

For off-diagonal matrix elements we have:

$$Z_{ij} = Z_{ji} = -\sum_k^{II} Z_k, \tag{6.87}$$

which means that an element on the intersection of the ith row and jth column of the coefficient matrix is the negative of the sum of all branch impedances common to meshes i and j.

For the source vector elements we have:

$$\hat{V}_s^{(i)} = \sum_k^{I} \hat{V}_{sk}, \tag{6.88}$$

which means that the ith element of the source vector is the algebraic sum of all voltage sources in the ith mesh. A voltage source is taken with a minus sign if the mesh current reference direction is pointing out of ("exiting") the negative terminal

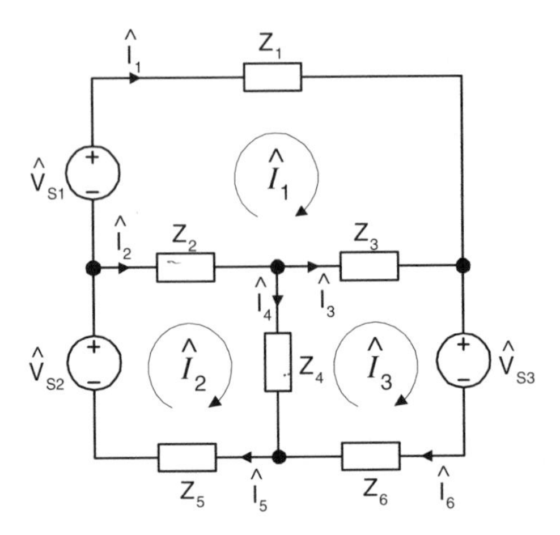

Figure 6.10: Example circuit for mesh current analysis.

of the voltage source, and it is taken with a plus sign if the reference direction of the mesh current is exiting the positive terminal of the voltage source.

The above three rules should be complemented by the previously discussed two rules which relate virtual mesh currents to the actual branch currents. We shall illustrate the previous discussion by the following example.

EXAMPLE 6.5 Consider the circuit shown in Figure 6.10. Our goal is to write the matrix mesh current equations and find all branch currents.

The circuit is given in phasor-impedance form as it is appropriate for mesh analysis. It is clear from the figure that the circuit is planar and contains three meshes. Therefore, we will need to set up a 3 × 3 impedance matrix. The resulting matrix equations are:

$$\begin{bmatrix} Z_1 + Z_2 + Z_3 & -Z_2 & -Z_3 \\ -Z_2 & Z_2 + Z_4 + Z_5 & -Z_4 \\ -Z_3 & -Z_4 & Z_3 + Z_4 + Z_6 \end{bmatrix} \begin{bmatrix} \hat{\mathcal{I}}_1 \\ \hat{\mathcal{I}}_2 \\ \hat{\mathcal{I}}_3 \end{bmatrix} = \begin{bmatrix} \hat{V}_{s1} \\ \hat{V}_{s2} \\ -\hat{V}_{s3} \end{bmatrix}. \tag{6.89}$$

These equations can now be solved by using the Gaussian elimination technique to find the mesh currents. Using the mesh currents and the two rules for relating mesh currents to actual branch currents, we can calculate the branch currents in the circuit:

$$\hat{I}_1 = \hat{\mathcal{I}}_1, \tag{6.90}$$

$$\hat{I}_2 = \hat{\mathcal{I}}_2 - \hat{\mathcal{I}}_1, \tag{6.91}$$

$$\hat{I}_3 = \hat{\mathcal{I}}_3 - \hat{\mathcal{I}}_1, \tag{6.92}$$

$$\hat{I}_4 = \hat{\mathcal{I}}_2 - \hat{\mathcal{I}}_3, \tag{6.93}$$

$$\hat{I}_5 = \hat{\mathcal{I}}_2, \tag{6.94}$$

$$\hat{I}_6 = \hat{\mathcal{I}}_3. \tag{6.95}$$

This concludes the solution of the problem. ■

In the above discussion of mesh current analysis we dealt only with circuits that contained voltage sources in series with impedances. We now consider three more possibilities which may arise: a current source in parallel with an impedance, a current source in series with an impedance, and a voltage source in parallel with an impedance.

The case of a current source in parallel with an impedance is relatively easy to remedy. This situation is illustrated on the left in Figure 6.11. Note that this connection can be simply construed as a nonideal current source, which can be equivalently transformed to a nonideal voltage source. Now, the source is in a form which we already know how to handle and we also end up with a circuit which has fewer meshes.

The second case is a bit more complicated to handle. If a branch contains a current source, then we already know the value of the branch current—it is the source current itself. Therefore, this current source either defines one of the mesh currents (if the branch is not shared by two meshes), or it defines the difference between two mesh currents (if the branch is common to two meshes). In any case, the current source will reduce the number of unknowns. As an example, consider the circuit shown in Figure 6.12, which is the same as the circuit in Figure 6.10 except for a current source connected in series with impedance Z_4. We know $\hat{I}_4$ is equal to the current source, and it is also equal to the difference between $\hat{\mathcal{I}}_2$ and $\hat{\mathcal{I}}_3$. So we have:

$$\hat{I}_4 = \hat{I}_s = \hat{\mathcal{I}}_2 - \hat{\mathcal{I}}_3, \tag{6.96}$$

$$\hat{\mathcal{I}}_3 = \hat{\mathcal{I}}_2 - \hat{I}_s. \tag{6.97}$$

Clearly, $\hat{\mathcal{I}}_2$ and $\hat{\mathcal{I}}_3$ are no longer independent so the actual number of unknowns is reduced by one. But what seems to present a problem is the unknown voltage $\hat{V}_x$ across the current source. However, this voltage can be eliminated when composing the mesh current equations. Consider the mesh equations set up with this voltage

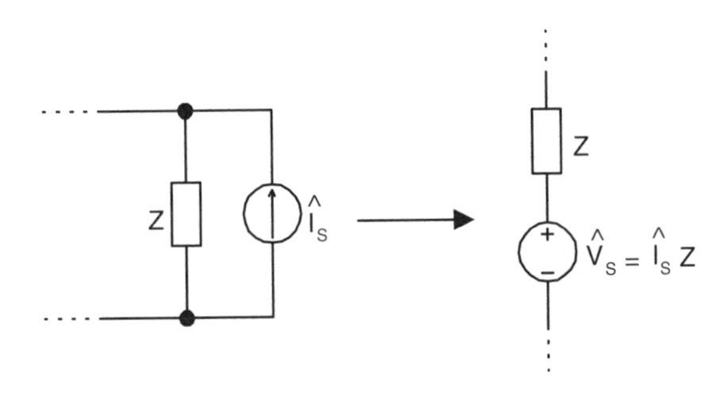

Figure 6.11: Current source in parallel with impedance.

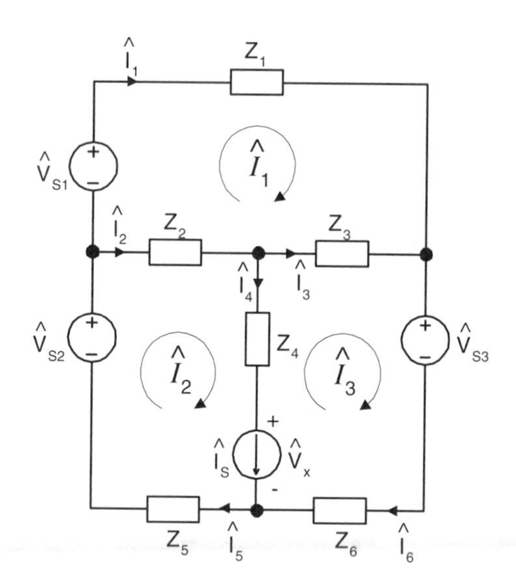

Figure 6.12: Current source added in series.

included as a voltage source. It then appears in the source vector:

$$\cdots = \begin{bmatrix} \hat{V}_{s1} \\ \hat{V}_{s2} - \hat{V}_x \\ -\hat{V}_{s3} + \hat{V}_x \end{bmatrix}. \tag{6.98}$$

(The coefficient matrix is no different than before; that is why it is omitted here.) We can easily eliminate this unknown voltage by summing up the bottom two equations. This results in two equations with no unknown voltage sources present. These equations are:

$$\begin{bmatrix} Z_1 + Z_2 + Z_3 & -Z_2 & -Z_3 \\ -(Z_2 + Z_3) & Z_2 + Z_5 & Z_3 + Z_6 \end{bmatrix} \begin{bmatrix} \hat{\mathcal{I}}_1 \\ \hat{\mathcal{I}}_2 \\ \hat{\mathcal{I}}_2 - \hat{I}_s \end{bmatrix} = \begin{bmatrix} \hat{V}_{s1} \\ \hat{V}_{s2} - \hat{V}_{s3} \end{bmatrix}. \tag{6.99}$$

By using simple transformations, equation (6.99) can be reduced to the following form:

$$\begin{bmatrix} Z_1 + Z_2 + Z_3 & -Z_2 - Z_3 \\ -Z_2 - Z_3 & Z_2 + Z_3 + Z_5 + Z_6 \end{bmatrix} \begin{bmatrix} \hat{\mathcal{I}}_1 \\ \hat{\mathcal{I}}_2 \end{bmatrix} = \begin{bmatrix} \hat{V}_{s1} - \hat{I}_s Z_3 \\ \hat{V}_{s2} - \hat{V}_{s3} + \hat{I}_s(Z_3 + Z_6) \end{bmatrix}. \tag{6.100}$$

These equations can now be solved by employing the conventional linear algebra techniques. It is worthwhile to note that the mesh current analysis of circuits with current sources is dual to the nodal analysis of circuits with voltage sources.

The final case of voltage sources in parallel with impedances is the simplest of the three. This is because the values for the currents through these impedances are immediate and these impedances do not affect any of the other mesh currents. This means, for the purpose of finding mesh currents, that we can treat parallel impedances

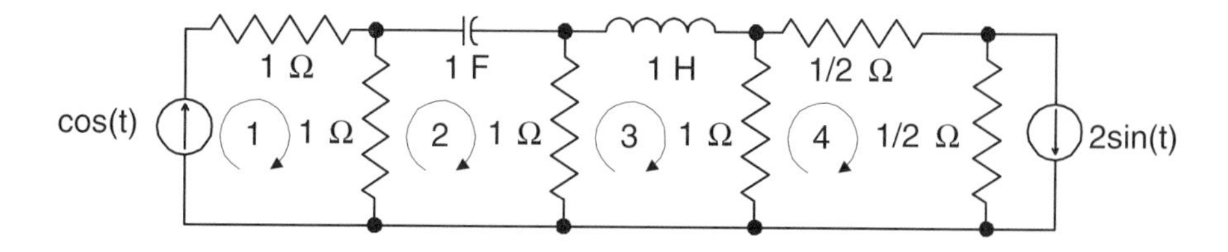

Figure 6.13: A circuit for mesh analysis.

as though they weren't there (i.e., replace them with an open branch). Still, these impedances do affect the circuit—they modify how much power is absorbed or supplied by the voltage sources.

EXAMPLE 6.6 Consider the circuit shown in Figure 6.13 with all source currents given in amps. Find all the phasor mesh currents in the circuit.

We start by converting the nonideal current source to the right of the fourth mesh into a nonideal voltage source and then combine the two $1/2\ \Omega$ resistors, which become connected in series. We then convert the sources to phasor form ($\omega = 1$), and find the impedances of all branches. As a result, we arrive at the circuit shown in Figure 6.14. By using the standard mesh analysis procedure, we obtain the matrix equation:

$$\begin{bmatrix} 2 & -1 & 0 & 0 \\ -1 & 2-j & -1 & 0 \\ 0 & -1 & 2+j & -1 \\ 0 & 0 & -1 & 2 \end{bmatrix} \begin{bmatrix} 1 \\ \hat{\mathcal{I}}_2 \\ \hat{\mathcal{I}}_3 \\ \hat{\mathcal{I}}_4 \end{bmatrix} = \begin{bmatrix} -\hat{V}_x \\ 0 \\ 0 \\ -j \end{bmatrix}. \tag{6.101}$$

Because the current source is contained only in the first mesh, $\hat{\mathcal{I}}_1$ is specified and we reduce this problem to three equations by simply ignoring the equation represented by the first row and by shifting the first column of the matrix (multiplied by 1 A) to

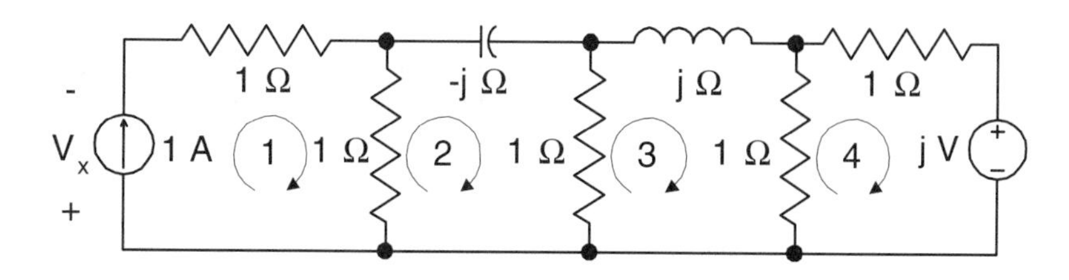

Figure 6.14: An equivalent circuit for the mesh analysis problem.

the source column:

$$\begin{bmatrix} 2-j & -1 & 0 \\ -1 & 2+j & -1 \\ 0 & -1 & 2 \end{bmatrix} \begin{bmatrix} \hat{\mathcal{I}}_2 \\ \hat{\mathcal{I}}_3 \\ \hat{\mathcal{I}}_4 \end{bmatrix} = \begin{bmatrix} 1 \\ 0 \\ -j \end{bmatrix}. \tag{6.102}$$

We use Gaussian elimination to solve for the mesh currents. The upper triangular matrix is:

$$\begin{bmatrix} 2-j & -1 & 0 \\ 0 & 4(2+j)/5 & -1 \\ 0 & 0 & (6+j)/4 \end{bmatrix} \begin{bmatrix} \hat{\mathcal{I}}_2 \\ \hat{\mathcal{I}}_3 \\ \hat{\mathcal{I}}_4 \end{bmatrix} = \begin{bmatrix} 1 \\ (2+j)/5 \\ 1/4-j \end{bmatrix}. \tag{6.103}$$

We complete the problem with back-substitution to find $\hat{\mathcal{I}}_4 = (2-25j)/37$, $\hat{\mathcal{I}}_3 = (4-13j)/37$, and $\hat{\mathcal{I}}_2 = (19+3j)/37$. This concludes the analysis of the problem. ■

EXAMPLE 6.7 Consider a circuit similar to the one shown in Figure 6.13 except that the current source is common to meshes 1 and 2 as shown in Figure 6.15. In this case the standard mesh analysis procedure produces:

$$\begin{bmatrix} 2 & -1 & 0 & 0 \\ -1 & 2-j & -1 & 0 \\ 0 & -1 & 2+j & -1 \\ 0 & 0 & -1 & 2 \end{bmatrix} \begin{bmatrix} \hat{\mathcal{I}}_1 \\ \hat{\mathcal{I}}_2 \\ \hat{\mathcal{I}}_3 \\ \hat{\mathcal{I}}_4 \end{bmatrix} = \begin{bmatrix} -V_x \\ +V_x \\ 0 \\ -j \end{bmatrix}. \tag{6.104}$$

Now we must add rows 1 and 2 to eliminate the unknown voltage:

$$\begin{pmatrix} 1 & 1-j & -1 & 0 \\ 0 & -1 & 2+j & -1 \\ 0 & 0 & -1 & 2 \end{pmatrix} \begin{bmatrix} \hat{\mathcal{I}}_1 \\ \hat{\mathcal{I}}_2 \\ \hat{\mathcal{I}}_3 \\ \hat{\mathcal{I}}_4 \end{bmatrix} = \begin{pmatrix} 0 \\ 0 \\ -j \end{pmatrix}. \tag{6.105}$$

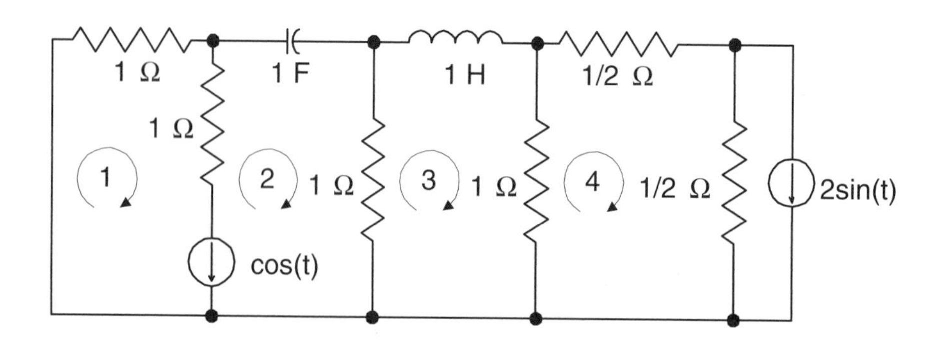

Figure 6.15: Another example circuit for mesh analysis.

We use the fact that $\hat{\mathcal{I}}_1 - \hat{\mathcal{I}}_2 = 1A$ to eliminate $\hat{\mathcal{I}}_1$:

$$\begin{pmatrix} 2-j & -1 & 0 \\ -1 & 2+j & -1 \\ 0 & -1 & 2 \end{pmatrix} \begin{pmatrix} \hat{\mathcal{I}}_2 \\ \hat{\mathcal{I}}_3 \\ \hat{\mathcal{I}}_4 \end{pmatrix} = \begin{pmatrix} -1 \\ 0 \\ -j \end{pmatrix}. \tag{6.106}$$

The application of Gaussian elimination to determine the mesh currents is left as an exercise for the reader. ■

6.4 MicroSim PSpice Simulations

In this section, we will use the PSpice circuit simulation code to explore the performance of three of the example circuits from this chapter. If you did not work through the examples discussed at the end of Chapter 4 with the PSpice code, we strongly urge you to go back and do that now, so that you can maximize your understanding of this simulation tool. If necessary, Appendix C lists several references which contain considerably more information about the code.

The first circuit that we consider is the three-phase network of Example 6.3. The circuit, as redrawn by the PSpice schematic generator, is shown in Figure 6.16. The three voltage sources represent household power, have the part name "VSIN," and are given magnitudes (VAMPL) of 169.7 V, offset voltages (VOFF) of zero, and frequencies (FREQ) of 60 Hz. The phases (PHASE) of v1, v2, and v3 are 0°, 120°, and 240°, respectively. The neutral wire is given a resistance of 10 Ω and the nominal resistance of the three loads is 1 kΩ. The ground (AGND) is connected to the second node as in the example.

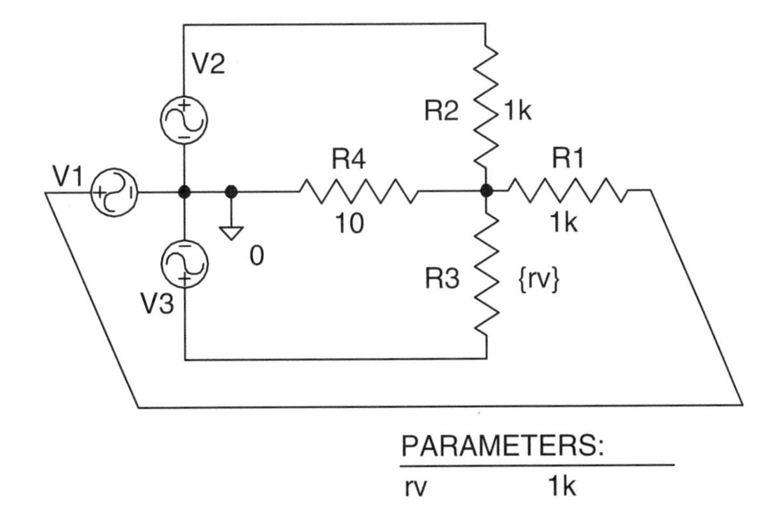

Figure 6.16: The schematic drawing of the three-phase network.

We will use PSpice to investigate the current through the neutral wire (R4) and the instantaneous power dissipation when the load is not balanced. A transient analysis is selected with a final time of 50 ms (3 cycles) and a print step of 0.2 ms. A parametric analysis will be used to vary the resistance of the third leg. The resistor R3 is assigned a value of "rv" and rv is defined to be a "Global Parameter" in the "Parametric..." section of the "Analysis Setup..." menu. "Value List" is chosen as the "Sweep Type" and we give rv the values of 790, 990, 1190, and 1390. We must also assign a value to R3 for the dc bias point analysis and this is done by typing "PARAM" into the "Add Part" dialog box. We place the resulting icon anywhere on the schematic in the usual way and double-click on the word "Parameter" to bring up the "PM1 Part Name: PARAM" dialog box. We set Name1 equal to rv and set Value1 equal to 1k (remember to hit the "Save Attr" button for each change). After selecting the "OK" button, the initial value of 1k for rv should be listed under "Parameters." After saving the schematic, PSpice can be run by selecting "Simulate" from the "Analysis" drop-down menu.

When Probe is launched at the end of the simulation, a dialog box will appear that allows the user to select which of the parametric results to plot. All four simulations can be plotted by selecting "All" and "OK." The current in the neutral wire (I(R4)) is plotted in Figure 6.17a. The dashed line is the 790 Ω case, the solid line corresponds to the 990 Ω case, the dot-dashed line is the 1190 Ω case, and the dotted line is the 1390 Ω case. The current is almost zero for the 990 Ω case since the load is nearly balanced. The peak value of about 1.66 mA (found by expanding the y-axis) agrees with (6.51) when the circuit parameters are inserted into the equation. Note also that the phase of the current flips by 180° when the resistance of R3 switches from being less than 1 kΩ to being greater than 1 kΩ.

The net power dissipated by the three voltage sources is plotted in Figure 6.17b. Again, the dashed line is the 790 Ω case, the solid line corresponds to the 990 Ω case, the dot-dashed line is the 1190 Ω case, and the dotted line is the 1390 Ω case. The power is found by multiplying the relevant voltages and currents for each of the power supplies and adding them together. The fact that the power is negative is a consequence of the passive sign convention. The net power is nearly a constant for the 990 Ω case and the variation with time is due mainly to the power supplied by the third voltage source. This can be seen by finding the analytic expression for the total power:

$$\begin{aligned} P(t) &= -V_0^2[\sin^2(\omega t)/R_1 + \sin^2(\omega t + 120°)/R_2 + \sin^2(\omega t + 240°)/R_3)] \\ &= -(V_0^2/R_1)[3/2 + \sin^2(\omega t + 240°)(1 - R_1/R_3)] \end{aligned} \tag{6.107}$$

where the second equality follows only if R1 = R2. For the perfectly matched case (R1= R2 = R3 = 1 kΩ) and the parameters of this problem ($V_0 = 169.7$ V and $\omega = 377$ rad/s), expression (6.107) yields $P(t) = -43.2$ W, as indicated in the figure.

The second circuit comes from Example 6.4 and the PSpice schematic is reproduced in Figure 6.18. The goal of the example is simply to find the current through the 1 Ω resistor, so we need only perform a dc bias point solution. The part names for the

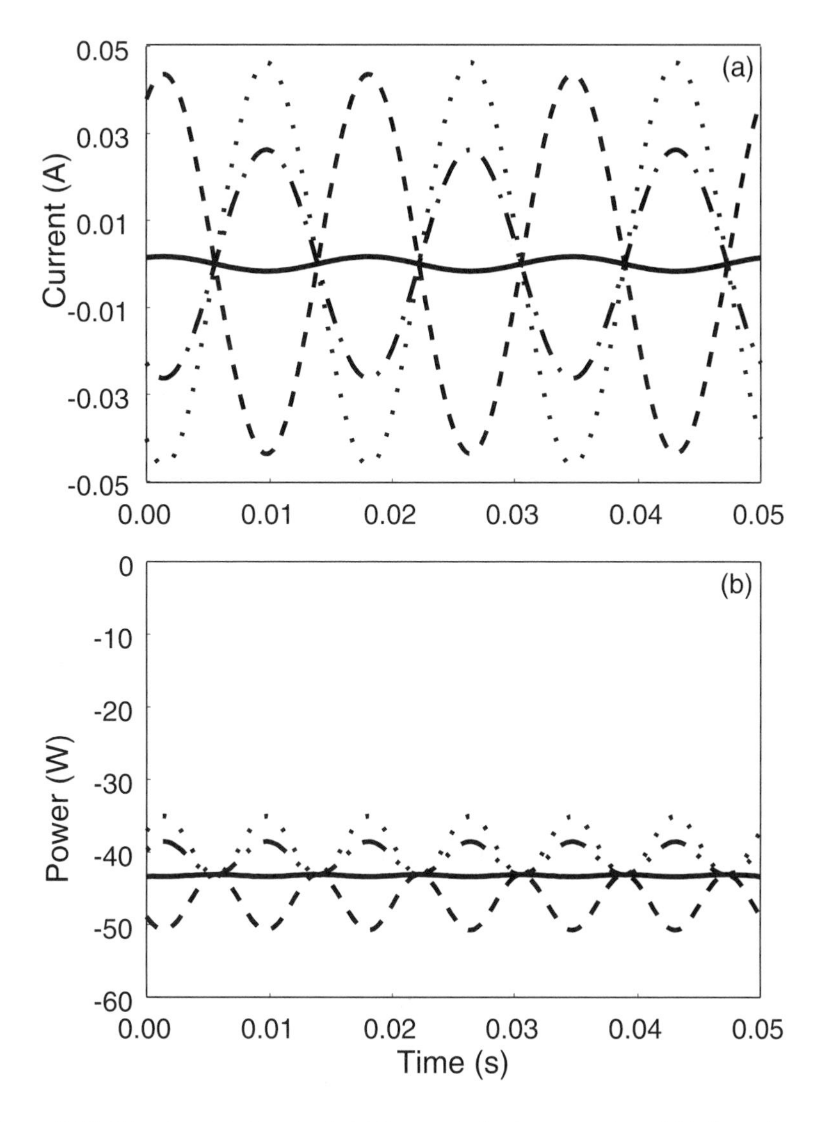

Figure 6.17: The simulated output: (a) the neutral wire current and (b) the net power supplied by three voltage sources.

voltage and current sources are "VSRC" and "ISRC," respectively. The 1 Ω resistor is R4 and the PSpice nodes \$N_0001 through \$N_0003 are chosen to correspond to nodes 1 through 3 in Figure 6.7. Selected lines from the output file are given in Table 6.1. The current through the 1 Ω resistor is found by taking the difference in node potentials across the resistor and dividing by the impedance. In this case, the resistor current in amperes is equivalent to the potential of node N\$_0002. The table reveals this potential to be 1.5 V, in agreement with the analytic result found in the example.

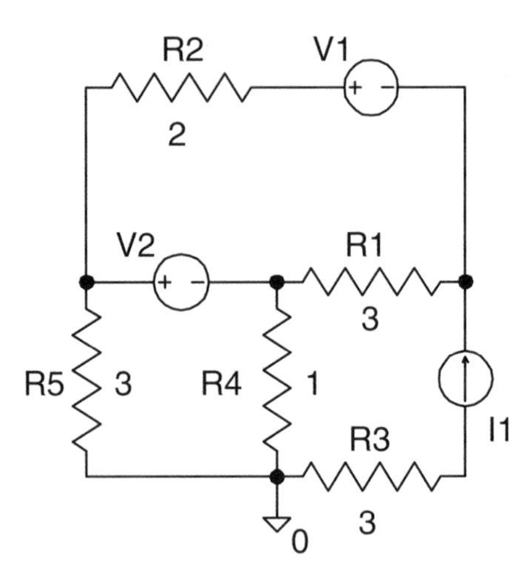

Figure 6.18: The PSpice schematic for the circuit in Example 6.4.

Table 6.1: Selected lines from the PSpice output file.

```
** Analysis setup **
.OP
R_R1 $N_0003 $N_0002 3
R_R2 $N_0001 $N_0004 2
R_R3 0 $N_0005 3
R_R4 0 $N_0003 1
R_R5 0 $N_0001 3
V_V1 $N_0004 $N_0002 DC 4 AC 0
V_V2 $N_0001 $N_0003 DC 3 AC 0
I_I1 $N_0005 $N_0002 DC 3 AC 0
.probe
.END
*** SMALL SIGNAL BIAS SOLUTION
NODE VOLTAGE
($N_0001) 4.5000
($N_0002) 4.5000
($N_0003) 1.5000
($N_0004) 8.5000
($N_0005) -9.0000
VOLTAGE SOURCE CURRENTS
NAME CURRENT
NAME CURRENT
V_V1 -2.000E+00
V_V2 5.000E-01
```

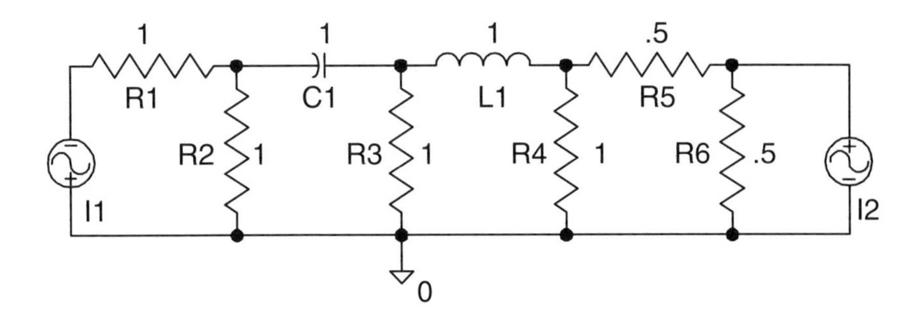

Figure 6.19: The PSpice schematic for the circuit in Example 6.6.

The remaining node potentials and the currents through the voltage sources are also given in the table. Note that the current through V2 is positive, indicating that power is being absorbed by the source.

The final PSpice simulation comes from the mesh analysis problem in Example 6.6. The PSpice schematic is redrawn in Figure 6.19. Note that PSpice uses a non-standard symbol to represent the ac current source (ISIN). The default names I1 and I2 indicate that the parts are current sources. Any doubt can be removed by double-clicking on the part to bring up the "ISIN" dialog box. We will use PSpice to find the current through the inductor and compare that result to the analytic steady-state solution found in the example:

$$i_{L_1}(t) = 0.3676 \cos(t - 72.897°). \tag{6.108}$$

A transient analysis is performed with a final time of 20 s and a print step of 100 ms. The analytic and simulated results are plotted in Figure 6.20 on page 200. There is a discrepancy in the curves at early times, but the results become nearly identical after the first cycle. The difference between the two curves involves the transient solution of the problem and is the subject of investigation in the next chapter.

6.5 Summary

The nodal and mesh current techniques allow one to write *by inspection* matrix equations which can be used to quickly analyze complicated ac steady-state problems. These techniques have been derived from the KCL and KVL equations by expressing them respectively in terms of node potentials and circulating mesh currents. These techniques have been presented for various source configurations, such as current or voltage sources connected in series or parallel with impedances. However, the solution method varies only slightly for each type of source configuration.

In the case of nodal analysis, we have first considered the standard $n + 1$ node circuit with admittances and current sources. The matrix equation for this circuit is

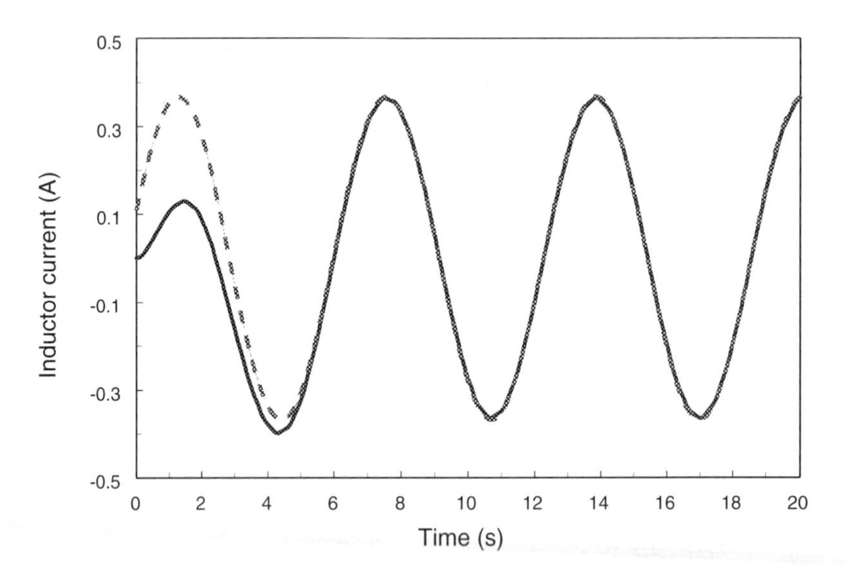

Figure 6.20: The analytic (dashed line) and the simulated (solid line) results for the current through the inductor in the circuit in Figure 6.19.

$\tilde{Y} \cdot \vec{v} = \vec{I}_s$, where $\tilde{Y}$ is the $n \times n$ admittance matrix, $\vec{v}$ is the column vector of node potentials, and $\vec{I}_s$ is the column vector of sources. The diagonal elements Y_{ii} of the admittance matrix are equal to the sum of all the admittances connected to the ith node. The off-diagonal elements $Y_{ij} = Y_{ji}$ are the negative sum of all the admittances which are connected to both the ith and jth nodes. The source vector component $I_s^{(j)}$ is equal to the algebraic sum of current sources connected to the jth node (the source currents flowing into the node are taken with positive signs).

In the case of mesh current analysis, we have first considered the standard planar circuit with m meshes which has only impedances and voltage sources. The matrix equation for this circuit is $\tilde{Z} \cdot \vec{\mathcal{I}} = \vec{V}_s$, where $\tilde{Z}$ is the $m \times m$ impedance matrix, $\vec{\mathcal{I}}$ is the column vector of fictitious mesh currents, and $\vec{V}_s$ is the column vector of sources. The diagonal elements Z_{ii} of the impedance matrix are equal to the sum of all the impedances in the ith mesh. The off-diagonal elements $Z_{ij} = Z_{ji}$ are equal to the negative sum of the impedances which are common to both the ith and jth meshes. The source vector component $V_s^{(j)}$ is found by taking the algebraic sum of voltage sources in the jth mesh (positive means the mesh current is flowing out of the positive terminal of the source).

The modifications of the described standard forms of nodal and mesh current equations for various source configurations have been discussed in the chapter and illustrated by specific examples.

6.6 Problems

1. Find all the node potentials in the circuit shown in Figure P6-1.

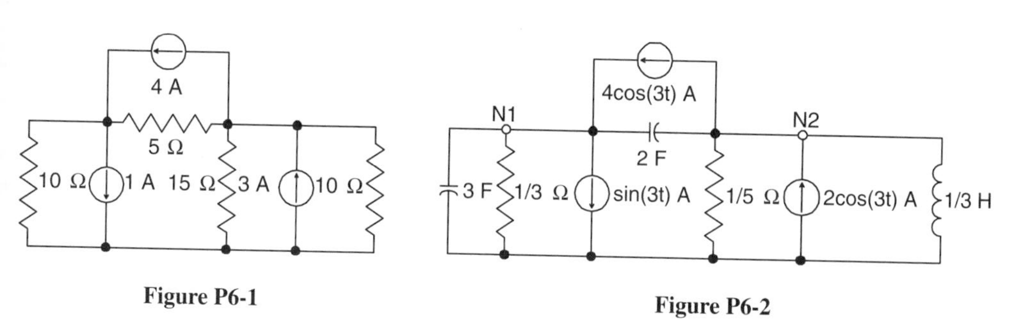

Figure P6-1 **Figure P6-2**

2. Find the time dependence of the second circuit node potential shown in Figure P6-2.
3. Set up, but do not solve, the nodal matrix equation for the circuit shown in Figure P6-3.
4. Set up, but do not solve, the nodal matrix equation for the circuit shown in Figure P6-4.

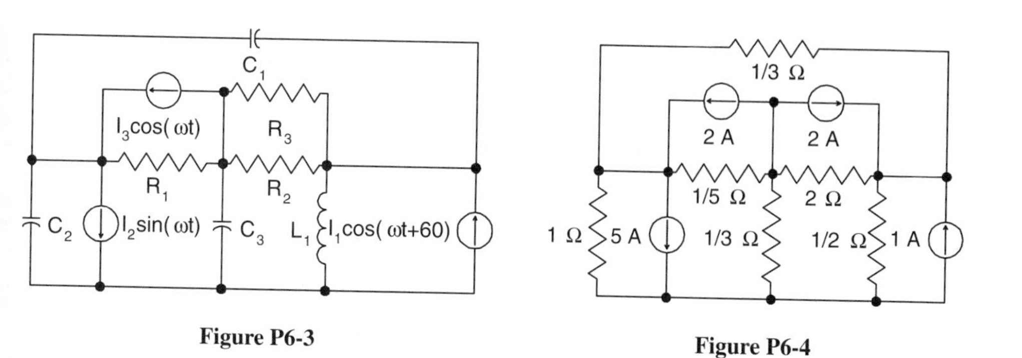

Figure P6-3 **Figure P6-4**

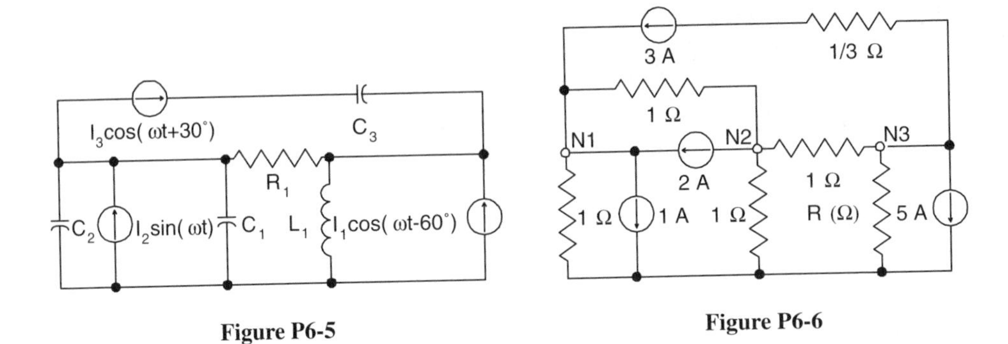

Figure P6-5

Figure P6-6

5. Find the current through C_1 in the circuit shown in Figure P6-5.
6. For the circuit shown in Figure P6-6, find the value of the resistance R that results in a third node potential of -2 V.
7. Find the phasor of the potential of the third node for the circuit shown in Figure P6-7.
8. Find all the node potentials for the circuit shown in Figure P6-8.

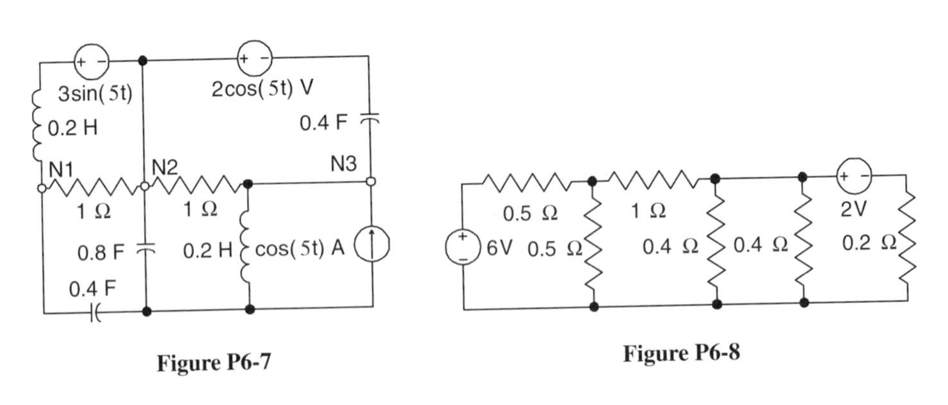

Figure P6-7

Figure P6-8

9. For the circuit shown in Figure P6-9, find the value of the resistance R that results in a third node potential of 5 V.
10. Find the current through the 10 Ω resistor in the circuit shown in Figure P6-10.

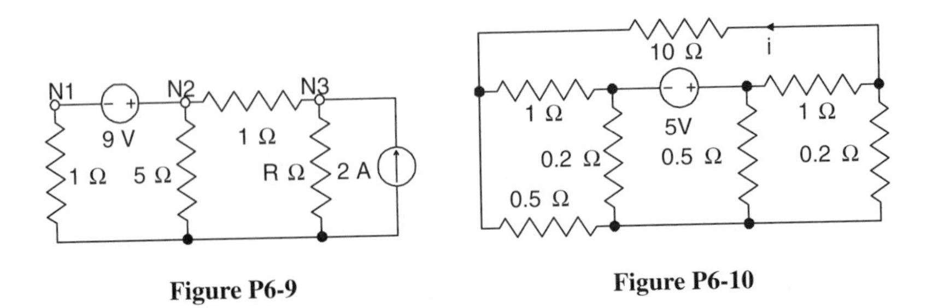

Figure P6-9

Figure P6-10

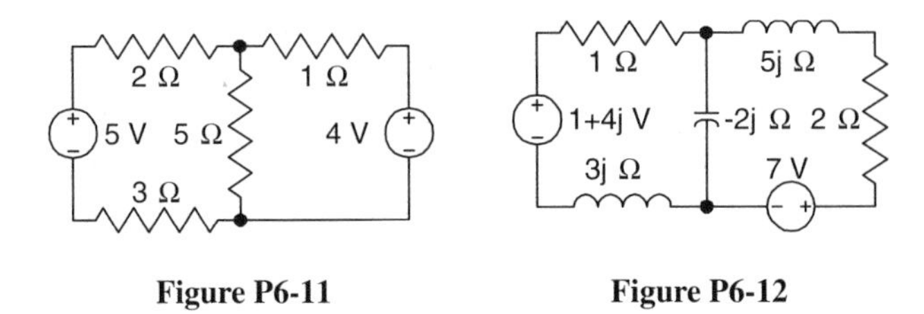

Figure P6-11 **Figure P6-12**

11. Find the power dissipated by the 5 Ω resister in the circuit shown in Figure P6-11.

12. Find the average power dissipated by the 1 Ω resister in the circuit shown in Figure P6-12.

13. Set up, but do not solve, the mesh analysis equations for the circuit shown in Figure P6-13.

14. Set up, but do not solve, the mesh analysis equations for the circuit shown in Figure P6-14.

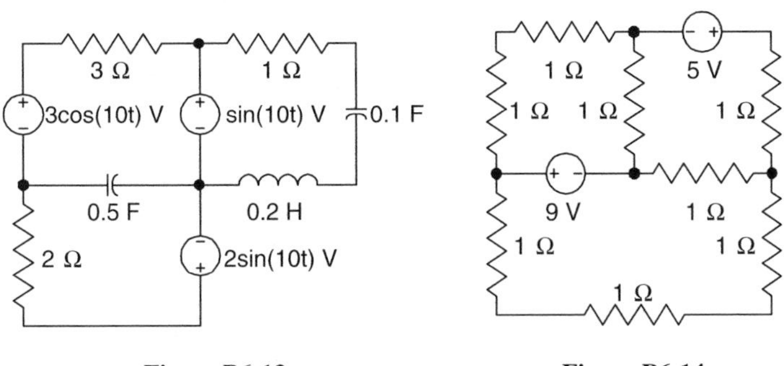

Figure P6-13 **Figure P6-14**

15. Find the current $i(t)$ through the 2 F capacitor in the circuit shown in Figure P6-15 with all source currents given in amps.

16. For the circuit shown in Figure P6-16, find the value of the resistance R that results in second mesh current of -2 A.

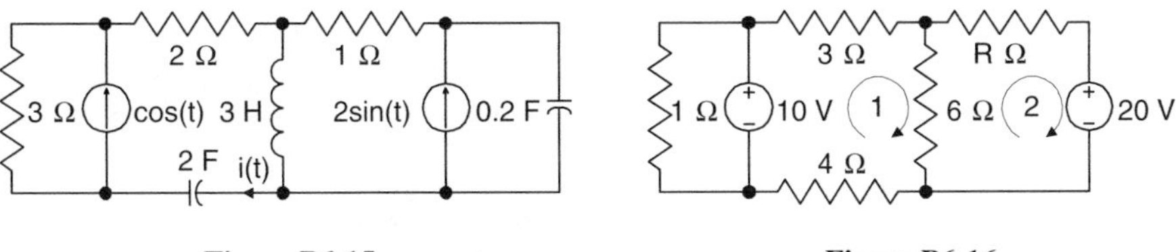

Figure P6-15 **Figure P6-16**

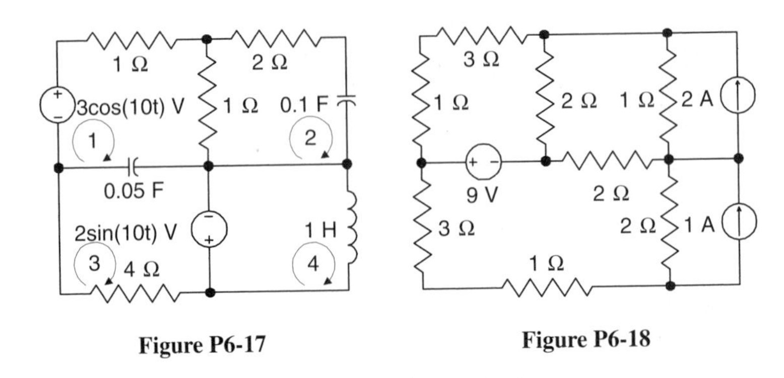

Figure P6-17 **Figure P6-18**

17. Find the phasor current of the third mesh for the circuit shown in Figure P6-17.
18. Find all the mesh currents for the circuit shown in Figure P6-18.
19. Find the phasor current through the 2 Ω resistor in the circuit shown in Figure P6-19.
20. For the circuit shown in Figure P6-20, find the value of the resistance R that results in a first mesh current of 1 A.

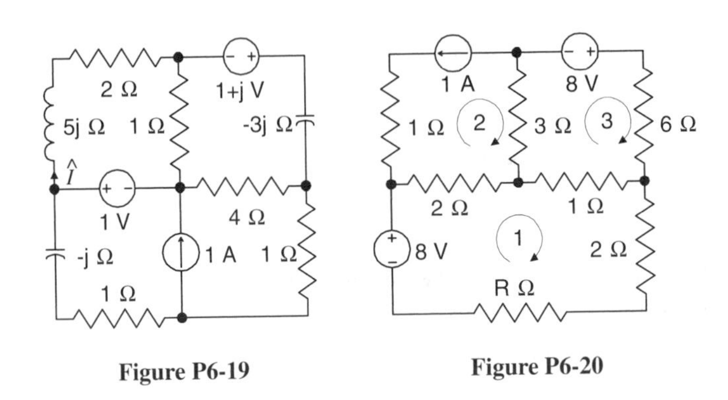

Figure P6-19 **Figure P6-20**

21. Find the current flowing through the neutral wire of the three-phase circuit shown in Figure P6-21.
22. Find the total instantaneous power dissipated in all three resistors of the three-phase circuit indicated in Figure P6-22.

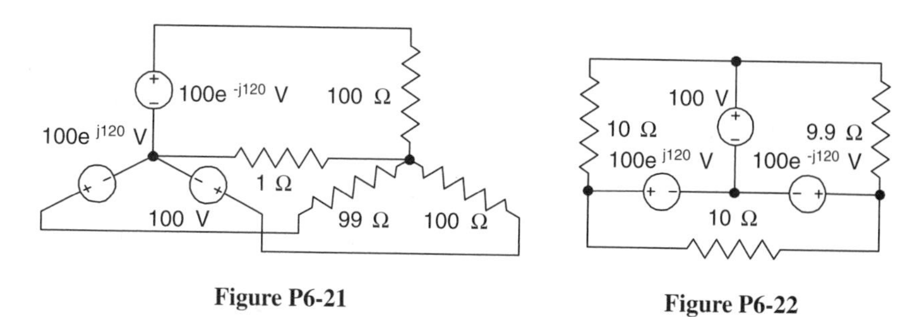

Figure P6-21

Figure P6-22

23. Consider the circuit shown in Figure P6-23. All voltages are in volts; all currents are in amps. Draw the phasor equivalent circuit so that it is ready for the general mesh analysis formalism. Clearly indicate all the circulating mesh currents. Write down but do not solve the general mesh equations in matrix form.

24. Consider the circuit shown in Figure P6-24. Solve for all the mesh currents.

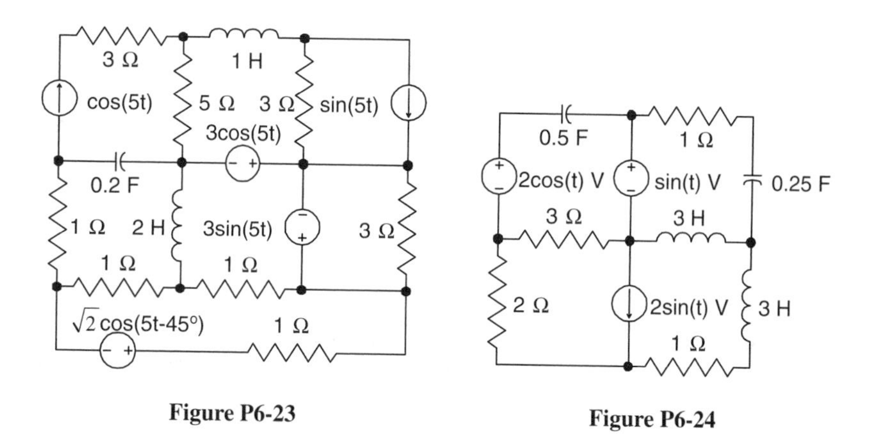

Figure P6-23

Figure P6-24

25. Consider the circuit shown in Figure P6-25. Write the matrix equations for the general node analysis problem. Find the value of the resistance R that results in a voltage of 3 V across the 2 Ω resistor.

26. Consider the circuit shown in Figure P6-25. Write the matrix form for the general mesh analysis problem. Find the value of the resistance R that results in a voltage of 3 V across the 2 Ω resistor.

27. Consider the circuit shown in Figure P6-27. All voltages are in volts; all currents are in amps. Write down but do not solve the general nodal equations in matrix form.

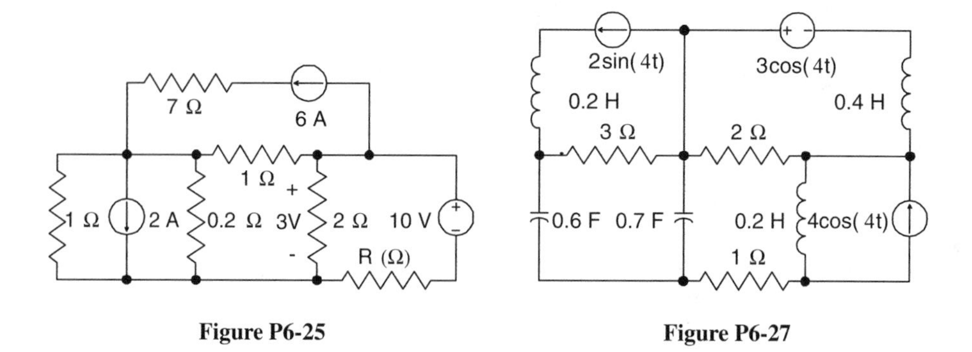

Figure P6-25 **Figure P6-27**

28. Use general mesh analysis to find the current through the 3 Ω resistor in the circuit shown in Figure P6-28.

29. Use nodal analysis to find the value of the current flowing through the $1/2\ \Omega$ resistor in the circuit shown in Figure P6-29.

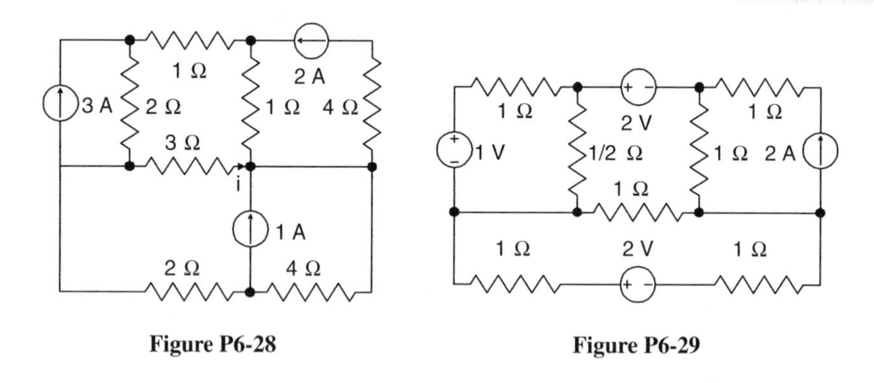

Figure P6-28 **Figure P6-29**

30. Use nodal analysis to find the time-dependent potentials of the two nodes (relative to the ground) in the circuit shown in Figure P6-30. All voltages are in volts; all currents are in amps.

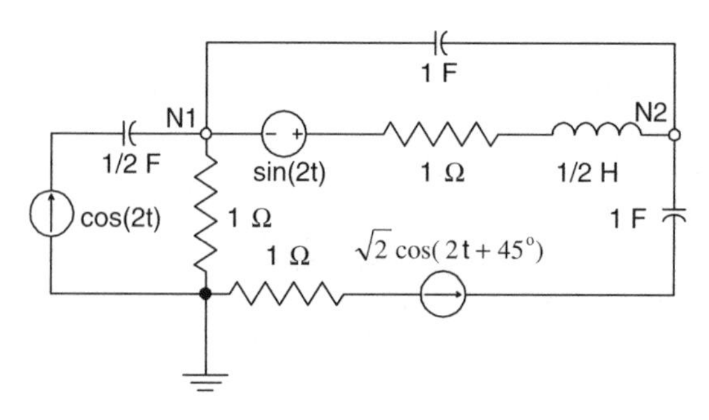

Figure P6-30

Use the PSpice code to solve Problems 31–38.

31. Find all the node potentials in the circuit shown in Figure P6-1.

32. Find the current through the 10 Ω resistor in the circuit shown in Figure P6-10.

33. Find the value of the current flowing through the 1/2 Ω resistor in the circuit shown in Figure P6-29.

34. For the circuit shown in Figure P6-6, find the value of the resistance R that results in a third node potential of -2 V.

35. For the circuit shown in Figure P6-20, find the value of the resistance R that results in a first mesh current of 1 A.

36. For the circuit shown in Figure P6-25, find the value of the resistance R that results in a voltage of 3 V across the 2 Ω resistor.

37. Plot the current through C1 in the circuit shown in Figure P6-5. Let $\omega = 377$ rad/s, $C_1 = C_2 = 2C_3 = 10\ \mu$F, $R_1 = 1\ k\Omega$, $L1 = 1$ mH, and $I_1 = 2I_2 = 3I_3 = 6$ A.

38. Plot the current flowing through the neutral wire (1 Ω) of the three-phase circuit shown in Figure P6-21.

Chapter 7

Transient Analysis

7.1 Introduction

Transients are non-steady-state time variations of voltages and currents which occur as a result of switching circuit elements and (or) changing interconnections of electric circuits. The transients occur because of the presence of energy storage elements (i.e., inductors and capacitors). We recall that a current through an inductor and a voltage across a capacitor are both continuous functions of time. This continuity determines initial conditions for energy storage elements at moments of switchings. When switchings occur, the energy storage elements usually have initial conditions that are different from new steady-state conditions. Due to the same continuity, it takes some time for the energy storage elements to reach the new steady-state conditions. This gradual (in time) adjustment of energy storage elements to new steady-state conditions results in transients in electric circuits.

In this chapter, we will study circuits with one or two energy storage elements. First, we will consider circuits with one energy storage element which are called *first-order circuits*. This is because the application of Kirchhoff's laws leads to a first-order differential equation which describes the time evolution of circuit variables. We first investigate transients in circuits which are excited by initial conditions and then move on to the discussion of circuits which are excited by sources. It will be shown that the method of phasors can be used to simplify the analysis of transients that are excited by ac sources. The analysis of first-order circuits will be concluded with a discussion of circuits that are excited by both initial conditions and sources.

Second-order circuits contain two energy storage elements and are described by second-order differential equations. First, we will develop solution techniques for circuits that are excited by initial conditions. Then, we will discuss circuits excited by sources. It will be shown that the nature of the transients depends strongly on the value of the resistance in the circuit. We will consider separately four cases: overdamped circuits, critically damped circuits, underdamped circuits, and undamped circuits. We shall also demonstrate how the method of phasors can be applied to the analysis of transients in second-order circuits driven by ac sources.

We will then present two powerful techniques that can be used to solve transient problems. First, we will define the transfer function of a circuit and present an algebraic approach to the solution of transient problems which utilizes the transfer function as a function of the complex frequency s. Next, convolution integrals will be introduced along with the concept of unit step response and unit impulse response. It will be demonstrated that the convolution integral provides a systematic approach for the calculation of circuit responses to arbitrary sources.

Finally, we introduce *diodes* and simple diode circuits and demonstrate how transient analysis can be useful for the analysis of diode circuits. The chapter is closed with a collection of PSpice examples.

7.2 First-Order Circuits

7.2.1 Circuits Excited by Initial Conditions

The first type of circuits we consider are those which have only one energy storage element and which are driven only by the initial conditions for these energy storage elements.

Consider the circuits shown in Figure 7.1. The drawing on the left shows the circuit before switching. Here, a capacitor is connected to a voltage source. At time $t = 0$ the switches are thrown, and the voltage source is disconnected leaving the charged capacitor connected in series with the resistor. This is shown by the drawing on the right. The capacitor now proceeds to discharge itself through the resistor, producing the transient process.

Now, we shall analyze this transient process mathematically. The analysis procedure generally involves two distinct steps:

1. Determination of the initial conditions for the energy storage elements. This step often requires analysis of the circuit before switching.
2. Analysis of the circuit after switching by using the previously found initial conditions. This step normally involves solving an initial value problem for differential equations.

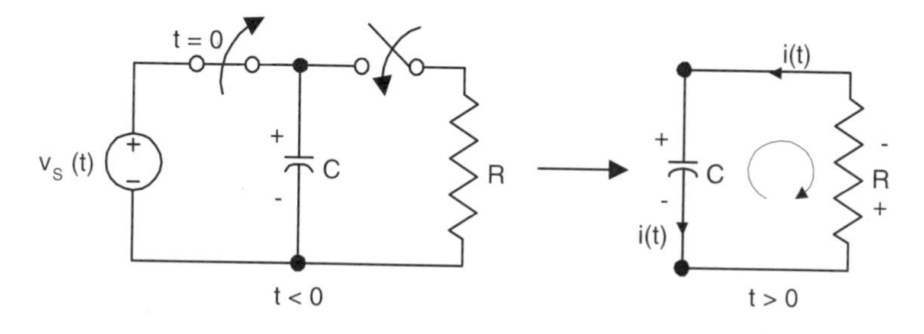

Figure 7.1: Example *RC* circuit driven by initial conditions.

We begin with step one: finding the initial conditions. We examine the circuit before switching ($t < 0$), shown on the left in Figure 7.1. In this case there is one capacitor in the circuit, so we need to find the initial voltage across the capacitor. We use KVL for the single loop, which results in:

$$v_C(t) - v_s(t) = 0, \qquad (t \leq 0), \tag{7.1}$$

$$v_C(t) = v_s(t), \quad (t \leq 0). \tag{7.2}$$

Thus, we know that the voltage across the capacitor at times $t \leq 0$ is the same as the source voltage at $t \leq 0$. Therefore, we can conclude that:

$$v_C(0_-) = v_s(0_-) = V_0. \tag{7.3}$$

Since a voltage across a capacitor is a continuous function of time, we have:

$$v_C(0_+) = v_C(0_-) = V_0. \tag{7.4}$$

Equation (7.4) is the initial condition for the posed problem. We shall now analyze the circuit after switching, that is, for $t > 0$. Applying KVL yields:

$$v_R(t) + v_C(t) = 0, \tag{7.5}$$

where $v_R(t)$ and $v_C(t)$ are the voltages across the resistor and the capacitor, respectively. By substituting in (7.5) the terminal relationship between voltage and current for a resistor, we find:

$$i(t)R + v_C(t) = 0. \tag{7.6}$$

The last equation has two unknowns, $i(t)$ and $v_C(t)$, so we need to eliminate $i(t)$. Since the same current $i(t)$ is flowing through both the resistor and the capacitor, we can express it in terms of the capacitor voltage by using the terminal relationship for a capacitor:

$$i(t) = C\frac{dv_C(t)}{dt}. \tag{7.7}$$

Substituting (7.7) into (7.6) produces the following differential equation:

$$RC\frac{dv_C(t)}{dt} + v_C(t) = 0. \tag{7.8}$$

Now, by combining (7.8) with (7.4), we end up with the following initial value problem: find the solution of the differential equation

$$RC\frac{dv_C(t)}{dt} + v_C(t) = 0 \tag{7.9}$$

subject to the initial condition:

$$v_C(0_+) = V_0. \tag{7.10}$$

The above equation is known as a first-order, linear homogeneous differential equation with constant coefficients and can be solved by using the following method. We look for the solution in the form:

$$v_C(t) = Ae^{st} \tag{7.11}$$

where A and s are some unknown constants. To find s, we substitute (7.11) and its first derivative into (7.9), which yields sequentially:

$$sRCAe^{st} + Ae^{st} = 0, \tag{7.12}$$

$$Ae^{st}(sRC + 1) = 0, \tag{7.13}$$

$$sRC + 1 = 0, \tag{7.14}$$

$$R + \frac{1}{sC} = 0. \tag{7.15}$$

Equation (7.14) is called the *characteristic equation* of the differential equation and can be solved for s:

$$s = -\frac{1}{RC}. \tag{7.16}$$

By the way, there is a quick way to find the characteristic equation just by looking at the differential equation. To do this, we simply replace the first derivative with the factor s and the function itself with 1.

If we write the characteristic equation as (7.15), it can be interpreted in terms of impedance. Indeed, setting $s = j\omega$ yields

$$R + \frac{1}{j\omega C} = Z(j\omega), \tag{7.17}$$

which is the familiar impedance of the series RC circuit. This justifies the notation:

$$R + \frac{1}{sC} = Z(s). \tag{7.18}$$

Thus, the solutions of the characteristic equation are zeros of the impedance as a function of complex frequency s:

$$Z(s) = 0. \tag{7.19}$$

Recall that the term "complex frequency" refers to the fact that s may assume arbitrary (not only imaginary) values.

Returning to the original problem and substituting (7.16) into (7.11), we find a general solution to equation (7.9):

$$v_C(t) = Ae^{-t/RC}. \tag{7.20}$$

This solution is called "general" because it satisfies (7.9) for *any* constant A. To find this constant, we employ initial condition (7.10), which yields:

$$v_C(0) = Ae^{-0/RC} = A = V_0, \tag{7.21}$$

$$A = V_0. \tag{7.22}$$

Thus, the solution to the initial value problem (7.9)–(7.10) is:

$$v_C(t) = V_0 e^{-t/RC}. \tag{7.23}$$

The last formula gives the expression for the voltage across the capacitor for all $t \geq 0$. It can be used to find the expressions for the voltage across the resistor and the current through the circuit:

$$v_R(t) = -v_C(t) = -V_0 e^{-t/RC}, \tag{7.24}$$

$$i(t) = C\frac{dv_C(t)}{dt} = -\frac{V_0}{R}e^{-t/RC}. \tag{7.25}$$

It is apparent from (7.24) and (7.25) that all circuit variables decay exponentially with time. These exponential decays can be characterized by the same *time constant* τ, which is defined by:

$$\tau = RC. \tag{7.26}$$

The time constant always has the unit of time. Indeed:

$$[\tau] = \Omega\frac{\text{s}}{\Omega} = \text{s}. \tag{7.27}$$

For exponential decay, τ is such that at time $t = \tau$ the quantity is decreased by a factor of $1/e$. Indeed, at time $t = \tau = RC$ equation (7.23) yields:

$$v_c(\tau) = V_0 e^{-\tau/RC} = V_0 e^{-RC/RC} = \frac{V_0}{e}. \tag{7.28}$$

In terms of the time constant, equation (7.23) is written as:

$$v_C(t) = V_0 e^{-t/\tau}. \tag{7.29}$$

The time constant is a measure of exponential decay. Indeed, since τ is the time needed for a quantity to decay by a factor of $1/e$, we conclude that the smaller the time constant the faster the decay. The graph presented in Figure 7.2 shows the decay of $v_C(t)$ for various values of τ.

Another way to interpret the relationship between τ and resistance R is by using power considerations. For our circuit, power dissipation is defined by:

$$p(t) = \frac{v_C(t)^2}{R}. \tag{7.30}$$

It is clear from the last equation that a smaller value of R will lead to a larger power dissipation, which will result in faster decay. This is consistent with the expression of $\tau = RC$. A smaller R means a smaller τ, which also corresponds to faster decay.

It is also clear from (7.23) that with time (theoretically speaking, with infinite time) the voltage across the capacitor will adjust to its new zero steady-state condition.

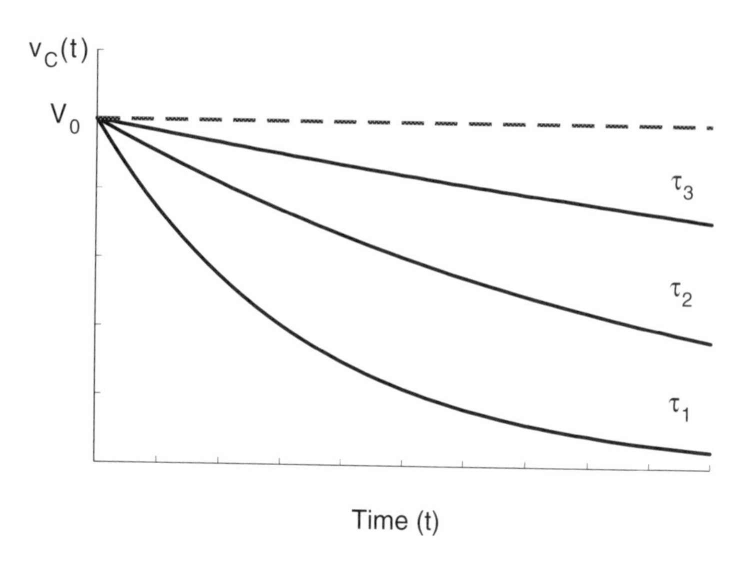

Figure 7.2: Graph of exponential decay: $\tau_1 < \tau_2 < \tau_3$.

EXAMPLE 7.1 In Figure 7.1, let $C = 1\ \mu$F, $R = 10$ kΩ, and $V_0 = 10$ V. We want to find the time constant for the circuit and the time it takes the capacitor to discharge to 1 V. Also, find the instantaneous power dissipated in the resistor when $t = 1$ ms.

The time constant for the circuit is $\tau = RC = 1\ \mu\text{F} \times 10\ \text{k}\Omega = .01\ \text{s} = 10$ ms.

If the capacitor has discharged to 1 V by the time t_0, then equation (7.29) yields $v_C(t_0) = 1 = 10e^{-t_0/.01}$, which can be solved for t_0: $t_0 = .01 \times \ln(10)s = 23.03$ ms. Note that we used the fact that $\ln(1/x) = -\ln(x)$ to get the final answer. Finally, from equation (7.30) we find that the instantaneous power dissipated in the resistor is:

$$p(t) = (V_0 e^{-t/\tau})^2/R = \frac{V_0^2}{R} e^{-2t/\tau} = 0.01e^{-200t}\ \text{W} = 0.01e^{-0.2}\ \text{W} = 8.19\ \text{mW}. \tag{7.31}$$

■

EXAMPLE 7.2 The previous example shows how to analyze first-order circuits excited by initial conditions that contain one capacitor and one resistor. Now, we consider a slightly more complicated problem, shown in Figure 7.3; let us find expressions for the current in all the resistors.

This problem is similar to the first *RC* circuit problem; however, it contains more than one resistor. We start in the usual manner, by finding the initial condition. Since this circuit before switching is identical to the first one we considered, the initial condition is also the same:

$$v_C(0_+) = V_0. \tag{7.32}$$

After the circuit is switched, it looks like the one shown on the right in Figure 7.3. Now, the circuit appears to be somewhat different from the first *RC* circuit we

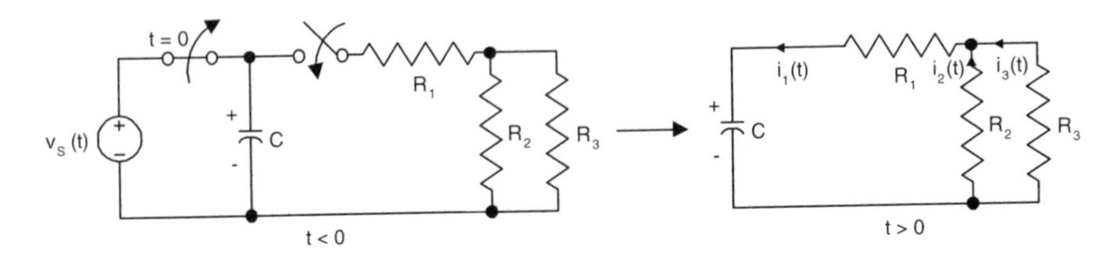

Figure 7.3: A more complicated *RC* circuit.

considered. However, in actuality it is very similar. We simply have to use equivalent transformations to combine all the resistors into one equivalent resistor R_e. From Figure 7.3 it is evident that the equivalent resistance (shown in Figure 7.4) is:

$$R_e = R_1 + \frac{R_2 R_3}{R_2 + R_3}. \tag{7.33}$$

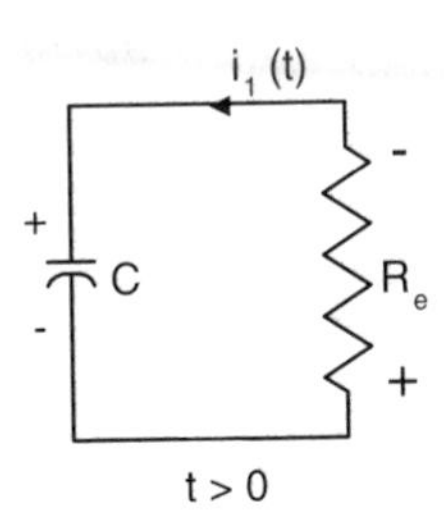

Figure 7.4: An equivalent circuit when all resistors are combined into one.

Now, the circuit is mathematically identical to the first *RC* circuit and has the same solution except that the R is replaced by R_e:

$$v_C(t) = V_0 e^{-t/R_e C}. \tag{7.34}$$

Next, all the currents in the circuit can be found. First, we calculate $i_1(t)$:

$$i_1(t) = C\frac{dv_C(t)}{dt} = -\frac{V_0}{R_e} e^{-t/R_e C}. \tag{7.35}$$

Using $i_1(t)$ and the current divider rule, the currents $i_2(t)$ and $i_3(t)$ are found:

$$i_2(t) = -\frac{V_0}{R_e}\frac{R_3}{R_2 + R_3} e^{-t/R_e C}, \tag{7.36}$$

$$i_3(t) = -\frac{V_0}{R_e}\frac{R_2}{R_2 + R_3} e^{-t/R_e C}. \tag{7.37}$$

■

EXAMPLE 7.3 We now consider the analysis of a circuit with an inductor and a resistor excited by initial conditions. The two configurations of the circuit, before switching and after switching, are illustrated in Figure 7.5. Notice that after switching

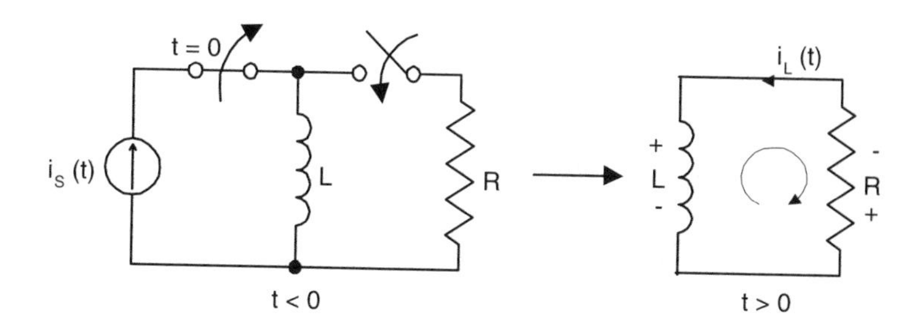

Figure 7.5: Example of an *RL* circuit driven by initial conditions.

there are no sources in the circuit—it is driven solely by the energy stored in the inductor.

We begin the analysis by determining the initial conditions for the energy storage element. In particular, we want to find the current through the inductor at time $t = 0$. From the circuit on the left in Figure 7.5 it is obvious that the current through the inductor is equal to the source current since they are connected in series. So we know that at any time $t \leq 0$,

$$i_L(t) = i_s(t), \quad (t \leq 0). \tag{7.38}$$

Therefore, we find

$$i_L(0_-) = i_s(0_-) = I_0. \tag{7.39}$$

Since the current through an inductor is a continuous function of time, we obtain the initial condition

$$i_L(0_+) = i_L(0_-) = I_0. \tag{7.40}$$

Once the initial condition is found, we can analyze the circuit after switching, that is, the circuit shown on the right in Figure 7.5. An application of KVL yields

$$v_R(t) + v_L(t) = 0. \tag{7.41}$$

We would like to express this equation in terms of $i_L(t)$ since this is the function for which we have the initial condition. This is easy to do by using the terminal relationships for resistors and inductors. After substitution of these relationships, the last equation looks like:

$$L\frac{di_L(t)}{dt} + Ri_L(t) = 0. \tag{7.42}$$

This equation, along with the initial condition (7.40), makes up the initial value problem to be solved. The above equation is also a first-order, linear homogeneous differential equation with constant coefficients. Therefore, we look for a solution in the form:

$$i_L(t) = Ae^{st}. \tag{7.43}$$

By substituting (7.43) into (7.42), we find the characteristic equation:

$$Ls + R = 0. \tag{7.44}$$

Notice that the solution to the characteristic equation is once again a zero of the impedance as a function of complex frequency. This is clear if we let $s = j\omega$; then:

$$Z(j\omega) = R + j\omega L. \tag{7.45}$$

This justifies the notation:

$$Ls + R = Z(s). \tag{7.46}$$

Consequently, the characteristic equation (7.44) can be written as follows:

$$Z(s) = 0. \tag{7.47}$$

The characteristic equation is easily solved for s:

$$s = -\frac{R}{L}. \tag{7.48}$$

Now, the solution of equation (7.42) is:

$$i_L(t) = Ae^{-\frac{R}{L}t} \tag{7.49}$$

and finding the constant A is all that is required to complete the solution. This constant is found from the initial condition. Substituting $t = 0$ into equation (7.49) and setting it equal to the initial condition (7.40) gives the value for A:

$$i_L(0_+) = I_0 = A. \tag{7.50}$$

Now, the solution is known:

$$i_L(t) = I_0 e^{-\frac{R}{L}t}. \tag{7.51}$$

From equation (7.51), the voltage across the inductor and the resistor can be found by using the voltage-current relationships for these elements:

$$v_R(t) = Ri_L(t) = I_0 Re^{-\frac{R}{L}t}, \tag{7.52}$$

$$v_L(t) = L\frac{di_L(t)}{dt} = -I_0 Re^{-\frac{R}{L}t}. \tag{7.53}$$

From the above expressions, we find the time constant:

$$\tau = \frac{L}{R}. \tag{7.54}$$

Although it is defined differently than in the previous problem, the time constant still has the unit of time, as it always must:

$$[\tau] = \frac{\Omega \cdot \text{s}}{\Omega} = \text{s}. \tag{7.55}$$

In terms of the time constant, equation (7.51) can be written as:

$$i_L(t) = I_0 e^{-t/\tau}. \tag{7.56}$$

In this case, the larger the value of R, the smaller τ and the faster the decay. Once again, this can also be seen by considering the power dissipation. For this circuit, the power dissipation is given by the expression:

$$p(t) = i_L^2(t)R, \tag{7.57}$$

which shows that the larger the resistance, the greater the power dissipation and the faster the decay. ■

EXAMPLE 7.4 A nonideal inductor, with an inductance of 10 mH and a series resistance of 10 Ω, was connected to a current source of 1 A for $t < 0$ and then short-circuited at t = 0. How long will it take for the current to decay to 1 mA? How long will it take before it decays to 1 μA? How long would it take to reach those currents if the series resistance was only 10 mΩ?

A model for the shorted nonideal inductor is given by the rightmost circuit in Figure 7.5. The instantaneous current in this inductor can be found from equation (7.56). When we plug in the numbers for the first part of our problem, we find $i_L(t_0) = 0.001 = 1e^{-1000t_0}$. Next, we invert the equation to get $t_0 = .001 \times \ln(1000)$ s $= 6.91$ ms. For $i_L(t_0) = 1$ μA, we get $t_0 = .001 \times \ln(10^6)$ s $= 13.82$ ms. We observe that it takes only twice as long to get to 1 μA as it did to get to 1 mA. If the resistance is only 10 mΩ, the time constant is 1000 times larger than in the first case, so the circuit would take 1000 times longer to decay. Thus, 1 mA would be achieved in 6.91 s and 1 μA would be left after 13.82 s. ■

7.2.2 Circuits Excited by Sources

Now, we consider electric circuits excited by sources. In situations where there is a dc or ac source, the circuit response eventually reaches the appropriate steady-state condition. This is because the transient part of the response driven by an initial condition always decays to zero, while the steady-state (forced) component of the response which is driven by a source remains.

Consider, as the first example, the *RL* circuit pictured in Figure 7.6. The problem is approached in the same way as any problem involving transients—the initial conditions must be found first. Examining the circuit before it is switched (left circuit in Figure 7.6) suggests that the initial current through the inductor is zero. This is because the circuit is open before switching and the current through the inductor is a continuous function of time. Therefore, the initial condition is:

$$i_L(0_-) = i_L(0_+) = 0. \tag{7.58}$$

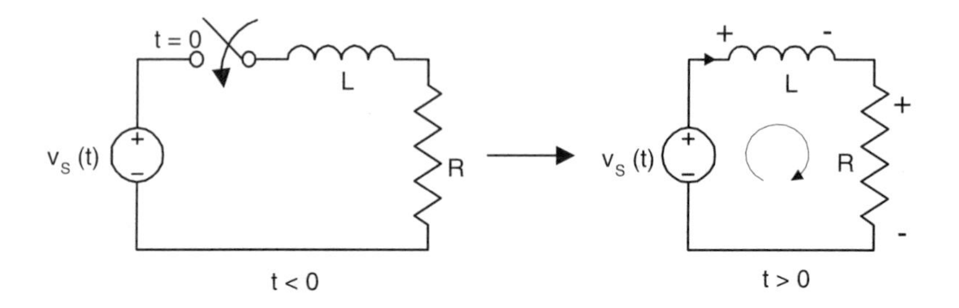

Figure 7.6: Example of an *RL* circuit excited by a source.

Now, we shall analyze the circuit after switching (right circuit in Figure 7.6). Writing KVL for this circuit yields:

$$v_L(t) + v_R(t) - v_s(t) = 0. \tag{7.59}$$

We want to express equation (7.59) in terms of the current through the inductor, because this is the physical quantity for which we have the initial condition. To this end, we use the terminal relationships:

$$v_L(t) = L\frac{di_L(t)}{dt}, \tag{7.60}$$

$$v_R(t) = Ri_L(t) \tag{7.61}$$

and substitute them into (7.59). This yields:

$$L\frac{di_L(t)}{dt} + Ri_L(t) = v_s(t). \tag{7.62}$$

Differential equation (7.62) combined with initial condition (7.58) constitutes the initial value problem.

The differential equation (7.62) is a first-order linear differential equation with constant coefficients, but it is no longer homogeneous (i.e., the right-hand side of the equation is not zero). In fact, this equation is similar to the equation we derived earlier for the *RL* circuit excited only by initial conditions, except for the appearance of the voltage source on the right-hand side. In general, circuits excited by sources will lead to nonhomogeneous differential equations because their sources will always appear on the right-hand side of the equations.

It is known from mathematics that a complete solution of a nonhomogeneous linear differential equation can be represented as a sum of a general solution of the corresponding homogeneous equation and a particular solution of the nonhomogeneous equation:

$$i_L(t) = i_h(t) + i_p(t), \tag{7.63}$$

where $i_h(t)$ stands for the general solution of the homogeneous equation

$$L\frac{di_h(t)}{dt} + Ri_h(t) = 0, \tag{7.64}$$

while $i_p(t)$ represents the particular solution of the nonhomogeneous equation

$$L\frac{di_p(t)}{dt} + Ri_p(t) = v_s(t). \tag{7.65}$$

Equation (7.64) is mathematically identical to (7.42). Consequently, its solution can be written in the form:

$$i_h(t) = Ae^{-\frac{R}{L}t}. \tag{7.66}$$

Please note that the constant A needs to be determined. However, we must delay solving for A until we have found the particular solution, *because the initial condition must be applied to the complete solution (7.63).*

The method of finding the particular solution (7.65) depends on the mathematical form of $v_s(t)$ on the right-hand side of the nonhomogeneous equation. We begin with the simple problem, when our circuit is excited by a dc voltage source:

$$v_s(t) = V_0 = \text{ constant}. \tag{7.67}$$

In this case, equation (7.65) has the form:

$$L\frac{di_p(t)}{dt} + Ri_p(t) = V_0. \tag{7.68}$$

The particular solution of this equation can be found by using the method of undetermined coefficients. The basic idea of this method is to look for the solution in the form which mimics the form of the right-hand side. It means that in the case of equation (7.68), we have to look for the solution in the form:

$$i_p(t) = I_0 = \text{ constant}. \tag{7.69}$$

Now, we substitute (7.69) into equation (7.68), which results in:

$$RI_0 = V_0, \tag{7.70}$$

$$i_p(t) = I_0 = \frac{V_0}{R}. \tag{7.71}$$

Combining the homogeneous and particular solutions gives the complete solution to equation (7.62):

$$i_L(t) = Ae^{-\frac{R}{L}t} + \frac{V_0}{R}. \tag{7.72}$$

However, the constant A is still unknown. To find it, we evoke the initial condition (7.58). According to the initial condition:

$$i_L(0_+) = 0 = A + \frac{V_0}{R}, \tag{7.73}$$

$$A = -\frac{V_0}{R}. \tag{7.74}$$

So, the final solution to the initial value problem is:

$$i_L(t) = \frac{V_0}{R} - \frac{V_0}{R} e^{-\frac{R}{L}t}. \tag{7.75}$$

It is clear from equation (7.75) that there are two components of the solution: component V_0/R corresponds to the steady-state solution and does not decay with time, while component $-(V_0/R)e^{-Rt/L}$ corresponds to the transient response, and it does decay with time. As time goes to infinity, the transient goes to zero, and only the steady-state response remains.

Thus, we can draw the following conclusion (which is, by the way, very general in nature): *the homogeneous solution has the physical meaning of transient (free) response, while the particular solution has the physical meaning of steady-state (forced) response*. The latter suggests that we can find the particular solution without invoking the method of undetermined coefficients but by using techniques for steady-state analysis of electric circuits. For instance, in our example circuit, inductor L has zero voltage drop at dc steady state; consequently, the source voltage is applied across the resistor R and the dc steady-state current is equal to

$$I_0 = \frac{V_0}{R}, \tag{7.76}$$

which coincides with the particular solution.

We can also find the voltages across the resistor and the inductor using equation (7.75):

$$v_R(t) = Ri_L(t) = V_0 - V_0 e^{-\frac{R}{L}t} \tag{7.77}$$

$$v_L(t) = L\frac{di_L(t)}{dt} = V_0 e^{-\frac{R}{L}t} \tag{7.78}$$

One way to get a good feeling for how the circuit behaves over time is to graph its voltages and currents. Figure 7.7 shows the graphs of the current, the voltage across the resistor, and the voltage across the inductor. These graphs show that the current through and the voltage across the resistor both start at zero and then become larger until they reach their steady-state values. On the other hand, the voltage across

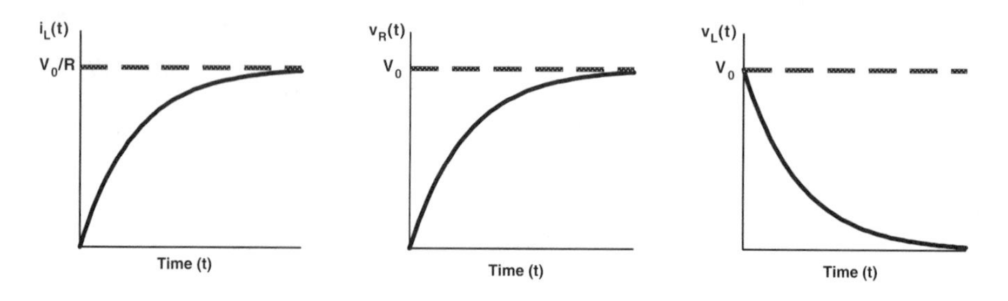

Figure 7.7: Graphs of $i_L(t)$, $v_R(t)$, and $v_L(t)$.

the inductor starts at its maximum value and then decays to zero at steady-state conditions. This is a good illustration of the properties of an inductor. Initially the inductor behaves as an *open circuit* and it has all the source voltage applied across it. This is because there is no current flowing through the inductor as a result of zero initial condition. When the dc steady-state condition is reached, the inductor acts as a *short circuit* and its voltage drop is zero and the current flows through it unrestricted.

We shall see later that an uncharged capacitor behaves in the opposite manner. At time $t = 0$ the capacitor acts as a short circuit, while at time $t = \infty$ it acts as an open circuit.

EXAMPLE 7.5 Consider the *RL* circuit again, but this time with a sinusoidal voltage source. This source has the form:

$$v_s(t) = V_m \cos(\omega t + \phi_s). \tag{7.79}$$

The differential equation is nearly the same as before—only the right-hand side of the equation is different:

$$L\frac{di_L(t)}{dt} + Ri_L(t) = V_m \cos(\omega t + \phi_s). \tag{7.80}$$

Clearly the "homogeneous" solution to this equation will be identical to the previous one (see (7.66)), so we need only consider the particular solution. To use the method of undetermined coefficients for this problem, we have to assume the form of the particular solution which mimics the form of the right-hand side of the equation (7.80):

$$i_p(t) = I_m \cos(\omega t + \phi_I). \tag{7.81}$$

However, it is not necessary to use the method of undetermined coefficients to solve for the particular solution when the source is sinusoidal. We recall that the particular solution corresponds to the ac steady-state response of the circuit. For this reason, in order to find the particular solution we can employ the phasor technique, which has been studied in detail in previous chapters.

As always with the phasor technique, the first step is to convert the source into phasor form:

$$\hat{V}_s = V_m e^{j\phi_s}. \tag{7.82}$$

Then, we represent the unknown current (7.81) in the phasor form as well:

$$\hat{I} = I_m e^{j\phi_I}. \tag{7.83}$$

The total impedance of our circuit is

$$Z = R + j\omega L. \tag{7.84}$$

To find the unknown current, we use the relationship between the phasors of voltage and current:

$$\hat{I} = \frac{\hat{V}_s}{Z}. \tag{7.85}$$

Thus, we can find the peak value I_m and phase angle ϕ_I of the unknown current as follows:

$$I_m = \frac{V_m}{|Z|} = \frac{V_m}{\sqrt{R^2 + \omega^2 L^2}}, \tag{7.86}$$

$$\phi_I = \phi_s - \phi, \tag{7.87}$$

where

$$\phi = \arctan \frac{\omega L}{R}. \tag{7.88}$$

By substituting (7.86) and (7.87) into (7.81), we obtain the particular solution:

$$i_p(t) = \frac{V_m}{\sqrt{R^2 + \omega^2 L^2}} \cos(\omega t + \phi_s - \phi). \tag{7.89}$$

Combining this expression with the "homogeneous" solution yields the expression for the complete solution to the differential equation:

$$i_L(t) = \frac{V_m}{\sqrt{R^2 + \omega^2 L^2}} \cos(\omega t + \phi_s - \phi) + Ae^{-\frac{R}{L}t}. \tag{7.90}$$

Now, the problem is nearly done and only the unknown constant A remains to be found. This is accomplished by using the initial condition (the same one as in the previous RL circuit):

$$i_L(0_+) = \frac{V_m}{\sqrt{R^2 + \omega^2 L^2}} \cos(\phi_s - \phi) + A = 0, \tag{7.91}$$

$$A = -\frac{V_m}{\sqrt{R^2 + \omega^2 L^2}} \cos(\phi_s - \phi). \tag{7.92}$$

By substituting this value for A into (7.90), we arrive at the final solution:

$$i_L(t) = \frac{V_m}{\sqrt{R^2 + \omega^2 L^2}} \cos(\omega t + \phi_s - \phi) - \frac{V_m e^{-\frac{R}{L}t}}{\sqrt{R^2 + \omega^2 L^2}} \cos(\phi_s - \phi). \tag{7.93}$$

Two instructive observations can be made concerning the solution (7.93).

Consider such an initial phase ϕ_s of the voltage source that

$$\phi_s - \phi = \phi_s - \arctan \frac{\omega L}{R} = \frac{\pi}{2}. \tag{7.94}$$

Then, it is clear from (7.92) that

$$A = -\frac{V_m}{\sqrt{R^2 + \omega^2 L^2}} \cos(\phi_s - \phi) = -\frac{V_m}{\sqrt{R^2 + \omega^2 L^2}} \cos \frac{\pi}{2} = 0. \tag{7.95}$$

From (7.93), (7.94), and (7.95) we find:

$$i_L(t) = \frac{V_m}{\sqrt{R^2 + \omega^2 L^2}} \cos(\omega t + \frac{\pi}{2}) = i_p(t). \tag{7.96}$$

This means that if the initial phase ϕ_s is properly adjusted [see formula (7.94)] there will be no transient response. The reason for this is that:

$$i_p(0) = 0 = i_L(0_+). \tag{7.97}$$

In other words, there is no transient response because the initial condition for the inductor coincides with its new steady-state condition.

The above discussion clearly reveals that the physical cause for transients is a mismatch between initial conditions for energy storage elements and their new steady-state conditions. This mismatch drives transients.

Another observation is related to the interpretation of the complete solution (7.90) in terms of impedance $Z(s) = sL + R$ as a function of "complex" frequency s. It is clear from (7.90) (as well as from (7.86)–(7.89)) that the value of this impedance at the frequency of excitation $s = j\omega$ completely determines the steady-state response of the circuit. It is also clear from (7.44) that zeros of this impedance coincide with the exponents of the transient (free) response. Thus, the impedance $Z(s)$ provides the complete information concerning both steady-state and transient responses. For this reason, the impedance as a function of complex frequency is extensively used in the advanced theory of electric circuits. ■

EXAMPLE 7.6 Consider the specific example of an *RL* circuit excited by an ac source. In this case $R = 1/2\ \Omega$, $L = 1/2$ H, and $v_s(t) = 2\cos(\sqrt{3}t)$. Find the inductor current.

First, we find the impedance magnitude

$$|Z| = \sqrt{R^2 + \omega^2 L^2} = \sqrt{(1/2)^2 + (\sqrt{3}/2)^2} = 1\ \Omega.$$

For the impedance phase we use equation (7.88): $\phi = \arctan(\omega L/R) = \arctan(\sqrt{3}) = 60^\circ$. Plugging these values into equation (7.89), we find the particular solution to be $i_p(t) = 2\cos(\sqrt{3}t - 60^\circ)$. The unknown constant in the homogeneous solution is found from equation (7.92): $A = -2\cos(-60^\circ) = -1$. The time constant, L/R, for the homogeneous solution is unity. Combining the two solutions and performing some algebra, we find the final answer: $i_L(t) = \cos(\sqrt{3}t) + \sqrt{3}\sin(\sqrt{3}t) - e^{-t}$ A. As a useful exercise, the reader may wish to try to recover the same result by using the method of undetermined coefficients. ■

Now, we consider the excitation of *RC* circuits by sources. We first examine the excitation of the circuit shown in Figure 7.8 by the dc current source:

$$i_s(t) = I_o = \text{constant}. \tag{7.98}$$

First, we find the initial condition. In this case, we need to determine the initial voltage across the capacitor. Since the circuit is not excited before the switch closes, we conclude that the initial voltage across the capacitor must be zero. Therefore, the initial condition is:

$$v(0_+) = 0. \tag{7.99}$$

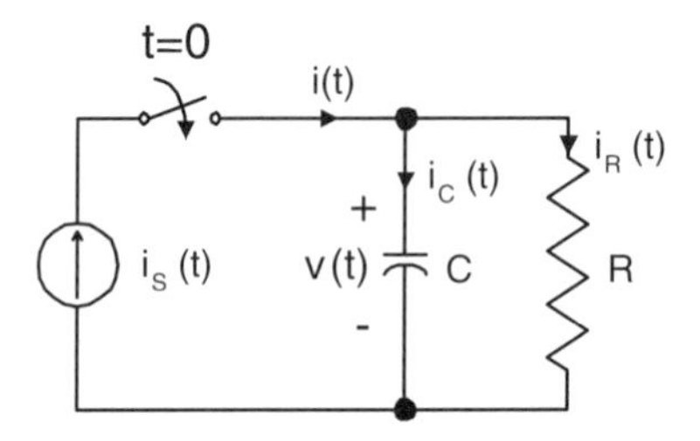

Figure 7.8: *RC* circuit excited by a current source.

Now, we analyze the circuit for $t > 0$. Applying KCL yields:

$$i_s(t) = i_C(t) + i_R(t), \tag{7.100}$$

where $i_C(t)$ and $i_R(t)$ are the currents through the capacitor and the resistor, respectively. We need to express the KCL equation in terms of the voltage across the capacitor, because this voltage is a physical quantity for which we have the initial condition. To this end, we evoke the terminal relationships:

$$i_C(t) = C\frac{dv(t)}{dt}, \tag{7.101}$$

$$i_R(t) = \frac{v(t)}{R}. \tag{7.102}$$

Substituting these expressions into (7.100) gives the differential equation in the form we need. This equation together with initial condition (7.99) leads to the following initial value problem:

$$C\frac{dv(t)}{dt} + \frac{v(t)}{R} = i_s(t), \tag{7.103}$$

$$v(0_+) = 0. \tag{7.104}$$

This initial value problem can be solved in the standard manner.

We know that the complete solution of differential equation (7.103) is a sum of the homogeneous solution and the particular solution:

$$v(t) = v_h(t) + v_p(t). \tag{7.105}$$

The homogeneous solution is in the form

$$v_h(t) = Ae^{st}, \tag{7.106}$$

where s is the solution of the following characteristic equation:

$$sC + \frac{1}{R} = 0, \tag{7.107}$$

which can also be rewritten as

$$Y(s) = \frac{1}{R} + sC = 0. \tag{7.108}$$

The above equation leads to:

$$s = -\frac{1}{RC}. \tag{7.109}$$

Therefore,

$$v_h(t) = Ae^{-t/RC}. \tag{7.110}$$

Now, the particular solution is required. For $t > 0$, the right-hand side of (7.103) is a constant. Using the method of undetermined coefficients, we look for the particular solution in the same form as the right-hand side of the differential equation:

$$v_p(t) = V_0 = \text{ constant.} \tag{7.111}$$

Substituting $v_p(t)$ into (7.103) we find

$$0 + \frac{V_0}{R} = I_0, \tag{7.112}$$

$$V_0 = I_0 R. \tag{7.113}$$

So, as should be expected, the particular solution coincides with the dc steady-state response of the circuit. Thus, the complete solution is the sum of the homogeneous and particular solutions:

$$v(t) = RI_0 + Ae^{-t/RC}. \tag{7.114}$$

We next find the constant A by using initial condition (7.104):

$$v(0_+) = RI_0 + A = 0, \tag{7.115}$$

$$A = -RI_0, \tag{7.116}$$

which leads to the final answer:

$$v(t) = RI_0 - RI_0 e^{-t/RC}. \tag{7.117}$$

Now that we have the expression for the voltage across the capacitor, we can easily find expressions for the currents through the capacitor and the resistor:

$$i_C(t) = C\frac{dv(t)}{dt} = I_0 e^{-t/RC}, \tag{7.118}$$

$$i_R(t) = \frac{v(t)}{R} = I_0 - I_0 e^{-t/RC}. \tag{7.119}$$

The graphs of these two functions (Figure 7.9) reveal some interesting properties of the capacitor. At time $t = 0$, there is no voltage drop across the capacitor, so it acts like a short circuit and *shunts* the current source. In other words, all the current flows

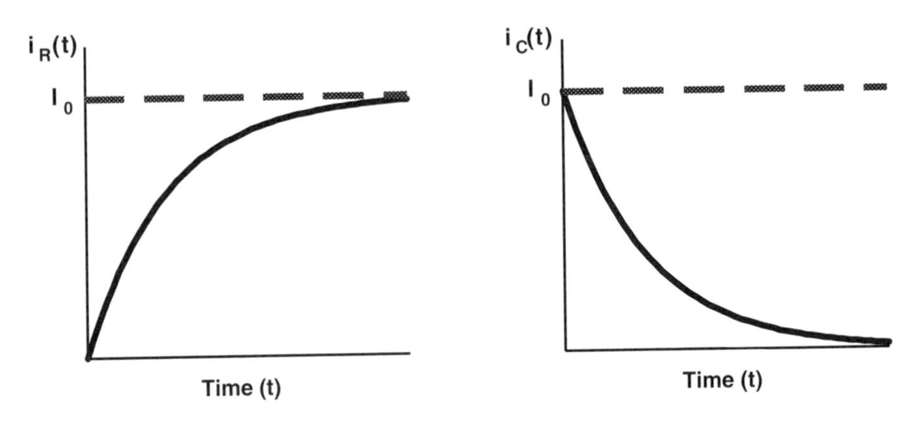

Figure 7.9: Graphs of the current through the resistor and capacitor.

through the branch with the capacitor and none flows through the branch with the resistor. However, at dc steady state ($t = \infty$) the capacitor acts like an open circuit. None of the current flows through it because there is no time variation of the voltage. All the current flows through the branch with the resistor.

The property of a capacitor to act as a short circuit and to absorb most of the transient current during the initial stage of transients is utilized in practice. Namely, capacitors are often connected in parallel with expensive devices (equipment such as computers, etc.) to protect these devices from large (and destructive) transient currents which usually occur during initial stages of transients.

EXAMPLE 7.7 Consider the case of an ac current source:

$$i_s(t) = I_{ms} \cos(\omega t + \phi_s) \tag{7.120}$$

driving the circuit in Figure 7.8. Find the expression for $v(t)$.

In this case, differential equation (7.103) has the form:

$$C\frac{dv(t)}{dt} + \frac{v(t)}{R} = I_{ms} \cos(\omega t + \phi_s). \tag{7.121}$$

The homogeneous solution for this equation is the same as before (see (7.110)). To find the particular solution to equation (7.121), we take advantage of the fact that this solution has the physical meaning of the ac steady-state response of the RC circuit. For this reason, we shall employ the phasor technique to find the particular solution.

In the time domain, this particular solution has the form:

$$v_p(t) = V_m \cos(\omega t + \phi_V). \tag{7.122}$$

The phasor form of this solution is:

$$\hat{V} = V_m e^{j\phi_V}. \tag{7.123}$$

Similarly, the phasor of the current source is:

$$\hat{I}_s = I_{ms} e^{j\phi_s}. \tag{7.124}$$

It is clear that the voltage phasor, $\hat{V}$, and the current source phasor, $\hat{I}_s$, are related to one another through the admittance, Y:

$$\hat{I}_s = Y\hat{V}, \tag{7.125}$$

where

$$Y = \frac{1}{R} + j\omega C. \tag{7.126}$$

From (7.125), we find:

$$\hat{V} = \frac{\hat{I}_s}{Y}, \tag{7.127}$$

which according to (7.126) leads to:

$$V_m = \frac{I_{ms}}{\sqrt{\frac{1}{R^2} + \omega^2 C^2}}, \tag{7.128}$$

$$\phi_V = \phi_s - \phi, \tag{7.129}$$

where

$$\phi = \arctan \omega RC. \tag{7.130}$$

Thus, the particular solution (7.122) is:

$$v_p(t) = \frac{I_{ms} R}{\sqrt{1 + \omega^2 C^2 R^2}} \cos(\omega t + \phi_s - \phi). \tag{7.131}$$

Combining (7.131) with (7.110), we find the complete solution of differential equation (7.121):

$$v(t) = \frac{I_{ms} R}{\sqrt{1 + \omega^2 C^2 R^2}} \cos(\omega t + \phi_s - \phi) + Ae^{-t/RC}. \tag{7.132}$$

To find the constant, A, we shall invoke the initial condition (7.104), which leads to:

$$\frac{I_{ms} R}{\sqrt{1 + \omega^2 C^2 R^2}} \cos(\phi_s - \phi) + A = 0. \tag{7.133}$$

Consequently:

$$A = -\frac{I_{ms} R}{\sqrt{1 + \omega^2 C^2 R^2}} \cos(\phi_s - \phi). \tag{7.134}$$

By substituting (7.134) into (7.132), we end up with the final expression for the response of the RC circuit excited by ac current source:

$$v(t) = \frac{I_{ms} R}{\sqrt{1 + \omega^2 C^2 R^2}} \cos(\omega t + \phi_s - \phi) - \frac{I_{ms} R e^{-t/RC}}{\sqrt{1 + \omega^2 C^2 R^2}} \cos(\phi_s - \phi), \tag{7.135}$$

where, according to (7.130):

$$\phi = \arctan \omega RC. \tag{7.136}$$

Formulas (7.135) and (7.136) explicitly express the circuit response in terms of ac excitation and circuit parameters. ■

Two instructive observations can be made concerning the solution (7.135). It can be shown that the initial phase, ϕ_s, of the current source can be chosen in such a way that there will be no transient. Indeed, if

$$\phi_s = \phi + \frac{\pi}{2} = \arctan \omega RC + \frac{\pi}{2}, \tag{7.137}$$

then from (7.134) and (7.135) we find:

$$A = 0, \tag{7.138}$$

$$v(t) = \frac{I_{ms}R}{\sqrt{1 + \omega^2 C^2 R^2}} \cos(\omega t + \phi_s - \phi) = v_p(t). \tag{7.139}$$

It is clear from (7.139) (as well as from (7.131)) that if the initial phase of the ac current source is chosen according to (7.137) then the initial condition for the voltage across the capacitor coincides with the ac steady-state condition for this voltage:

$$v_p(0) = v(0_+) = 0. \tag{7.140}$$

The above discussion again underlines the fact that the physical cause for transients is a mismatch between initial conditions for energy storage elements and their new steady-state conditions. This mismatch drives transients.

Another observation, which is appropriate here, is that the admittance, $Y(s)$, as a function of complex frequency fully determines the form of the complete solution (7.132). Indeed, the value of this admittance at the frequency of excitation, $s = j\omega$, completely determines the steady-state response of the circuit, while zeros of this admittance [see (7.108)–(7.109)] are the exponents of the transient (free) response.

We recall that we reached a similar conclusion with respect to impedance $Z(s)$ when we analyzed the RL circuit. The natural question is why in one case we deal with impedance $Z(s)$ while in the other case we deal with admittance $Y(s)$. A related question is: “Is it possible to achieve some uniformity in this matter so that we shall deal with the same quantity regardless of the type of electric circuit?”

The answers to the posed questions are based on the notion of **transfer functions**, which are discussed in detail in Section 7.4.

7.2.3 Circuits Excited by Initial Conditions and Sources

We now consider the case of a circuit excited by both initial conditions and sources.

EXAMPLE 7.8 Consider a circuit shown in Figure 7.10, where $v_s(t)$ is a dc source:

$$v_s(t) = V_0 = \text{constant}. \tag{7.141}$$

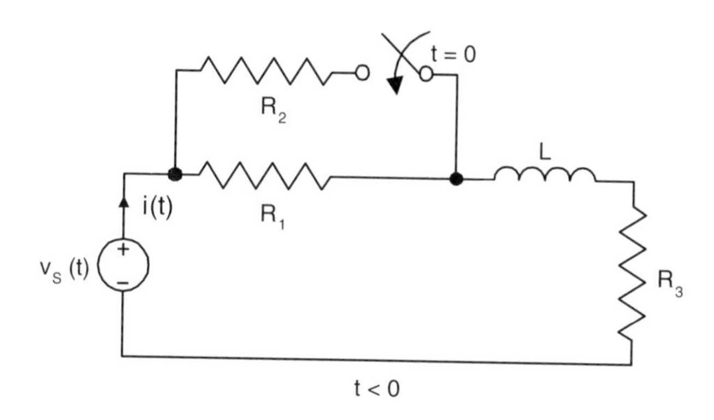

Figure 7.10: Circuit excited by initial conditions and sources.

At time $t < 0$ the circuit is at dc steady state. Then at time $t = 0$ the switch is thrown and the circuit changes its configuration—specifically, a new resistor R_2 is connected. As a result, some transients will be generated and it will take some time for the circuit to readjust to new steady-state conditions corresponding to the new configuration.

We approach the analysis in the same way as with any problem involving transients. First, the initial condition is to be found. In this case, we need the initial value of the current through the inductor. Examining the circuit before switching, we see that the current is given by the expression:

$$i(t) = \frac{V_0}{R_1 + R_3}. \tag{7.142}$$

This is because immediately before switching, the circuit is at dc steady-state conditions. This means that the inductor acts as a short circuit because its drop in voltage is zero, and, consequently, the current through the inductor is given by equation (7.142). Therefore, the initial condition is:

$$i(0_+) = i(0_-) = \frac{V_0}{R_1 + R_3}. \tag{7.143}$$

After switching, the circuit can be represented in the new form shown in Figure 7.11. Here, to simplify the calculations, the resistors are combined into one equivalent resistor:

$$R_e = R_3 + \frac{R_1 R_2}{R_1 + R_2}. \tag{7.144}$$

We can now write the differential equation for this circuit, which is similar to those for the *RL* circuits we have considered. By combining this differential equation with the initial condition, we end up with the following initial value problem:

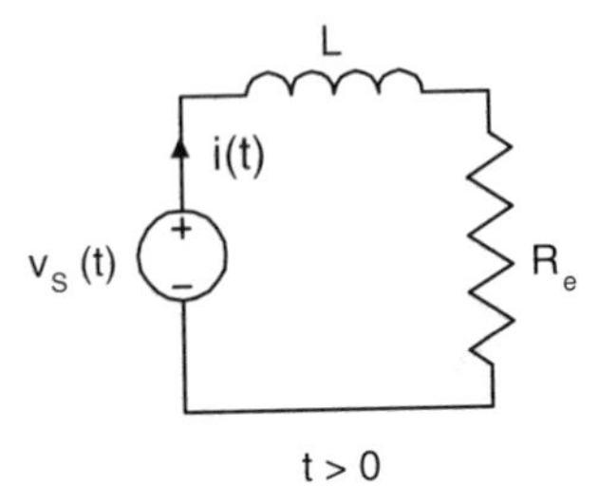

Figure 7.11: The circuit after switching.

$$L\frac{di(t)}{dt} + R_e i(t) = V_0, \tag{7.145}$$

$$i(0_+) = \frac{V_0}{R_1 + R_3}. \tag{7.146}$$

This problem can be solved by using the methods outlined in the previous sections. The complete solution will be a sum of the homogeneous and particular solutions:

$$i(t) = i_h(t) + i_p(t). \tag{7.147}$$

The homogeneous solution will have the form

$$i_h(t) = Ae^{st}, \tag{7.148}$$

where A is to be found from the initial condition and s is the solution to the characteristic equation. The characteristic equation for this differential equation is:

$$Z(s) = sL + R_e = 0, \tag{7.149}$$

and it is clear that

$$s = -\frac{R_e}{L}. \tag{7.150}$$

Therefore, the homogeneous solution is

$$i_h(t) = Ae^{-(R_e/L)t}. \tag{7.151}$$

Now, the particular solution must be found. Since the right-hand side of (7.145) is a constant, we assume the particular solution to be a constant as well:

$$i_p(t) = I_0 = \text{ constant}. \tag{7.152}$$

By substituting $i_p(t)$ into equation (7.145), we find:

$$0 + R_e I_0 = V_0, \tag{7.153}$$

$$I_0 = \frac{V_0}{R_e}. \tag{7.154}$$

As expected, the particular solution coincides with the new steady-state response.

Now, the complete solution can be written as follows:

$$i(t) = \frac{V_0}{R_e} + Ae^{-(R_e/L)t}. \tag{7.155}$$

The constant A is found by using the initial condition (7.146):

$$i(0_+) = \frac{V_0}{R_e} + A = \frac{V_0}{R_1 + R_3}, \tag{7.156}$$

$$A = \frac{V_0}{R_1 + R_3} - \frac{V_0}{R_e}. \tag{7.157}$$

Substituting A back into (7.155) gives the final solution to the problem:

$$i(t) = \frac{V_0}{R_e} + \left(\frac{V_0}{R_1 + R_3} - \frac{V_0}{R_e} \right) e^{-(R_e/L)t}. \tag{7.158}$$

As before, this solution has two components: the particular solution which corresponds to the steady-state response and the homogeneous solution which corresponds to the transient response. ■

EXAMPLE 7.9 Consider the same circuit excited by an ac voltage source (see Figure 7.10):

$$v_s(t) = V_{ms} \cos(\omega t + \phi_s). \tag{7.159}$$

To find the initial condition, we have to first analyze the ac steady-state response which existed before switching. This task can be accomplished by using the phasor technique. The impedance of the circuit before switching is equal to:

$$Z_- = (R_1 + R_3) + j\omega L, \tag{7.160}$$

where the subscript "−" indicates the impedance of the circuit for negative times.

Now, by using the voltage source phasor

$$\hat{V}_s = V_{ms} e^{j\phi_s}, \tag{7.161}$$

we find the current phasor:

$$\hat{I}_- = \frac{\hat{V}_s}{Z_-} = I_{m-} e^{j(\phi_s - \phi_-)}, \tag{7.162}$$

where as before the subscript "−" means the ac steady-state current value before switching. From (7.160), (7.161), and (7.162), we derive:

$$I_{m-} = \frac{V_{ms}}{\sqrt{(R_1 + R_3)^2 + \omega^2 L^2}}, \quad \phi_- = \arctan \frac{\omega L}{R_1 + R_3}. \tag{7.163}$$

Thus, in the time domain, the ac steady-state current prior to switching can be written as follows:

$$i_-(t) = \frac{V_{ms}}{\sqrt{(R_1 + R_3)^2 + \omega^2 L^2}} \cos(\omega t + \phi_s - \phi_-). \tag{7.164}$$

From expression (7.164) we can find the initial condition for the current through the inductor:

$$i(0_+) = i(0_-) = \frac{V_{ms}}{\sqrt{(R_1 + R_3)^2 + \omega^2 L^2}} \cos(\phi_s - \phi_-). \tag{7.165}$$

Now, the initial value problem (7.145)–(7.146) can be reformulated as follows:

$$L\frac{di(t)}{dt} + R_e i(t) = V_{ms}\cos(\omega t + \phi_s), \tag{7.166}$$

$$i(0_+) = \frac{V_{ms}}{\sqrt{(R_1 + R_3)^2 + \omega^2 L^2}} \cos(\phi_s - \phi_-). \tag{7.167}$$

The homogeneous solution for equation (7.166) is the same as before (see (7.151)). The particular solution for equation (7.166) has the physical meaning of the ac steady-state response of the circuit shown in Figure 7.11 and it can be found by using again the phasor technique. The final result is as follows:

$$i_p(t) = \frac{V_{ms}}{\sqrt{R_e^2 + \omega^2 L^2}} \cos(\omega t + \phi_s - \phi), \tag{7.168}$$

where

$$\phi = \arctan\frac{\omega L}{R_e}. \tag{7.169}$$

Thus, the complete solution to differential equation (7.166) is:

$$i(t) = \frac{V_{ms}}{\sqrt{R_e^2 + \omega^2 L^2}} \cos(\omega t + \phi_s - \phi) + Ae^{-(R_e/L)t}. \tag{7.170}$$

To find the unknown constant A, we invoke the initial condition (7.167), which leads to:

$$\frac{V_{ms}\cos(\phi_s - \phi_-)}{\sqrt{(R_1 + R_3)^2 + \omega^2 L^2}} = \frac{V_{ms}\cos(\phi_s - \phi)}{\sqrt{R_e^2 + \omega^2 L^2}} + A. \tag{7.171}$$

This results in:

$$A = \frac{V_{ms}\cos(\phi_s - \phi_-)}{\sqrt{(R_1 + R_3)^2 + \omega^2 L^2}} - \frac{V_{ms}\cos(\phi_s - \phi)}{\sqrt{R_e^2 + \omega^2 L^2}}. \tag{7.172}$$

By substituting (7.172) into (7.170), we find the final expression:

$$\begin{aligned} i(t) = {} & \frac{V_{ms}}{\sqrt{R_e^2 + \omega^2 L^2}} \cos(\omega t + \phi_s - \phi) \\ & + \left(\frac{V_{ms}\cos(\phi_s - \phi_-)}{\sqrt{(R_1 + R_3)^2 + \omega^2 L^2}} - \frac{V_{ms}\cos(\phi_s - \phi)}{\sqrt{R_e^2 + \omega^2 L^2}} \right) e^{-(R_e/L)t}. \end{aligned} \tag{7.173}$$

■

7.3 Second-Order Circuits

All the circuits that we have discussed thus far dealing with transients have been first-order circuits, because they contained only one energy storage element. For this reason, they led to first-order differential equations. We now consider circuits which are described by second-order differential equations. These circuits generally feature two energy storage elements. We consider circuits excited by initial conditions before examining circuits driven by sources.

7.3.1 Circuits Excited by Initial Conditions

Consider the circuit shown on the left in Figure 7.12. Notice that it contains both an inductor and a capacitor. The approach to the analysis of this circuit is basically the same as to the analysis of first-order circuits: namely, we have to first find the initial condition and then analyze the circuit after switching. However, when there are two energy storage elements in the circuit, two initial conditions must be found. In this case, we need to find the initial current through the inductor and the initial voltage across the capacitor.

Using KVL for the circuit before switching, we can see that the voltage across the capacitor is equal to the source voltage:

$$v_s(t) - v_C(t) = 0, \tag{7.174}$$

$$v_C(t) = v_s(t). \tag{7.175}$$

The initial current through the inductor is zero, because before switching, the inductor is in an open branch with no current flowing. Thus, the initial conditions are

$$v_C(0_+) = v_C(0_-) = v_s(0) = V_0, \tag{7.176}$$

$$i_L(0_+) = i_L(0_-) = 0. \tag{7.177}$$

After switching, the circuit has the configuration shown on the right in Figure 7.12. We now use KVL to set up the differential equation for this circuit:

$$v_C(t) + v_L(t) + v_R(t) = 0. \tag{7.178}$$

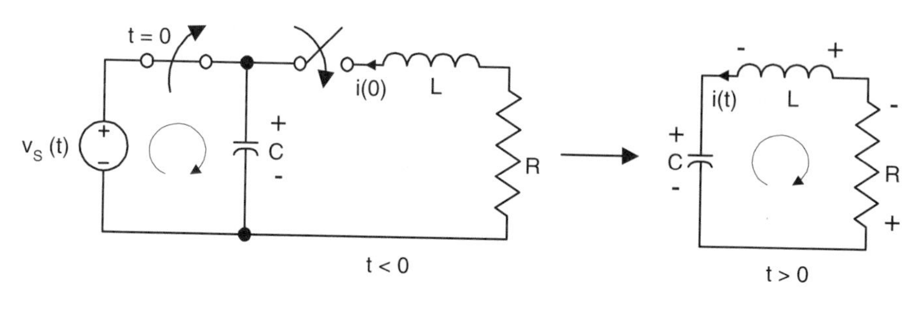

Figure 7.12: Second-order circuit.

Ultimately, we want to get the differential equation only in terms of the voltage across the capacitor. To this end we first use the terminal relationships for the inductor and the resistor. This leads to the following form of equation (7.178):

$$v_C(t) + L\frac{di(t)}{dt} + Ri(t) = 0. \tag{7.179}$$

Since all three elements are connected in series, the same current flows through all these elements. This current can be expressed in terms of the voltage across the capacitor:

$$i(t) = C\frac{dv_C(t)}{dt}. \tag{7.180}$$

By substituting (7.180) into (7.179), we obtain:

$$LC\frac{d^2v_C(t)}{dt^2} + RC\frac{dv_C(t)}{dt} + v_C(t) = 0. \tag{7.181}$$

Now the initial condition (7.177) must be expressed in terms of $v_C(t)$ as well. This can be accomplished by combining (7.177) with (7.180), which leads to:

$$i(0_+) = C\frac{dv_C}{dt}(0_+) = 0, \tag{7.182}$$

$$\frac{dv_C}{dt}(0_+) = 0. \tag{7.183}$$

Thus, we arrive at the following initial value problem: find the solution of the equation:

$$LC\frac{d^2v_C(t)}{dt^2} + RC\frac{dv_C(t)}{dt} + v_C(t) = 0 \tag{7.184}$$

subject to the initial conditions:

$$v_C(0_+) = V_0, \tag{7.185}$$

$$\frac{dv_C}{dt}(0_+) = 0. \tag{7.186}$$

Differential equation (7.184) is a linear, homogeneous second-order differential equation with constant coefficients. As with any linear homogeneous differential equation with constant coefficients, we look for a solution in the form:

$$v_C(t) = Ae^{st}. \tag{7.187}$$

By substituting (7.187) into equation (7.184), we find:

$$s^2LCAe^{st} + sRCAe^{st} + Ae^{st} = 0. \tag{7.188}$$

By factoring out Ae^{st}, we obtain:

$$Ae^{st}(s^2LC + sRC + 1) = 0. \tag{7.189}$$

From (7.189) we conclude that the function in (7.187) will be a solution of the differential equation only if the exponent s satisfies the characteristic equation:

$$s^2LC + sRC + 1 = 0. \tag{7.190}$$

The last equation can be written in the equivalent form:

$$Z(s) = R + sL + \frac{1}{sC} = 0 \tag{7.191}$$

which again shows that the zeros of the characteristic equation are the zeros of the impedance as a function of complex frequency. The characteristic equation is a quadratic equation which has two roots (solutions) given by the formula:

$$s_{1,2} = -\frac{R}{2L} \pm \sqrt{\frac{R^2}{4L^2} - \frac{1}{LC}}. \tag{7.192}$$

As far as the roots of the characteristic equation are concerned, three different cases can be clearly distinguished. These cases depend on the *discriminant*, which is the quantity under the radical:

$$d = \frac{R^2}{4L^2} - \frac{1}{LC}. \tag{7.193}$$

If the discriminant is greater than zero, then s_1 and s_2 are both real and different. If it is equal to zero, then s_1 and s_2 are real and equal. If the discriminant is less than zero, then s_1 and s_2 are complex conjugates. Note that since $LC > 0$, the real part of both roots will always be negative (or zero if $R = 0$). It is well known from the theory of differential equations that the above three cases correspond to three distinct forms of the solutions of equation (7.184). These three forms of the solution are given below:

$$v_C(t) = \begin{cases} A_1 e^{s_1 t} + A_2 e^{s_2 t}, & \text{if } d > 0,\ (s_1 < 0, s_2 < 0), \\ A_1 e^{st} + A_2 t e^{st}, & \text{if } d = 0,\ (s_1 = s_2 = s), \\ e^{-\sigma t}(A_1 \cos \omega t + A_2 \sin \omega t), & \text{if } d < 0, \end{cases} \tag{7.194}$$

where σ and ω in (7.194) are determined by the expressions:

$$\sigma = \frac{R}{2L}, \tag{7.195}$$

$$\omega = \sqrt{\frac{1}{LC} - \frac{R^2}{4L^2}}. \tag{7.196}$$

Since in our problem we do not assume specific values for the circuit elements, we shall discuss the above three cases separately.

Case a): $d > 0$.

In this case, the solution to equation (7.184) and its derivative have the forms:

$$v_C(t) = A_1 e^{s_1 t} + A_2 e^{s_2 t}, \tag{7.197}$$

$$\frac{dv_C(t)}{dt} = s_1 A_1 e^{s_1 t} + s_2 A_2 e^{s_2 t}. \tag{7.198}$$

By setting $t = 0$ in these expressions and invoking the initial conditions (7.185) and (7.186), we obtain:

$$A_1 + A_2 = V_0, \tag{7.199}$$

$$s_1 A_1 + s_2 A_2 = 0. \tag{7.200}$$

By solving this simple system of equations, we find the values for A_1 and A_2:

$$A_1 = \frac{s_2 V_0}{s_2 - s_1}, \tag{7.201}$$

$$A_2 = \frac{s_1 V_0}{s_1 - s_2}. \tag{7.202}$$

By substituting (7.201) and (7.202) into (7.197), we arrive at the final solution:

$$v_C(t) = V_0 \left(\frac{s_2}{s_2 - s_1} e^{s_1 t} + \frac{s_1}{s_1 - s_2} e^{s_2 t} \right). \tag{7.203}$$

Next consider the case b): $d = 0$.

In this case the solution and its derivative take on the forms:

$$v_C(t) = A_1 e^{st} + A_2 t e^{st}, \tag{7.204}$$

$$\frac{dv_C(t)}{dt} = s A_1 e^{st} + A_2 e^{st} + s A_2 t e^{st}. \tag{7.205}$$

Now, by setting $t = 0$ and invoking the initial conditions (7.185) and (7.186), we obtain:

$$A_1 = V_0, \tag{7.206}$$

$$s A_1 + A_2 = 0, \tag{7.207}$$

$$A_2 = -s V_0. \tag{7.208}$$

Therefore, the final solution is:

$$v_C(t) = V_0 e^{st}(1 - st). \tag{7.209}$$

Finally, consider the case c): $d < 0$

For this case, the solution and its derivative have the forms:

$$v_C(t) = e^{-\sigma t}(A_1 \cos \omega t + A_2 \sin \omega t), \tag{7.210}$$

$$\frac{dv_C(t)}{dt} = -\sigma e^{-\sigma t}(A_1 \cos \omega t + A_2 \sin \omega t) + e^{-\sigma t}(-\omega A_1 \sin \omega t + \omega A_2 \cos \omega t). \tag{7.211}$$

By setting $t = 0$ and by invoking again the initial conditions, we get:

$$V_0 = A_1, \tag{7.212}$$

$$-\sigma A_1 + \omega A_2 = 0 \tag{7.213}$$

$$A_2 = \frac{\sigma V_0}{\omega}. \tag{7.214}$$

Thus, the final solution is:

$$v_C(t) = V_0 e^{-\sigma t}\left(\cos \omega t + \frac{\sigma}{\omega}\sin \omega t\right). \tag{7.215}$$

It is helpful to examine the plots of the solution for the above three different cases in order to gain some feeling for its behavior. The plot in Figure 7.13a corresponds to positive discriminant; and it is called the *overdamped* case. This means that ohmic losses due to the resistor do not allow for any oscillation in the solution. The plot in Figure 7.13b corresponds to zero discriminant. This is the borderline case, sometimes called *critically damped*. This means that losses due to the resistor are just enough to prevent any oscillation. Note that in general critically damped solutions decay more rapidly than overdamped solutions. The final plot (Figure 7.13c) corresponds to negative discriminant; it is called an *underdamped* solution, and it exhibits *damped oscillations*. This means that the solution oscillates while it decays.

Note that in each of the cases, the response always decays. This is the result of irreversible losses of energy in the resistor.

EXAMPLE 7.10 Consider the specific case of the circuit shown in Figure 7.12 when $L = 1$ H, $R = 2\ \Omega$, and $V_0 = 10$ V. Find the expression for the voltage across the capacitor (a) when $C = 4/3$ F, (b) when $C = 1$ F, and (c) when $C = 4/5$ F. Compare the time evolution of the capacitor voltage for each of the three cases above.

From equation (7.193), the discriminant for case (a) is found to be: $d = 2^2/4 - 3/4 = 1/4 > 0$. Thus, the circuit is overdamped and the two roots are given by equation (7.192):

$$s_1 = -1 + 1/2 = -1/2\ \text{s}^{-1}; \qquad s_2 = -1 - 1/2 = -3/2\ \text{s}^{-1}. \tag{7.216}$$

The expression for the capacitor voltage comes from equation (7.203):

$$v_C(t) = 10[3/2e^{-t/2} - 1/2e^{-3t/2}]\ \text{V}. \tag{7.217}$$

The discriminant for case (b) is $d = 2^2/4 - 1 = 0$ and the circuit is critically damped. Both roots are then equal to negative one and the capacitor voltage can be found from equation (7.209):

$$v_C(t) = 10(1 + t)e^{-t}\ \text{V}. \tag{7.218}$$

Finally, for case (c) we have $d = 2^2/4 - 5/4 = -1/4 < 0$ and the circuit is underdamped. From equations (7.195) and (7.196) we see that $\sigma = 1$ and $\omega = 1/2$. The expression for the capacitor voltage comes from equation (7.215):

$$v_C(t) = 10e^{-t}[\cos(t/2) + 2\sin(t/2)]\ \text{V}. \tag{7.219}$$

Sample values of the capacitor voltages at some selected times are given in Table 7.1. You can see that the overdamped case decays considerably more slowly than the other two cases. That decay is limited by the smallest of the two roots. In fact, when

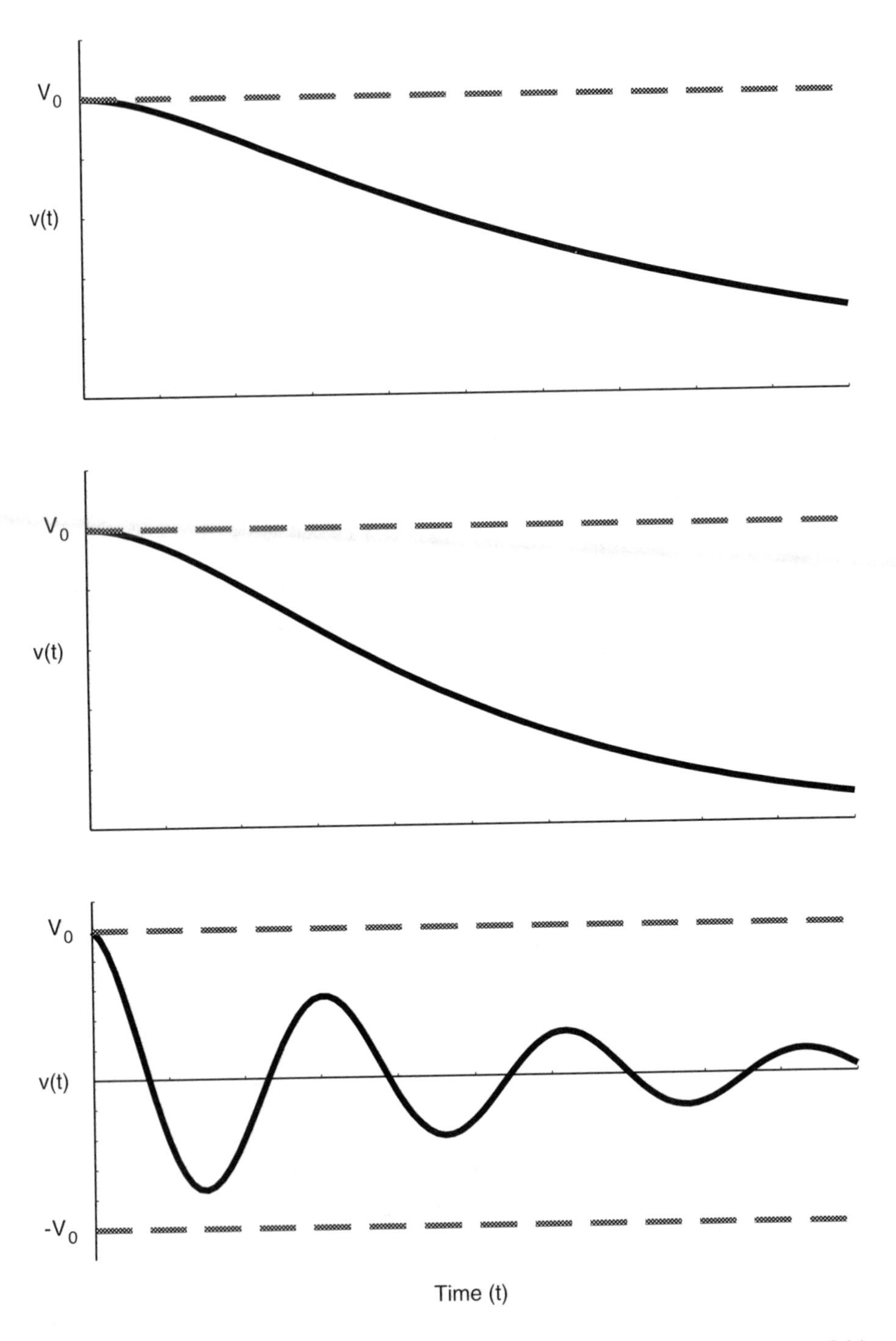

Figure 7.13: Plots for: (a) overdamped ($d > 0$), (b) critically damped ($d = 0$), and (c) underdamped ($d < 0$) solutions.

Table 7.1: The capacitor voltages $v_C(t)$ V at specified times for the above three cases.

t(s)	**case (a)**	**case (b)**	**case (c)**
0	10.00	10.00	10.00
1	7.98	7.36	6.76
2	5.27	4.06	3.01
5	1.23	0.404	.027
10	0.101	0.005	$-.74 \times 10^{-3}$
20	0.68×10^{-3}	0.43×10^{-6}	$-.40\times10^{-7}$

$t \geq 5$ s, the second term in the expression for the overdamped case can essentially be ignored. The underdamped circuit clearly oscillates (goes negative) and decays most rapidly of all the cases. ■

EXAMPLE 7.11 Although there can never be a real-world circuit that has *no* energy losses, these inevitable losses can be reduced to a very small, negligible amount. This justifies the consideration of a circuit which contains only an inductor and a capacitor (see Figure 7.14).

This circuit is a particular case of the circuit we considered earlier ($R = 0$) and is called an undamped circuit. From equation (7.192), we can see that:

$$s_{1,2} = \pm\frac{j}{\sqrt{LC}}. \tag{7.220}$$

The discriminant in this case is less than zero so the roots s_1 and s_2 are complex conjugates. Because there is no resistor, their real parts are zero:

$$\sigma = 0, \tag{7.221}$$

and:

$$\omega = \frac{1}{\sqrt{LC}}. \tag{7.222}$$

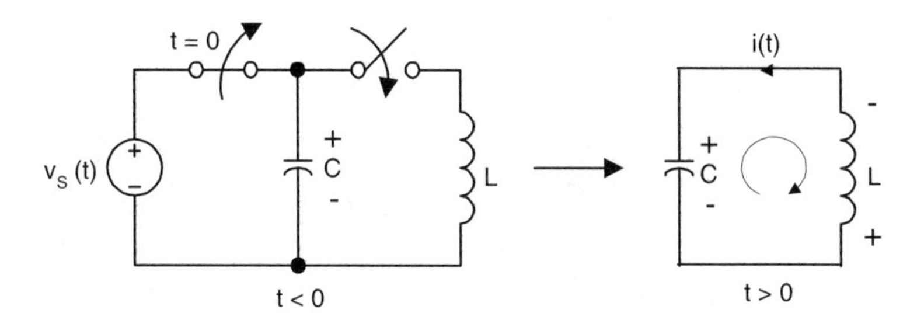

Figure 7.14: An *LC* circuit with no resistors.

From (7.215), we find that:

$$v_C(t) = V_0 \cos \omega t. \tag{7.223}$$

From equation (7.223), we see that the response of this circuit never decays, it just continuously oscillates between V_0 and $-V_0$. The frequency of the oscillation, $\omega = 1/\sqrt{LC}$, is called the *natural frequency* of the circuit. This is because this frequency is not caused by a source (indeed, there is no source), but rather it is due to the structure of the circuit itself. Thus, a simple LC circuit can be considered as a generator of harmonic oscillations, and the frequency of these oscillations can be chosen by the appropriate selection of L and C. In practice, however, there are always some losses. To maintain the oscillations, the lost energy should be continuously replenished without disrupting the oscillations. This can be achieved by using LC circuits in combination with active elements—transistors.

The expression for the current through the circuit can be found from the previous expression for the voltage across the capacitor:

$$i(t) = C\frac{dv_C(t)}{dt} = -\omega C V_0 \sin \omega t. \tag{7.224}$$

By using the above expressions for voltage and current, we can make an interesting observation concerning the energy stored in the capacitor and the inductor. For the capacitor we have

$$w_e(t) = \frac{Cv_C^2(t)}{2} = \frac{CV_0^2 \cos^2 \omega t}{2}, \tag{7.225}$$

while for the inductor we find:

$$w_m(t) = \frac{Li^2(t)}{2} = \frac{CV_0^2 \sin^2 \omega t}{2}. \tag{7.226}$$

The voltage across the capacitor and the current through the inductor as well as the energy stored in the capacitor and the inductor are represented by the graphs in Figure 7.15. It is apparent from these graphs that as the current increases, the voltage

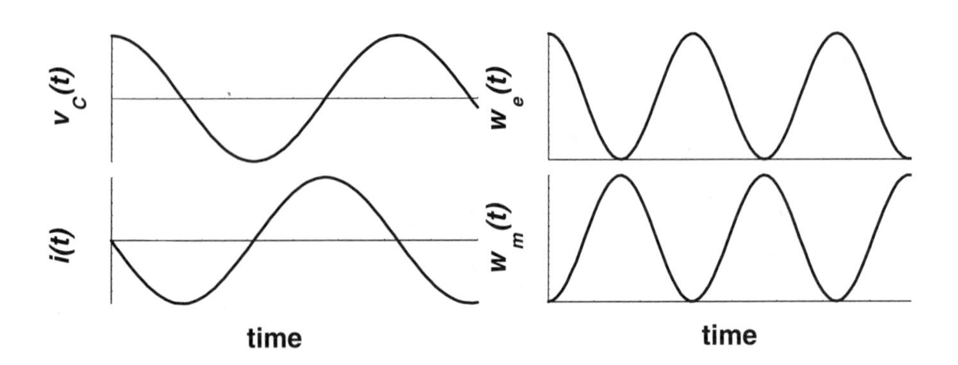

Figure 7.15: Graphs of voltage, current, and energy in the LC circuit.

across the capacitor decreases, and the energy transfers from the electric field of the capacitor to the magnetic field of the inductor. On the other hand, as the current decreases, the voltage across the capacitor increases, and the energy transfers from the magnetic field of the inductor to the electric field of the capacitor. This energy exchange between the capacitor and the inductor is the physical mechanism of the oscillations.

The *LC* circuits found practical application years ago in various electric devices generating electromagnetic oscillations at various frequencies. They performed this duty well, except at high frequencies when energy losses due to radiation and some other imperfections appeared. Nowadays, their role has been filled by modern circuits which avoid the use of inductors (difficult-to-work-with circuit elements that are both expensive and hard to miniaturize). ■

7.3.2 Circuits Excited by Sources

Now, we consider a second-order circuit that is excited by a source (Figure 7.16). We begin as usual with the initial conditions, which in this case are obvious:

$$v(0_+) = v(0_-) = 0, \tag{7.227}$$

$$i(0_+) = i(0_-) = 0, \tag{7.228}$$

where $v(t)$ stands for the voltage across the capacitor and $i(t)$ is the current through all of the elements. By using KVL and the terminal relationships for the inductor and resistor, we obtain:

$$v(t) + L\frac{di(t)}{dt} + Ri(t) = v_s(t). \tag{7.229}$$

Then, by invoking the expression

$$i(t) = C\frac{dv(t)}{dt}, \tag{7.230}$$

we arrive at the following initial value problem: find the solution of the equation:

$$LC\frac{d^2v(t)}{dt^2} + RC\frac{dv(t)}{dt} + v(t) = v_s(t) \tag{7.231}$$

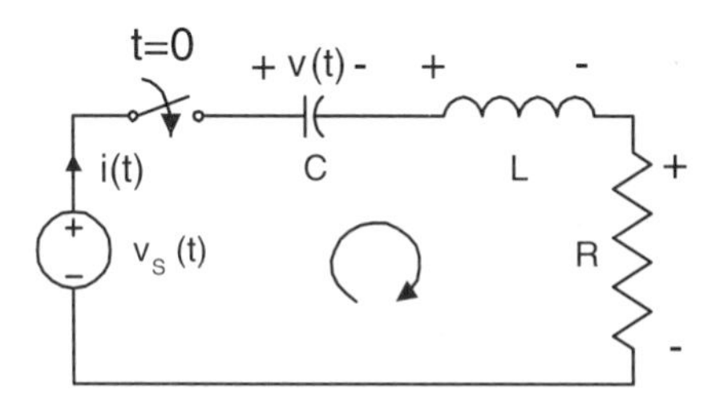

Figure 7.16: Second-order circuit with source.

subject to the initial conditions:

$$v(0_+) = 0, \tag{7.232}$$

$$\frac{dv}{dt}(0_+) = 0. \tag{7.233}$$

Since we deal here with a *nonhomogeneous* differential equation, its complete solution is a sum of the homogeneous and particular solutions:

$$v(t) = v_h(t) + v_p(t). \tag{7.234}$$

We already know how to find the homogeneous solution. First, we must solve the characteristic equation:

$$s^2LC + sRC + 1 = 0. \tag{7.235}$$

For the sake of being specific, we will assume that this characteristic equation has two real and different roots. This means the homogeneous solution has the form:

$$v_h(t) = A_1e^{s_1t} + A_2e^{s_2t}. \tag{7.236}$$

Now, the particular solution must be found. The form of this solution depends on the nature of the voltage source. We first consider a dc voltage source:

$$v_s(t) = V_0 = \text{ constant.} \tag{7.237}$$

In this case, the particular solution has the physical meaning of the dc steady-state voltage across the capacitor. Therefore, it is clear that:

$$v_p(t) = V_0. \tag{7.238}$$

Now, the complete solution to differential equation (7.231) is:

$$v(t) = V_0 + A_1e^{s_1t} + A_2e^{s_2t}. \tag{7.239}$$

Constants A_1 and A_2 remain to be found. By differentiating (7.239) and setting $t = 0$, from initial conditions (7.232) and (7.233) we find:

$$V_0 + A_1 + A_2 = 0, \tag{7.240}$$

$$s_1A_1 + s_2A_2 = 0. \tag{7.241}$$

These two equations can be easily solved to find A_1 and A_2:

$$A_1 = \frac{s_2V_0}{s_1 - s_2}, \tag{7.242}$$

$$A_2 = \frac{s_1V_0}{s_2 - s_1}. \tag{7.243}$$

Thus, the final solution is:

$$v(t) = V_0 \left(1 + \frac{s_2}{s_1 - s_2} e^{s_1 t} + \frac{s_1}{s_2 - s_1} e^{s_2 t}\right). \tag{7.244}$$

From the expression for the voltage, we can easily find the expression for the current through the circuit (see equation (7.230)):

$$i(t) = CV_0 \left(\frac{s_2 s_1}{s_1 - s_2} e^{s_1 t} + \frac{s_1 s_2}{s_2 - s_1} e^{s_2 t}\right). \tag{7.245}$$

EXAMPLE 7.12 Consider the same circuit excited by an ac voltage source (see Figure 7.16):

$$v_s(t) = V_{ms} \cos(\omega t + \phi_s). \tag{7.246}$$

In this case, differential equation (7.231) has the form:

$$LC \frac{d^2 v(t)}{dt^2} + RC \frac{dv(t)}{dt} + v(t) = V_{ms} \cos(\omega t + \phi_s). \tag{7.247}$$

The homogeneous solution for this equation is the same as before (see expression (7.194)). The particular solution for equation (7.247) has the physical meaning of the ac steady-state voltage across the capacitor in the circuit shown in Figure 7.16. For this reason, we shall employ the phasor technique to find the particular solution.

First, we transform the voltage (7.246) of the ac source into phasor form:

$$\hat{V}_s = V_{ms} e^{j\phi_s}. \tag{7.248}$$

Next, we can find the phasor $\hat{I}$ of the current through the circuit:

$$\hat{I} = \frac{\hat{V}_s}{Z} = \frac{\hat{V}_s}{R + j(\omega L - \frac{1}{\omega C})}. \tag{7.249}$$

Finally, we can find the phasor of the voltage across the capacitor:

$$\hat{V}_C = -\frac{j}{\omega C} \hat{I} = \hat{V}_s \frac{-\frac{j}{\omega C}}{R + j(\omega L - \frac{1}{\omega C})}. \tag{7.250}$$

From (7.250) we obtain:

$$\hat{V}_C = V_{mc} e^{j\phi_c}, \tag{7.251}$$

$$V_{mc} = V_{ms} \frac{\frac{1}{\omega C}}{\sqrt{R^2 + (\omega L - \frac{1}{\omega C})^2}}, \tag{7.252}$$

$$\phi_c = \phi_s - \phi - \frac{\pi}{2}, \tag{7.253}$$

where

$$\phi = \arctan \frac{\omega L - \frac{1}{\omega C}}{R}. \tag{7.254}$$

By transforming the phasor $\hat{V}_C$ into the time domain, we end up with the particular solution:

$$v_p(t) = V_{ms} \frac{\frac{1}{\omega C}}{\sqrt{R^2 + (\omega L - \frac{1}{\omega C})^2}} \cos\left(\omega t + \phi_s - \phi - \frac{\pi}{2}\right). \tag{7.255}$$

To be specific, we assume that the discriminant is positive and, consequently, the homogeneous solution has the form:

$$v_h(t) = A_1 e^{s_1 t} + A_2 e^{s_2 t}. \tag{7.256}$$

Then, the general solution of equation (7.247) is:

$$v(t) = A_1 e^{s_1 t} + A_2 e^{s_2 t} + V_{ms} \frac{\frac{1}{\omega C}}{\sqrt{R^2 + (\omega L - \frac{1}{\omega C})^2}} \cos\left(\omega t + \phi_s - \phi - \frac{\pi}{2}\right). \tag{7.257}$$

From (7.257), we obtain:

$$\frac{dv(t)}{dt} = s_1 A_1 e^{s_1 t} + s_2 A_2 e^{s_2 t} - V_{ms} \frac{\frac{1}{C}}{\sqrt{R^2 + (\omega L - \frac{1}{\omega C})^2}} \sin\left(\omega t + \phi_s - \phi - \frac{\pi}{2}\right). \tag{7.258}$$

By setting $t = 0$ in (7.257) and (7.258) and by using initial conditions (7.232) and (7.233), we arrive at the following equations for A_1 and A_2:

$$A_1 + A_2 = -V_{ms} \frac{\frac{1}{\omega C} \cos(\phi_s - \phi - \frac{\pi}{2})}{\sqrt{R^2 + (\omega L - \frac{1}{\omega C})^2}}, \tag{7.259}$$

$$s_1 A_1 + s_2 A_2 = V_{ms} \frac{\frac{1}{C} \sin(\phi_s - \phi - \frac{\pi}{2})}{\sqrt{R^2 + (\omega L - \frac{1}{\omega C})^2}}. \tag{7.260}$$

From (7.259) and (7.260), we derive:

$$A_1 = \frac{V_{ms}[\frac{s_2}{\omega C} \sin(\phi_s - \phi) - \frac{1}{C} \cos(\phi_s - \phi)]}{(s_1 - s_2)\sqrt{R^2 + (\omega L - \frac{1}{\omega C})^2}}, \tag{7.261}$$

$$A_2 = \frac{V_{ms}[\frac{s_1}{\omega C} \sin(\phi_s - \phi) - \frac{1}{C} \cos(\phi_s - \phi)]}{(s_2 - s_1)\sqrt{R^2 + (\omega L - \frac{1}{\omega C})^2}}. \tag{7.262}$$

By substituting (7.261) and (7.262) into (7.257) we obtain the explicit analytical form of the solution. ■

EXAMPLE 7.13 Consider the circuit shown in Figure 7.17, where the source current is given by:

$$i_s(t) = 10\cos(t) \text{ A}. \tag{7.263}$$

Assuming zero initial conditions for the current through the inductor and the voltage across the capacitor, derive the expression for the current through the inductor.

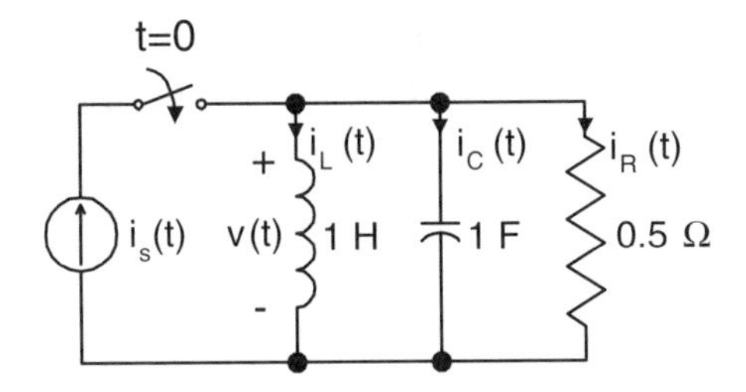

Figure 7.17: A second-order circuit example.

By using KCL, we find:

$$i_L(t) + i_C(t) + i_R(t) = i_s(t). \tag{7.264}$$

The voltage $v(t)$ across the inductor, capacitor, and resistor can be expressed in terms of the current through the inductor as follows:

$$v(t) = L\frac{di_L(t)}{dt}. \tag{7.265}$$

By using this expression for the voltage and the terminal relationships for capacitors and resistors, we express $i_C(t)$ and $i_R(t)$ in terms of $i_L(t)$:

$$i_C(t) = C\frac{dv(t)}{dt} = CL\frac{d^2 i_L(t)}{dt^2}, \tag{7.266}$$

$$i_R(t) = \frac{v(t)}{R} = \frac{L}{R}\frac{di_L(t)}{dt}. \tag{7.267}$$

Now, by substituting (7.266) and (7.267) into (7.264) and rearranging the terms, we derive the differential equation for $i_L(t)$:

$$CL\frac{d^2 i_L(t)}{dt^2} + \frac{L}{R}\frac{di_L(t)}{dt} + i_L(t) = i_s(t). \tag{7.268}$$

By substituting the expression for $i_s(t)$ and the values for R, L, and C, we obtain:

$$\frac{d^2 i_L(t)}{dt^2} + 2\frac{di_L(t)}{dt} + i_L(t) = 10\cos t. \tag{7.269}$$

By using zero initial conditions for the current through the inductor and the voltage across the capacitor along with (7.265), we obtain:

$$i_L(0_+) = 0, \tag{7.270}$$

$$\frac{di_L}{dt}(0_+) = 0. \tag{7.271}$$

Expressions (7.269), (7.270), and (7.271) constitute the initial value problem for $i_L(t)$. To solve this problem we first consider the characteristic equation for (7.269):

$$s^2 + 2s + 1 = 0. \tag{7.272}$$

From this equation we find:

$$s_1 = s_2 = -1 \text{ s}^{-1}. \tag{7.273}$$

Consequently, the transient (free) response has the form:

$$i_{L_h}(t) = A_1 e^{-t} + A_2 t e^{-t}. \tag{7.274}$$

Next, we shall use the phasor technique to find the particular solution of equation (7.269). It is clear that:

$$\hat{I}_s = 10 \text{ A}, \tag{7.275}$$

$$Y_{in} = \frac{1}{j} + j + 2 = 2\ \mho. \tag{7.276}$$

From (7.275) and (7.276) we find the phasor $\hat{V}$ of the voltage across the inductor:

$$\hat{V} = \frac{\hat{I}_s}{Y_{in}} = 5 \text{ V}. \tag{7.277}$$

Now, we can find the phasor $\hat{I}_L$ of the current through the inductor:

$$\hat{I}_L = \frac{\hat{V}}{j\omega L} = \frac{5}{j} = -j5 \text{ A}. \tag{7.278}$$

By using (7.278), we can write the particular solution $i_{L_p}(t)$ for (7.269) as follows:

$$i_{L_p}(t) = 5\cos\left(t - \frac{\pi}{2}\right) = 5\sin t \text{ A}. \tag{7.279}$$

By combining (7.274) and (7.279), we obtain the total response:

$$i_L(t) = 5\sin t + A_1 e^{-t} + A_2 t e^{-t}. \tag{7.280}$$

To find A_1 and A_2, we use the initial conditions (7.270) and (7.271), which yield:

$$i_L(0_+) = A_1 = 0, \tag{7.281}$$

$$\frac{di_L}{dt}(0_+) = 5 + A_2 - A_1 = 0. \tag{7.282}$$

By taking into account (7.281) in (7.282), we obtain:

$$A_2 = -5. \tag{7.283}$$

By substituting (7.281) and (7.283) into (7.280), we get the final result

$$i_L(t) = 5\sin t - 5te^{-t} \text{ A}. \tag{7.284}$$

■

7.4 Transfer Functions and Their Applications

It is clear from the previous discussion that the most involved part of transient analysis is the derivation of the differential equation for the appropriate circuit variable. This difficulty can be circumvented by using the notion of a transfer function. This notion is central to the system and signal areas and it also permeates many different branches of electrical engineering. As far as transient analysis is concerned, the machinery of the transfer function allows one to avoid completely the derivation of differential equations and to solve transient problems by using mostly algebraic manipulations on phasors.

By definition, transfer functions are the ratios of output to input signals. In the case of electric circuits, the input can be defined as a source which drives an electric circuit, while the output is the circuit variable which we are interested in calculating. For example, in the case of the *RL* circuit shown in Figure 7.18a, the input is the voltage source, while the output is the current through the circuit. So, the transfer function is:

$$H(t) = \frac{i_o(t)}{v_s(t)}. \tag{7.285}$$

In the case of the *RC* circuit shown in Figure 7.18b, the input is the current source, $i_s(t)$, while the output is the voltage, $v_o(t)$, across the capacitor. Thus, the transfer function is:

$$H(t) = \frac{v_o(t)}{i_s(t)}. \tag{7.286}$$

Transfer functions are generally used not in the time domain but rather in the frequency domain. If we consider ac excitation, then we can define the transfer function as the ratio of the output phasor to the input phasor. In this case, expressions (7.285) and (7.286) become, respectively:

$$\hat{H}(j\omega) = \frac{\hat{I}_o(j\omega)}{\hat{V}_s} = \frac{1}{Z(j\omega)} = \frac{1}{R + j\omega L}, \tag{7.287}$$

$$\hat{H}(j\omega) = \frac{\hat{V}_o(j\omega)}{\hat{I}_s} = \frac{1}{Y(j\omega)} = \frac{1}{\frac{1}{R} + j\omega C}. \tag{7.288}$$

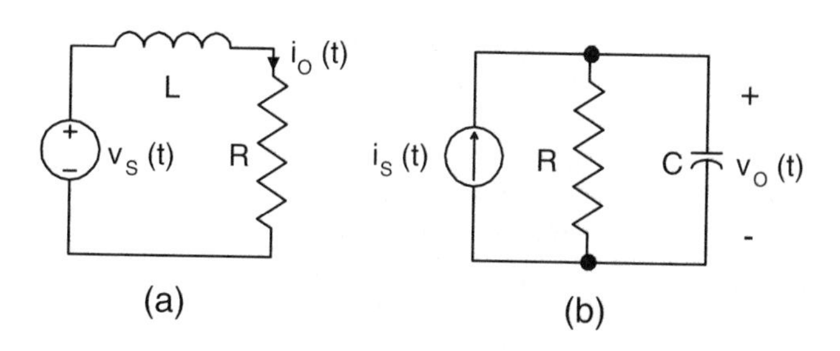

Figure 7.18: Two first-order circuits.

If the above circuits are driven by sources of complex frequency s, then the transfer function $\hat{H}$ will be a function of s as well and expressions (7.287) and (7.288) can be written as follows:

$$\hat{H}(s) = \frac{1}{Z(s)} = \frac{1}{R + sL}, \tag{7.289}$$

$$\hat{H}(s) = \frac{1}{Y(s)} = \frac{1}{\frac{1}{R} + sC}. \tag{7.290}$$

The circuits shown in Figure 7.18a and b have been discussed in Examples 7.5 and 7.7, respectively, and we have reached the following conclusion: zeros of $Z(s)$ and $Y(s)$ are the exponents of transient responses while the values of $Z(s)$ and $Y(s)$ at the frequency of excitation determine the steady-state responses. By using expressions (7.289) and (7.290), this conclusion can be rephrased in the following way: poles of the transfer functions are exponents of free (transient) responses, while the values of the transfer functions at the excitation frequency determine steady-state responses.

This conclusion applies to circuits with more than one energy storage element. Indeed, let us recall the *RLC* series circuit discussed in Example 7.12. From expression (7.250), we can find the transfer function:

$$H(j\omega) = \frac{\hat{V}_C}{\hat{V}_s} = \frac{\frac{1}{j\omega C}}{R + j\omega L + \frac{1}{j\omega C}}. \tag{7.291}$$

If we consider this transfer function as the function of complex frequency s, we arrive at:

$$\hat{H}(s) = \frac{\frac{1}{sC}}{R + sL + \frac{1}{sC}} = \frac{1}{s^2LC + sRC + 1}. \tag{7.292}$$

Now, we again observe that:
The values of the transfer functions at the frequency of excitation, $s = j\omega$, completely determine the ac steady-state response of linear electric circuits, while the poles of the transfer function coincide with the roots of the characteristic equation (see (7.235)) and, consequently, they determine the form of the transient (free) response.

The stated fact is very general in nature and is extensively used in linear system theory. The above discussion suggests that the machinery of the transfer function can be very powerful in the analysis of transients in electric circuits. This analysis can proceed as follows. A circuit variable, which we are interested in, can be identified as the output. By using the phasor technique, we can find the transfer function $\hat{H}(j\omega)$ for this variable and, consequently, the ac steady-state response. Then, we can find the poles of the transfer function $\hat{H}(s)$ which determine the form of the transient (free) response. The unknown constants in this free response should be determined from the initial conditions for the chosen circuit variable. The described approach is algebraic in nature; it completely avoids the derivation and solution of differential equations and it fully exploits the machinery of the phasor technique.

The assertion that the poles of the transfer function coincide with complex frequencies of the free (transient) response can be in general explained by using the following reasoning.

Consider a circuit which is excited by a voltage source $v_s(t)$ with complex frequency s. We are interested in a voltage $v_k(t)$ across the kth branch. According to the definition of the transfer function, we have:

$$\hat{V}_k = \hat{H}_k(s)\hat{V}_s, \tag{7.293}$$

where $\hat{V}_s$ and $\hat{V}_k$ are the phasors of the voltage source and the voltage across the kth branch, respectively.

Now, we shall reduce the peak value of the voltage source (and, consequently, $\hat{V}_s$) to zero. According to the last expression, $\hat{V}_k$ can be nonzero only for such complex frequencies s that $\hat{H}_k(s) = \infty$. It is clear that such complex frequencies are the poles of the transfer function. It is also clear that for these complex frequencies, the "sourceless" voltage $v_k(t)$ across branch number k can exist. Since this is a sourceless voltage, it has the physical meaning of free (transient) response.

It is interesting to note that by using the transfer function we can also find the dc steady-state response. For example, consider the RLC series circuit shown in Figure 7.16. In the case of $v_s(t) = V_0 =$ constant, from (7.292) we derive:

$$v_p(t) = H(0)v_s(t) = V_0, \tag{7.294}$$

which corresponds to the result given in (7.238).

Next, we shall demonstrate the transfer function approach by considering the following examples.

EXAMPLE 7.14 Recall the RLC parallel circuit considered in Example 7.13 (shown in Figure 7.17), for which we have found the inductor current by deriving and solving a second-order differential equation. Now we show that, by using the transfer function approach, we can arrive at the same result without having to perform any calculus.

From the definition of the transfer function and the current divider rule, we obtain:

$$\hat{H}(s) = \frac{\hat{I}_L}{\hat{I}_s} = \frac{Y_L}{Y_L + Y_C + Y_R} = \frac{\frac{1}{sL}}{\frac{1}{sL} + sC + \frac{1}{R}}. \tag{7.295}$$

By substituting the values $R = 0.5\ \Omega$, $L = 1$ H and $C = 1$ F and making simple transformations, we derive:

$$\hat{H}(s) = \frac{\frac{1}{s}}{\frac{1}{s} + s + 2} = \frac{1}{s^2 + 2s + 1}. \tag{7.296}$$

Next, we find the poles of the transfer function:

$$s^2 + 2s + 1 = 0, \tag{7.297}$$

$$s_1 = s_2 = -1\ \mathrm{s}^{-1}. \tag{7.298}$$

Consequently, the free (transient) response has the form:

$$i_{L_h}(t) = A_1 e^{-t} + A_2 t e^{-t}. \tag{7.299}$$

To find the forced (ac steady state) response, we evaluate $\hat{H}(s)$ at $s = j \cdot 1$:

$$\hat{H}(j \cdot 1) = \frac{1}{j^2 + 2j + 1} = \frac{1}{2j} = -\frac{j}{2}. \tag{7.300}$$

From (7.275) and (7.300), we obtain:

$$\hat{I}_L = \hat{H}(j \cdot 1)\hat{I}_s = -\frac{j}{2} \cdot 10 = -j5 \text{ A}. \tag{7.301}$$

By transforming phasor $\hat{I}_L$ into the time domain, we end up with the forced response:

$$i_{L_p}(t) = 5\cos\left(t - \frac{\pi}{2}\right) = 5\sin t \text{ A}. \tag{7.302}$$

By combining (7.299) and (7.302), we obtain the total response:

$$i_L(t) = 5\sin t + A_1 e^{-t} + A_2 t e^{-t}, \tag{7.303}$$

which coincides with (7.280) as it must. What is left is to use the initial conditions (7.270) and (7.271) in order to find A_1 and A_2. However, this part of the solution process is literally the same as before. That is why it is omitted here.

It is clear from the above example that the transfer function approach is purely algebraic in nature and completely avoids the derivation of any differential equations.

■

EXAMPLE 7.15 Consider the electric circuit shown in Figure 7.19. This circuit contains two energy storage elements (capacitors C_1 and C_2); that is why this is a second-order circuit. We want to find the voltage $v_2(t)$ across the capacitor C_2.

First, we shall find the transfer function $\hat{H}(s)$, which in our case is defined as the ratio of the phasor $\hat{V}_2$ of the voltage $v_2(t)$ to the phasor $\hat{V}_s$ of the source voltage:

$$\hat{H}(s) = \frac{\hat{V}_2}{\hat{V}_s}. \tag{7.304}$$

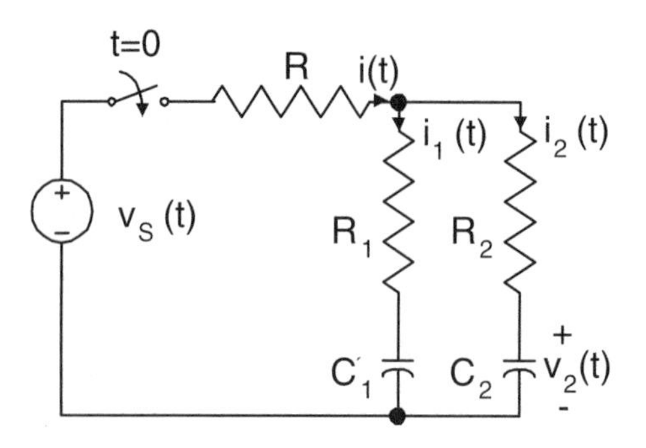

Figure 7.19: A second-order circuit example.

To find $\hat{V}_2$, we characterize each branch of the circuit by impedance. There are three branches in the circuit and their impedances as functions of complex frequency s are given by the following expressions:

$$Z = R, \quad Z_1 = R_1 + \frac{1}{sC_1}, \quad Z_2 = R_2 + \frac{1}{sC_2}. \tag{7.305}$$

By using the above impedances, we derive the following expression for the equivalent input impedance:

$$Z_{eq} = Z + \frac{Z_1 Z_2}{Z_1 + Z_2} = \frac{Z_1 Z + Z_2 Z + Z_1 Z_2}{Z_1 + Z_2}. \tag{7.306}$$

We next find the phasor $\hat{I}$ of the total current $i(t)$:

$$\hat{I} = \frac{\hat{V}_s}{Z_{eq}} = \hat{V}_s \frac{Z_1 + Z_2}{Z_1 Z + Z_2 Z + Z_1 Z_2}. \tag{7.307}$$

By using the current divider rule, we derive:

$$\hat{I}_2 = \hat{I} \frac{Z_1}{Z_1 + Z_2} = \hat{V}_s \frac{Z_1}{Z_1 Z + Z_2 Z + Z_1 Z_2}. \tag{7.308}$$

Now, we can obtain the expression for $\hat{V}_2$:

$$\hat{V}_2 = \hat{I}_2 \frac{1}{sC_2} = \hat{V}_s \frac{Z_1}{sC_2(Z_1 Z + Z_2 Z + Z_1 Z_2)}. \tag{7.309}$$

From (7.304) and (7.309), we conclude:

$$\hat{H}(s) = \frac{Z_1}{sC_2(Z_1 Z + Z_2 Z + Z_1 Z_2)}. \tag{7.310}$$

By substituting (7.305) into (7.310) and by using simple algebraic transformations, we derive:

$$\hat{H}(s) = \frac{sR_1 C_1 + 1}{s^2 C_1 C_2 (RR_1 + RR_2 + R_1 R_2) + s(C_2 R + C_1 R + C_1 R_1 + C_2 R_2) + 1}. \tag{7.311}$$

Now, we can find the poles of the transfer function. They are the roots of the equation:

$$s^2 C_1 C_2 (RR_1 + RR_2 + R_1 R_2) + s(C_2 R + C_1 R + C_1 R_1 + C_2 R_2) + 1 = 0. \tag{7.312}$$

To be specific, we will assume that this characteristic equation has two real and distinct roots s_1 and s_2. This means that the free (transient) response has the form:

$$v_{2h}(t) = A_1 e^{s_1 t} + A_2 e^{s_2 t}. \tag{7.313}$$

In the cases when the roots of the characteristic equation are real and identical or complex and conjugate, the free response will be given by expressions similar to (7.204) and (7.210), respectively.

Next, we assume that the circuit is excited by an ac voltage source:

$$v_s(t) = V_{ms}\cos(\omega t + \phi_s). \tag{7.314}$$

This means that

$$\hat{V}_s = V_{ms}e^{j\phi_s}. \tag{7.315}$$

To find the forced (ac steady state) response, we evaluate $\hat{H}(s)$ at $s = j\omega$:

$$\hat{H}(j\omega) = |\hat{H}(j\omega)|e^{j\phi_H}. \tag{7.316}$$

From (7.304), (7.315), and (7.316) we find:

$$\hat{V}_2 = V_{ms}|\hat{H}(j\omega)|e^{j(\phi_s+\phi_H)}. \tag{7.317}$$

By transforming phasor $\hat{V}_2$ into the time domain, we end up with the forced response:

$$v_{2p}(t) = V_{ms}|\hat{H}(j\omega)|\cos(\omega t + \phi_s + \phi_H). \tag{7.318}$$

By combining (7.313) and (7.318), we obtain the total response:

$$v_2(t) = A_1e^{s_1t} + A_2e^{s_2t} + V_{ms}|\hat{H}(j\omega)|\cos(\omega t + \phi_s + \phi_H). \tag{7.319}$$

The previous discussion clearly illustrates how the machinery of the transfer function can be used in order to find the full response of the electric circuit.

The unknown constants in (7.319) should be determined from the initial conditions for $v_2(t)$. We assume that before switching ($t < 0$) the capacitors C_1 and C_2 were not charged. This means that

$$v_1(0_+) = 0, \tag{7.320}$$

$$v_2(0_+) = 0. \tag{7.321}$$

Expression (7.321) gives us one initial condition for $v_2(t)$. In order to find the second initial condition for $v_2(t)$, we consider the circuit shown in Figure 7.19 **at the initial instant of time** $t = 0$.

By using (7.320) and (7.321), this circuit can be redrawn in the way shown in Figure 7.20.

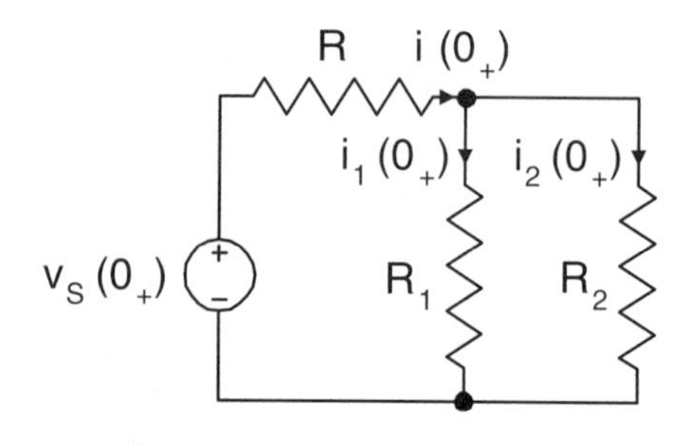

Figure 7.20: The circuit at $t = 0$.

This is because at time $t = 0$, the uncharged capacitors can be replaced by short circuits. The circuit shown in Figure 7.20 can be interpreted as a dc resistive circuit, which makes its analysis quite simple. It is clear from the above circuit that

$$i(0_+) = \frac{v_s(0_+)}{R_{eq}}, \tag{7.322}$$

where

$$R_{eq} = R + \frac{R_1 R_2}{R_1 + R_2} = \frac{R_1 R + R_2 R + R_1 R_2}{R_1 + R_2}. \tag{7.323}$$

By using the current divider rule, we find

$$i_2(0_+) = i(0_+)\frac{R_1}{R_1 + R_2}. \tag{7.324}$$

From (7.322), (7.323), and (7.324), we obtain:

$$i_2(0_+) = v_s(0_+)\frac{R_1}{R_1 R + R_2 R + R_1 R_2}. \tag{7.325}$$

Now, according to the circuit shown in Figure 7.19, we have:

$$i_2(t) = C_2 \frac{dv_2(t)}{dt}. \tag{7.326}$$

From (7.325) and (7.326), we find:

$$\frac{dv_2}{dt}(0_+) = \frac{v_s(0_+) R_1}{C_2(R_1 R + R_2 R + R_1 R_2)}. \tag{7.327}$$

Thus, we have determined the second initial condition for $v_2(t)$. The way we have achieved this is quite general in nature and can be summarized as follows.

To find the initial values for electric currents in a circuit, we redraw this circuit for the initial instant of time $t = 0$ by replacing uncharged capacitors by short circuits and "currentless" inductors by open circuits. As a result, we obtain a dc resistive circuit which can be used to find initial values for all currents and voltages. Then, by employing terminal relationships for capacitors and inductors, we determine additional initial conditions for energy storage elements.

In the case of nonzero initial conditions, capacitors and inductors should be "replaced" by dc voltage and current sources, respectively. As a result, we arrive at dc resistive circuits with several sources. Analysis of these circuits yields initial values for all currents and voltages.

To finish the solution of our problem, we shall use initial conditions (7.321) and (7.327) to find the unknown constants A_1 and A_2 in (7.319). From (7.319), (7.321), and (7.327) we obtain:

$$A_1 + A_2 = -V_{ms}|\hat{H}(j\omega)|\cos(\phi_s + \phi_H), \tag{7.328}$$

$$s_1 A_1 + s_2 A_2 = \omega V_{ms}|\hat{H}(j\omega)|\sin(\phi_s + \phi_H) + \frac{V_{ms} R_1 \cos\phi_s}{C_2(R_1 R + R_2 R + R_1 R_2)} \tag{7.329}$$

where we have used the fact that according to (7.314) $v_s(0) = V_{ms}\cos\phi_s$. Simultaneous equations (7.328) and (7.329) can be solved for A_1 and A_2. By substituting the found values for A_1 and A_2 into (7.319), we end up with the final expression for $v_2(t)$. This completes the solution of the problem. ■

EXAMPLE 7.16 Consider the circuit shown in Figure 7.21. It is assumed that the initial voltages across the capacitors C_1 and C_2 are equal to zero. We intend to find the voltage $v_1(t)$ across resistor R_1 for the following values of circuit parameters:

$$C_1 = 10^{-6}\text{ F},\ R_1 = 10^6\ \Omega,\ C_2 = 10^{-6}\text{ F},\ R_2 = 0.5\times 10^6\ \Omega.$$

By using the definition of transfer function and the voltage divider rule, we derive:

$$\hat{H}(s) = \frac{\hat{V}_1}{\hat{V}_s} = \frac{R_1}{R_1 + \frac{1}{sC_1} + \frac{R_2\cdot\frac{1}{sC_2}}{R_2+\frac{1}{sC_2}}}. \tag{7.330}$$

By using simple algebraic transformations, we find:

$$\hat{H}(s) = \frac{R_1}{R_1 + \frac{1}{sC_1} + \frac{R_2}{sR_2C_2+1}}, \tag{7.331}$$

$$\hat{H}(s) = \frac{sR_1C_1(sR_2C_2+1)}{s^2R_1R_2C_1C_2 + s(R_1C_1 + R_2C_2 + R_2C_1) + 1}. \tag{7.332}$$

By substituting the values of C_1, C_2, R_1, and R_2 into (7.332), we obtain:

$$\hat{H}(s) = \frac{s(s+2)}{s^2+4s+2}. \tag{7.333}$$

Next, we find the poles of the transfer function:

$$s^2 + 4s + 2 = 0, \tag{7.334}$$

$$s_1 = -2 + \sqrt{2} = -0.59\ \text{s}^{-1}, \tag{7.335}$$

$$s_2 = -2 - \sqrt{2} = -3.41\ \text{s}^{-1}. \tag{7.336}$$

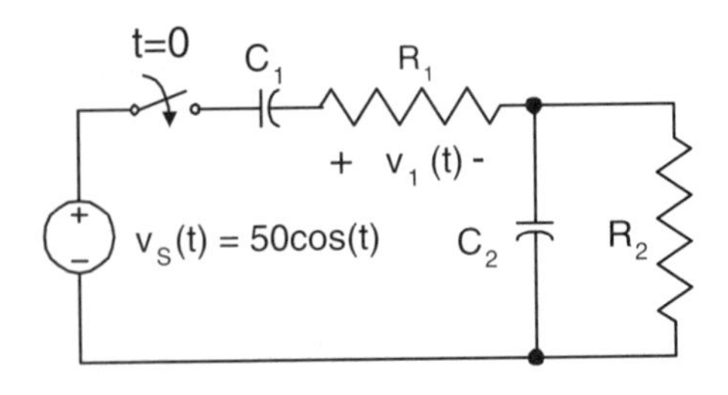

Figure 7.21: A second-order circuit example.

Consequently, the free transient response has the form:

$$v_{1_h}(t) = A_1 e^{-0.59t} + A_2 e^{-3.41t}. \tag{7.337}$$

To find the forced (ac steady state) response, we evaluate $\hat{H}(s)$ at $s = j \cdot 1$:

$$\hat{H}(j \cdot 1) = \frac{j(2+j)}{j^2 + 4j + 2} = \frac{-1+2j}{1+4j} = \frac{7+j6}{17}, \tag{7.338}$$

$$\hat{H}(j) = 0.54e^{j41^\circ}. \tag{7.339}$$

From (7.330), (7.339), and the fact that $\hat{V}_s = 50$, we obtain:

$$\hat{V}_1 = 0.54e^{j41^\circ} \cdot 50 = 27e^{j41^\circ} \text{ V}. \tag{7.340}$$

By transforming phasor $\hat{V}_1$ into the time domain, we end up with the forced response:

$$v_{1p}(t) = 27\cos(t + 41^\circ) \text{ V}. \tag{7.341}$$

By combining (7.337) and (7.341), we obtain the total response:

$$v_1(t) = 27\cos(t + 41^\circ) + A_1 e^{-0.59t} + A_2 e^{-3.41t}. \tag{7.342}$$

To find A_1 and A_2, we need the initial conditions $v_1(0_+)$ and $\frac{dv_1}{dt}(0_+)$. To this end, we consider the circuit shown in Figure 7.21 at the initial instant of time $t = 0$. Since at this instant of time voltages across the capacitors C_1 and C_2 are equal to zero, these capacitors can be replaced by short-circuit branches. This leads to the circuit shown in Figure 7.22. From this figure, we find:

$$v_1(0_+) = 50 \text{ V}, \tag{7.343}$$

$$i(0_+) = 50/10^6 \text{ A}. \tag{7.344}$$

It is clear that $i(0_+) = i_{C_1}(0_+) = i_{C_2}(0_+)$. Consequently,

$$i_{C_1}(0_+) = i_{C_2}(0_+) = 50 \cdot 10^{-6}. \tag{7.345}$$

Next, we consider again the circuit shown in Figure 7.21 and write KVL for the loop consisting of the source, capacitor C_1, resistor R_1, and capacitor C_2:

$$50\cos t = v_{C_1}(t) + v_1(t) + v_{C_2}(t). \tag{7.346}$$

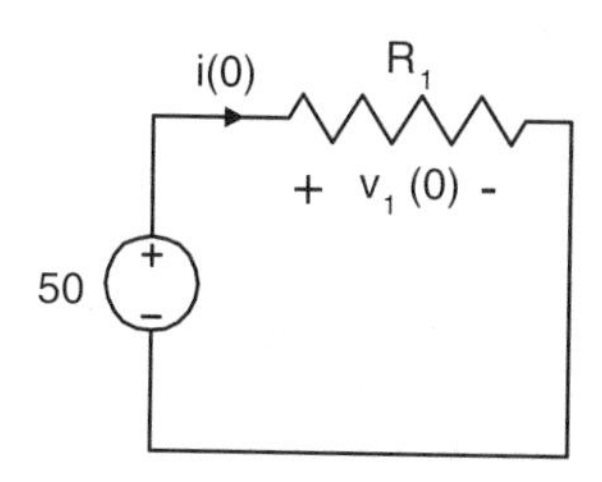

Figure 7.22: The previous circuit at $t = 0$.

By differentiating the last equation, we obtain:

$$-50 \sin t = \frac{dv_{C_1}(t)}{dt} + \frac{dv_1(t)}{dt} + \frac{dv_{C_2}(t)}{dt}. \tag{7.347}$$

The last equation can be rewritten as follows:

$$-50 \sin t = \frac{i_{C_1}(t)}{C_1} + \frac{dv_1(t)}{dt} + \frac{i_{C_2}(t)}{C_2}. \tag{7.348}$$

By setting $t = 0$ and using (7.345), we derive:

$$\frac{dv_1}{dt}(0_+) = -100 \text{ V/s}. \tag{7.349}$$

Having found the initial conditions (7.343) and (7.349), we can determine A_1 and A_2 in (7.342). Indeed, the above initial conditions lead to the equations:

$$50 = 27 \cos 41^\circ + A_1 + A_2, \tag{7.350}$$

$$-100 = -27 \sin 41^\circ - 0.59 A_1 - 3.41 A_2. \tag{7.351}$$

By solving these equations for A_1 and A_2, we find:

$$A_1 = 6.64, \qquad A_2 = 22.98. \tag{7.352}$$

From (7.352) and (7.342), we finally obtain:

$$v_1(t) = 27 \cos(t + 41^\circ) + 6.64 e^{-0.59t} + 22.98 e^{-3.41t} \text{ V}. \tag{7.353}$$

To better appreciate the benefits of the transfer function approach, it is suggested that the reader try to solve this problem by using the differential equation approach. ■

7.5 Impulse Response and Convolution Integral

Up until now, we have discussed electric circuits which contain only dc or ac sources. Using the method of undetermined coefficients, we can also analyze many circuits that have sources described by other mathematical functions. But, how can we calculate a response of an electric circuit to an *arbitrary* source? This can be accomplished by using a special technique called the *convolution integral technique*. The convolution integral gives the response of a circuit to any voltage or current source. In this case, there is no separation of the response into transient and steady-state components because with an arbitrary source there may not be any steady-state response to speak of.

To introduce the convolution integral, we shall first consider a specific circuit, namely the *RL* circuit used in our earlier discussions. Later we will show how to derive the convolution integral for any linear circuit.

7.5.1 Convolution Integral for an *RL* Circuit

The differential equation for the *RL* circuit shown in Figure 7.6 was given in (7.62) as:

$$L\frac{di(t)}{dt} + Ri(t) = v_s(t), \tag{7.354}$$

where we have omitted the subscript "L" for the sake of convenience. Multiplying (7.354) by $e^{\frac{R}{L}t}$ and dividing by L yields:

$$e^{\frac{R}{L}t}\frac{di(t)}{dt} + \frac{R}{L}e^{\frac{R}{L}t}i(t) = \frac{e^{\frac{R}{L}t}}{L}v_s(t). \tag{7.355}$$

By using the rule for the differentiation of the product of two functions, this equation can be rewritten as:

$$\frac{d}{dt}[e^{\frac{R}{L}t}i(t)] = \frac{e^{\frac{R}{L}t}}{L}v_s(t). \tag{7.356}$$

By integrating the last equation from 0 to t, and by using initial condition (7.58), we obtain:

$$e^{\frac{R}{L}t}i(t) = \int_0^t \frac{e^{\frac{R}{L}\tau}}{L}v_s(\tau)d\tau. \tag{7.357}$$

Next, we multiply both sides of (7.357) by $e^{-\frac{R}{L}t}$ which yields:

$$i(t) = \int_0^t \frac{e^{-\frac{R}{L}(t-\tau)}}{L}v_s(\tau)d\tau. \tag{7.358}$$

Expression (7.358) is the convolution integral for our particular circuit. It will give the response (current) of the circuit for any voltage source. Indeed, since $v_s(t)$ is nowhere defined in this problem, we can see that the convolution integral will work for *arbitrary* sources.

In general, integrals which fit the following form

$$i(t) = \int_0^t h(t-\tau)v_s(\tau)d\tau \tag{7.359}$$

are called convolution integrals.

In equation (7.359), t is known as *observation time*, τ is *integration time*, and $h(t-\tau)$ is called the *kernel of convolution*. In the expression (7.358), the kernel of convolution is

$$h(t-\tau) = \frac{e^{-\frac{R}{L}(t-\tau)}}{L}. \tag{7.360}$$

Expression (7.358) was derived by using the integrating factor technique which is applicable to solving first-order differential equations.

EXAMPLE 7.17 Consider the *RL* circuit in Figure 7.6 which is excited by the non-periodic source

$$v_s(t) = \begin{cases} 2t \text{ V}, & 0 \leq t \leq 1 \text{ s}, \\ 2 \text{ V}, & t > 1 \text{ s}. \end{cases} \quad (7.361)$$

Let $R = 1\ \Omega$ and $L = 2$ H. Use the convolution integral to find an expression for the time dependence of the current through the inductor.

Since the voltage source is specified differently for two different time intervals, we must consider separately these two intervals. The kernel of convolution is the same for both time intervals, since it depends only on the circuit structure and, according to equation (7.360), it is equal to

$$h(t-\tau) = e^{-\frac{1}{2}(t-\tau)}/2. \quad (7.362)$$

For times between zero and one second, we find the convolution integral (7.358) to yield:

$$i_L(t) = \int_0^t \frac{e^{-\frac{1}{2}(t-\tau)}}{2} 2\tau d\tau = e^{-t/2} \int_0^t e^{\tau/2} \tau d\tau. \quad (7.363)$$

This integral can be evaluated by using integration by parts, which leads to

$$i_L(t) = 2t - 4 + 4e^{-t/2}. \quad (7.364)$$

For times greater than one second, the convolution integral becomes:

$$i_L(t) = \int_1^t \frac{e^{-\frac{1}{2}(t-\tau)}}{2} 2d\tau + e^{-t/2} \int_0^1 e^{\tau/2} \tau d\tau. \quad (7.365)$$

The first integral can be found in a straightforward way, while the second integral can be calculated in the same way as for the first time interval. Thus, the final answer can be expressed as:

$$i_L(t) = \begin{cases} 2t - 4 + 4e^{-t/2} \text{ A}, & 0 \leq t \leq 1 \text{ s}, \\ 2(1 - 2e^{(1-t)/2}) + 4e^{-t/2} \text{ A}, & t > 1 \text{ s}. \end{cases} \quad (7.366)$$

There are two important points to note here. First, $i_L(1_-) = i_L(1_+)$, as it must be according to the continuity of electric current through an inductor. Second, $i_L(t) \to 2$ A as $t \to \infty$. This also should be expected because the inductor will act like a short circuit after the transient dies away (that is, at dc steady-state). ■

Later, we shall show that the convolution integral can be derived for any linear circuit. To achieve this, we shall need some facts concerning the unit step function and the unit impulse function. These facts will also be instrumental in arriving at the physical interpretation of the convolution kernel (7.360).

The *unit step function*, denoted as $u(t)$, is defined as follows:

$$u(t) = \begin{cases} 0, & \text{if } t < 0, \\ 1, & \text{if } t > 0. \end{cases} \quad (7.367)$$

This expression means that when the argument (expression in parentheses) of the unit step function is less than zero, this function is equal to zero, and when its argument is greater than zero, the unit step function is equal to 1 (see Figure 7.23). The unit

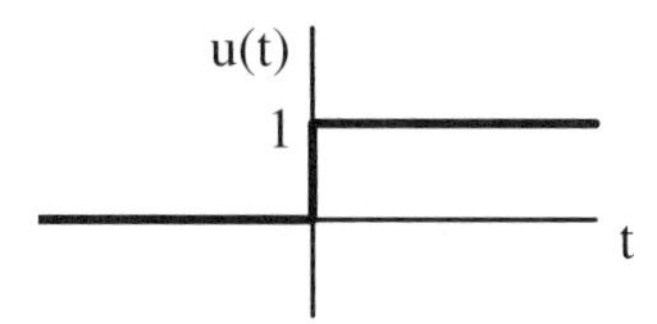

Figure 7.23: Graph of the unit step function.

step function is very useful for mathematical descriptions of switching of sources. For example, if a source is defined as

$$v_s(t)u(t) = \begin{cases} 0, & \text{if } t < 0, \\ v_s(t), & \text{if } t > 0, \end{cases} \tag{7.368}$$

this means that the source is "off" when $t < 0$ and the source is "on" when $t > 0$ (see Figure 7.24). The unit step function can also be used to describe the switching off of sources. From the very definition of the unit step function, we find that:

$$u(-t) = \begin{cases} 1, & \text{if } t < 0, \\ 0, & \text{if } t > 0. \end{cases} \tag{7.369}$$

This version of the unit step function can be used to turn sources off at $t = 0$. Indeed,

$$v_s(t)u(-t) = \begin{cases} v_s(t), & \text{if } t < 0, \\ 0, & \text{if } t > 0, \end{cases} \tag{7.370}$$

which means that the source is on when $t < 0$ and the source is off when $t > 0$. Next, we introduce the shifted unit step function $u(t - t_0)$. It is clear from the definition of the unit step function that:

$$u(t - t_0) = \begin{cases} 0, & \text{if } t < t_0, \\ 1, & \text{if } t > t_0. \end{cases} \tag{7.371}$$

It is easy to see that:

$$v_s(t)u(t - t_0) = \begin{cases} 0, & \text{if } t < t_0, \\ v_s(t), & \text{if } t > t_0. \end{cases} \tag{7.372}$$

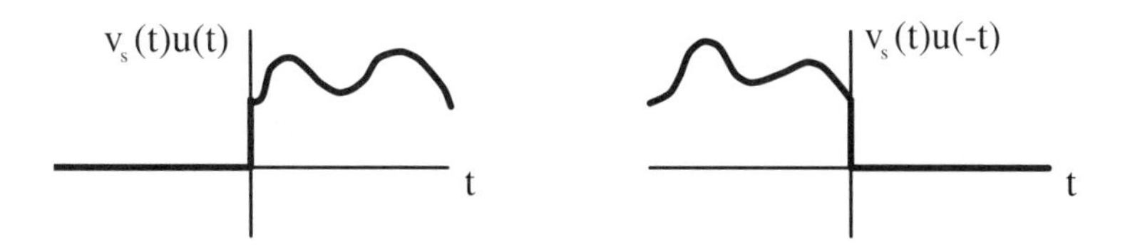

Figure 7.24: Description of a source switched on and off by using the unit step function.

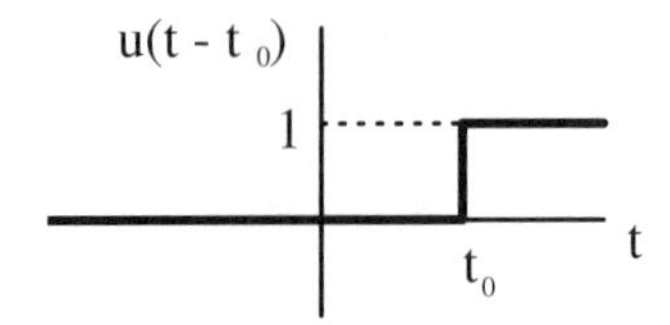

Figure 7.25: Shifted unit step function.

Consequently, $v_s(t)u(t - t_0)$ can be interpreted as a voltage source which is turned on at time t_0 (see Figure 7.25).

One of the important applications of the unit step function is the staircase approximation of continuous functions. First, we note that staircase (or stepwise) functions can be represented as sums of shifted step functions:

$$f(t) = \sum_k \alpha_k u(t - t_k), \tag{7.373}$$

where the t_k's form an ordered sequence ($t_{k+1} > t_k$). These functions look like the one shown in Figure 7.26. It is clear from this figure that the α_k have the meaning of the increments of $f(t)$ at times t_k ($k = 1, 2, \ldots$). These stepwise functions can in turn be used to approximate continuous functions. If $v_s(t)$ is a continuous source function, then it can be approximated with step functions by using the expression:

$$v_s(t) \approx v_s(0)u(t) + \sum_k \Delta v_s^{(k)} u(t - t_k) \tag{7.374}$$

where $\Delta v_s^{(k)}$ are the increments of the source function on successive time intervals. A graph of a source function and its stepwise approximation is shown in Figure 7.27.

The second important function which must be covered to fully understand the convolution integral is the *unit impulse function*. This function, denoted $\delta(t)$, is formally defined as the derivative of the unit step function:

$$\delta(t) = \frac{du(t)}{dt} = \begin{cases} 0, & \text{if } t < 0, \\ \infty, & \text{if } t = 0, \\ 0, & \text{if } t > 0. \end{cases} \tag{7.375}$$

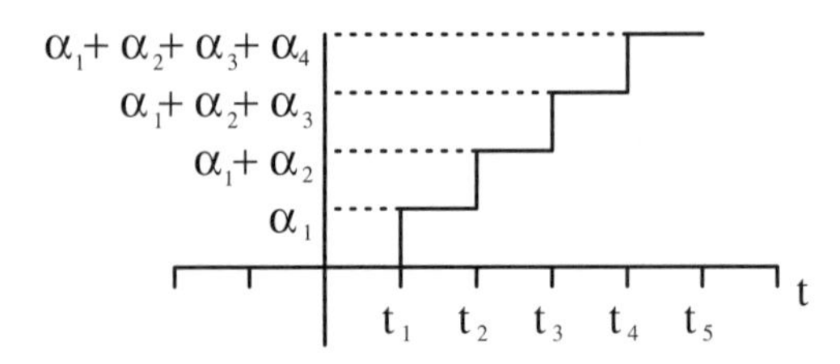

Figure 7.26: Staircase function represented by step functions.

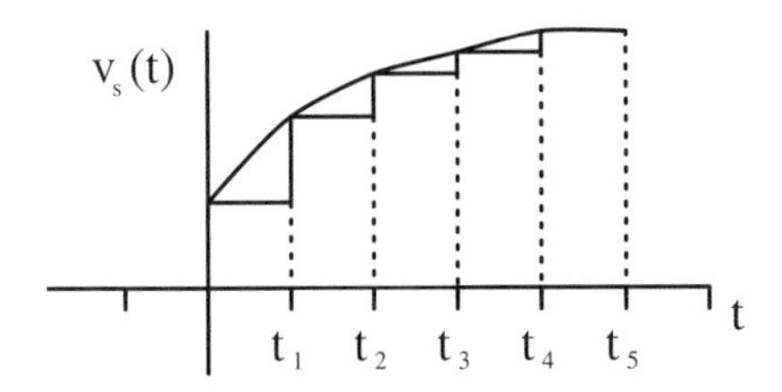

Figure 7.27: Approximating a continuous function with step functions.

This means that the unit impulse function is infinite for $t = 0$ and zero for all other values of t. We can also evaluate the integral of the unit impulse function by using its definition:

$$\int_b^a \delta(t)dt = \int_b^a \frac{du(t)}{dt}dt = u(a) - u(b). \tag{7.376}$$

From this expression we can see that the integral of the unit impulse function is equal to either one or zero, depending on whether or not zero falls inside or outside the limits of integration:

$$\int_b^a \delta(t)dt = \begin{cases} 1, & \text{if } 0 \in (b, a), \\ 0, & \text{if } 0 \notin (b, a). \end{cases} \tag{7.377}$$

Equation (7.377), along with the expression $\delta(t) = 0$ if $t \neq 0$, is sometimes used as another definition of the unit impulse function.

Now that we know the unit step function and the unit impulse function, we can elucidate the physical meaning of the convolution integral. Consider again the simple *RL* circuit excited by a voltage source shown in Figure 7.28. This circuit is basically identical to the other *RL* circuits considered previously; however, it has a unit step voltage source. The current produced by a unit step voltage source, $i_u(t)$, is called the unit step response. We can write the differential equation for this circuit as before (the only difference is the source term):

$$L\frac{di_u(t)}{dt} + Ri_u(t) = u(t). \tag{7.378}$$

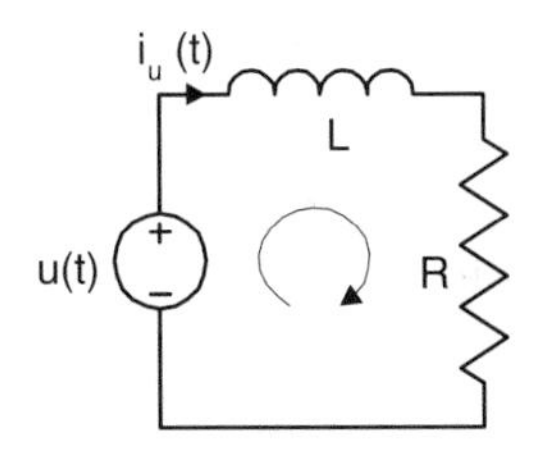

Figure 7.28: *RL* circuit with unit step voltage source.

The solution to this equation is the same as the one for the constant dc voltage source. Indeed, for $t < 0$ the circuit is switched off. For $t > 0$ the source is switched on, and we just treat the unit step voltage source as a constant voltage source with a value of 1. Therefore, according to (7.75), the solution to this differential equation is

$$i_u(t) = \frac{1}{R} - \frac{1}{R}e^{-\frac{R}{L}t}. \tag{7.379}$$

We now return to equation (7.378) and differentiate both sides with respect to time:

$$L\frac{d}{dt}\left[\frac{di_u(t)}{dt}\right] + R\frac{di_u(t)}{dt} = \frac{du(t)}{dt}. \tag{7.380}$$

Now, we define the unit impulse response, denoted $i_\delta(t)$, to be the derivative of the unit step function response:

$$i_\delta(t) = \frac{di_u(t)}{dt}. \tag{7.381}$$

Substituting (7.381) into (7.380) and recalling (7.375) yields:

$$L\frac{di_\delta(t)}{dt} + Ri_\delta(t) = \delta(t). \tag{7.382}$$

This equation has the same mathematical form as the original differential equation (7.378) and can be interpreted as the equation describing a current in the circuit excited by a unit impulse voltage source (see Figure 7.29). Thus, the unit impulse response can be also defined as a response produced by a unit impulse source.

We next want to find $i_\delta(t)$. Since $i_u(t)$ is already known (equation (7.379)), calculating the unit impulse function response is straightforward. According to (7.381), we simply take the derivative of (7.379):

$$i_\delta(t) = \frac{di_u(t)}{dt} = \frac{e^{-\frac{R}{L}t}}{L}. \tag{7.383}$$

Notice that the expression for $i_\delta(t-\tau)$ is identical to the kernel of convolution we derived earlier for this circuit (see formula (7.360)). Thus, the convolution integral can be written as follows:

$$i(t) = \int_0^t i_\delta(t-\tau)v_s(\tau)d\tau. \tag{7.384}$$

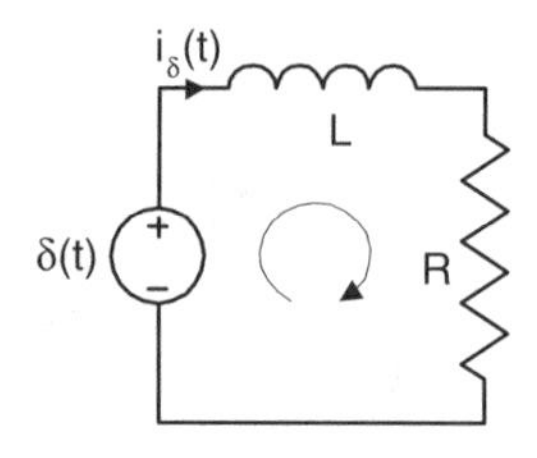

Figure 7.29: *RL* circuit with unit impulse function source.

Therefore, **we have established that the kernel of the convolution integral has the physical meaning of the unit impulse response**. Thus, if the unit impulse response is known, then the response to any source can be found by using the convolution integral. In this sense, the unit impulse response gives a complete characterization of a linear circuit. Of course, we have discussed only the *RL* circuit and this does not prove our general statement. That task is left to the next section.

7.5.2 Convolution Integral for Arbitrary Linear Circuits

We want to prove that if the unit step function response (and consequently the unit impulse response) of a circuit is known, then the response for any source can be found. We begin with an arbitrary linear circuit driven by an arbitrary voltage source, shown on the left in Figure 7.30. The same circuit excited by the unit step function source is shown on the right in the same figure. We know from the previous section that any continuous function can be approximated by using linear combinations of shifted step functions:

$$v_s(t) \approx v_s(0)u(t) + \sum_k \Delta v_s^{(k)} u(t - t_k). \tag{7.385}$$

Each term in (7.385) can be interpreted as a voltage source which is turned on at time t_k. Consequently, the last expression has a physical interpretation, illustrated in Figure 7.31. All these voltage sources are connected in series and contribute to the response current. Since the circuit is *linear* (and time invariant), if a voltage source $u(t)$ results in a response current $i_u(t)$, a source $\Delta v_s^{(k)} u(t - t_k)$ will result in a response $\Delta v_s^{(k)} i_u(t - t_k)$. According to the *superposition* principle, we can add these separate components of the response current together to find the total current. This current is given by the expression

$$i(t) \approx v_s(0)i_u(t) + \sum_k \Delta v_s^{(k)} i_u(t - t_k). \tag{7.386}$$

Using elementary calculus, $\Delta v_s^{(k)}$ can be replaced with:

$$\Delta v_s^{(k)} \approx \left. \frac{dv_s(t)}{dt} \right|_{t=t_k} \cdot \Delta t_k. \tag{7.387}$$

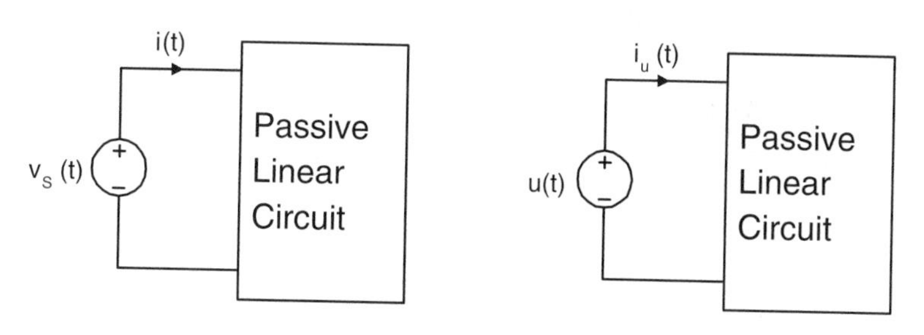

Figure 7.30: Arbitrary circuit and its unit step response.

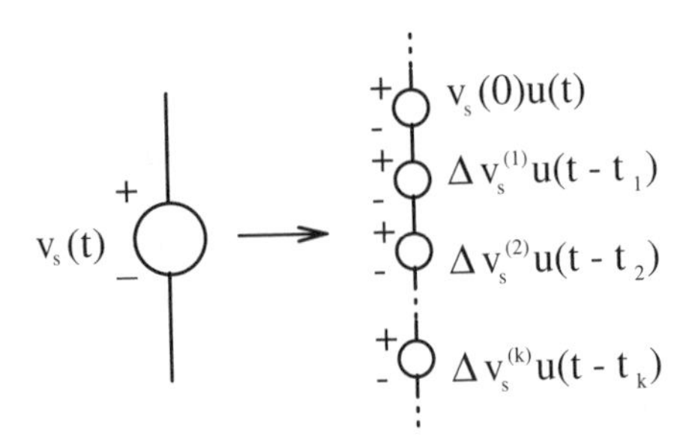

Figure 7.31: A series connection of voltage sources.

Substitution of (7.387) into (7.386) yields:

$$i(t) \approx v_s(0)i_u(t) + \sum_k i_u(t - t_k) \left. \frac{dv_s(t)}{dt} \right|_{t_k} \cdot \Delta t_k. \qquad (7.388)$$

As the time intervals Δt_k get smaller and smaller the last expression gets more and more accurate. In the limit of $\Delta t_k \rightarrow 0$, the last expression becomes an exact equality, while the sum in this expression becomes an integral. Thus, we obtain:

$$i(t) = v_s(0)i_u(t) + \int_0^t i_u(t - \tau)\frac{dv_s(\tau)}{d\tau}d\tau. \qquad (7.389)$$

This integral is called the *superposition* integral, and it is interesting in its own right. It shows that if we know the unit step function response $i_u(t - \tau)$ of an electric circuit, then we can find the response of this circuit to *any* voltage source.

The last integral can be transformed through integration by parts:

$$i(t) = v_s(0)i_u(t) + i_u(t - \tau)v_s(\tau)|_{\tau=0}^{\tau=t} - \int_0^t \frac{di_u(t - \tau)}{d\tau}v_s(\tau)d\tau. \qquad (7.390)$$

By substituting the limits of integration, we find:

$$i(t) = v_s(0)i_u(t) + i_u(0)v_s(t) - i_u(t)v_s(0) - \int_0^t \frac{di_u(t - \tau)}{d\tau}v_s(\tau)d\tau. \qquad (7.391)$$

By canceling the identical terms, we derive:

$$i(t) = i_u(0)v_s(t) - \int_0^t \frac{di_u(t - \tau)}{d\tau}v_s(\tau)d\tau. \qquad (7.392)$$

Next, we recall (see (7.381)) that:

$$\frac{di_u(t - \tau)}{d\tau} = -\frac{di_u(t - \tau)}{dt} = -i_\delta(t - \tau). \qquad (7.393)$$

By substituting the last expression into (7.392), we end up with the convolution integral:

$$i(t) = i_u(0)v_s(t) + \int_0^t i_\delta(t-\tau)v_s(\tau)d\tau. \tag{7.394}$$

For many circuits that we will encounter, we will find that $i_u(0) = 0$. For those circuits (7.394) reduces to:

$$i(t) = \int_0^t i_\delta(t-\tau)v_s(\tau)d\tau. \tag{7.395}$$

This is the result we sought to prove.

Previously, we have stressed the importance of the convolution and superposition integrals from the computational point of view. Namely, we have emphasized that, by computing the unit impulse response or the unit step function response, we can then find the response to an arbitrary source just by performing certain integrations. The calculation of the unit step function response (as well as the unit impulse response) requires the analysis of an electric circuit for the simplest source excitation. This underlines the computational efficiency of convolution and superposition integrals.

These integrals are also important from the experimental point of view. The reason is that the unit step function response and the unit impulse response can be measured by using relatively simple tests. Then, these measured responses can be utilized to predict responses of an electric circuit to arbitrary source excitations.

7.5.3 Applications of the Convolution Integral

In this section we will demonstrate the usefulness of the convolution integral technique by applying it to a number of circuit problems.

EXAMPLE 7.18 In this example, we will use the convolution integral technique to solve the problem considered previously in Example 7.7. The circuit under consideration is redrawn in Figure 7.32. Note that no explicit switches are drawn in this circuit. Instead, the turning "on" of the source is described by using the unit step

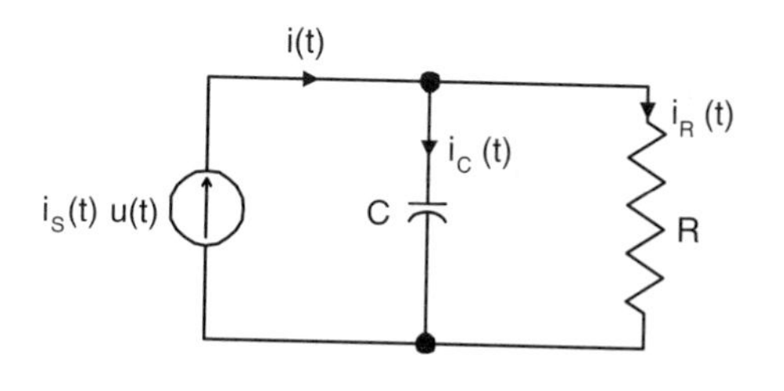

Figure 7.32: *RC* circuit excited by a current source.

function as:

$$i_s(t)u(t). \tag{7.396}$$

In order to find the circuit response by using the convolution techniques, we must first find the unit step response of the circuit. Just prior to Example 7.7, the expressions (7.117), (7.118), and (7.119) were derived for the case of excitation of the RC circuit by a dc current source. By putting $I_0 = 1$ in (7.117), we obtain the unit step response $v_u(t)$:

$$v_u(t) = R - Re^{-t/RC}. \tag{7.397}$$

From the last expression, we can find the unit impulse response, $v_\delta(t)$:

$$v_\delta(t) = \frac{dv_u(t)}{dt} = \frac{1}{C}e^{-t/RC}. \tag{7.398}$$

By using this unit impulse response, we can employ the convolution integral and find the response to an arbitrary current source:

$$v(t) = \int_0^t v_\delta(t-\tau)i_s(\tau)d\tau = \int_0^t \frac{e^{-\frac{t-\tau}{RC}}}{C} i_s(\tau)d\tau. \tag{7.399}$$

In Example 7.7, we considered an ac current source:

$$i_s(t) = I_{ms}\cos(\omega t + \phi_s). \tag{7.400}$$

By substituting this expression for the current source into the convolution integral (7.399), we obtain:

$$v(t) = \frac{1}{C}\int_0^t e^{-(t-\tau)/RC} I_{ms}\cos(\omega\tau + \phi_s)d\tau. \tag{7.401}$$

We can simplify the integration by recalling that $\mathrm{Re}(e^{jx}) = \cos(x)$. After using this fact and rearranging terms we arrive at:

$$v(t) = \mathrm{Re}\left\{\frac{I_{ms}}{C}e^{-t/RC}\int_0^t e^{\tau/RC}e^{j(\omega\tau+\phi_s)}d\tau\right\}. \tag{7.402}$$

Now we can easily perform the integration and rearrange terms to get:

$$v(t) = \mathrm{Re}\left\{\frac{RI_{ms}}{\sqrt{1+(\omega RC)^2}}\left[\frac{1-j\omega RC}{\sqrt{1+(\omega RC)^2}}\right]e^{j\phi_s}(e^{j\omega t} - e^{-t/RC})\right\}. \tag{7.403}$$

The bracketed term is just $e^{-j\phi}$ where $\phi = \arctan(\omega RC)$. Thus we can take the real part to arrive at the final answer:

$$v(t) = \frac{RI_{ms}}{\sqrt{1+(\omega RC)^2}}[\cos(\omega t + \phi_s - \phi) - e^{-t/RC}\cos(\phi_s - \phi)]. \tag{7.404}$$

This agrees with equation (7.135) as it must. ■

EXAMPLE 7.19 Consider the *RLC* parallel circuit of Examples 7.13 and 7.14, which is redrawn in Figure 7.33. Use the convolution integral technique to find the current through the inductor.

We can find the required differential equation by replacing the source current in (7.268) with the unit step function and plugging in the component values:

$$\frac{d^2 i_L}{dt^2} + 2\frac{di_L}{dt} + i_L(t) = u(t). \tag{7.405}$$

The particular solution is simply:

$$i_{L_p}(t) = 1. \tag{7.406}$$

The homogeneous solution was found in (7.274) to be:

$$i_{L_h}(t) = A_1 e^{-t} + A_2 t e^{-t}. \tag{7.407}$$

From (7.270) and (7.271) we have:

$$i_L(0_+) = 0 = \frac{di_L}{dt}(0_+), \tag{7.408}$$

and a few algebraic manipulations yields the unit step response:

$$i_u(t) = 1 - e^{-t} - te^{-t}. \tag{7.409}$$

The unit impulse response is given by the derivative of (7.409):

$$i_\delta(t) = \frac{di_u}{dt} = te^{-t}. \tag{7.410}$$

The convolution integral in our case (when the source and output quantities are both currents) takes on the form:

$$i(t) = \int_0^t i_\delta(t-\tau) i_s(\tau) d\tau \tag{7.411}$$

which for this example becomes:

$$i(t) = 10\,\text{Re}\left\{\int_0^t (t-\tau)e^{-(t-\tau)}e^{j\tau}d\tau\right\}\ \text{A}, \tag{7.412}$$

$$i(t) = 10e^{-t}\,\text{Re}\left\{\int_0^t \frac{e^{\tau(1+j)}}{1+j}d\tau\right\} - 5te^{-t}\ \text{A}, \tag{7.413}$$

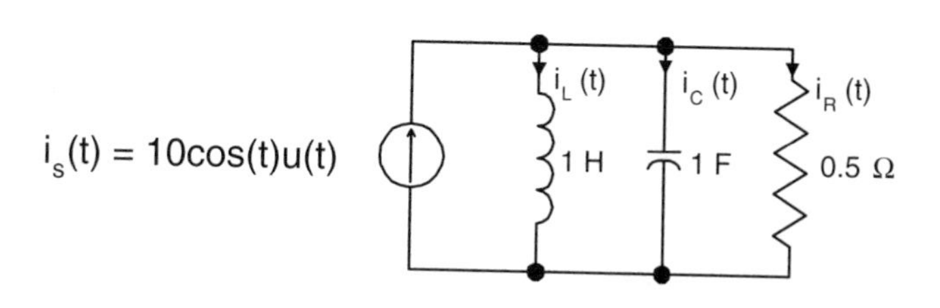

Figure 7.33: *RLC* parallel circuit excited by a current source.

$$i(t) = 5\sin(t) - 5te^{-t} \text{ A.} \tag{7.414}$$

Equation (7.414) is identical to (7.284), as it must be. This concludes this example. ■

EXAMPLE 7.20 Consider the circuit shown in Figure 7.34. Find the voltage $v(t)$ across the 2 F capacitor.

For the above circuit, the differential equation which describes the time evolution of $v(t)$ is quite difficult to derive directly from KVL, KCL, and the terminal relationships. However, this equation is needed in order to find the particular solution for the given voltage source:

$$v_s(t) = te^{-t}. \tag{7.415}$$

The above difficulty can be circumvented by combining the convolution integral and transfer function techniques. The convolution integral technique requires us to find the unit impulse response, which can be done if we know the unit step response. Because dc sources are a limiting case of ac sources, the method of the transfer function can be used to find the unit step response and we can proceed with the convolution technique from there.

The transfer function approach for this circuit was considered in Example 7.15. For the parameters given in Figure 7.34, the transfer function (7.311) becomes:

$$\hat{H}(s) = \frac{s+1}{6s^2 + 6s + 1}. \tag{7.416}$$

The free response can be found from the poles of (7.416):

$$s_{1,2} = -\frac{1}{2}(1 \pm 1/\sqrt{3})\ \text{s}^{-1}. \tag{7.417}$$

The forced response can be found from (7.294) and (7.416) to be:

$$v_p(t) = H(0)u(t) = 1. \tag{7.418}$$

Thus, the complete response is:

$$v_u(t) = 1 + A_1 e^{-0.211t} + A_2 e^{-0.789t}. \tag{7.419}$$

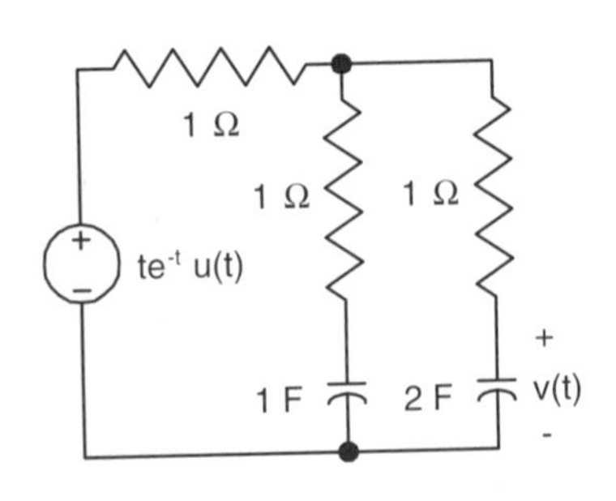

Figure 7.34: A second-order circuit excited by a voltage source.

The initial conditions are found from (7.321) and (7.327) to be:

$$v_u(0_+) = 0 \tag{7.420}$$

$$\frac{dv_u}{dt}(0_+) = \frac{1}{6}. \tag{7.421}$$

By using these two initial conditions and (7.419), we can solve for A_1 and A_2 and arrive at the expression for the unit step response:

$$v_u(t) = 1 - 1.077e^{-0.211t} + 0.077e^{-0.789t}. \tag{7.422}$$

The unit impulse response in found in the usual way by differentiating (7.422):

$$v_\delta(t) = 0.227e^{-0.211t} - 0.061e^{-0.789t}. \tag{7.423}$$

When the source and output signals are both voltages, the convolution integral takes the form:

$$v(t) = \int_0^t v_\delta(t-\tau)v_s(\tau)d\tau. \tag{7.424}$$

Plugging the source voltage (7.415) and the unit impulse response (7.423) into the convolution integral (7.424) yields the solution:

$$v(t) = e^{-t} + 0.366e^{-0.211t} - 1.366e^{-0.789t}\ \mathrm{A} \tag{7.425}$$

after integrating by parts. This concludes the example. ■

7.6 Circuits with Diodes (Rectifiers)

In this section, we shall apply the transient analysis techniques developed in previous sections to the analysis of steady-state responses of special electric circuits—rectifiers.

Many electric devices require dc power supplies. Utility companies supply ac power. Thus, there is a need for special circuits which convert ac voltage sources into dc voltage sources. Such circuits are called rectifiers and the main element of these circuits is a diode. The diode can be defined as a two-terminal element whose resistance depends on the polarity of applied voltage. The circuit notation for the diode is shown in Figure 7.35.

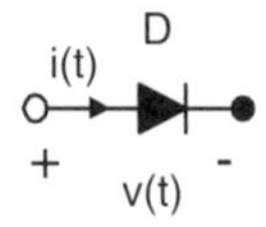

Figure 7.35: The circuit notation for a diode.

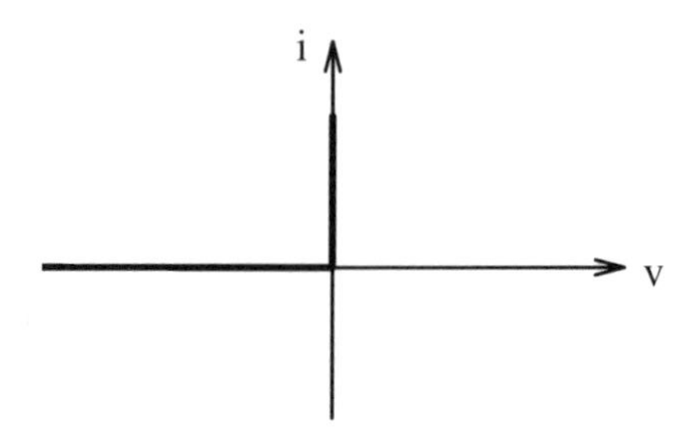

Figure 7.36: The i-v curve of an ideal diode.

We shall consider only ideal diodes. Such diodes are characterized by the i-v curve shown in Figure 7.36. According to this figure, the diode resistance is equal to zero when the voltage across the diode is positive, and the diode resistance is equal to infinity if the voltage across the diode is negative. In other words, the diode acts as a short circuit if the applied voltage is positive, and it acts as an open circuit if the applied voltage is negative. This means that an electric current through the diode can flow only in one direction, and this is implied by the circuit notation for the diode. Actual (real-world) diodes deviate somewhat from the ideal properties described above. However, as a first approximation, these deviations can be neglected. This helps to understand the effect of diodes on the operation of electric circuits.

The diodes are usually constructed by using semiconductors. One of the most frequently used diodes is the p–n junction diode, which is studied in detail in courses on semiconductor devices.

The fact that the diode is a polarity-dependent element suggests that diodes can be used for rectification of ac power sources. As an example of such rectification, let us first consider the circuit shown in Figure 7.37. Here, the circuit is excited by ac voltage source $v_s(t)$:

$$v_s(t) = V_{ms} \sin \omega t, \tag{7.426}$$

the graph of which is depicted in Figure 7.38.

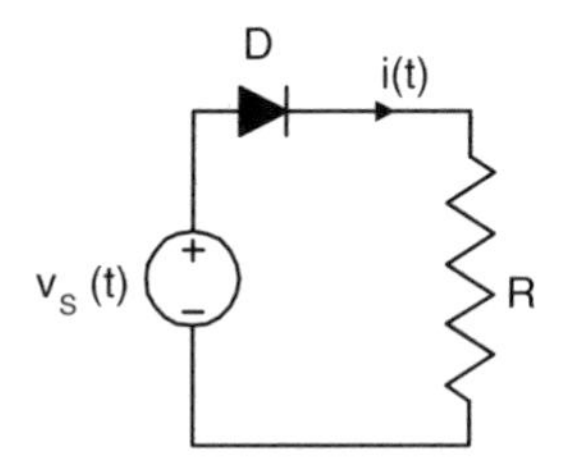

Figure 7.37: A half-wave rectifier circuit.

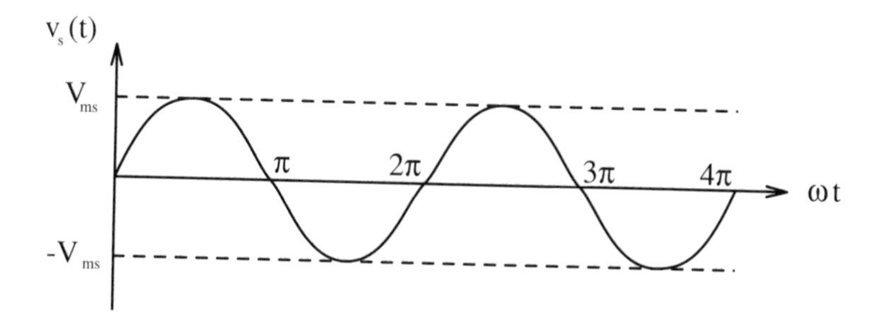

Figure 7.38: The input voltage source.

During the first half-period ($0 < \omega t < \pi$), a voltage of positive polarity is applied across the diode D. Thus, this diode is "closed" and the total source voltage is applied across the resistive load R. During the next half-period, a voltage of negative polarity is applied across the diode. As a result, the diode is "open" and no voltage is applied across the resistive load. In the subsequent half-periods, the situation repeats itself. Thus, it can be concluded that the resistive load is subject to the rectified voltage $v^{(r)}(t)$ shown in Figure 7.39.

According to Ohm's law $i(t) = v(t)/R$ and the shape of the load current mimics the shape of the rectified voltage (see Figure 7.40).

Thus, we have achieved rectification. However, in the circuit shown in Figure 7.37 the current is being conducted during only positive half-periods. For this reason,

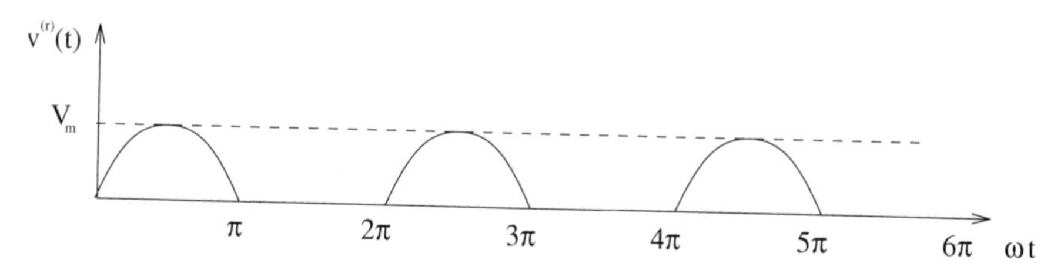

Figure 7.39: Voltage across the resistor.

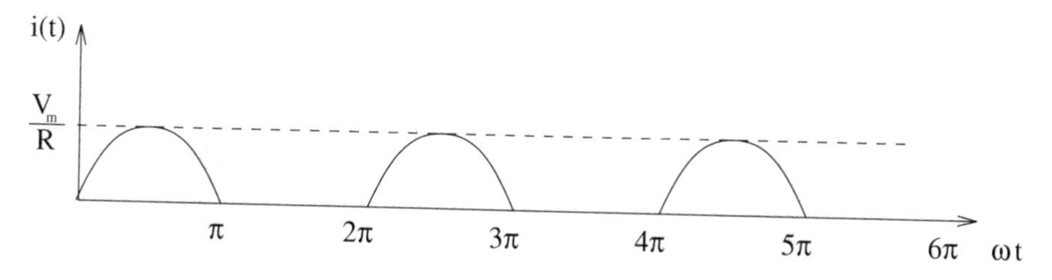

Figure 7.40: Current through the resistor.

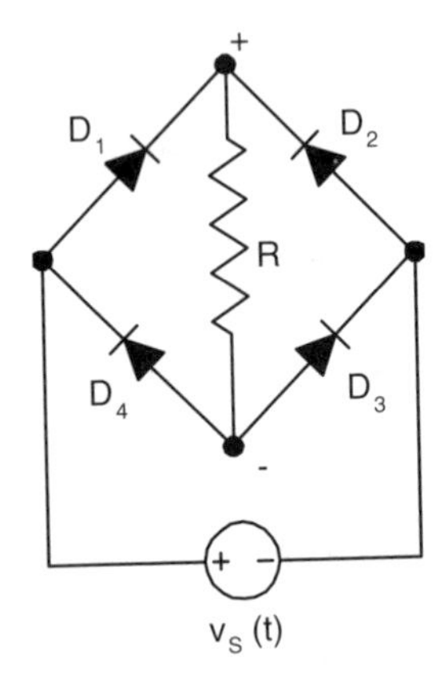

Figure 7.41: A full-wave rectifier circuit.

this circuit is called a **half-wave** rectifier. To achieve the current conduction during positive and negative half-periods, **full-wave** rectifiers are designed. These rectifiers usually employ the diode bridge circuit shown in Figure 7.41. In this circuit, during the positive half-periods, a voltage of positive polarity is applied across diodes D_1 and D_3, while a voltage of negative polarity is applied across diodes D_2 and D_4. For this reason, diodes D_1 and D_3 are closed, while diodes D_2 and D_4 are open. Thus, the total voltage source is applied across the resistive branch with positive polarity being applied to the upper terminal of R and negative polarity being applied to the lower terminal of R.

During the negative half-periods, the polarity of the ac voltage source $v_s(t)$ is reversed. For this reason, a voltage of positive polarity is applied across diodes D_2 and D_4, while a voltage of negative polarity is applied across diodes D_1 and D_3. As a result, diodes D_2 and D_4 are closed, while diodes D_1 and D_3 are open. Again, the total voltage source is applied across the resistive branch with positive polarity being applied to the upper terminal of R and negative polarity being applied to the lower terminal of R.

Thus, it can be concluded that the resistive branch is subject to the rectified voltage shown in Figure 7.42. It is clear that the shape of the electric current mimics the shape of the rectified voltage. In this way, full-wave rectification is achieved.

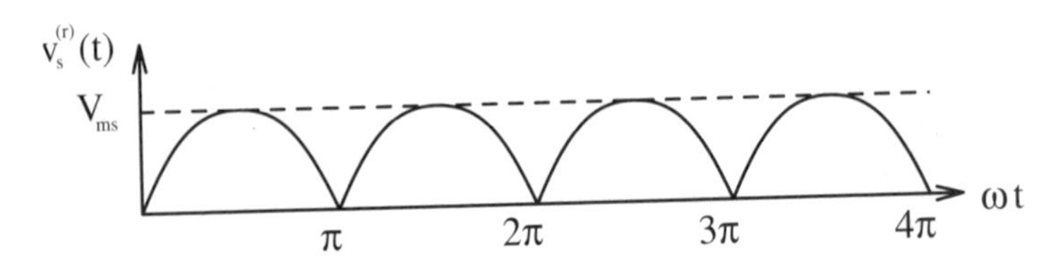

Figure 7.42: Voltage across the diagonal branch.

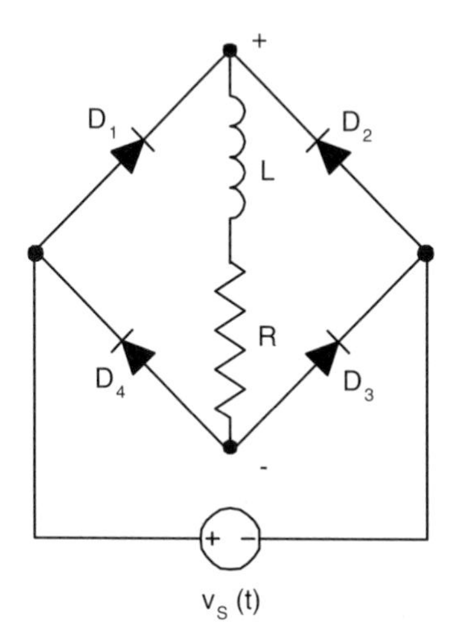

Figure 7.43: A diode bridge circuit with an inductor.

However, the rectified voltage and current have a substantial amount of ripple. **To reduce the ripple level**, diode circuits with energy storage elements (inductors and capacitors) are employed. One of these circuits is shown in Figure 7.43 and will be analyzed below. It is clear from the previous discussion that the rectified voltage (shown in Figure 7.42) is applied across the *LR* branch. Consequently, the circuit shown in Figure 7.43 can be replaced by the equivalent circuit shown in Figure 7.44. This circuit is equivalent to the one shown in Figure 7.43 as far as the current through the *LR* branch as well as the voltages across L and R are concerned.

As is evident from Figure 7.42, the rectified voltage source, $v_s^{(r)}(t)$, is periodic with period equal to π/ω. This implies that the *steady-state* current, $i(t)$, is periodic as well and has the same period. For this reason, it suffices to consider the circuit shown in Figure 7.44 during one period:

$$0 \leq t \leq \frac{\pi}{\omega}. \tag{7.427}$$

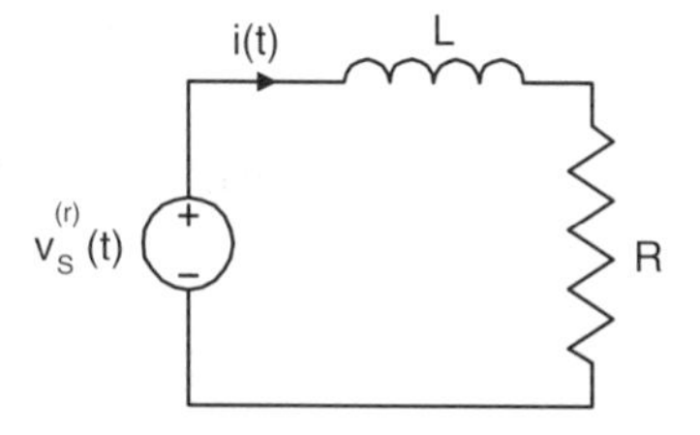

Figure 7.44: Equivalent circuit for the full-wave rectifier with inductor.

During this period, the current, $i(t)$, satisfies the differential equation:

$$L\frac{di(t)}{dt} + Ri(t) = V_{ms}\sin\omega t \tag{7.428}$$

and the following periodic boundary condition:

$$i(0) = i\left(\frac{\pi}{\omega}\right). \tag{7.429}$$

As far as the differential equation is concerned, it just follows from KVL. Indeed, the first term in this equation is the voltage drop across the inductor, the second term is the voltage drop across the resistor, and the right-hand side is the expression for the rectified voltage source during the first period specified by (7.427). (Please note that this expression is not valid for the second period.) As far as the periodic boundary condition (7.429) is concerned, it is just a consequence of the periodicity of the steady-state current $i(t)$ in our circuit.

Differential equation (7.428) is identical to the differential equation for the *RL* circuit excited by an ac voltage source. Consequently, the solution to equation (7.428) can be found by using the same technique as before. (This is the main reason why we consider circuits with diodes in this chapter.) The general solution to equation (7.428) has the form:

$$i(t) = \frac{V_{ms}}{\sqrt{R^2 + \omega^2 L^2}}\sin(\omega t - \phi) + Ae^{-\frac{R}{L}t}, \tag{7.430}$$

where

$$\phi = \arctan\frac{\omega L}{R}. \tag{7.431}$$

The first term is the particular solution of equation (7.428), which is found by using the phasor technique. The second term is the homogeneous solution with unknown constant A. When we discussed transients for the *RL* circuit, this constant was found from initial conditions. In our problem, and this is the main mathematical difference, this constant should be found from the periodic boundary condition (7.429). By setting $t = 0$ and $t = \pi/\omega$ in (7.430) and by using (7.429), we obtain the following equation for A:

$$\frac{-V_{ms}\sin\phi}{\sqrt{R^2 + \omega^2 L^2}} + A = \frac{V_{ms}\sin(\pi - \phi)}{\sqrt{R^2 + \omega^2 L^2}} + Ae^{-\frac{\pi R}{\omega L}}. \tag{7.432}$$

By solving (7.432) for A, we arrive at:

$$A = \frac{2V_{ms}\sin\phi}{(1 - e^{-\frac{\pi R}{\omega L}})\sqrt{R^2 + \omega^2 L^2}}. \tag{7.433}$$

By substituting (7.433) into (7.430), we find the final solution:

$$i(t) = \frac{V_{ms}}{\sqrt{R^2 + \omega^2 L^2}}\sin(\omega t - \phi) + \frac{2V_{ms}\sin\phi}{(1 - e^{-\frac{\pi R}{\omega L}})\sqrt{R^2 + \omega^2 L^2}}e^{-\frac{R}{L}t}. \tag{7.434}$$

From (7.434) we obtain:

$$v_R(t) = \frac{V_{ms}R}{\sqrt{R^2 + \omega^2 L^2}} \sin(\omega t - \phi) + \frac{2V_{ms}R \sin\phi}{(1 - e^{-\frac{\pi R}{\omega L}})\sqrt{R^2 + \omega^2 L^2}} e^{-\frac{R}{L}t}. \tag{7.435}$$

Let us examine the last equation. First, let us evaluate the average value of $v_R(t)$:

$$\overline{v_R(t)} = \frac{\omega}{\pi} \int_0^{\frac{\pi}{\omega}} v_R(t)dt. \tag{7.436}$$

By substituting (7.435) into (7.436) and by performing the integration, we derive:

$$\overline{v_R(t)} = \frac{2V_{ms}R \cos\phi}{\pi\sqrt{R^2 + \omega^2 L^2}} + \frac{2V_{ms}R \sin\phi}{\pi(1 - e^{-\frac{\pi R}{\omega L}})\sqrt{R^2 + \omega^2 L^2}} \frac{\omega L}{R}(1 - e^{-\frac{\pi R}{\omega L}}). \tag{7.437}$$

By performing some cancellations in the second term of (7.437) and by recalling (7.431), we transform (7.437) as follows:

$$\begin{aligned}
\overline{v_R(t)} &= \frac{2V_{ms}R}{\pi\sqrt{R^2 + \omega^2 L^2}} \left(\cos\phi + \frac{\omega L}{R}\sin\phi\right) \\
&= \frac{2V_{ms}R}{\pi\sqrt{R^2 + \omega^2 L^2}}(\cos\phi + \tan\phi \sin\phi) \\
&= \frac{2V_{ms}R}{\pi\sqrt{R^2 + \omega^2 L^2}\cos\phi} = \frac{2V_{ms}R}{\pi\sqrt{R^2 + \omega^2 L^2}} \frac{\sqrt{R^2 + \omega^2 L^2}}{R} = \frac{2V_{ms}}{\pi}.
\end{aligned} \tag{7.438}$$

So, we arrive at the following remarkable result:

$$\overline{v_R(t)} = \frac{2}{\pi}V_{ms}. \tag{7.439}$$

It is remarkable because *the average voltage across the resistor does not depend on the value of inductance or resistance.* It is the same for any value of the inductance, L, and resistance, R. However, the value of L will affect ripples of the voltage across the resistor. To demonstrate this, we consider two limiting cases: $L \longrightarrow 0$ and $L \longrightarrow \infty$. In the first case, from (7.435) and (7.431) we find:

$$v_R(t) = V_{ms} \sin\omega t = v_s^{(r)}(t). \tag{7.440}$$

Thus, for very small L, the voltage across the resistance has the same (substantial) amount of ripple as the rectified voltage source shown in Figure 7.42.

For very large L, we have:

$$1 - e^{-\frac{\pi R}{\omega L}} \approx \frac{\pi R}{\omega L}, \tag{7.441}$$

$$\sqrt{R^2 + \omega^2 L^2} \approx \omega L, \tag{7.442}$$

$$\sin\phi = \frac{\omega L}{\sqrt{R^2 + \omega^2 L^2}} \approx 1. \tag{7.443}$$

By using (7.441), (7.442), and (7.443) in (7.435), in the second limiting case we find:

$$v_R(t) = \frac{2}{\pi} V_{ms} = \overline{v_R(t)}. \tag{7.444}$$

Thus, for very large L, the voltage across the resistor has a very small amount of ripple. Physically, this can be explained as follows. For large L, the second term in (7.435) is almost constant and equal to the average value, $\overline{v_R(t)}$, while the first term in (7.435) is quite small and mostly responsible for the ripple.

The voltage $v_R(t)$ across the resistor in the circuit shown in Figure 7.43 can be viewed as the output voltage. It is clear from the above discussion that for large L this output voltage can be considered as a high-quality dc voltage source. It is important to note that the voltage of this dc source does not depend on the resistive load. This follows from expression (7.439), which shows that the average voltage across the resistor does not depend on the value of R. All this suggests that for large L the circuit shown in Figure 7.43 is a good rectifier circuit. The problem is that it is technically difficult to realize a sufficiently large inductance. This inductance also results in very long transients. For this reason, many different rectifier circuits which avoid the use of inductors have been developed.

One example of such a circuit is shown in Figure 7.45. For the sake of analysis, this circuit can be replaced by the equivalent one shown in Figure 7.46. Here, $v_s^{(r)}(t)$ is the rectified voltage source shown in Figure 7.42.

The circuit shown in Figure 7.46 is equivalent to the circuit shown in Figure 7.45 only when $i(t) > 0$, that is, when the current $i(t)$ flows from node 1 to node 2. The opposite flow of the current is prohibited by the diode bridge circuit. Let us consider the time interval $0 < t < \pi/\omega$ and find when the condition $i(t) > 0$ is violated.

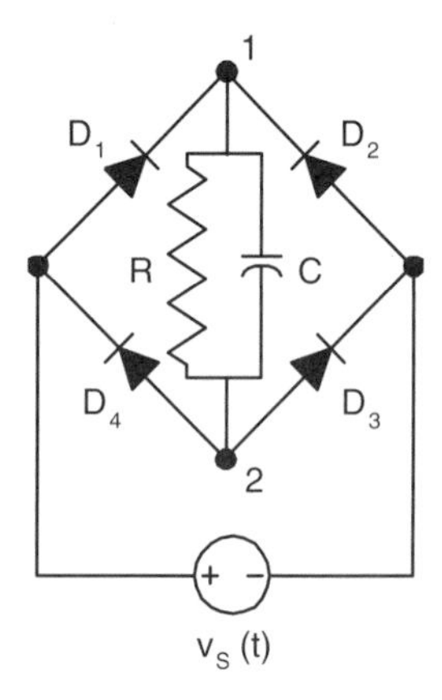

Figure 7.45: Full-wave rectifier with capacitor.

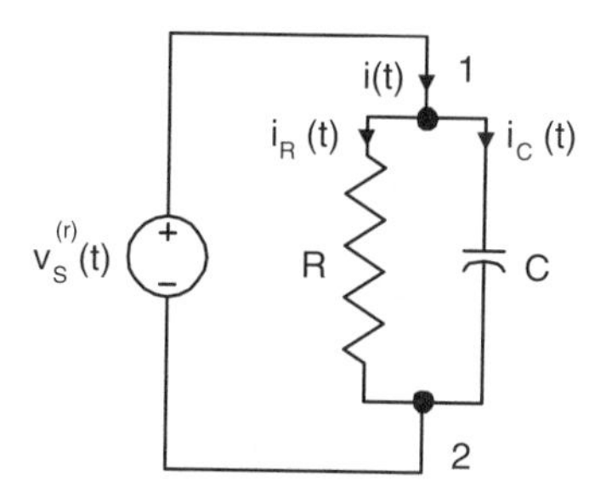

Figure 7.46: Equivalent circuit for the full-wave rectifier with capacitor.

From the circuit shown in Figure 7.46 we infer:

$$i(t) = i_R(t) + i_C(t), \tag{7.445}$$

$$i_R(t) = \frac{V_{ms}}{R} \sin \omega t, \tag{7.446}$$

$$i_C(t) = C\frac{dv_s^{(r)}}{dt} = V_{ms}\omega C \cos \omega t. \tag{7.447}$$

By substituting (7.446) and (7.447) into (7.445), we obtain:

$$i(t) = \frac{V_{ms}}{R} \sin \omega t + V_{ms}\omega C \cos \omega t. \tag{7.448}$$

Let us find the instant of time t_1 when the current $i(t)$ reaches zero:

$$i(t_1) = \frac{V_{ms}}{R} \sin \omega t_1 + V_{ms}\omega C \cos \omega t_1 = 0. \tag{7.449}$$

The last expression yields

$$\tan \omega t_1 = -\omega RC. \tag{7.450}$$

There is only one solution of equation (7.450) in the time interval $0 < t < \pi/\omega$ and this solution is given by:

$$t_1 = \frac{\pi}{\omega} - \frac{1}{\omega} \tan^{-1}(\omega CR). \tag{7.451}$$

It is clear from (7.448) that if $t_1 < t < \pi/\omega$ then $i(t) < 0$. However, the negative current values are prohibited by the diode bridge circuit. This means that the current $i(t)$ should be equal to zero for $t_1 < t < \pi/\omega$. As soon as the current $i(t)$ ceases to flow, the capacitor C starts to discharge through the resistor R. This means that the equivalent circuit takes the form shown in Figure 7.47. This is exactly the circuit that we discussed at the very beginning of the chapter. We found that the capacitor discharge is described by the equation:

$$v_C(t) = Ae^{-\frac{t}{RC}}, \tag{7.452}$$

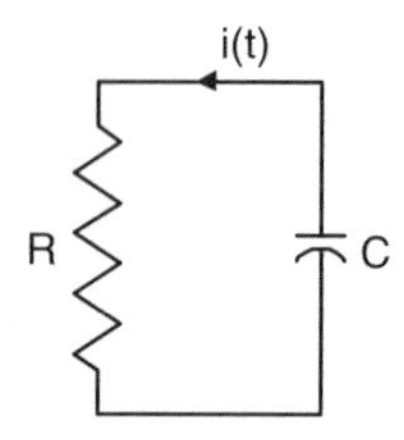

Figure 7.47: Equivalent circuit for the capacitor discharge.

where the constant A should be found from the initial condition. In our case, this condition is:

$$v_C(t_1) = V_{ms} \sin \omega t_1, \tag{7.453}$$

which is immediately clear from the circuit shown in Figure 7.46.

From (7.452) and (7.453), we find:

$$A = V_{ms} e^{\frac{t_1}{RC}} \sin \omega t_1. \tag{7.454}$$

By substituting (7.454) into (7.452), we obtain:

$$v_C(t) = V_{ms} e^{\frac{t_1 - t}{RC}} \sin \omega t_1. \tag{7.455}$$

This capacitor discharge will continue until the voltage across the capacitor becomes equal to the rectified voltage source $v_s^{(r)}(t_2)$ at some instant of time t_2 in the interval $\pi/\omega < t < 2\pi/\omega$ (see Figure 7.48). This leads to the following equation:

$$V_{ms} \sin \omega t_1 e^{\frac{t_1 - t_2}{RC}} = v_s^{(r)}(t_2). \tag{7.456}$$

Since the voltage $v_C(t)$ is periodic with the period equal to π/ω, we find that:

$$t_2 = t_0 + \frac{\pi}{\omega}, \tag{7.457}$$

where time t_0 is indicated on Figure 7.48. It is also clear that

$$v_s^{(r)}(t_2) = v_s^{(r)}(t_0) = V_{ms} \sin \omega t_0. \tag{7.458}$$

By substituting (7.457) and (7.458) into (7.456), we end up with the following equation for t_0:

$$e^{\frac{\omega(t_1 - t_0) - \pi}{\omega RC}} \sin \omega t_1 = \sin \omega t_0, \tag{7.459}$$

which can be reduced to:

$$e^{\frac{t_0}{RC}} \sin \omega t_0 = e^{\frac{\omega t_1 - \pi}{\omega RC}} \sin \omega t_1. \tag{7.460}$$

Note that t_1 is given by expression (7.451). Consequently, expression (7.460) can be construed as a nonlinear algebraic equation which can be solved (by using graphical techniques or a computer) for t_0. After t_0 is found, the expression for $v_C(t)$ is given

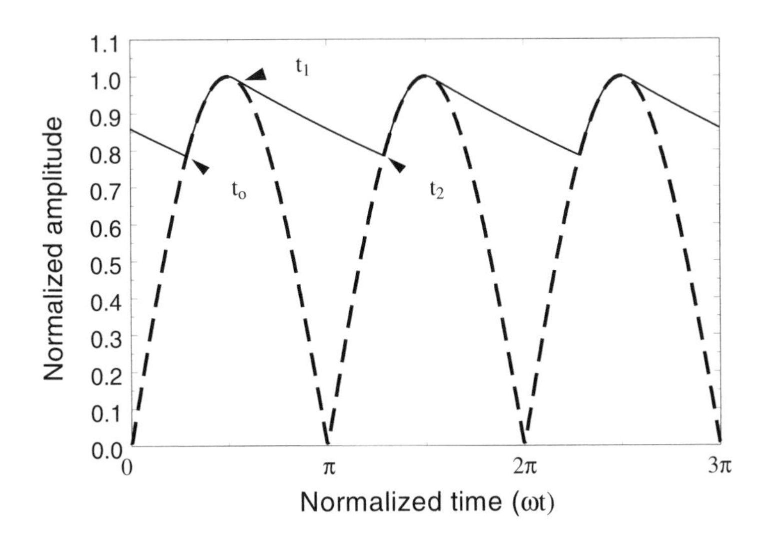

Figure 7.48: Voltage across the capacitor.

by:

$$v_C(t) = \begin{cases} V_{ms} \sin \omega t, & \text{if } t_0 \le t \le t_1, \\ V_{ms} \sin \omega t_1 e^{\frac{t_1 - t}{RC}}, & \text{if } t_1 \le t \le t_0 + \frac{\pi}{\omega}. \end{cases} \tag{7.461}$$

This voltage is shown by the continuous line in Figure 7.48, while the full-wave rectified source voltage is presented by a dashed line. It is clear from this figure that the ripple level for the voltage $v_C(t)$ (which by the way is equal to $v_R(t)$) is substantially smaller than the ripple level for the full-wave rectified source voltage. This figure also clearly reveals the physical mechanism of the variation of the voltage $v_C(t)$. During the time interval $t_0 \le t \le t_1$, the capacitor is being charged by the rectified voltage source, while during the time interval $t_1 \le t \le t_0 + \pi/\omega$ the capacitor is being discharged through the resistor R. This pattern periodically repeats itself.

7.7 MicroSim PSpice Simulations

The first PSpice simulation that we will consider in this section is for the source-driven first-order *RL* circuit in Example 7.5. The circuit, redrawn by the PSpice schematic generator, is shown in Figure 7.49. The values of the passive elements are indicated in the figure and the voltage source has been given the time dependence via the "VSIN" dialog box.

The part name for the switch is "Sw_tClose," which can be accessed by typing this name in the "Add Part" dialog box. If one is uncertain of the name of a circuit model or even uncertain if a model exists, one can press the "Browse... " button

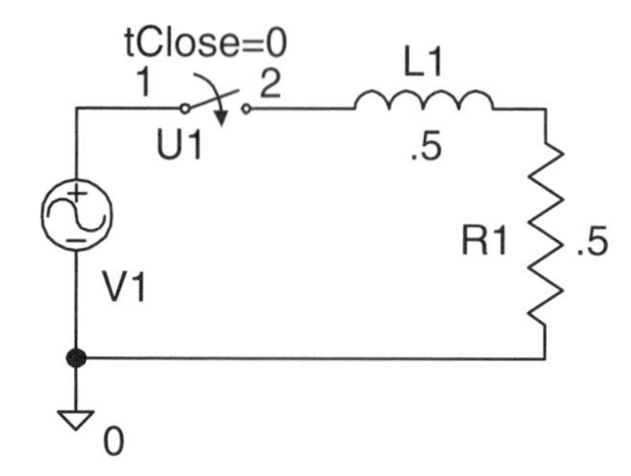

Figure 7.49: The PSpice schematic for the circuit in Example 7.5.

in the "Add Part" dialog box for help. That button brings up the "Get Part" dialog box that lists the various libraries that contain the part models and lists the parts contained in whichever library is highlighted. The switch required for this example can be found by clicking on the "eval.slb" library (using the scroll bars if necessary) and scrolling down the parts list until the name "Sw_tClose" is visible. Click on the name to highlight it. This name will appear in the "Part Name" box at the top of the window and the description "Switch: closes at tClose=?" will appear directly below the part name box. Pressing the "OK" button will select the switch and it can then be placed on the schematic in the usual way.

The properties of the switch can be found by double-clicking on the switch after it has been placed in the circuit. There are five properties of this model that can be adjusted. The first property is the Package Reference Designator "PKGREF" (i.e., name) which uniquely identifies each part that is placed in the circuit. An ideal switch is modeled by infinite open resistance and zero closed resistance. However, for numerical reasons and because some switches are nonideal, the switch model has the parameters Ropen and Rclosed to model the nonideal open and closed resistances of the switch, respectfully. The default values are given in the dialog box and can be changed if necessary. In particular, care should be taken if either of these values is the same order of magnitude as other resistances in the circuit. The "tClose" variable indicates the time when the switch begins to change from its open state to its closed state. The final parameter is "ttran," which is the time it takes to make the transition from the high-impedance state to the low-impedance state. All of the switch default values were used for this simulation.

A transient analysis is called for with a final time of 10 s and a print step of 50 ms. The simulated current through the inductor is plotted in Figure 7.50. The analytic solution is also plotted in the figure. For the circuit parameters stated above, equation (7.93) gives the analytic solution as:

$$i_{L_1}(t) = 2\cos(\sqrt{3}t - 60^\circ) - e^{-t}\text{ A}. \tag{7.462}$$

As expected, the two curves are virtually identical for the entire simulation period. The differences stem from the resistance of the switch and the inaccuracies of the PSpice numerical methods.

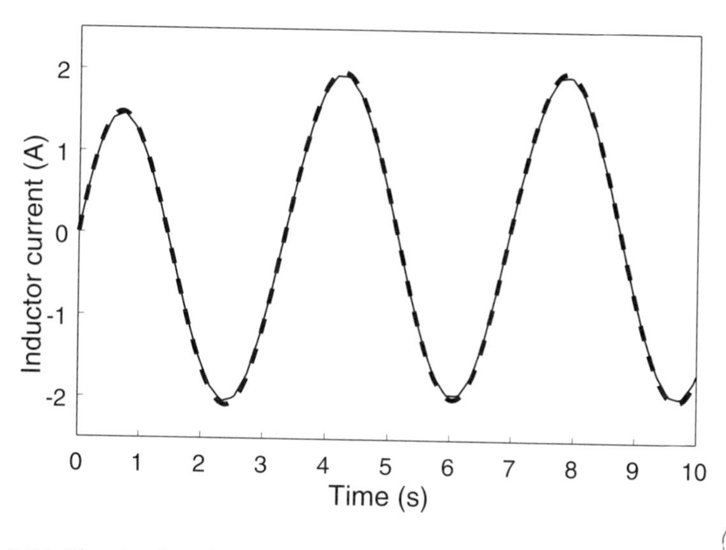

Figure 7.50: The simulated (solid line) and analytic (dashed line) solutions for the inductor current in the circuit of Figure 7.49.

The next PSpice simulation revolves around the second-order circuit of Example 7.10 which is excited by initial conditions. The numerical values in the schematic shown in Figure 7.51 are identical to the values given in the example. A parametric sweep of the capacitance is selected (with List Values: C = 0.1, 0.8, 1.0, and 1.333 F) as described in Section 4.7. The only new part type is the opening switch, which has the name "Sw_tOpen" and can be found in the "eval.slb" library.

The results of the simulation for the voltage across the capacitor are plotted in Figure 7.52 for all four cases. As expected, the voltage decays more slowly for the overdamped case (C = 1.333 F) than for the critically damped case (C = 1F). The underdamped cases (C = 0.1, 0.8 F) decay the most rapidly. The voltage oscillations

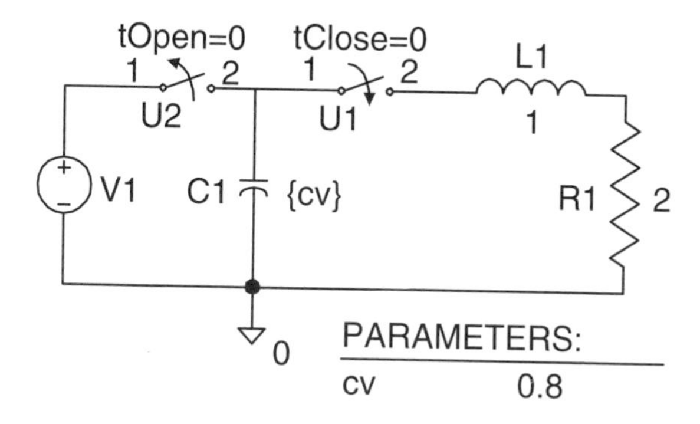

Figure 7.51: The PSpice schematic for the circuit in Example 7.10.

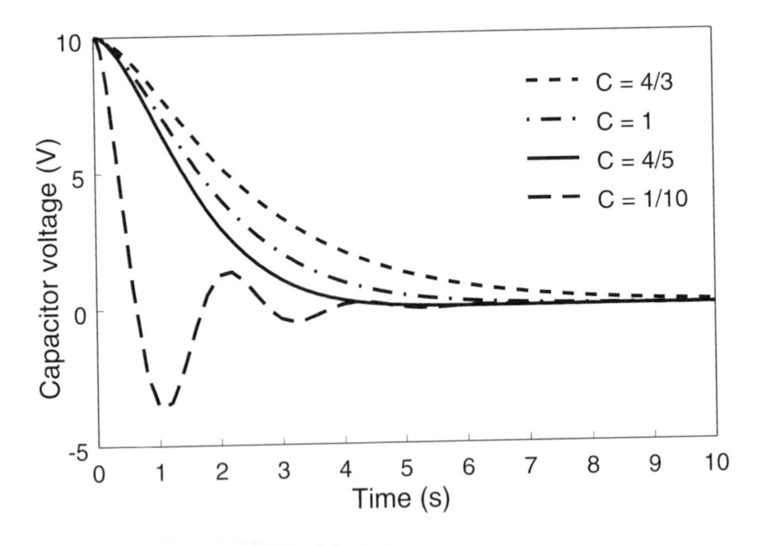

Figure 7.52: The simulated capacitor voltage in the circuit of Figure 7.51.

for the C = 0.1 F case are quite evident, but the capacitor voltage for the C = 0.8 F case goes only slightly negative. This is difficult to see on the figure because the damping is so strong, but can be seen in Probe by expanding the y-axis. A comparison of the figure with the numerical values listed in Table 7.1 shows good agreement.

The third PSpice simulation demonstrates the power and utility of PSpice by showing the ease with which we generate the solution to a difficult second-order transient problem. The schematic is shown in Figure 7.53 and uses the source-excited dual capacitor circuit of Example 7.15. The parameters for the passive elements are indicated in the figure. The voltage source is a damped sinusoidal with an amplitude

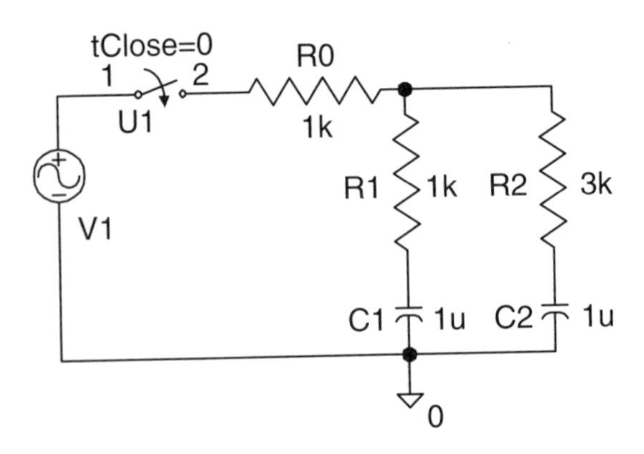

Figure 7.53: The PSpice schematic for the circuit in Example 7.15.

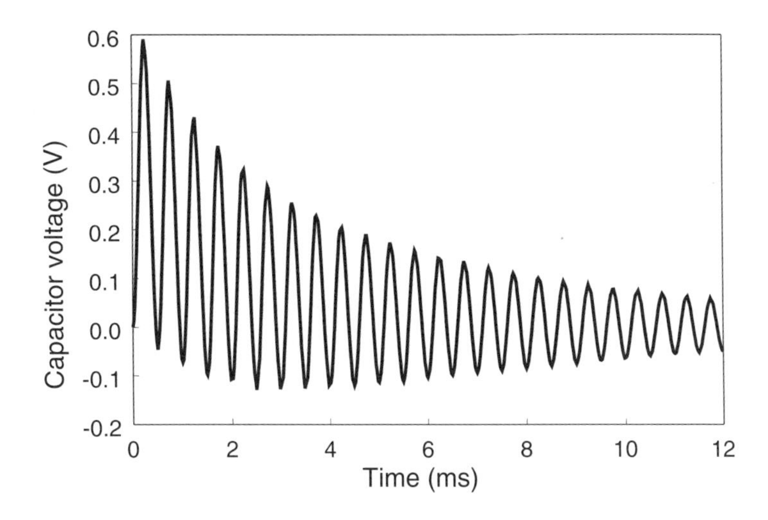

Figure 7.54: The simulated capacitor voltage in the circuit of Figure 7.53.

of 10 V, a frequency of 2000 Hz, an offset voltage of 0 V, a phase of 0 (i.e., a $\sin(\omega t)$ dependence), and a damping factor (DF) of 150. The damping factor is simply related to the complex frequency, which for this case becomes:

$$s = \sigma + j\omega = -150 + j2\pi(2000)\ \mathrm{s}^{-1}. \tag{7.463}$$

The voltage across the capacitor C1 is plotted as a function of time in Figure 7.54. The reader who still doubts the usefulness of a circuit simulator is encouraged to reproduce the analytic result for this circuit and compare it to the PSpice simulation.

The schematic shown in Figure 7.55 is the PSpice realization of the second-order transient problem that was analyzed in Example 7.16. The numerical values indicated in the figure are the same as in the example. Likewise, the sinusoidal voltage source (VSIN) has a magnitude of 50 V and an angular frequency of 1 rad/s. Remember that you must use "MEG" for the resistances to get the proper values.

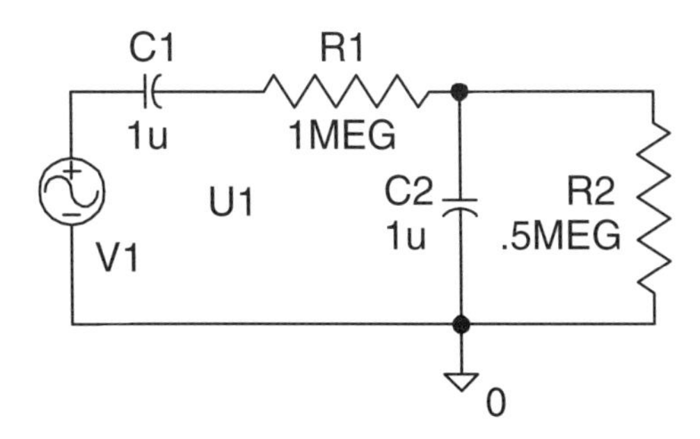

Figure 7.55: The PSpice schematic for the circuit in Example 7.16.

A transient analysis is selected with a "Print Step" of 50 ms and a "Final Time" of 20 s. In the same "Transient" dialog box that is used to select these times, there are two additional options at the bottom of the "Transient Analysis" section of the window. The left option calls for a "Detailed Bias Pt." and the right option reads "Use Init. Conditions." To get the desired simulation results, place an "x" in the box to the immediate left of the right option by clicking on either it or the text. The "bias point detail" is normally selected by default in the "Analysis Setup" dialog box, so there is no need to activate the left option. The analytic expression for the voltage across R1, which is given in (7.353), is indicated by the upper trace in Figure 7.56. The lower trace in that figure displays the result for the simulation. Once again, the results are virtually identical.

The final simulation uses PSpice to analyze the filtered, full-wave diode rectifier circuit considered in the previous section (see Figure 7.45). As seen in the text, this problem is quite challenging and requires the solution of a transcendental equation. However, the PSpice solution is no more difficult than any of the previous problems. The schematic is drawn in Figure 7.57 and has a 1 kΩ resistive load. The capacitor has been parameterized (C = 300 pF, 3 μF, 30 μF, and 300 μF) to demonstrate the

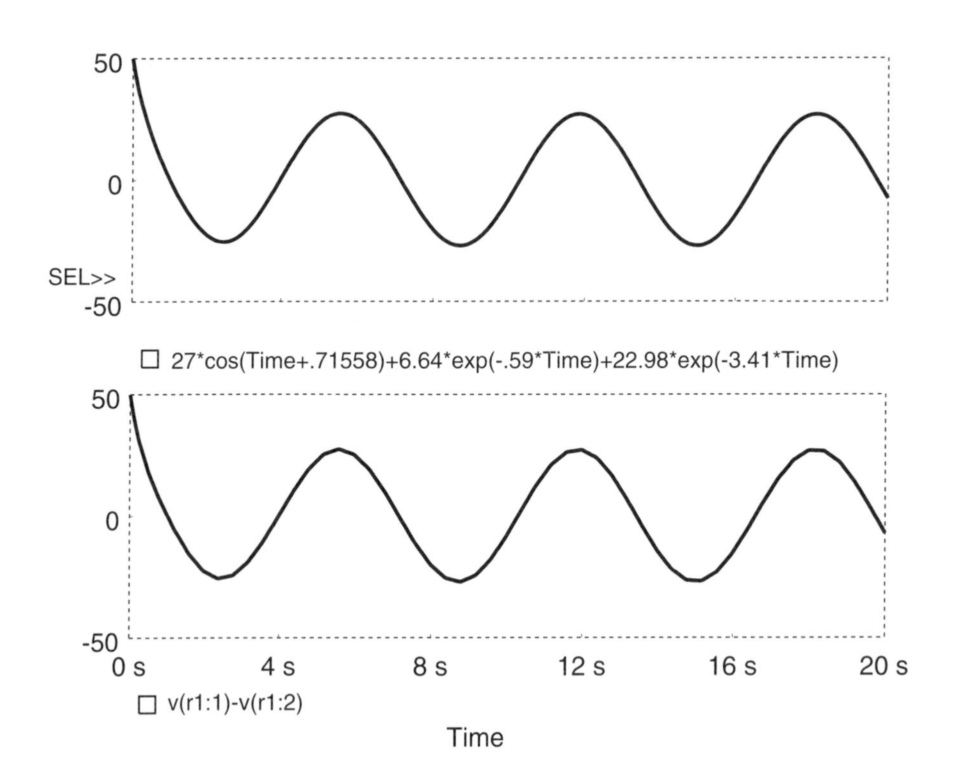

Figure 7.56: The analytic expression (upper trace) and simulated result (lower trace) for the voltage across R1 in the circuit of Figure 7.55.

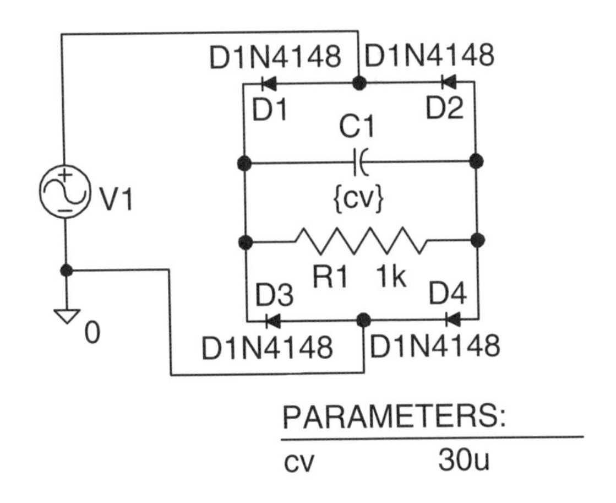

Figure 7.57: The PSpice schematic for the filtered full-wave diode rectifier.

effect of its magnitude on filtering. We assume a household voltage source with an amplitude of 169.7 V, a phase of 0°, and a frequency of 60 Hz.

The only elements in this circuit that we have not yet considered in previous examples are the four identical diodes, which have the part name "D1N4148" and can be found in the "eval.slb" library. The modeling of this element is a subject for an electronics course, but we need to make one simple modification nonetheless. It suffices to say that while an ideal diode has zero current for any voltage applied across it in the "wrong" direction, a real diode has a maximum reverse breakdown voltage that should not be exceeded. For this diode, the model has this breakdown voltage set to Bv = 100 V. Since we are using a voltage source with a maximum voltage of 169.7 V, we will change the breakdown voltage of the model to 200 V. *(Note: if we were to use real D1N4148's in this bridge, exceeding the reverse breakdown voltage would most likely cause them to burn out with potentially dire consequences!)* Click on a diode to select it and from the "Edit" drop-down menu select "Model." In the "Edit Model" dialog box that appears, push the "Edit Instance Model" button. A "Model Editor" dialog box will appear that will list all the parameters of the model. Change the "Bv" variable from 100 to 200 and then push the "OK" button. Repeat this procedure for the other three diodes. (The procedure required to create entirely new models is beyond the scope of this text but is described in some of the references listed in Appendix C.)

The voltage across the resistor/capacitor load is shown in Figure 7.58 for the four different cases. The lowest capacitance does virtually nothing and the voltage is the same as for an unfiltered full-wave rectifier. The time constant for the 3 μF capacitor is a few times smaller than the source period and the ripple is quite large. The ripple is much reduced for the 30 μF capacitor, which has a time constant nearly twice that of the source period. The ripple for the 300 μF capacitor is the lowest, of course, but the time constant is so long that the voltage has not yet reached steady state after three periods of the source (six periods of the effective rectified source).

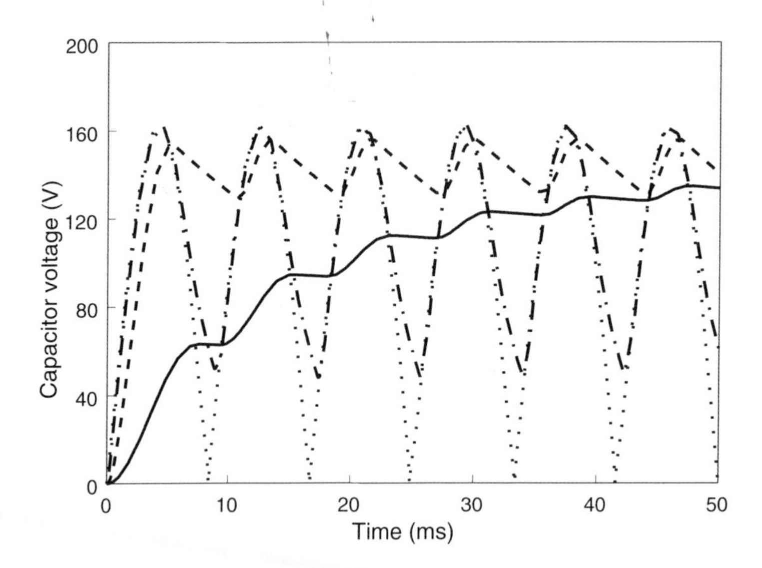

Figure 7.58: The simulated capacitor voltage in the circuit of Figure 7.57. The solid line corresponds to the C = 300 μF case, the dashed line is for C = 30 μF, the dot-dash line is for C = 3 μF, and the dotted line is for C = 300 pF.

7.8 Summary

In this chapter, we systematically developed various techniques for the analysis of transients in first- and second-order electric circuits. The most important facts and results discussed in the chapter can be summarized as follows:

- Transients in electric circuits occur due to the presence of energy storage elements (i.e., inductors and capacitors).
- Transients in electric circuits can be excited by initial conditions, by sources, or by both.
- Analysis of transients can be broken down into two major steps:
    1. Determination of initial conditions for the energy storage elements by using the continuity of voltage across a capacitor and the continuity of current through an inductor.
    2. Analysis of electric circuits after switching. This step normally involves the solution of initial value problems for ordinary differential equations.
- Analysis of transients excited by initial conditions requires the solution of **homogeneous** differential equations subject to nonzero initial conditions.
- Analysis of transients excited by sources requires the solution of **nonhomogeneous** differential equations subject to **zero** initial conditions.

- Analysis of transients excited by initial conditions and sources requires the solution of **nonhomogeneous** equations subject to **nonzero** initial conditions.

- A complete solution of a nonhomogeneous linear differential equation can be represented as a sum of a particular solution of the nonhomogeneous equation and a general solution of the corresponding homogeneous equation.

- In the case of excitation by ac sources, the particular solution of the nonhomogeneous equation can be found by using the phasor technique for the calculation of the steady-state response. The particular solution has the physical meaning of **forced response**.

- A general solution of the corresponding homogeneous equation has the physical meaning of **free (transient) response**. Its calculation requires the solution of characteristic equations and determination of unknown constants from initial conditions.

- The nature of the transient (free) response of second-order circuits is determined by the roots of the characteristic equation. There are four distinct cases:

  a) overdamped response when the two roots are real, negative, and distinct;

  b) critically damped response when the two roots are real, negative, and identical;

  c) underdamped response when the two roots are complex and conjugate;

  d) undamped response when the two roots are imaginary and conjugate.

- The machinery of transfer functions is a very powerful tool in the analysis of transients in electric circuits. This analysis proceeds as follows. A circuit variable, which we are interested in, is identified as the output. The transfer function is defined as the ratio of the output phasor to the input phasor, i.e., excitation source. By using the phasor technique, the transfer function $\hat{H}(s)$ is found as the function of complex frequency s. The value of this function at the excitation frequency ($s = j\omega$) fully determines the ac steady-state (forced) response, while the poles of the transfer function determine the exponents and, consequently, the form of the free response. The described approach is algebraic in nature; it completely avoids the derivation and solution of differential equations and fully exploits the machinery of the phasor technique.

- To calculate the response of an electric circuit to an arbitrary source, the **convolution integral** can be used. The convolution integral has the form:

$$i(t) = \int_0^t i_\delta(t - \tau) v_s(\tau) d\tau, \tag{7.464}$$

  where $i(t)$ is the response (current) caused by the excitation by the voltage source $v_s(t)$, while $i_\delta(t)$ is the unit impulse response caused by the unit impulse source excitation. The unit impulse response can be found as the time derivative of unit step response, which in turn can be found from the transient analysis of the electric circuit excited by a unit dc source. Thus, the analysis of an electric circuit by using the convolution integral consists of two major steps: a) calculation of unit impulse response; b) evaluation of convolution integral for an arbitrary (but given) voltage source $v_s(t)$. Expressions similar to (7.464) hold for the convolution integrals when the desired circuit response is a voltage and/or the circuit excitation is a current source.

- A diode is a two-terminal element whose resistance depends on the polarity of the applied voltage. Ideal diodes act as short circuits if applied voltages are positive, and they act as open circuits if applied voltages are negative. Diodes are used in rectifier circuits to convert ac voltage sources into dc voltage sources. A single diode can be used to construct a **half-wave** rectifier. A diode bridge circuit can be used to construct a **full-wave rectifier**. Energy storage elements are employed in rectifier circuits to reduce the level of ripples. **The techniques for transient analysis of electric circuits can be used for the steady-state analysis of rectifiers**.

7.9 Problems

1. A nonideal capacitor, after being charged to an initial voltage V_o, will slowly discharge. For the capacitances, conductances, and initial voltages given in each row of the following table, find the time it takes for the capacitor to discharge to 1 V.

Circuit	V_o (V)	C (μF)	G_C (m℧)
a	2	.001	100.
b	5	2.2	10.
c	10	470	1.

2. An initial current, I_o, in a shorted nonideal inductor will slowly decay with time. For the inductances, resistances, and initial currents given in each row of the following table, find the time it takes for the inductor current to reduce to 10 mA.

Circuit	I_o (A)	L (μH)	R_L ($\mu\Omega$)
a	.02	1	1.
b	5	10	10,000.
c	.06	470	100.

3. Plot the current through the 10 Ω resistor in the circuit shown in Figure P7-3 for $t > 0$.

4. Plot the voltage across the 10 μH inductor in the circuit shown in Figure P7-4 for $t > 0$.

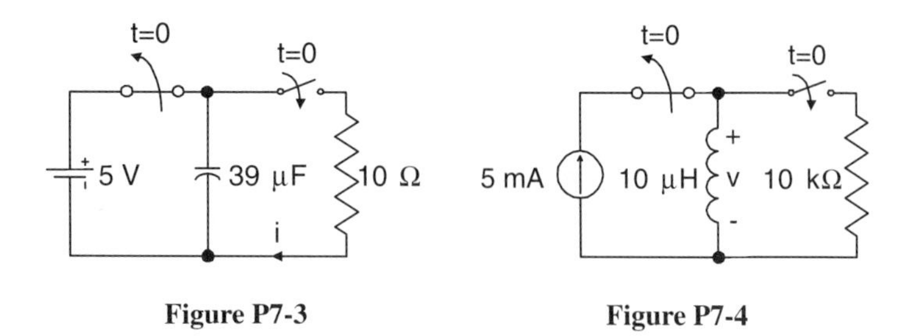

Figure P7-3 **Figure P7-4**

5. For the circuit shown in Figure P7-5, find the voltage across R_3 for $t > 0$.

6. For the circuit shown in Figure P7-6, find the voltage across C_2 for $t > 0$.

7. For the circuit shown in Figure P7-6, let $V_o = 20$ V, $R_1 = 6$ kΩ, $R_2 = 2$ kΩ, and $C_1 = C_2 = 470$ μF. Find the value of R_3 required so that the voltage across C_2 decays to 1 V in 1 second.

8. For the circuit shown in Figure P7-5, let $V_o = 10$ V, $R_1 = 100$ Ω, $R_2 = 25$ Ω, and $L_1 = L_2 = 10$ mH. Find the value of R_3 required so that the current through L_2 decays to 2 mA in 2 ms.

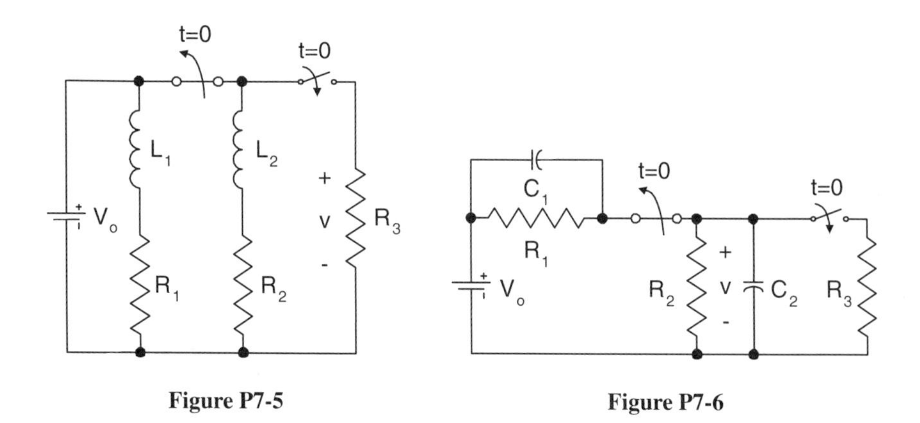

Figure P7-5 **Figure P7-6**

9. Find the current through the inductor in the circuit shown in Figure P7-9 (a) by the method of phasors and (b) by the method of undetermined coefficients. Assume $v_s(t) = 3\cos(2t + \pi/8)$ V and $i_L(0_-) = 0$.

10. Find the voltage across the capacitor in the circuit shown in Figure P7-10 (a) by the method of phasors and (b) by the method of undetermined coefficients. Assume $i_s(t) = 10\sin(10^6 t)$ mA and $v_C(0_-) = 0$.

11. Assume that $i_s(t) = 3t$ mA when $t > 0$ in the circuit shown in Figure P7-11 and plot the current through the inductor for $t > 0$. All quantities are zero for $t < 0$.

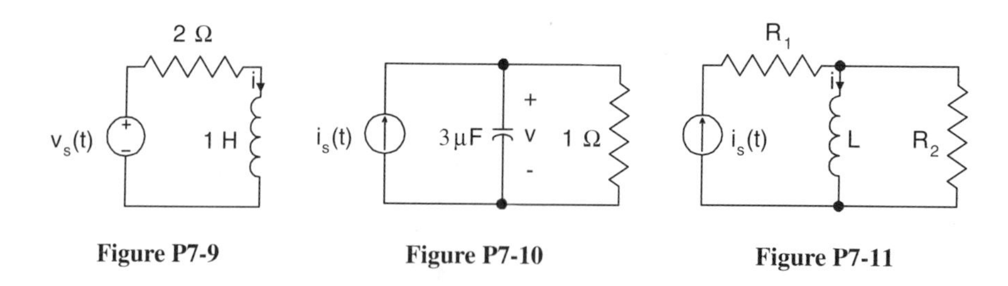

Figure P7-9 **Figure P7-10** **Figure P7-11**

12. Assume that $v_s(t) = 2e^{2t/t_o}$ V when $t > 0$ in the circuit shown in Figure P7-12 and plot the voltage across the capacitor for $t > 0$. All quantities are zero for $t < 0$.

13. Find the current through the resistor in the circuit shown in Figure P7-10 if $i_s(t) = 2(t/t_o)^2 u(t - t_o) - (t/t_o)u(t - 3t_o)\mu$A and $t_o = 1\ \mu$s.

14. Find the voltage across the resistor in the circuit shown in Figure P7-9 if $v_s(t) = e^{(t-1)}u(t - 1)$ kV.

15. Use the convolution integral to find the voltage across the resistor R_1 in the circuit shown in Figure P7-12 if the source voltage is $v_s(t) = \cos(t)$ V and $v_C(0_-) = 0$.

16. Use the convolution integral to find the current through the resistor R_2 in the circuit shown in Figure P7-11 if the source current is $i_s(t) = \sin(t)$ A and $i_L(0_-) = 0$.

17. Find the unit step and impulse responses for the current through the 1Ω resistor in the circuit shown in Figure P7-17.

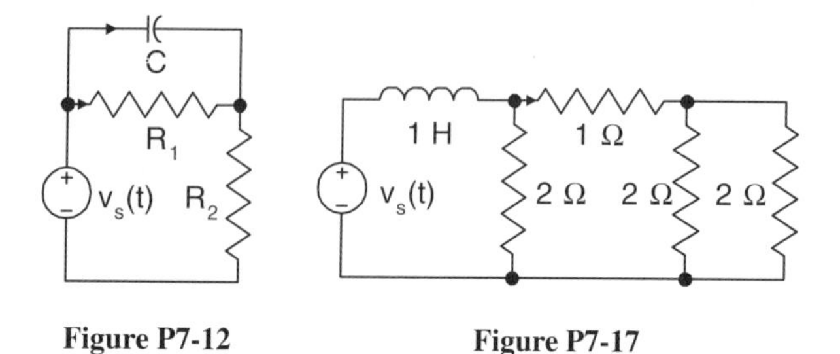

Figure P7-12 **Figure P7-17**

18. Find the unit step and impulse responses for the voltage across the capacitor in the circuit shown in Figure P7-18.

19. Find the current through the resistor R_1 in the circuit shown in Figure P7-18. Assume the initial voltage across the capacitor is $v_C(0_-) = 2$ V, $v_s(t) = \cos(t)$ V, $C = 1$ F, and $R_2 = 2R_1 = 2\ \Omega$.

20. Find the voltage across the $1\ \Omega$ resistor in the circuit shown in Figure P7-17. Assume the initial current in the inductor is $i_L(0_-) = 1$ mA and $v_s(t) = 3u(t)$ mV.

21. Find the voltage across the resistor in the circuit shown in Figure P7-21 for $t > 0$. Assume that $v_1(t) = \cos(10^8 t)$ V and $v_2(t) = \cos(2 \times 10^8 t)$ V.

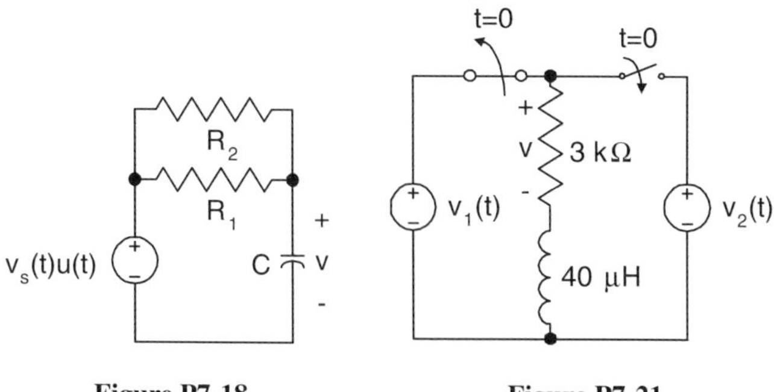

Figure P7-18 **Figure P7-21**

22. Find the voltage across the 10 Ω resistor in the circuit shown in Figure P7-22 for $t > 0$.

23. Consider the circuit shown in Figure P7-23. Draw the circuit for $t < 0$ and find the relevant initial conditions. Draw the circuit for $t > 0$ and find the differential equation for the current through the inductor. Find the solution for the inductor current for $t > 0$. Plot the power through the inductor. (Be careful to use an appropriate time scale.)

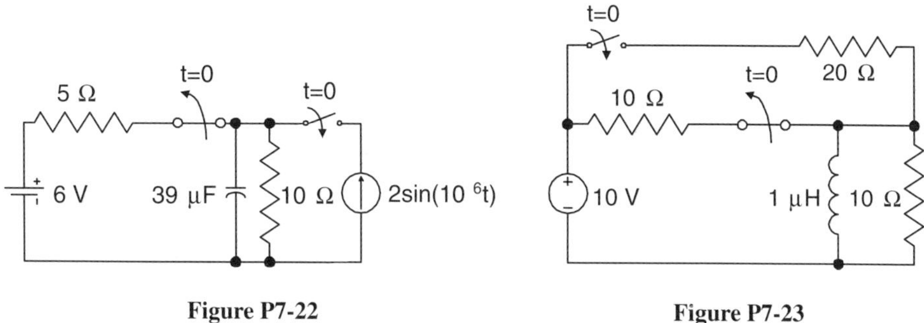

Figure P7-22 **Figure P7-23**

24. Consider the circuit shown in Figure P7-24. Find the voltage across the capacitor for $t > 0$.

25. Consider the circuit shown in Figure P7-25. For the capacitances, resistances, inductances, and initial voltages given in each row of the following table, find the time it takes for the capacitor to discharge to 1 V.

Circuit	V_o (V)	C (F)	L (H)	R (Ω)
a	2	10^{-8}	10^{-3}	470
b	10	0.2	1	20
c	100	0.125	2	8

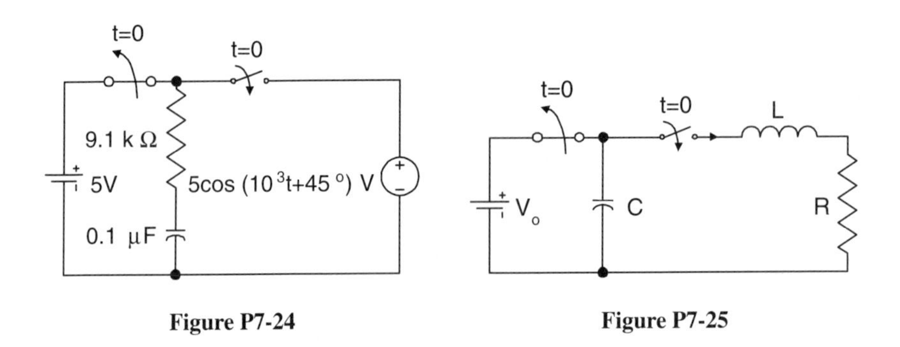

Figure P7-24 Figure P7-25

26. Consider the circuit shown in Figure P7-25. If $L = 1$ mH, $C = 470$ mF, $R = 2$ kΩ, and $V_o = 5$ V, plot the current through the inductor when $t > 0$.

27. Consider the circuit shown in Figure P7-27. If $L = 1\ \mu$H, $C = 1\ \mu$F, $R = 10\ \Omega$, and $I_o = 1$ A, plot the current through the resistor when $t > 0$.

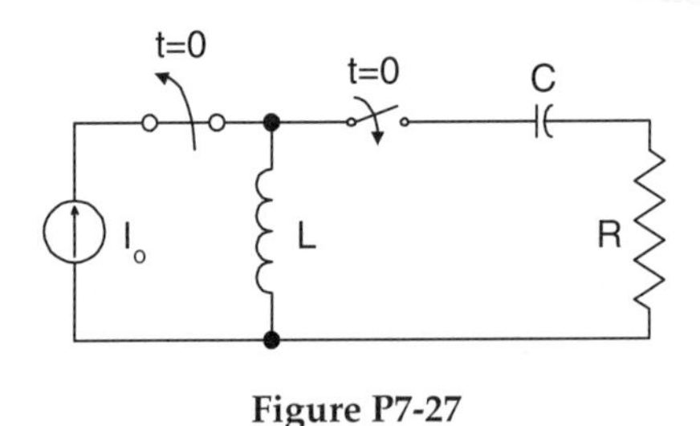

Figure P7-27

28. Consider the circuit shown in Figure P7-27. For the capacitances, resistances, inductances, and initial currents given in each row of the following table, find the time it takes for the inductor current to reduce to 10 mA.

Circuit	I_o (A)	L (H)	C (F)	R (Ω)
a	.02	10^{-3}	10^{-7}	200
b	0.2	10^{-4}	2×10^{-6}	20
c	2.	.01	10^{-7}	2000

29. Plot the current through the resistor in the circuit shown in Figure P7-29 for $t > 0$. Assume $R = 1/2\ \Omega$, $C = 1$ F, $L = 1$ H, and $V_o = 5$ V.

30. Plot the voltage across the inductor in the circuit shown in Figure P7-30 for $t > 0$. Assume $R = 1/2\ \Omega$, $C = 1$ F, $L = 1$ H, and $I_o = 5$ A.

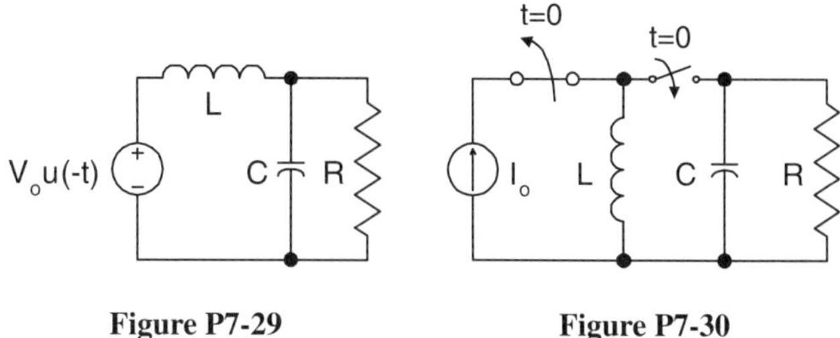

Figure P7-29 **Figure P7-30**

31. For the circuit shown in Figure P7-31, use the convolution integral to find the voltage across the resistor for $t > 0$. Assume that $R = 4\ \Omega$, $L = 1$ H, $C = 0.25$ F, and $v_s(t) = (2t - 3t^2)u(t)$ V.

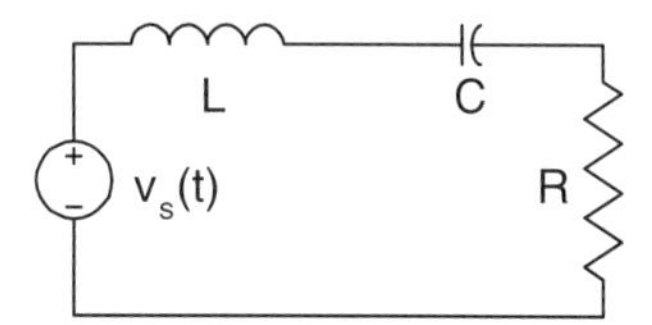

Figure P7-31

32. For the circuit shown in Figure P7-31, use the convolution integral to find the voltage across the capacitor for $t > 0$. Assume that $R = 10\ \Omega$, $L = 0.1$ H, $C = 0.1$ F, and $v_s(t) = (1 - e^{-2t})u(t)$ V.

33. Find the current through the inductor in the circuit shown in Figure P7-31. Assume $v_s(t) = 2\cos(20t + \pi/6)u(t)$ V, $L = 0.1$H, $C = 0.02$ F, and $R = 5\ \Omega$.

34. Find the voltage across the capacitor in the circuit shown in Figure P7-31. Assume $v_s(t) = 5\sin(10t + 30°)u(t)$ kV, $L = 0.5$ H, $C = 0.2$ F, and $R = \sqrt{10}\ \Omega$.

35. Find the unit step response and the unit impulse response of the current through the inductor in the circuit shown in Figure P7-35. Let $L = 1.2$ H, $C = 0.3$ F, and $R = 1\ \Omega$.

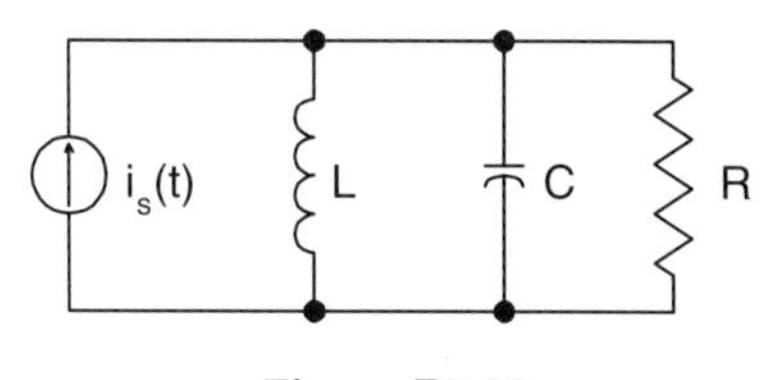

Figure P7-35

36. Find the unit step response and the unit impulse response of the voltage across the resistor in the circuit shown in Figure P7-35. Assume that $(1/2RC)^2 - 1/(LC) < 0$.

37. Find the voltage across the resistor in the circuit shown in Figure P7-35 if $i_s(t) = I_m \cos(\omega t + \phi)u(t)$ A. Assume that $(1/2RC)^2 > 1/(LC)$.

38. Consider that $i_s(t) = 7tu(t)$mA in the circuit shown in Figure P7-35 and plot the current through the inductor for $t > 0$. Let $L = 1$ H, $C = 1$ F, and $R = 3\ \Omega$.

39. Plot the power dissipated in the capacitor and inductor in the circuit shown in Figure P7-25 as a function of time if $R = 0\ \Omega$, $L = 1$ mH, $C = 1$ mF, and $V_o = 1$ kV.

40. Plot the power dissipated in the capacitor and inductor in the circuit shown in Figure P7-30 as a function of time if $R \rightarrow \infty\ \Omega$, $L = 91\ \mu$F, $C = 22\ \mu$F, and $I_o = 1$ mA.

41. Plot the current through the source ($v_s(t) = V_{ms} \cos(\omega t + \phi_s)u(t)$) in the circuit shown in Figure P7-41 over one period.

42. Find the steady-state voltage across the capacitor in the circuit shown in Figure P7-42 when $i_s(t) = I_{ms}|\sin(\omega t)|$.

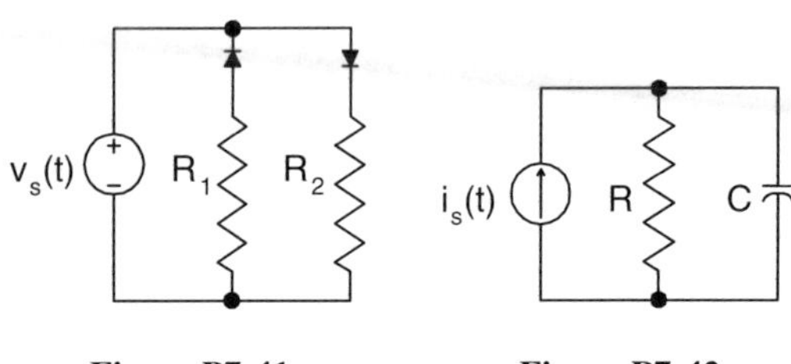

Figure P7-41 **Figure P7-42**

43. Use the convolution integral to find the output current indicated in the circuit shown in Figure P7-43 when $v_s(t) = (1 + 2t)u(t)$. Clearly explain all major steps of the analysis.

44. Use the convolution integral to find the output current indicated in the circuit shown in Figure P7-44 when $i_s(t) = [1 + \cos(t)]u(t)$ A. Clearly explain all major steps of the analysis.

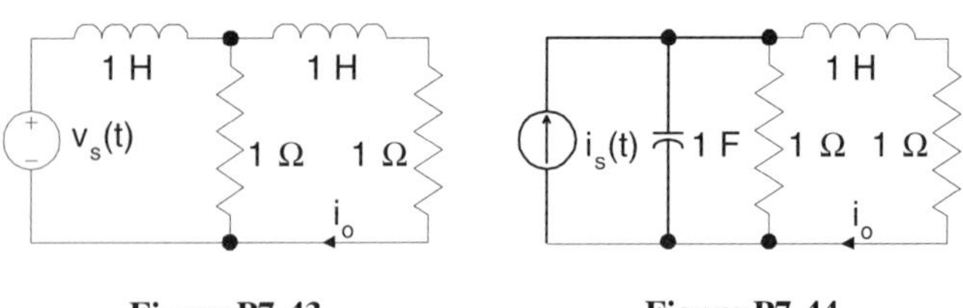

Figure P7-43 **Figure P7-44**

45. Consider the circuit shown in Figure P7-45. Find the equation which describes the evolution of $v_o(t)$. If all parameter values are unity and $v_s(t) = \cos(3t)u(t)$, find $v_o(t)$. (Hint: you may want to consider using transfer functions.)

46. Consider the circuit shown in Figure P7-45. If $v_s(t) = 10u(-t)$ V, $C_1 = 1\ \mu$F, $C_2 = 3\ \mu$F, $R_1 = 2$ kΩ, and $R_2 = 1$ kΩ, find the current through C_2 as a function of time.

47. Consider the circuit shown in Figure P7-47. If $C = 470\ \mu$F, $L = 10\ \mu$H, $R_1 = 10$ kΩ, and $R_2 = 2$ kΩ, find the unit step and impulse responses of $v_o(t)$.

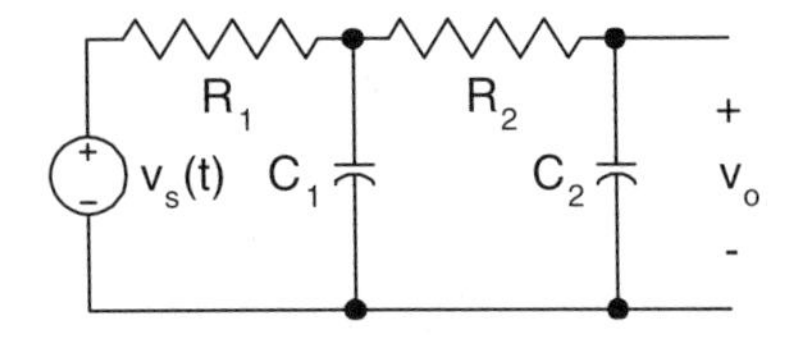

Figure P7-45

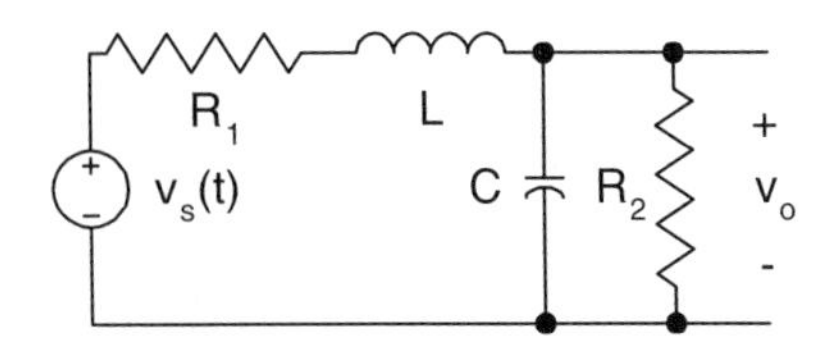

Figure P7-47

48. Consider the circuit shown in Figure P7-47. If all parameter values are unity and $v_s(t) = \cos(4t)u(t)$, find $v_o(t)$. (Hint: you may want to consider using transfer functions.)

49. Consider the circuit shown in Figure P7-49. Find the differential equation which describes the evolution of $i_o(t)$. If $R_1 = 1/2\ \Omega$, $R_2 = 1\ \Omega$, $L = 2$ H, $C = 1$ F, and $i_s(t) = \cos(2t + 45°)u(t)$, find $i_o(t)$.

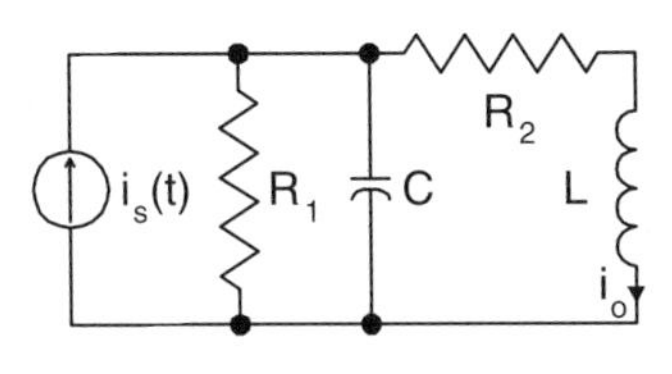

Figure P7-49

50. Consider the circuit shown in Figure P7-49. If $L = 1$ mH, $C = 1$ nF, and $R_1 = 1$ kΩ, find the value of R_2 that will result in critical damping of $i_o(t)$.

In Problems 51–60, use PSpice to plot the circuit parameters indicated.

51. Plot the current through the 10 Ω resistor in the circuit shown in Figure P7-3 for $t > 0$.

52. Plot the voltage across the 10 μH inductor in the circuit shown in Figure P7-4 for $t > 0$.

53. Plot the voltage across the resistor in the circuit shown in Figure P7-21 for $t > 0$. Assume that $v_1(t) = \cos(10^8 t)$ V and $v_2(t) = \cos(2 \times 10^8 t)$ V.

54. Plot the current through the resistor in the circuit shown in Figure P7-29 for $t > 0$. Assume $R = 1/2\ \Omega$, $C = 1$ F, $L = 1$ H, and $V_o = 5$ V.

55. Plot the voltage across the inductor in the circuit shown in Figure P7-30 for $t > 0$. Assume $R = 1/2\ \Omega$, $C = 1$ F, $L = 1$ H, and $I_o = 5$ A.

56. Plot the power dissipated in the capacitor and inductor in the circuit shown in Figure P7-25 as a function of time if $R = 0\ \Omega$, $L = 1$ mH, $C = 1$ mF, and $V_o = 1$ kV.

57. Plot the power dissipated in the capacitor and inductor in the circuit shown in Figure P7-30 as a function of time if $R \to \infty\ \Omega$, $L = 91\ \mu$F, $C = 22\ \mu$F, and $I_o = 1$ mA.

58. Plot the current through the source ($v_s(t) = V_{ms} \cos((t + \phi s)u(t))$) in the circuit shown in Figure P7-41 over one period.

59. Consider the circuit shown in Figure P7-45. If all parameter values are unity and $v_s(t) = \cos(3t)u(t)$, plot $v_o(t)$.

60. Consider the circuit shown in Figure P7-49. If $i_s(t) = 10u(-t)$ mA, $C = 1\ \mu$F, $L = 1$ H, $R_1 = 2$ kΩ, and $R_2 = 1$ kΩ, plot the current through the inductor as a function of time.

Chapter 8

Dependent Sources and Operational Amplifiers

8.1 Introduction

In previous chapters, we have considered two basic sources: current and voltage sources. These sources were always assumed to be *independent*. That meant that their values did not depend on other quantities in circuits they were connected to. In this chapter, we study *dependent* sources. *These are the sources whose values do depend on other quantities in the circuit.* They usually appear as linear models for transistors and other semiconductor devices.

After defining the four different types of dependent sources, we give a brief introduction to the two most important types of transistors. We introduce a dependent source model for the MOSFET transistor and demonstrate its validity. We then revisit the formalisms for general nodal analysis, general mesh analysis, and Thevenin's theorem and show how the presence of dependent sources affects the procedures that we have previously developed to analyze electric circuits. We demonstrate that for each technique only minor modifications are required to accommodate the presence of dependent sources.

In the second part of this chapter we discuss an important type of dependent voltage source with very special properties. This dependent voltage source is called an *operational amplifier* (also called an *op-amp* for short) and is realized via an integrated circuit which contains many transistors, resistors, and capacitors. We will not get into the details of the internal structure–that is left for a future course on electronics. Instead, we characterize the op-amp in terms of its terminal properties and use them to analyze many circuits containing op-amps. These circuits can be used to isolate loads from sources, amplify voltage or current signals, invert signals, add signals together, integrate signals, or differentiate them. The chapter is concluded with a demonstration that op-amp circuits can act as analog computers and sine wave generators.

8.2 Dependent Sources as Linear Models for Transistors

There are four types of dependent sources: voltage-controlled voltage sources (VCVS), current-controlled voltage sources (ICVS), voltage-controlled current sources (VCIS), and current-controlled current sources (ICIS). The dependent sources are represented by diamond notations instead of circles as with independent sources. The circuit notations for the four sources are shown in Figure 8.1. In this figure, v_x and i_x are controlling voltages and currents, respectively. In these notations, controlling branches are shown adjacent to dependent sources. This may not be the case for actual circuits! The constants A, G, and R are factors which couple controlling quantities with sources. Usually, A has the meaning of amplification factor and is dimensionless, while G and R have the dimensions of conductance and resistance, respectively. (They are often called transconductances and transresistances.) This explains the notations for the coupling factors. In summary, in order to specify a dependent source we have to specify a controlling branch and a coupling factor.

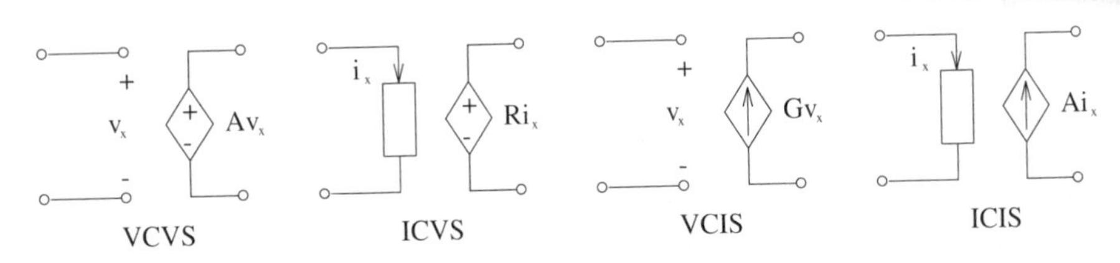

Figure 8.1: Circuit notation for various dependent sources.

Where do dependent sources come from? The answer is: *dependent sources usually appear in electric circuits as linear models for transistors.* To understand how they fulfill this role, we must briefly examine transistors themselves.

There are two most important types of transistors. The first kind is the bipolar junction transistor, or BJT. A diagram of the BJT is shown in Figure 8.2. The BJT is made out of a semiconductor which has three distinct regions (n^+, p, n^-), each one connected to an ohmic contact. The three ohmic contacts are called the emitter, the collector, and the base and they are specified in Figure 8.2.

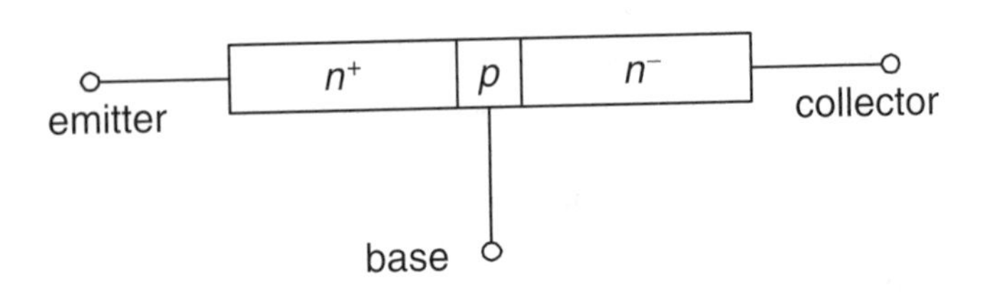

Figure 8.2: Bipolar junction transistor.

To understand these notations, we recall that in semiconductors there are two types of current carriers: electrons and (positively-charged) holes. Usually, a semiconductor is not very useful if the amounts of electrons and holes are balanced. For this reason, this balance is deliberately upset by introducing impurities into semiconductors. This process is called *doping*. After doping, if the amount of electrons is greater than the amount of holes, the semiconductor is called an n type. If the opposite is true, then the semiconductor is a p type. The semiconductor of the BJT is doped to create the three distinct regions. The emitter region is *heavily* doped and, as a result, the density of electrons is *much* larger than density of holes. This is indicated by the superscript "+." The base region is doped in such a way that the density of holes is larger than the density of electrons. For this reason, the base is a p region. Finally, the collector region is *lightly* n doped and this is indicated by a superscript "–." When two oppositely doped regions are adjacent to one another they form what is called a $p-n$ junction. Two such junctions are shown in Figure 8.2. This explains why this transistor is called the bipolar junction transistor. The middle area, the base, is made *very narrow* to allow for interactions between the two junctions. This is essential for the proper operation of the BJT.

The BJT is a current-controlled device. Indeed, it can be shown that injection of holes into the base causes large numbers of electrons to move from the emitter to the collector. Thus, the base current controls the emitter-collector current. For this reason, the BJT can be modeled as a current-controlled source.

At one time, the BJT was the most important device in electronics, but now it has been somewhat eclipsed by the MOSFET transistor. The MOSFET transistor is the main building block of VLSI (very large scale integrated) circuits, which are essential components of modern computers. The acronym MOSFET stands for metal oxide semiconductor field effect transistor. It is pictured in Figure 8.3. The MOSFET transistor is also made of a semiconductor with three distinctly doped regions. It has four ohmic contacts: the source, the drain, the gate, and the substrate. The first three letters of MOSFET describe the composition of the device. Indeed, the words metal oxide semiconductor refer to the metallic gate, the SiO_2 region and the bulk semiconductor region. The last words, field effect transistor, are related to the principle of operation of the device. Unlike the BJT, the MOSFET transistor is a voltage-controlled device. When a positive voltage is applied to the gate, a vertical electric field is created. This field is indicated in Figure 8.3 by arrows directed from the gate toward the p region. The positively charged holes in the p region are repelled by this electric field. This process is called *depletion*. As the gate voltage and, consequently, the field are further increased, electrons are introduced. This is called *inversion* because the applied gate voltage has inverted the region, making electrons more numerous than holes. This inversion creates an n channel between the source and the drain, allowing current to flow between them when some source-to-drain voltage is applied.

The current i_d between the source and drain (which is usually called a drain current) depends on the voltage, v_{sd}, between source and drain as well as on the voltage, v_{sg}, between source and gate. Mathematically, it can be written as follows:

$$i_d = f(v_{sd}, v_{sg}). \tag{8.1}$$

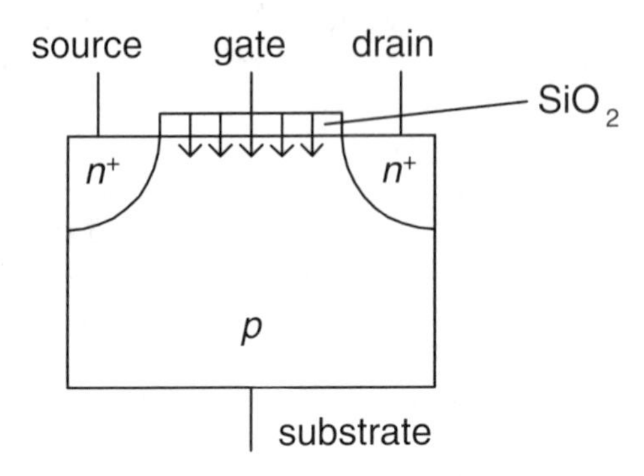

Figure 8.3: MOSFET transistor.

This dependence can be experimentally measured and plotted as a family of curves shown in Figure 8.4. In this figure, the plots of i_d vs. v_{sd} are given for monotonically increasing values of gate voltage v_{sg}:

$$v_{sg}^{(1)} < v_{sg}^{(2)} < v_{sg}^{(3)} < v_{sg}^{(4)}. \tag{8.2}$$

It is clear from the above figure that for the same fixed drain voltage, v_{sd}, the drain current, i_d, is increased as the gate voltage, v_{sg}, is increased. The last fact is transparent from the physical point of view: the larger the gate voltage, the larger the cross section of the inversion channel, the smaller its resistance, and the larger the drain current for the same drain voltage. This clearly suggests that the drain current is controlled by the gate voltage. For this reason, the MOSFET transistor is a voltage-controlled device and can be modeled as a voltage-controlled source.

In order to arrive at such models, we consider the case in which the applied drain and gate voltages have two distinct components:

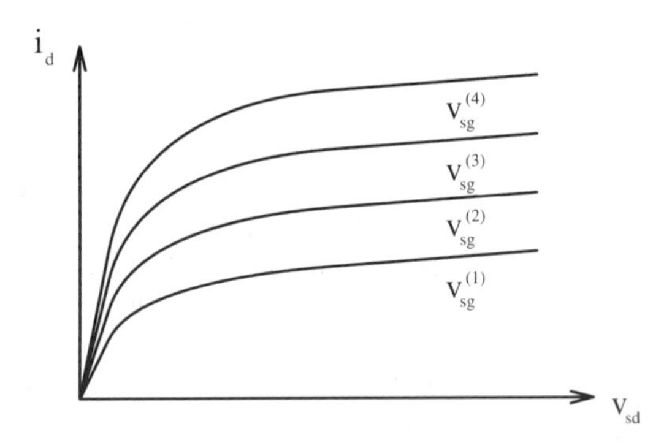

Figure 8.4: Drain current as a function of source-to-drain voltage for different values of gate voltage v_{sg}.

$$v_{sd} = v_{sd}^{(0)} + \Delta v_{sd}, \tag{8.3}$$

$$v_{sg} = v_{sg}^{(0)} + \Delta v_{sg}. \tag{8.4}$$

Here, $v_{sd}^{(0)}$ and $v_{sg}^{(0)}$ are constant (quiescent) components of the drain and gate voltages, which are usually called *bias* voltages, while Δv_{sd} and Δv_{sg} are small time-varying components called signals.

The bias voltages determine the bias current through the transistor, which according to (8.1) is given by:

$$i_d^{(0)} = f(v_{sd}^{(0)}, v_{sg}^{(0)}). \tag{8.5}$$

The "signal" voltages cause the time-varying (signal) component of the drain current, which according to (8.1) can be defined as follows:

$$\Delta i_d = f(v_{sd}^{(0)} + \Delta v_{sd}, v_{sg}^{(0)} + \Delta v_{sg}) - f(v_{sd}^{(0)}, v_{sg}^{(0)}). \tag{8.6}$$

Since the signal voltages are assumed to be small, we can use in (8.6) the Taylor expansion and retain only the first two terms. This leads to:

$$\Delta i_d = \left.\frac{\partial f}{\partial v_{sd}}\right|_{\text{bias}} \Delta v_{sd} + \left.\frac{\partial f}{\partial v_{sg}}\right|_{\text{bias}} \Delta v_{sg}. \tag{8.7}$$

Notation "$|_{\text{bias}}$" means that the corresponding derivatives in (8.7) are evaluated at bias voltages $v_{sd}^{(0)}$ and $v_{sg}^{(0)}$. Expression (8.7) is a local *linearization* of transistor characteristics (8.1) around an operating point of the transistor which is determined by the bias conditions. This linearization leads to *linear* models for the MOSFET transistor.

Now, we introduce new notations for signal components:

$$\Delta i_d = I_d, \quad \Delta v_{sd} = V_{sd}, \quad \Delta v_{sg} = V_{sg}, \tag{8.8}$$

as well as the following notations for the derivatives:

$$\left.\frac{\partial f}{\partial v_{sd}}\right|_{\text{bias}} = \frac{1}{R}, \quad \left.\frac{\partial f}{\partial v_{sg}}\right|_{\text{bias}} = G. \tag{8.9}$$

In the new notations, expression (8.7) can be written in the form:

$$I_d = \frac{V_{sd}}{R} + GV_{sg}. \tag{8.10}$$

The last expression clearly suggests that as far as the relationship between the signal components of the drain current, drain voltage, and gate voltage are concerned, the MOSFET transistor can be modeled as the nonideal voltage-controlled current source shown in Figure 8.5. In this figure, source, drain, and gate terminals are marked by "s,""d," and "g," respectively, and gate and drain voltages are identified. It is clear from this figure that the drain current, I_d, is the sum of two currents: the current through the resistor R, which is equal to V_{sd}/R, and the current GV_{sg} of the dependent current source. Consequently, the drain current, I_d, is given by (8.10). This

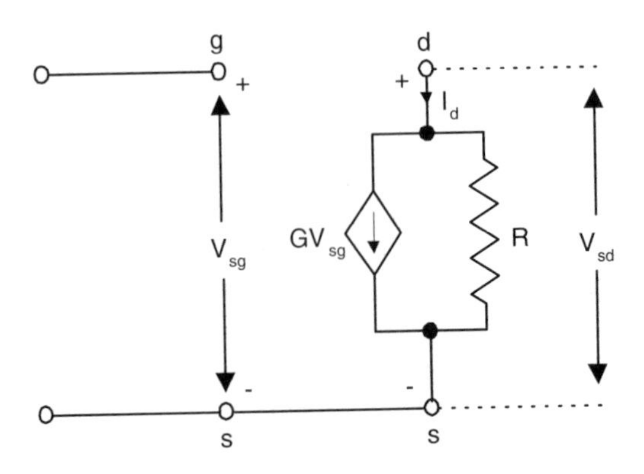

Figure 8.5: Small-signal MOSFET model.

proves our assertion that the voltage-controlled current source shown in Figure 8.5 is the appropriate model for the MOSFET transistor.

By using the equivalent transformation of a nonideal current source into a nonideal voltage-source, we end up with the nonideal voltage-controlled voltage source model for the MOSFET transistor. This model is shown in Figure 8.6. Here, the following expression is valid for the amplification factor A:

$$A = RG. \tag{8.11}$$

Thus, we have *demonstrated that dependent sources appear as linear models for transistors.*

It is important to note that the technique which we used in order to arrive at the voltage-controlled source models for the MOSFET is called *small-signal analysis*. This technique is extensively used for the analysis of electric circuits and is discussed in detail in courses on electronics.

It is also important to note that our discussion of bipolar and MOSFET transistors has been very superficial. We intended only to give some very general ideas concern-

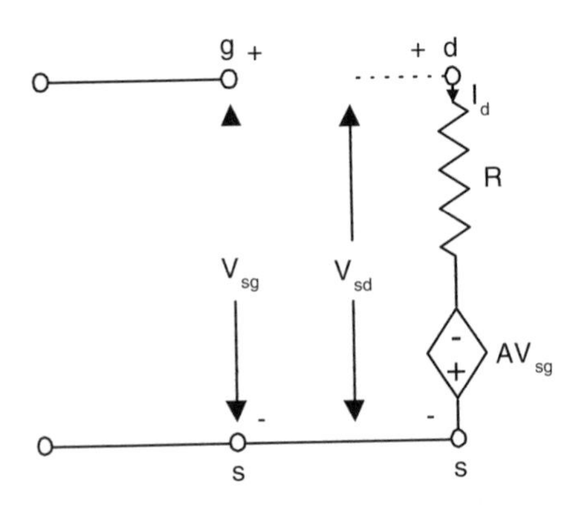

Figure 8.6: VCVS model of a MOSFET.

ing the structures and principles of operation of these devices and to demonstrate that there are real-world devices behind the notion of dependent sources. Extensive discussions of these transistors can be found in books on semiconductor devices.[1]

8.3 Analysis of Circuits with Dependent Sources

Analysis of circuits with dependent sources is similar to the analysis of circuits with independent sources. This is true for nodal analysis, mesh current analysis, and Thevenin's theorem. There are, however, some slight differences, which will be discussed below.

8.3.1 Nodal Analysis

The application of nodal analysis to electric circuits with dependent sources will be illustrated by the following examples.

EXAMPLE 8.1 Consider the circuit shown in Figure 8.7. It is assumed that Y_1, Y_2, Y_3, $\hat{I}_{s1}$ are given. The dependent sources are specified as follows:

$$\hat{I}_{s2}^{(d)} = G_2 \hat{V}_{23}, \tag{8.12}$$

$$\hat{I}_{s3}^{(d)} = G_3 \hat{V}_{13}. \tag{8.13}$$

Thus, the first dependent source depends on the voltage between nodes 2 and 3, and the second source depends on the voltage between nodes 1 and 3. Note that the controlling branches are not adjacent to the dependent sources.

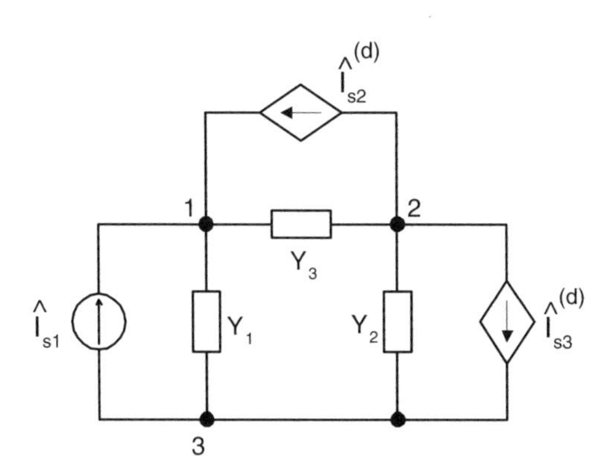

Figure 8.7: Circuit with dependent sources.

[1]See, for example, *Electronic Design* by C. J. Savant, M. S. Roden, and G. L. Carpenter (Redwood City: Benjamin Cummings, 1991) or *Microelectronic Circuits* by A. S. Sedra and K. C. Smith (New York: Holt, Rinehart and Winston, 1987).

The first step of using nodal analysis with dependent sources is to treat all the circuit sources as if they were independent and write the corresponding matrix equations. For our circuit these equations are:

$$\begin{bmatrix} Y_1 + Y_3 & -Y_3 \\ -Y_3 & Y_2 + Y_3 \end{bmatrix} \begin{bmatrix} \hat{v}_1 \\ \hat{v}_2 \end{bmatrix} = \begin{bmatrix} \hat{I}_{s1} + \hat{I}_{s2}^{(d)} \\ -\hat{I}_{s2}^{(d)} - \hat{I}_{s3}^{(d)} \end{bmatrix}, \tag{8.14}$$

where node 3 has been chosen as a reference node.

The second step is to express the dependent sources in terms of the node potentials. In this example, the dependent sources are voltage-controlled so they can be rewritten as

$$\hat{I}_{s2}^{(d)} = G_2(\hat{v}_2 - \hat{v}_3) = G_2\hat{v}_2, \tag{8.15}$$

$$\hat{I}_{s3}^{(d)} = G_3(\hat{v}_1 - \hat{v}_3) = G_3\hat{v}_1. \tag{8.16}$$

After the sources have been expressed in terms of the node potentials, the third and final step is to substitute the new expressions into the matrix equations and modify them accordingly. After substitution we have:

$$\begin{bmatrix} Y_1 + Y_3 & -Y_3 \\ -Y_3 & Y_2 + Y_3 \end{bmatrix} \begin{bmatrix} \hat{v}_1 \\ \hat{v}_2 \end{bmatrix} = \begin{bmatrix} \hat{I}_{s1} + G_2\hat{v}_2 \\ -G_2\hat{v}_2 - G_3\hat{v}_1 \end{bmatrix}. \tag{8.17}$$

Now, all unknowns must be moved to the left-hand side. This leads to the following final matrix equation:

$$\begin{bmatrix} Y_1 + Y_3 & -Y_3 - G_2 \\ -Y_3 + G_3 & Y_2 + Y_3 + G_2 \end{bmatrix} \begin{bmatrix} \hat{v}_1 \\ \hat{v}_2 \end{bmatrix} = \begin{bmatrix} \hat{I}_{s1} \\ 0 \end{bmatrix}. \tag{8.18}$$

The last matrix equation can be solved for $\hat{v}_1$ and $\hat{v}_2$, and afterward all currents can be easily found. ■

EXAMPLE 8.2 Consider a circuit with more complicated dependent sources (see Figure 8.8). It is assumed that Y_1, Y_2, Y_3, Y_4, Y_5, Y_6, $\hat{V}_{s1}$, are given as well as the following specifications for the dependent sources:

$$\hat{V}_{s2}^{(d)} = R_2\hat{I}_2, \tag{8.19}$$

$$\hat{V}_{s3}^{(d)} = A_3\hat{V}_{12}. \tag{8.20}$$

As before, the first step is to write the node equations as if all sources were independent:

$$\begin{bmatrix} Y_1 + Y_2 + Y_3 + Y_4 & -Y_3 - Y_4 \\ -Y_3 - Y_4 & Y_3 + Y_4 + Y_5 + Y_6 \end{bmatrix} \begin{bmatrix} \hat{v}_1 \\ \hat{v}_2 \end{bmatrix} = \begin{bmatrix} Y_1\hat{V}_{s1} + Y_4\hat{V}_{s2}^{(d)} \\ -Y_4\hat{V}_{s2}^{(d)} + Y_6\hat{V}_{s3}^{(d)} \end{bmatrix}, \tag{8.21}$$

where again node 3 is chosen as a reference node.

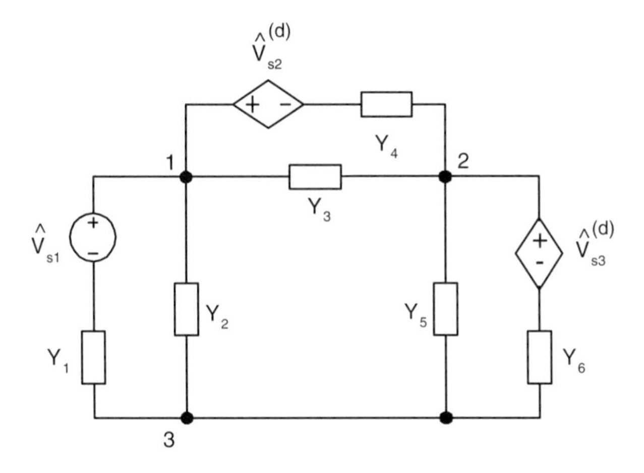

Figure 8.8: Example circuit with dependent sources.

Next, we express the dependent sources in terms of the unknowns:

$$\hat{V}_{s2}^{(d)} = R_2\hat{I}_2 = R_2Y_2\hat{V}_{13} = R_2Y_2(\hat{v}_1 - \hat{v}_3) = R_2Y_2\hat{v}_1, \tag{8.22}$$

$$\hat{V}_{s3}^{(d)} = A_3\hat{V}_{12} = A_3(\hat{v}_1 - \hat{v}_2). \tag{8.23}$$

Finally, we substitute these expressions into the previous node equations and modify the matrix of these equations by moving all unknowns to the left-hand side:

$$\begin{bmatrix} Y_1 + Y_2 + Y_3 + Y_4 - Y_4Y_2R_2 & -Y_3 - Y_4 \\ -Y_3 - Y_4 + R_2Y_2Y_4 - Y_6A_3 & Y_3 + Y_4 + Y_5 + Y_6 + A_3Y_6 \end{bmatrix} \begin{bmatrix} \hat{v}_1 \\ \hat{v}_2 \end{bmatrix} = \begin{bmatrix} Y_1\hat{V}_{s1} \\ 0 \end{bmatrix}. \tag{8.24}$$

Now, these equations can be solved for $\hat{v}_1$ and $\hat{v}_2$ and then all circuit variables can be found. ■

8.3.2 Mesh Current Analysis

The major steps for using mesh current analysis of circuits with dependent sources are similar to those used in nodal analysis. These steps are as follows.

1. Write the matrix equations as if all the sources were independent.
2. Express the dependent sources in terms of the unknowns in the matrix equations. For mesh current analysis, this means expressing the dependent sources in terms of mesh currents.
3. Substitute the expressions for the dependent sources into the matrix equations. Then modify the coefficient matrix by moving expressions for dependent sources in terms of mesh currents to the left-hand side of the matrix equation.

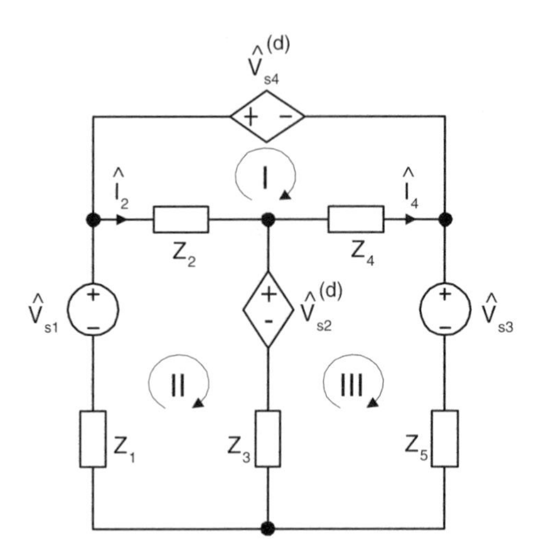

Figure 8.9: Example circuit with dependent sources.

EXAMPLE 8.3 Consider the circuit shown in Figure 8.9. Our goal is to write the mesh current equations. It is assumed that Z_1, Z_2, Z_3, Z_4, Z_5, $\hat{V}_{s1}$, $\hat{V}_{s2}$ are given as well as the following specifications for the dependent sources:

$$\hat{V}_{s2}^{(d)} = A\hat{V}_2, \tag{8.25}$$

$$\hat{V}_{s4}^{(d)} = R\hat{I}_4, \tag{8.26}$$

where $\hat{V}_2$ is the voltage across the impedance Z_2, while $\hat{I}_4$ is the current through the impedance Z_4.

First, we write the mesh current equations treating all sources as independent:

$$\begin{bmatrix} Z_2 + Z_4 & -Z_2 & -Z_4 \\ -Z_2 & Z_1 + Z_2 + Z_3 & -Z_3 \\ -Z_4 & -Z_3 & Z_3 + Z_4 + Z_5 \end{bmatrix} \begin{bmatrix} \hat{\mathcal{I}}_1 \\ \hat{\mathcal{I}}_2 \\ \hat{\mathcal{I}}_3 \end{bmatrix} = \begin{bmatrix} -\hat{V}_{s4}^{(d)} \\ -\hat{V}_{s2}^{(d)} + \hat{V}_{s1} \\ -\hat{V}_{s3} + \hat{V}_{s2}^{(d)} \end{bmatrix}. \tag{8.27}$$

Now, we express the dependent source in terms of the mesh currents:

$$\hat{V}_{s2}^{(d)} = A\hat{V}_2 = A\hat{I}_2 Z_2 = AZ_2(\hat{\mathcal{I}}_2 - \hat{\mathcal{I}}_1), \tag{8.28}$$

$$\hat{V}_{s4}^{(d)} = R\hat{I}_4 = R(\hat{\mathcal{I}}_3 - \hat{\mathcal{I}}_1). \tag{8.29}$$

Substituting the expressions for the dependent sources in terms of mesh currents into the matrix equations and moving them to the left-hand side yields the final set of equations:

$$\begin{bmatrix} Z_2 + Z_4 - R & -Z_2 & -Z_4 + R \\ -Z_2 - AZ_2 & Z_1 + Z_2 + Z_3 + AZ_2 & -Z_3 \\ -Z_4 + AZ_2 & -Z_3 - AZ_2 & Z_3 + Z_4 + Z_5 \end{bmatrix} \begin{bmatrix} \hat{\mathcal{I}}_1 \\ \hat{\mathcal{I}}_2 \\ \hat{\mathcal{I}}_3 \end{bmatrix} = \begin{bmatrix} 0 \\ +\hat{V}_{s1} \\ -\hat{V}_{s3} \end{bmatrix}. \tag{8.30}$$

By solving the last matrix equation, we find the mesh currents, which can then be used to calculate the actual currents in the circuit. ■

8.3.3 Thevenin's Theorem

We first recall that the use of Thevenin's theorem for analysis of electric circuits with independent sources requires the following three steps:

1. Remove the branch in question and determine the open-circuit voltage.
2. Set all sources in the active circuit to zero and find the input impedance with respect to the open terminals. When we set sources to zero, we replace current sources by open branches and voltage sources by short-circuit branches.
3. Use the equation

$$\hat{I} = \frac{\hat{V}_{oc}}{Z_{in} + Z} \tag{8.31}$$

to find the current through the branch in question.

The primary obstacle in the use of Thevenin's theorem for circuits with dependent sources is the second step. This is because the values of dependent sources are determined by controlling voltages and currents in other branches and, for this reason, dependent sources cannot independently be set to zero without affecting controlling branches. However, this obstacle can be removed by using an alternative approach. We recall that the input impedance can also be defined as:

$$Z_{in} = \frac{\hat{V}_{oc}}{\hat{I}_{sc}}. \tag{8.32}$$

Therefore, to find the input impedance we need to find the open-circuit voltage and the short-circuit current, both of which can be found for circuits containing dependent sources. The Thevenin theorem is especially effective when the branch in question is a *controlling branch*.

EXAMPLE 8.4 Consider the circuit shown in Figure 8.10. It is assumed that Z_1, Z_2, Z_3, $\hat{V}_{s1}$ are given as well as the following specifications for the dependent sources:

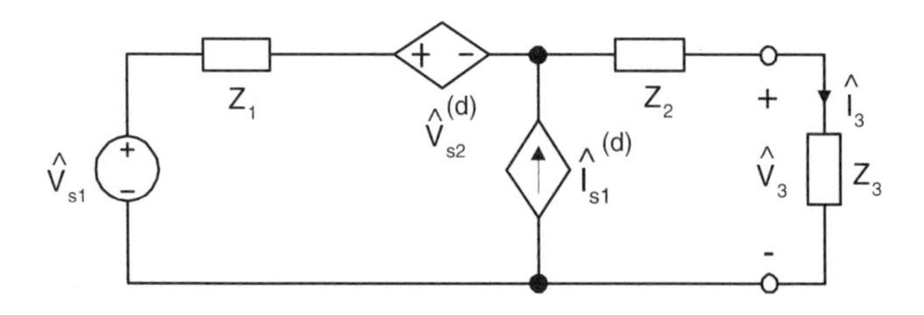

Figure 8.10: Circuit containing dependent sources.

$$\hat{V}_{s2}^{(d)} = A\hat{V}_3, \tag{8.33}$$

$$\hat{I}_{s1}^{(d)} = G\hat{V}_3. \tag{8.34}$$

In this circuit, branch 3 is the controlling branch because all of the dependent sources depend on the voltage across this branch. We want to solve for the current $\hat{I}_3$ through this branch.

We begin by removing this branch and analyzing the resulting circuit in order to find the open-circuit voltage. Figure 8.11 shows the circuit with the branch removed. Now we make the following observations. First of all, the current through the only loop is equal to the source current. Second, voltage $\hat{V}_3$ which controls the current source is equal to the voltage across the current source. This is because there is no current through Z_2 and hence no voltage drop across it. With these observations in mind, we set up the KVL for the loop:

$$-\hat{V}_{s1} + \hat{V}_3 + \hat{V}_{s2}^{(d)} - \hat{I}_{s1}^{(d)} Z_1 = 0. \tag{8.35}$$

Substituting in the expressions for the dependent sources in terms of $\hat{V}_3$ yields:

$$-\hat{V}_{s1} + \hat{V}_3 + A\hat{V}_3 - GZ_1\hat{V}_3 = 0. \tag{8.36}$$

Now, we can solve for $\hat{V}_3$, which is by the way equal to the open-circuit voltage:

$$\hat{V}_3 = \hat{V}_{oc} = \frac{\hat{V}_{s1}}{1 + A - GZ_1}. \tag{8.37}$$

Next, we short-circuit the branch in question in order to find the short-circuit current. Here we must make another important observation. Short-circuiting branch 3 means that the voltage drop across this branch is zero, $\hat{V}_3 = 0$. Thus, all the sources controlled by this branch are also zero. This is the trick which simplifies the calculation of short-circuit currents through controlling branches. We now have the circuit shown in Figure 8.12. It is trivial to solve this circuit for the short-circuit current. This current is given by:

$$\hat{I}_{sc} = \frac{\hat{V}_{s1}}{Z_1 + Z_2}. \tag{8.38}$$

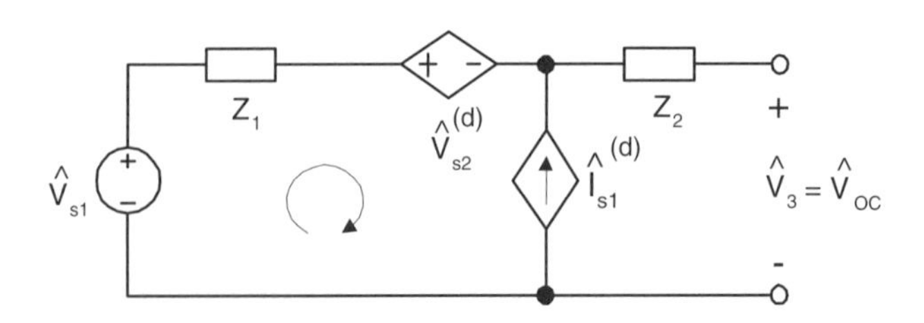

Figure 8.11: Branch is removed, creating open circuit.

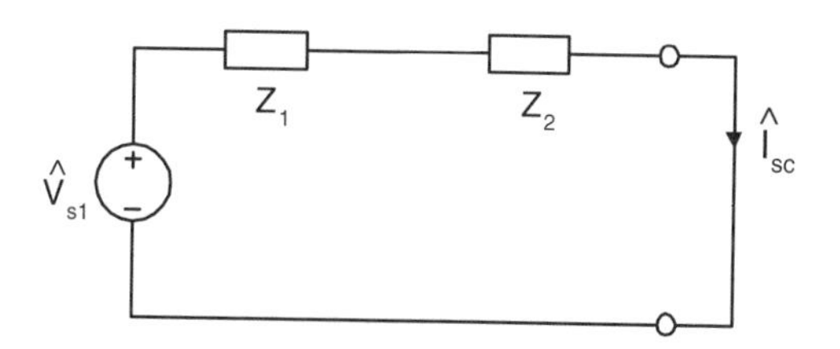

Figure 8.12: Branch is short-circuited—dependent sources shut off.

With the open-circuit voltage and the short-circuit current in hand, we can calculate the input impedance:

$$Z_{in} = \frac{\hat{V}_{oc}}{\hat{I}_{sc}} = \frac{Z_1 + Z_2}{1 + A - GZ_1}. \tag{8.39}$$

Using the last formula in (8.31), we find the current through branch 3 to be:

$$\hat{I}_3 = \frac{\hat{V}_{oc}}{Z_{in} + Z_3} = \frac{\hat{V}_{s1}}{Z_1 + Z_2 + Z_3(1 + A - GZ_1)}. \tag{8.40}$$

■

Another method, based on the principle of equivalent transformations, can be used to find both parameters of a Thevenin or Norton equivalent circuit with a single circuit calculation. In this method, we attach a "probing" independent voltage or current source across the circuit terminals for which we are seeking the equivalent circuit.

When using a probing voltage source, as shown in Figure 8.13a, a circuit analysis is performed to find the current through that source. This current, $\hat{I}_p$, will have two components, one constant and one proportional to $\hat{V}_p$: $\hat{I}_p = A - B\hat{V}_p$. For the equivalent circuit shown in Figure 8.13b, we can use KVL to find: $\hat{I}_p = \hat{V}_{TH}/Z_{TH} - \hat{V}_p/Z_{TH}$. Because these circuits are equivalent, the expressions for $\hat{I}_p$ must be identical

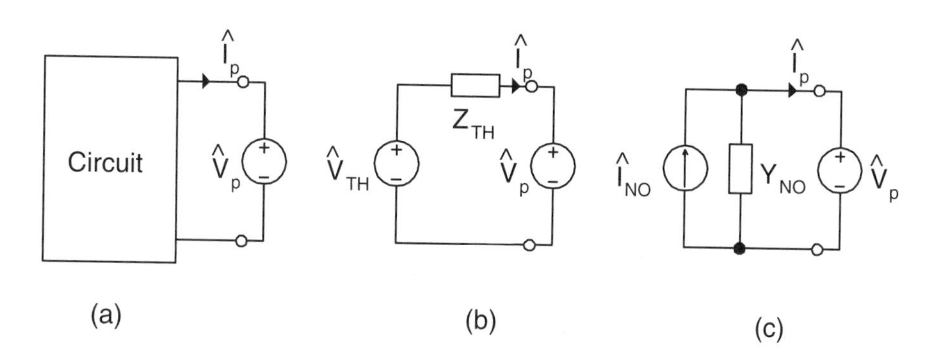

Figure 8.13: Application of the probing voltage method to find an equivalent circuit.

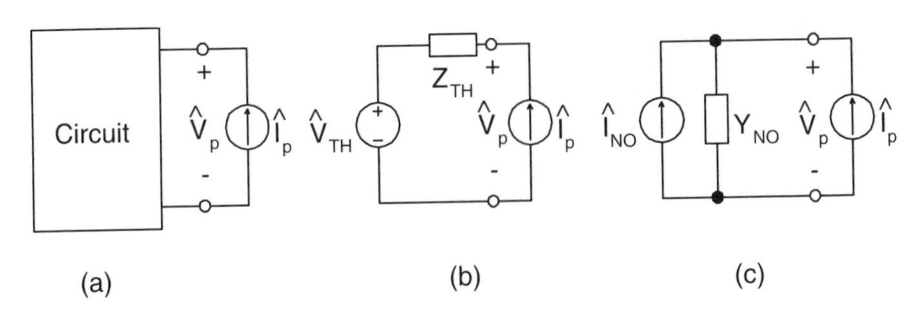

Figure 8.14: Application of the probing current method to find an equivalent nonideal source.

for any value of $\hat{V}_p$. This will be true only if $Z_{TH} = 1/B$ and $\hat{V}_{TH} = A/B$. Therefore, by solving for $\hat{I}_p$ in terms of $\hat{V}_p$ and the other circuit parameters, we can read off the Thevenin equivalent voltage and impedance. Similarly, using a probing voltage source to find a Norton equivalent circuit, as shown in Figure 8.13c, would result in the expressions: $\hat{I}_{NO} = A$ and $Y_{NO} = B$.

If a circuit appears to be easier to analyze with general nodal analysis, a probing current source should be used and the voltage across this source is the circuit variable which must be obtained. The application of KVL to the Thevenin equivalent circuit in Figure 8.14b yields $\hat{V}_p = \hat{V}_{TH} + Z_{TH}\hat{I}_p$. Nodal analysis of the circuit shown in Figure 8.14a yields $\hat{V}_p = A + B\hat{I}_p$. By following the same line of reasoning as in the above paragraph, we conclude that $\hat{V}_{TH} = A$ and $Z_{TH} = B$. An application of KCL to the circuit of Figure 8.14c will result in similar expressions for the Norton equivalent circuit.

EXAMPLE 8.5 Let's use a probing source to find the Thevenin equivalent circuit from the previous example depicted in Figure 8.11. If we convert the series combination of $\hat{V}_s$ and Z_1 to a nonideal current source and attach a probe current source, we arrive at the circuit shown in Figure 8.15, where $Y_1 = 1/Z_1$ and $Y_2 = 1/Z_2$. The

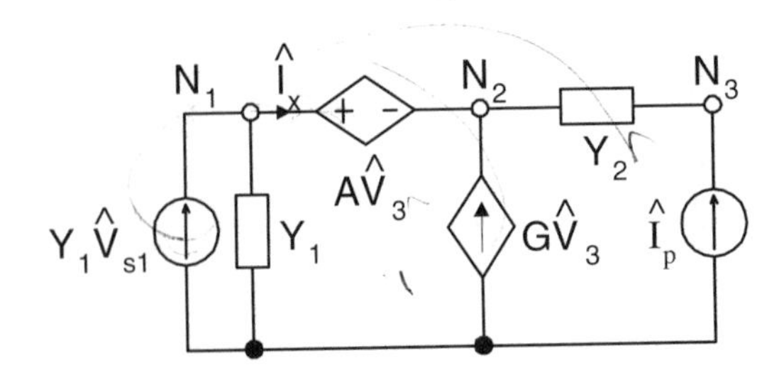

Figure 8.15: The circuit for the previous example with a probing current source.

general nodal analysis matrix equation is:

$$\begin{pmatrix} Y_1 & 0 & 0 \\ 0 & Y_2 & -Y_2 \\ 0 & -Y_2 & Y_2 \end{pmatrix} \begin{pmatrix} \hat{v}_1 \\ \hat{v}_2 \\ \hat{v}_3 \end{pmatrix} = \begin{pmatrix} \hat{V}_{s_1} Y_1 - \hat{I}_x \\ \hat{I}_x + G\hat{v}_3 \\ \hat{I}_p \end{pmatrix}. \tag{8.41}$$

By adding the first two equations and using the fact that $\hat{v}_1 = A\hat{v}_3 + \hat{v}_2$, we reduce the order of the matrix equation by one:

$$\begin{pmatrix} Y_1 + Y_2 & Y_1 A - G - Y_2 \\ -Y_2 & Y_2 \end{pmatrix} \begin{pmatrix} \hat{v}_2 \\ \hat{v}_3 \end{pmatrix} = \begin{pmatrix} \hat{V}_{s_1} Y_1 \\ \hat{I}_p \end{pmatrix}. \tag{8.42}$$

The node voltage $\hat{v}_3$ is then found to be

$$\hat{v}_3 = \frac{\hat{V}_{s_1}}{1 + A - Z_1 G} + \left(\frac{Z_1 + Z_2}{1 + A - Z_1 G} \right) \hat{I}_p, \tag{8.43}$$

so

$$\hat{V}_{TH} = \frac{\hat{V}_{s1}}{1 + A - Z_1 G} \quad \text{and} \quad Z_{TH} = \frac{Z_1 + Z_2}{1 + A - Z_1 G}. \tag{8.44}$$

These results are identical to the previous solution given by equations (8.37) and (8.39), as they must be. ■

EXAMPLE 8.6 Let us find the Norton equivalent nonideal source for the circuit shown in Figure 8.9 at the terminals formed by the removal of Z_2. The new circuit, complete with a probing voltage source, is shown in Figure 8.16. This source voltage

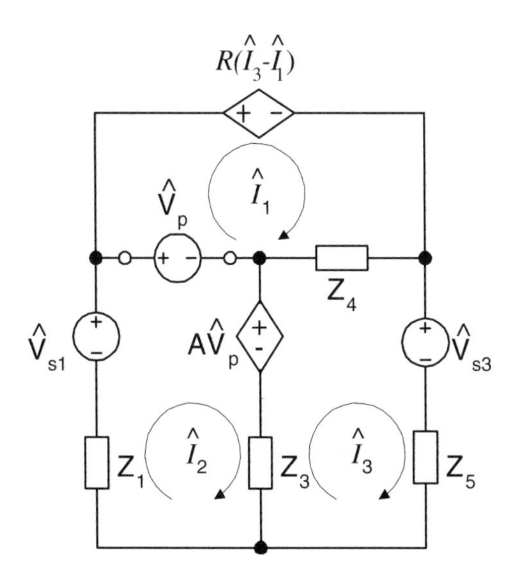

Figure 8.16: Example circuit with probing voltage source.

is equal to the controlling voltage $\hat{V}_2$ as reflected in the figure (see equation (8.25)). The general mesh equations are given by:

$$\begin{bmatrix} Z_4 & 0 & -Z_4 \\ 0 & Z_1 + Z_3 & -Z_3 \\ -Z_4 & -Z_3 & Z_3 + Z_4 + Z_5 \end{bmatrix} \begin{bmatrix} \hat{\mathcal{I}}_1 \\ \hat{\mathcal{I}}_2 \\ \hat{\mathcal{I}}_3 \end{bmatrix} = \begin{bmatrix} \hat{V}_p - R(\hat{\mathcal{I}}_3 - \hat{\mathcal{I}}_1) \\ \hat{V}_{s1} - \hat{V}_p - A\hat{V}_p \\ A\hat{V}_p - \hat{V}_{s3} \end{bmatrix}. \quad (8.45)$$

By moving the unknowns to the left-hand side of the equation we arrive at

$$\begin{bmatrix} Z_4 - R & 0 & R - Z_4 \\ 0 & Z_1 + Z_3 & -Z_3 \\ -Z_4 & -Z_3 & Z_3 + Z_4 + Z_5 \end{bmatrix} \begin{bmatrix} \hat{\mathcal{I}}_1 \\ \hat{\mathcal{I}}_2 \\ \hat{\mathcal{I}}_3 \end{bmatrix} = \begin{bmatrix} \hat{V}_p \\ \hat{V}_{s1} - (1 + A)\hat{V}_p \\ A\hat{V}_p - \hat{V}_{s3} \end{bmatrix}. \quad (8.46)$$

The current through the probe source is $\hat{I}_p = \hat{\mathcal{I}}_2 - \hat{\mathcal{I}}_1$, so we must use Gaussian elimination and back-substitution to find all the mesh currents. The equivalent upper-triangular matrix is found to be:

$$\begin{bmatrix} Z_4 - R & 0 & R - Z_4 \\ 0 & Z_1 + Z_3 & -Z_3 \\ 0 & 0 & Z_5 + \frac{Z_1 Z_3}{Z_1 + Z_3} \end{bmatrix} \begin{bmatrix} \hat{\mathcal{I}}_1 \\ \hat{\mathcal{I}}_2 \\ \hat{\mathcal{I}}_3 \end{bmatrix}$$

$$= \begin{bmatrix} \hat{V}_p \\ \hat{V}_{s1} - (1 + A)\hat{V}_p \\ \frac{Z_3}{Z_1 + Z_3}\hat{V}_{s1} - \hat{V}_{s3} + \left(\frac{Z_4}{Z_4 - R} + \frac{Z_1 A - Z_3}{Z_1 + Z_3}\right)\hat{V}_p \end{bmatrix}. \quad (8.47)$$

Solving for $\hat{I}_p = \hat{\mathcal{I}}_2 - \hat{\mathcal{I}}_1$, after some algebraic manipulations we arrive at:

$$\hat{I}_p = \frac{Z_5 \hat{V}_{s1} + Z_1 \hat{V}_{s3}}{(Z_1 + Z_3)(Z_5 + Z_e)} - \left\{ \frac{1}{Z_4 - R}\left[1 + \frac{Z_e R}{Z_3(Z_5 + Z_e)}\right] + \frac{Z_e(1 + A)}{Z_1 Z_3}\left[1 + \frac{Z_1 Z_e}{Z_3(Z_5 + Z_e)}\right] \right\} \hat{V}_p \quad (8.48)$$

where $Z_e = Z_1 Z_3/(Z_1 + Z_3)$ is the parallel combination of Z_1 and Z_3. Thus,

$$\hat{I}_{NO} = \frac{Z_5 \hat{V}_{s1} + Z_1 \hat{V}_{s3}}{(Z_1 + Z_3)(Z_5 + Z_e)} \quad (8.49)$$

and

$$Y_{NO} = \left\{ \frac{1}{Z_4 - R}\left[1 + \frac{Z_e R}{Z_3(Z_5 + Z_e)}\right] + \frac{Z_e(1 + A)}{Z_1 Z_3}\left[1 + \frac{Z_1 Z_e}{Z_3(Z_5 + Z_e)}\right] \right\}. \quad (8.50)$$

This concludes the analysis of this problem. ■

8.4 Operational Amplifiers

An operational amplifier (op-amp) is a voltage-controlled voltage source with very special properties which make this device very useful in numerous applications. The circuit notation for an operational amplifier is shown in Figure 8.17.

It is clear from this figure that the operational amplifier is a four-terminal device. One terminal (shown at the bottom of the figure) is common to both input voltages v_1 and v_2 and the output voltage v_0. This terminal is commonly called the ground terminal (ground lead) and it can be used as a natural zero reference node. The output voltage is related to input voltages v_1 and v_2 by the expression:

$$v_0 = A(v_2 - v_1), \tag{8.51}$$

where A is an amplification factor (sometimes called the open-loop gain). It is apparent from (8.51) that the output voltage is controlled by the *difference* of two input voltages. For this reason, this amplifier is also called a *differential* amplifier. The controlling voltage is $v_2 - v_1$, which explains why v_1 is called the "inverting" input, while v_2 is called the "noninverting" input. In the above figure, the inverting terminal 1 is marked by a "–," while the noninverting terminal 2 is marked by a "+."

Operational amplifiers are characterized by the following three properties:

1. A very high amplification factor A which typically has the following range of values:

$$10^4 \leq A \leq 10^7. \tag{8.52}$$

2. A very large input resistance R_i (i.e., the resistance between terminals 1 and 2):

$$10^5\ \Omega \leq R_i \leq 10^{15}\ \Omega. \tag{8.53}$$

3. A relatively small output resistance R_o (that is the resistance between output and ground terminals):

$$1\ \Omega \leq R_o \leq 10^3\ \Omega. \tag{8.54}$$

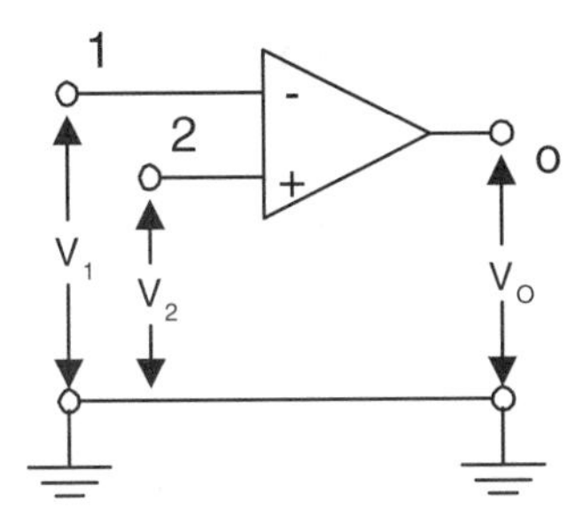

Figure 8.17: A circuit notation for operational amplifier.

It is customary to consider an ideal operational amplifier with the following idealized properties:

$$A = \infty, \quad R_i = \infty, \quad R_o = 0. \tag{8.55}$$

It is these three properties which make the operational amplifier a very valuable circuit device. Internally, an operational amplifier is a complicated active circuit which contains many transistors, resistors, and capacitors and must be connected to one or two power sources. However, details of the internal structure of the operational amplifier are immaterial as far as the effect of the operational amplifier on an external circuit is concerned. *This effect is completely determined by the above three terminal properties.* For this reason, in the analysis of electric circuits with operational amplifiers we can completely ignore the internal structure of these amplifiers and use only their terminal properties described above. These terminal properties can be used as the *definition* of the operational amplifier as a *circuit element*. This *axiomatic* approach is also justified by the extremely wide use of operational amplifiers. In fact, these amplifiers are encountered in modern circuits almost as often as resistors and capacitors. This justifies the elevation of the operational amplifier to the level of a *basic circuit element*, the element which is defined by the terminal properties (8.55).

Since the amplification factor is assumed to be infinite, no signal (no voltage) is applied between terminals 1 and 2. Usually, only one of these terminals is directly biased, while the potential of the other terminal adjusts itself in order to keep the output voltage, v_0, finite. According to expression (8.51), the infinite gain A and finite output voltage v_0 imply that:

$$v_1 = v_2. \tag{8.56}$$

The last formula is the first consequence of the terminal properties of the ideal operational amplifier. This formula significantly simplifies the analysis of circuits with operational amplifiers and it will be used time and time again in our subsequent discussions. For actual operational amplifiers equality (8.56) is fulfilled only approximately. However, deviations from this equality are almost always negligible.

Operational amplifiers are used with some circuitry around them in order to achieve desired effects. One constant element of this circuitry is a *feedback branch* which connects the output terminal with the inverting input terminal. Depending on the composition of this feedback branch, the operational amplifiers can perform many functions, which we proceed to discuss in the following sections.

8.4.1 Voltage Follower–Buffer Amplifier

An electric circuit of a voltage follower is shown in Figure 8.18. A voltage follower has a short-circuit feedback branch which connects the output terminal with the inverting input terminal. Consequently:

$$v_1(t) = v_o(t). \tag{8.57}$$

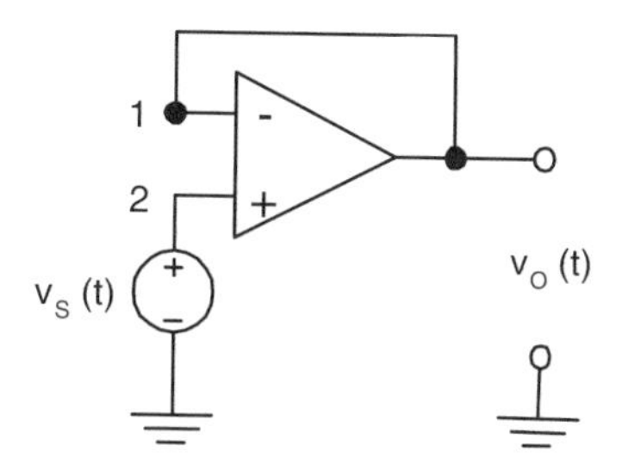

Figure 8.18: Voltage follower.

On the other hand, the voltage source $v_s(t)$ is connected between the ground terminal and the noninverting terminal. This means that:

$$v_2(t) = v_s(t). \tag{8.58}$$

By substituting (8.57) and (8.58) into (8.56), we obtain:

$$v_o(t) = v_s(t), \tag{8.59}$$

which explains why the above circuit is called a voltage follower. The practical utility of the voltage follower is that it provides isolation of a voltage source from a load. Indeed, if some load is connected between the output and ground terminals, this load will "feel" the source voltage. At the same time, the actual voltage source will not feel the load, because the voltage source faces the infinite input resistance (impedance) of the amplifier. Thus, the load will not affect the current through the voltage source and, consequently, it will not affect the voltage across the input terminals. This means that *any* load will experience the *same* source voltage appearing across the output and ground terminals. In other words, a *nonideal* voltage source connected between the noninverting and ground terminals will act as an *ideal* voltage source between the output and ground terminals. Due to the described isolation of a voltage source from a load, the voltage follower is also called a buffer amplifier.

It is instructive to examine what effect finite values of A, R_i, and R_o will have on the performance of the voltage follower. In other words, we would like to investigate how the voltage follower will perform if the operational amplifier is not ideal. If the operational amplifier is not ideal, then the voltage follower can be represented by an electric circuit shown in Figure 8.19. In this figure, R_i is an input resistance of the amplifier (that is resistance between inverting and noninverting terminals), R_0 is an output resistance (this is the resistance which can be measured between the output and ground terminals when no signal is applied to the amplifier and no load is connected between the ground and output terminals), and R_L is a load resistance. The amplifier itself is represented as a voltage-controlled voltage source with amplification factor (gain) A and controlling voltage $v_x(t)$:

$$v_x(t) = v_1(t) - v_2(t). \tag{8.60}$$

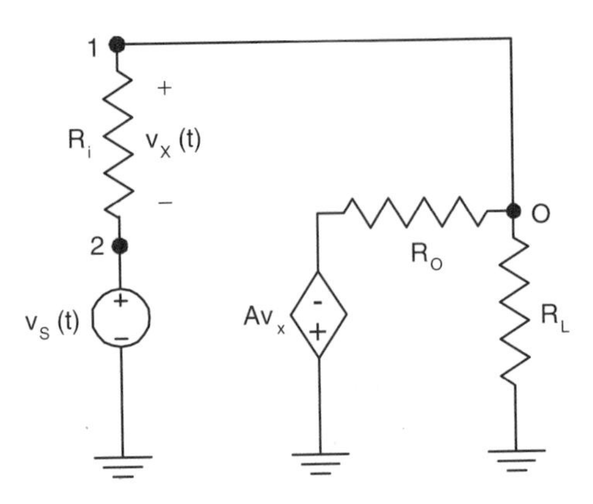

Figure 8.19: Electric circuit of a voltage follower in the case of a nonideal op-amp.

According to expressions (8.51) and (8.60), the voltage of the controlled voltage source is equal to:

$$-Av_x(t) = A(v_2(t) - v_1(t)), \tag{8.61}$$

which is reflected in polarity of this voltage source. Finally, the short-circuit feedback branch of the voltage follower is represented in Figure 8.19 by the short-circuit branch which connects the output and inverting terminals.

We shall apply the nodal analysis to the circuit in Figure 8.19, namely, the version of this analysis which deals with circuits containing voltage sources in series with admittances. According to this version of nodal analysis, we obtain the following equation for potential v_o of the output terminal:

$$\left(\frac{1}{R_i} + \frac{1}{R_o} + \frac{1}{R_L}\right) v_o = \frac{1}{R_i} v_s - \frac{1}{R_o} A v_x. \tag{8.62}$$

From Figure 8.19, we also find:

$$v_1 = v_o, \quad v_2 = v_s. \tag{8.63}$$

By substituting (8.63) into (8.60), we obtain:

$$v_x = v_o - v_s. \tag{8.64}$$

By substituting (8.64) into nodal equation (8.62), we derive the following equation for v_o:

$$\left(\frac{1}{R_i} + \frac{1}{R_o} + \frac{1}{R_L} + \frac{A}{R_o}\right) v_o = v_s \left(\frac{1}{R_i} + \frac{A}{R_o}\right). \tag{8.65}$$

By solving the last equation for v_o, we arrive at:

$$v_o = v_s \frac{\frac{1}{R_i} + \frac{A}{R_o}}{\frac{1}{R_i} + \frac{1+A}{R_o} + \frac{1}{R_L}}. \tag{8.66}$$

Consider the "worst" case (see (8.52), (8.53), and (8.54)) when $A = 10^4, R_i = 10^5\ \Omega$, $R_o = 10^3\ \Omega$, and $R_L = 10^3\ \Omega$. By substituting these values into (8.66), we find:

$$v_o(t) = 0.9998 v_s(t). \tag{8.67}$$

By comparing (8.67) with (8.59) we can see that the "nonideal nature" of the operational amplifier results in a very small deviation in the performance of the voltage follower. This conclusion is of general nature and applicable to all other circuits discussed below.

8.4.2 Noninverting Amplifier

An electric circuit for this amplifier is shown in Figure 8.20. The noninverting amplifier has a feedback branch with resistance R_f which connects the output terminal with the inverting terminal. The current, i, which flows through this branch is given by:

$$i = \frac{v_o - v_1}{R_f}. \tag{8.68}$$

The inverting terminal is connected to the ground terminal by the branch with resistance R_1. Since the input resistance of the operational amplifier is assumed to be infinite, the same current flows through R_1. This current can be expressed as follows:

$$i = \frac{v_1}{R_1}. \tag{8.69}$$

Finally, the voltage source $v_s(t)$ is connected between the noninverting and ground terminals. Consequently:

$$v_2 = v_s. \tag{8.70}$$

Thus, we have expressed all electric connections in the circuit shown in Figure 8.20 in terms of mathematical equations (8.68), (8.69), and (8.70). We shall next combine

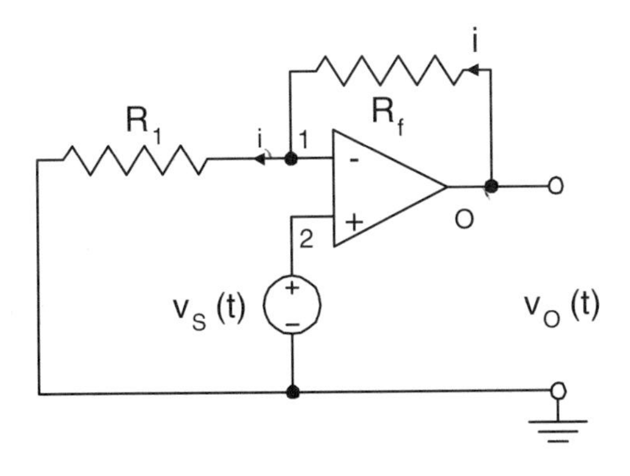

Figure 8.20: Noninverting amplifier.

these equations with (8.56) in order to find the expression for v_o in terms of v_s. From (8.56) and (8.70), we find:

$$v_1 = v_s. \tag{8.71}$$

By substituting (8.71) into (8.69) and (8.68) and by excluding i from the last two equations, we obtain:

$$\frac{v_o - v_s}{R_f} = \frac{v_s}{R_1}. \tag{8.72}$$

By solving the last equation for v_0, we obtain:

$$v_o(t) = \left(1 + \frac{R_f}{R_1}\right) v_s(t). \tag{8.73}$$

It is clear from (8.73) that at any instant of time the output voltage has the same polarity as the source voltage and it is amplified by the factor:

$$\frac{v_o(t)}{v_s(t)} = A_f = 1 + \frac{R_f}{R_1}. \tag{8.74}$$

That is the reason why the circuit shown in Figure 8.20 is called a noninverting amplifier. The utility of this circuit is that, by varying the resistances R_f and R_1, we can control the amplification factor (gain) of the amplifier.

8.4.3 Inverting Amplifier

An electric circuit for this amplifier is shown in Figure 8.21. As before, the inverting amplifier has a feedback branch with resistance R_f which connects the output terminal with the inverting terminal. The current, i, through this branch is given by:

$$i = \frac{v_1 - v_o}{R_f}. \tag{8.75}$$

Voltage source $v_s(t)$ is connected to the inverting terminal through resistance R_1. Since the input resistance (impedance) of the operational amplifier is assumed to be

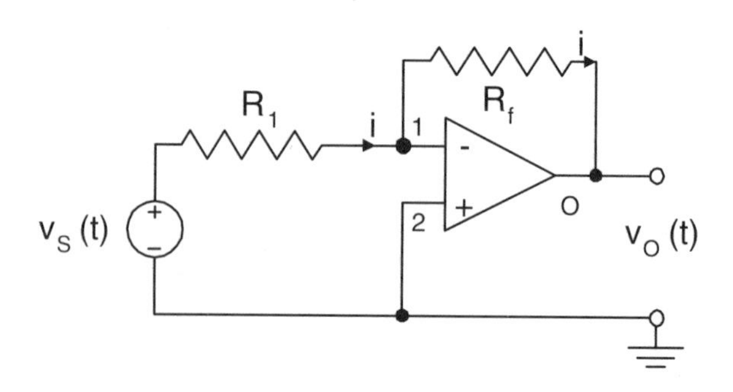

Figure 8.21: Inverting amplifier.

infinite, the same current, i, flows through resistance R_1. This current can be expressed as follows:

$$i = \frac{v_s - v_1}{R_1}. \tag{8.76}$$

Since the noninverting terminal is connected to the ground terminal, we find:

$$v_2 = 0. \tag{8.77}$$

We have expressed all electrical connections in the circuit shown in Figure 8.21 in terms of mathematical equations (8.75), (8.76), and (8.77). We shall combine these equations with (8.56) in order to find the expression for v_o in terms of v_s.

From (8.77) and (8.56), we find:

$$v_1 = 0. \tag{8.78}$$

By substituting (8.78) into (8.75) and (8.76) and by excluding the current i from the last two equations, we obtain:

$$-\frac{v_o}{R_f} = \frac{v_s}{R_1}, \tag{8.79}$$

which leads to

$$v_o(t) = -\frac{R_f}{R_1} v_s(t). \tag{8.80}$$

It is clear from (8.80) that at any instant of time the output voltage has a polarity opposite to the polarity of the source voltage. It is also clear that the amplification factor (sometimes called the closed-loop gain) of the amplifier is given by:

$$A_f = -\frac{R_f}{R_1}. \tag{8.81}$$

That is why this amplifier is called an inverting amplifier or inverter. The gain of this amplifier can be controlled by varying resistances R_f and R_1.

8.4.4 Adder (Summer) Circuit

This circuit is shown in Figure 8.22. As before, there is a feedback branch with resistance R_f which connects the output terminal with the inverting terminal. The current, i, through this branch is given by:

$$i = \frac{v_1 - v_o}{R_f}. \tag{8.82}$$

There is a group of n parallel branches and this group is connected between the inverting and ground terminals. Since the input resistance (impedance) of the operational amplifier is assumed to be infinite, the overall current of the above group of parallel branches is equal to the current, i, through the feedback branch. Thus:

$$i = \sum_{k=1}^{n} i_k. \tag{8.83}$$

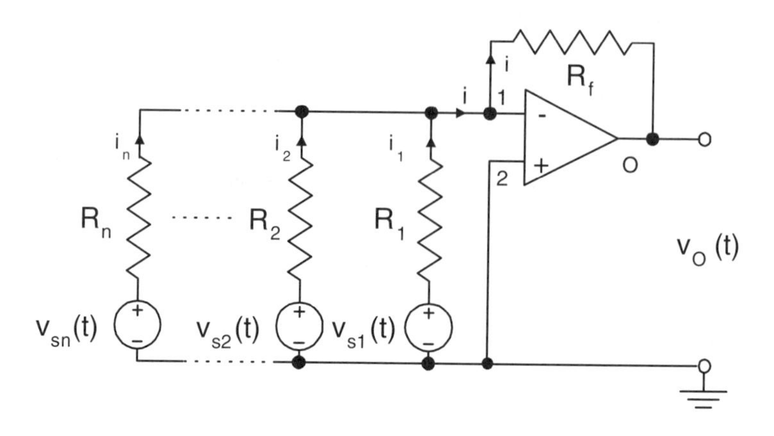

Figure 8.22: Adder circuit.

Each parallel branch contains a resistance R_k in series with a voltage source $v_{sk}(t)$. Consequently:

$$i_k = \frac{v_{sk} - v_1}{R_k}, \quad (k = 1, 2, ..., n). \tag{8.84}$$

Finally, the noninverting terminal is connected to the ground terminal, which means that:

$$v_2 = 0. \tag{8.85}$$

We have expressed all electrical connections in the circuit shown in Figure 8.22 in terms of mathematical equations (8.82), (8.83), (8.84), and (8.85). We shall next combine these equations with expression (8.56) in order to find the expression for the output voltage v_0 in terms of the source voltages v_{sk}. From (8.56) and (8.85), we find:

$$v_1 = 0. \tag{8.86}$$

We then substitute (8.86) into (8.84) and (8.82), which results in:

$$i_k = \frac{v_{sk}}{R_k}, \quad (k = 1, 2, ..., n), \tag{8.87}$$

$$i = -\frac{v_o}{R_f}. \tag{8.88}$$

From (8.87), (8.88), and (8.83), we obtain:

$$-\frac{v_o}{R_f} = \sum_{k=1}^{n} \frac{v_{sk}}{R_k}. \tag{8.89}$$

From (8.89), we arrive at:

$$v_o(t) = -\sum_{k=1}^{n} a_k v_{sk}(t), \tag{8.90}$$

where

$$a_k = \frac{R_f}{R_k}. \tag{8.91}$$

It is clear from (8.91) that the output is a "weighted" sum of source voltages. For this reason, the above circuit is called an adder circuit. It is also clear that the weighted coefficients can be controlled by varying resistances R_f and R_k $(k = 1, 2, ..., n)$.

The right-hand side of equation (8.90) can be written in the form of an inner product of two vectors. Indeed, by introducing the vector $\vec{v}_s$ of voltage sources

$$\vec{v}_s = (v_{s1}, v_{s2}, ...v_{sn}), \tag{8.92}$$

the vector $\vec{a}$ of weighted coefficients

$$\vec{a} = (a_1, a_2, ...a_n), \tag{8.93}$$

and the inner product

$$< \vec{a}, \vec{v}_s >= \sum_{k=1}^{n} a_k v_{sk}, \tag{8.94}$$

we can represent (8.90) as follows:

$$v_o(t) = - < \vec{a}, \vec{v}_s > . \tag{8.95}$$

For this reason, the circuit shown in Figure 8.22 can also be called an inner product circuit.

EXAMPLE 8.7 We would like to design a four-stage (binary) digital-to-analog (D/A) converter using two op-amps and any number of resistors with resistance values up to 10 kΩ.

In Chapter 4 we saw that a D/A converter is just a voltage summer where the weighted coefficients are powers of 2. For a four-stage system we need an output voltage that satisfies $v_{\text{out}} = 8v_4 + 4v_3 + 2v_2 + v_1$. The summer circuit described above yields a negative output voltage (for positive inputs), so we will have to connect it with a simple inverter op-amp circuit to restore the positive sign. The circuit schematic is shown in Figure 8.23 on page 322.

From equation (8.81), we see that $R/R_i = a_i = 2^{i-1}$ for $1 \leq i \leq 4$. If we take $R = 10\ k\Omega$, then we will need to have $R_1 = 10$ kΩ, $R_2 = 5$ kΩ, $R_3 = 2.5$ kΩ, and $R_4 = 1.25$ kΩ. The output of the first stage will be just the negative of the desired result. So, we see from equation (8.70) that we must have $R_A = R_B$. We might as well take both to be equal to 10 kΩ. ■

8.4.5 Integrator

An integrator circuit is shown in Figure 8.24. The operation of this circuit begins with the simultaneous closing of switch SW1 and opening of switch SW2 at time $t = 0$. The dc voltage source V_0, which for time $t < 0$ is connected in parallel with feedback capacitor C_f, will ensure the following initial condition for the voltage, $v_C(t)$, across

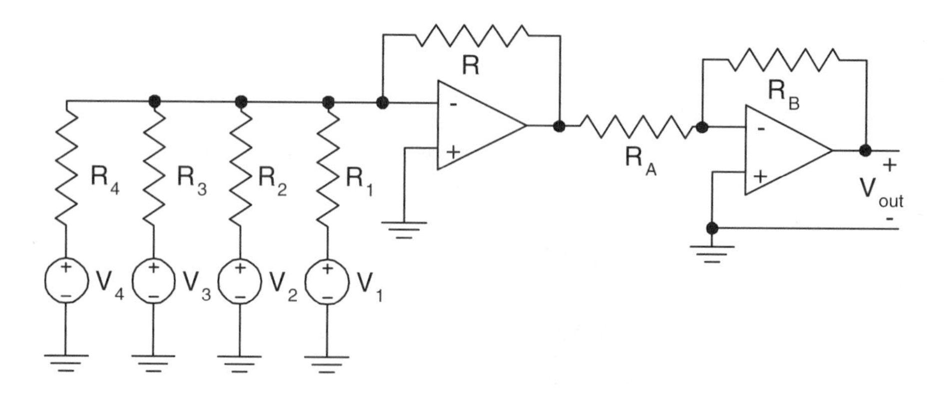

Figure 8.23: An op-amp D/A converter.

the capacitor:

$$v_C(0_+) = -V_0. \tag{8.96}$$

The feedback capacitor C_f is connected between the output and inverting terminals. Consequently:

$$v_C(t) = v_1(t) - v_o(t). \tag{8.97}$$

The noninverting terminal is connected to the ground terminal, which means that:

$$v_2 = 0. \tag{8.98}$$

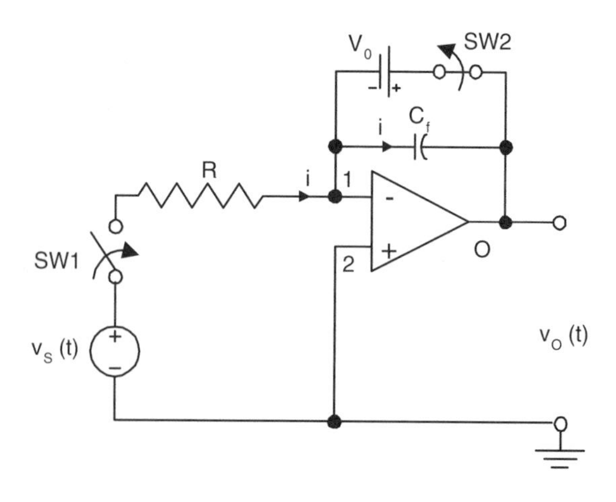

Figure 8.24: Integrator.

From (8.56) and (8.98), we find:

$$v_1(t) = 0. \tag{8.99}$$

By substituting (8.99) into (8.97), we have:

$$v_C(t) = -v_o(t). \tag{8.100}$$

From (8.100) and (8.96) we find the initial condition for $v_o(t)$:

$$v_o(0_+) = V_0. \tag{8.101}$$

For the current, $i(t)$, through the feedback branch we have:

$$i(t) = C_f \frac{dv_C(t)}{dt}, \tag{8.102}$$

which according to (8.100) leads to:

$$i(t) = -C_f \frac{dv_o(t)}{dt}. \tag{8.103}$$

Since the input resistance (impedance) of the operational amplifier is assumed to be infinite, the same current, $i(t)$, flows through resistance R. This current can be expressed as follows:

$$i(t) = \frac{v_s(t) - v_1(t)}{R}, \tag{8.104}$$

which according to (8.99) is tantamount to:

$$i(t) = \frac{v_s(t)}{R}. \tag{8.105}$$

From (8.103) and (8.105), we obtain:

$$\frac{dv_o(t)}{dt} = -\frac{1}{RC_f} v_s(t). \tag{8.106}$$

By integrating (8.106) and taking into account initial conditions (8.101), we arrive at:

$$v_o(t) = -\frac{1}{RC_f} \int_0^t v_s(\tau) d\tau + V_0. \tag{8.107}$$

It is clear from (8.107) that the output voltage is an integral (weighted integral) of the source voltage. For this reason, the circuit shown in Figure 8.24 is called an integrator. It is important to note that the above circuit also ensures the appropriate initial condition for the output voltage.

8.4.6 Differentiator

A differentiator circuit is shown in Figure 8.25. Here, there is a feedback branch with resistance R_f which connects the output terminal with the inverting terminal. The

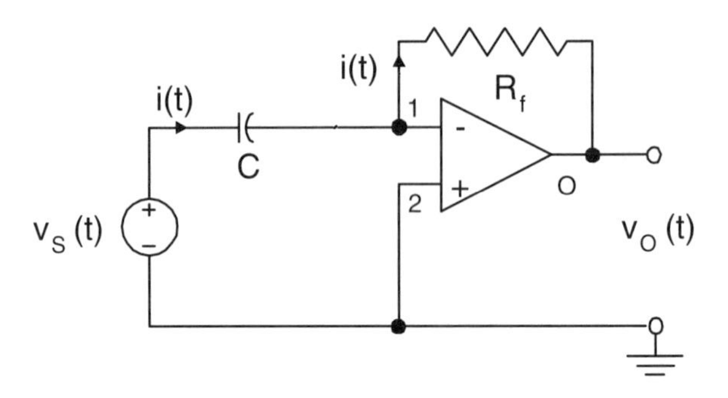

Figure 8.25: Differentiator.

current $i(t)$ through this feedback branch is given by:

$$i(t) = \frac{v_1(t) - v_o(t)}{R_f}. \tag{8.108}$$

Since the input resistance (impedance) of the operational amplifier is assumed to be infinite, the same current, $i(t)$, flows through capacitor C. The current through the capacitor can be expressed as follows:

$$i(t) = C\frac{d(v_s(t) - v_1(t))}{dt}. \tag{8.109}$$

Since the noninverting terminal is connected to the ground terminal, we have:

$$v_2(t) = 0. \tag{8.110}$$

From (8.56) and (8.110), we find:

$$v_1(t) = 0. \tag{8.111}$$

By substituting (8.111) into (8.108) and (8.109) and by eliminating the current $i(t)$ from these two equations, we obtain:

$$-\frac{v_o(t)}{R_f} = C\frac{dv_s(t)}{dt}. \tag{8.112}$$

From (8.112), we arrive at:

$$v_o(t) = -CR_f\frac{dv_s(t)}{dt}. \tag{8.113}$$

Thus, the output voltage is proportional to the time derivative of the voltage source. For this reason, the circuit shown in Figure 8.25 is called a differentiator.

8.4.7 Application of Operational Amplifiers to the Integration of Differential Equations (Analog Computer)

Consider the following initial value problem: find the solution to the equation

$$a_2 \frac{d^2 v(t)}{dt^2} + a_1 \frac{dv(t)}{dt} + a_0 v(t) = f(t) \tag{8.114}$$

subject to the initial conditions

$$v(0_+) = V_0, \tag{8.115}$$

$$\frac{dv}{dt}(0_+) = V_1. \tag{8.116}$$

In the above expressions a_2, a_1, a_0, $f(t)$, V_0, and V_1 are known.

Before constructing an electric circuit with operational amplifiers which integrates the initial value problem (8.114)–(8.116), we rewrite differential equation (8.114) as follows:

$$\frac{d^2 v(t)}{dt^2} = -\frac{a_1}{a_2}\frac{dv(t)}{dt} - \frac{a_0}{a_2} v(t) + \frac{f(t)}{a_2}. \tag{8.117}$$

Now, let us assume for the time being that we have a voltage equal to $d^2v(t)/dt^2$. This assumption is made just in order to start the design of our circuit. At the end of our design, we shall see how this voltage can be obtained and, consequently, our assumption will be removed. We apply voltage $d^2v(t)/dt^2$ as the input to the integrator circuit containing op-amp 1 shown in Figure 8.26. If R_1 and C_{f1} are chosen in such a way that $R_1 C_{f1} = 1$, then the output voltage of this integrator is equal to $-dv(t)/dt$. We also choose the dc voltage source V_{s1} (in parallel with C_{f1}) equal to $-V_1$, which guarantees initial condition (8.116) for $dv(t)/dt$. The first-stage output voltage of $-dv(t)/dt$ serves as the input for the second integrator shown in the same figure. If R_2 and C_{f2} are chosen such that $R_2 C_{f2} = 1$, then the output voltage of this integrator is equal to $v(t)$. We also choose the dc voltage source V_{s2} (in parallel with C_{f2}) equal to V_0. This guarantees initial condition (8.115) for $v(t)$. The circuit voltages equal to $-dv(t)/dt$ and $f(t)$ serve as the inputs for the adder circuit containing op-amp 3. We choose R_3, R_4, and R_{f3} such that

$$\frac{R_{f3}}{R_3} = \frac{1}{a_2}, \quad \frac{R_{f3}}{R_4} = \frac{a_1}{a_2}. \tag{8.118}$$

Then, the output voltage, $v_3(t)$, of this adder circuit is equal to:

$$v_3(t) = \frac{a_1}{a_2}\frac{dv(t)}{dt} - \frac{f(t)}{a_2}. \tag{8.119}$$

This voltage and the voltage equal to $v(t)$ serve as the inputs for the adder circuit containing op-amp 4. We choose R_5, R_6, and R_{f4} such that

$$\frac{R_{f4}}{R_5} = 1, \quad \frac{R_{f4}}{R_6} = \frac{a_0}{a_2}. \tag{8.120}$$

Then, the output voltage, $v_4(t)$, of this adder circuit is equal to:

$$v_4(t) = -v_3(t) - \frac{a_0}{a_2}v(t). \tag{8.121}$$

From (8.121) and (8.119), we find:

$$v_4(t) = -\frac{a_1}{a_2}\frac{dv(t)}{dt} - \frac{a_0}{a_2}v(t) + \frac{f(t)}{a_2}. \tag{8.122}$$

Now, we connect the output of adder circuit 4 with the input of integrator 1. In this way we force the equality

$$\frac{d^2v(t)}{dt^2} = v_4(t), \tag{8.123}$$

which according to (8.122) is equivalent to differential equation (8.117). Thus, we can conclude that the voltage $v(t)$ in the circuit shown in Figure 8.26 satisfies differential equation (8.114) and initial conditions (8.115) and (8.116). Consequently, if at time $t = 0$ we simultaneously open switches SW1 and SW2 and close switch SW3 and then measure the output, $v(t)$, of integrator 2 as a function of time, we find the solution of initial value problem (8.114)–(8.116). In other words, the electric circuit shown in Figure 8.26 can be considered as an analog computer which solves the initial value problem (8.114)–(8.116).

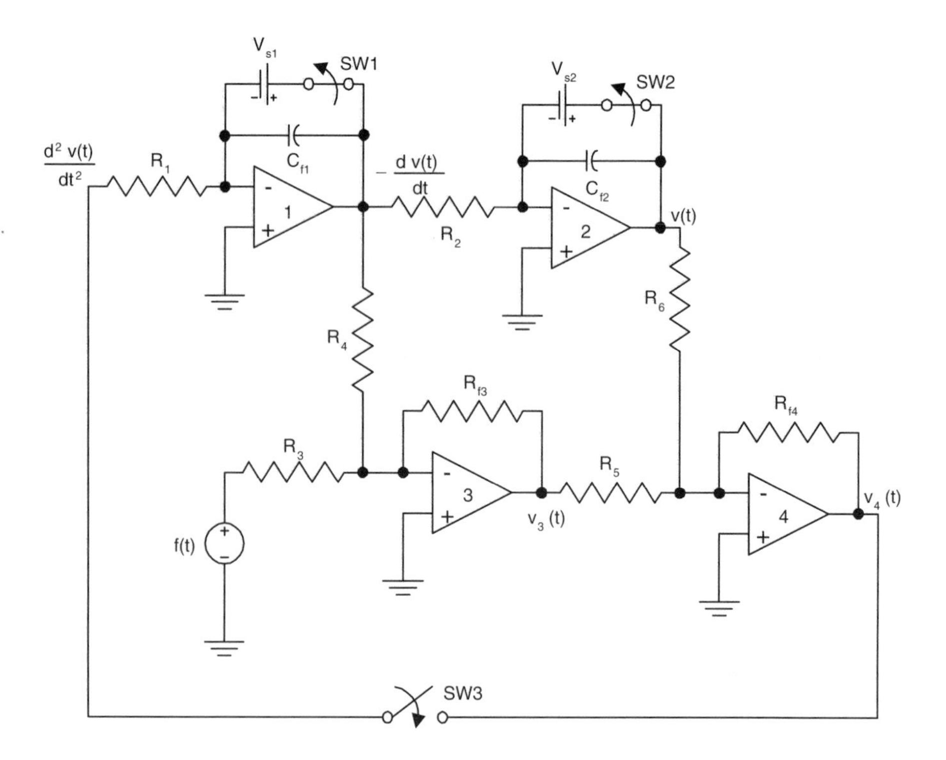

Figure 8.26: Circuit which integrates the initial value problem.

EXAMPLE 8.8 We shall design an electric circuit with operational amplifiers which solves the following initial value problem:

$$\frac{1}{\omega^2} \cdot \frac{d^2 v(t)}{dt^2} = -v(t), \tag{8.124}$$

$$v(0_+) = V_0, \tag{8.125}$$

$$\frac{dv}{dt}(0_+) = 0. \tag{8.126}$$

As before, we assume for the time being that we have a voltage equal to

$$\left(\frac{1}{\omega^2}\right) \frac{d^2 v(t)}{dt^2}.$$

This assumption will be removed at the end of our design. We apply the voltage $(1/\omega^2)d^2v(t)/dt^2$ as the input to integrator 1 shown in Figure 8.27. We choose R_1 and C_{f1} such that $R_1 C_{f1} = 1/\omega$, then the output voltage of this integrator is equal to $(-1/\omega)\, dv(t)/dt$. We also use a short-circuit branch in parallel with feedback capacitor C_{f1} in order to guarantee zero initial condition (8.126) for $dv(t)/dt$. The voltage of $(-1/\omega)dv/dt$ serves as the input for integrator 2. We choose R_2 and C_{f2} such that $R_2 C_{f2} = 1/\omega$, then the output voltage of this integrator is equal to $v(t)$. We also use the dc voltage source V_0 in parallel with feedback capacitor C_{f2} in order to guarantee initial condition (8.125) for the voltage $v(t)$. This voltage serves as the input for inverting amplifier 3. We choose R_3 and R_{f3} such that

$$\frac{R_{f3}}{R_3} = 1. \tag{8.127}$$

Then, the output voltage of inverting amplifier 3 is equal to $-v(t)$. Now, we connect the output of inverting amplifier 3 with the input of integrator 1. In this way, we force the equality (8.124). Thus, we can conclude that if at time $t = 0$ we simultaneously open switches SW1 and SW2 and close switch SW3, then the voltage $v(t)$ in the

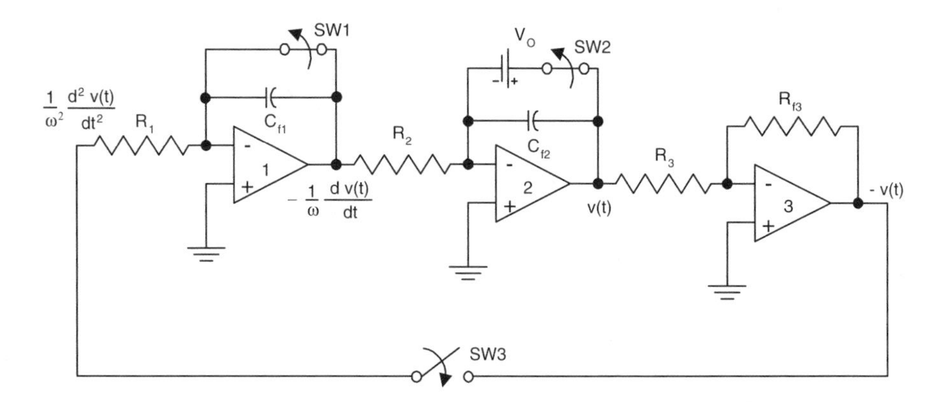

Figure 8.27: Generator circuit for harmonic electric oscillations.

circuit will satisfy differential equation (8.124) and initial conditions (8.125) and (8.126). It is easy to see that the solution of initial value problem (8.124)–(8.126) has the form:

$$v(t) = V_0 \cos \omega t. \tag{8.128}$$

Consequently, if we measure the voltage $v(t)$ between the output terminal of integrator 2 and the ground terminal, we find out that this voltage exhibits undamped harmonic oscillations. In other words, the circuit shown in Figure 8.27 can be considered as a generator of harmonic electric oscillations. The frequency of these oscillations can be controlled by varying the resistance of resistors R_1 or R_2.

The question can be asked how the circuit shown in Figure 8.27 can generate undamped oscillations when there are resistors and hence energy losses in the circuit. The answer is that these energy losses are replenished by the dc power supplies which are connected to the operational amplifiers.

A remarkable property of the above circuit is that it generates undamped harmonic oscillations without using inductors (compare with the *LC* circuit discussed in the previous chapter). This feature is typical for modern active *RC* circuits which can perform diverse functions without utilizing expensive and bulky inductors. As far as the cost of operational amplifiers is concerned, it is quite low due to the remarkable progress in semiconductor technology. ■

EXAMPLE 8.9 We want to design an op-amp circuit that generates the damped oscillation $v(t) = e^{-t} \sin(t)$. Use as few op-amps as possible; limit the resistors to 100 kΩ and make the capacitors as small as possible.

First, we need to determine the differential equation that describes this function. By taking two derivatives of $v(t)$ and performing a few algebraic manipulations, we quickly find that $d^2v/dt^2 = -2(dv/dt) - 2v$. We will need two op-amp integrator circuits, one summer circuit, and possibly one inverter circuit. Let's see if we can avoid the inverter by being a bit clever. Consider the circuit shown in Figure 8.28.

The key to avoiding the inverter is in the rightmost op-amp circuit. At first glance, it appears to be a circuit that we have not analyzed before. However, if

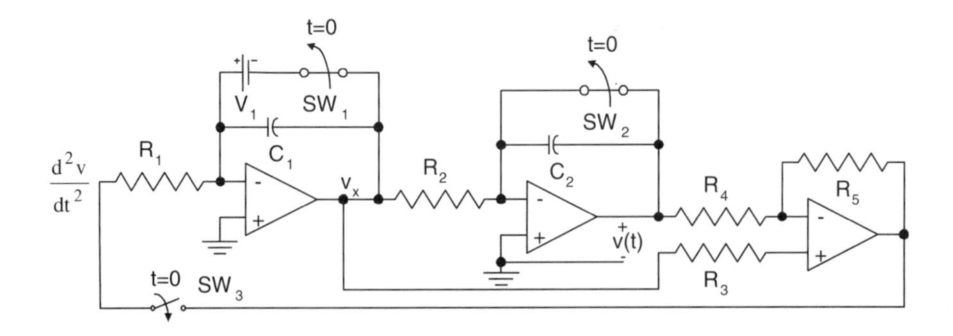

Figure 8.28: Circuit which generates the function $v(t) = e^{-t} \sin(t)$.

the voltage at the noninverting terminal were zero, it would be a simple inverting amplifier. Likewise, if $v(t)$ were set to zero, the op-amp circuit would be a simple noninverting amplifier. From the superposition principle, we see that the output of this circuit is just $v_x(1 + R_5/R_4) - v(t)R_5/R_4$. The differential equation requires that $R_5/R_4 = 2$, so we can take $R_5 = 10\ \text{k}\Omega$ and $R_4 = 5\ \text{k}\Omega$. This means that we must have $v_x = -\frac{2}{3}(dv/dt)$. Equation (8.97) tells us that R_1C_1 must be 3/2, so we can take $R_1 = 100\ \text{k}\Omega$ and $C_1 = 15\ \mu\text{F}$. The same equation places the condition $R_2C_2 = 2/3$, so we can try $R_2 = 100\ \text{k}\Omega$ and $C_2 = 6.66\ \mu F$. There is no direct expression for R_3 since no current flows through it, but in practice it is usually best to make it equal to R_4. The determination of V_1 is left as an exercise for the reader. ■

8.5 MicroSim PSpice Simulations

In this section, we will use PSpice simulations to examine the performance of six circuits which contain dependent sources and operational amplifiers. The first circuit that we will consider contains two dependent sources and was originally drawn in Figures 8.10 and 8.15. In the latter figure, one of the impedances (Z_3) was replaced by a probing current source to find the Thevenin equivalent nonideal voltage source at the open terminals. We will use PSpice to find the Thevenin voltage and impedance for the specific case of a dc circuit with Z_1 = R1 = 5 Ω, Z_2= R2 = 10 Ω, V_s1 = 5 V, A = 2, and G = 0.5 (these variables are defined in Example 8.5). For these values, (8.44) yields a Thevenin equivalent voltage of V_{TH} = 10 V and an impedance of Z_{TH} = 30 Ω.

The circuit, as drawn by the schematics editor, is shown in Figure 8.29. The independent voltage and current sources use "VSRC" and "ISRC" models, respectively, and are given dc values as indicated above. The voltage-controlled voltage

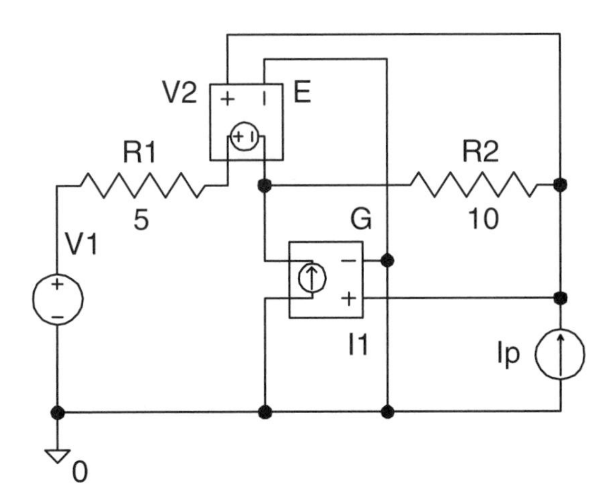

Figure 8.29: The schematic drawing of the circuit of Example 8.5.

source model is contained in the "analog.slb" library and can be selected by entering an "E" in the "Add Part" dialog box. Note that the VCVS part doesn't utilize the standard symbol but is given instead by a four-terminal box. This is done so that the controlling voltage can be specified simply by connecting the plus and minus "sensing" terminals to the appropriate circuit nodes via wires. The two terminals which are connected to the usual voltage source symbol contained within the box are the terminals of the dependent source. Double-clicking on the VCVS symbol brings up the model attributes dialog box. The gain should be set to 2 via the "Save Attr" button. The part name for the voltage-controlled current source is "G" and the model is also a four-port device that is contained in the "analog.slb" library. The gain for the VCIS should be set to 0.5 using the procedure outlined for the other dependent source.

Under "Setup..." in the Analysis drop-down menu we select a "DC Sweep." In the "DC Sweep" dialog box we select "Current source" for the "Swept Var. Type" and "Linear" for the "Sweep Type." We enter "Ip" under "Name" and select a start value of zero, an end value of two, and an increment of 0.1. The results of the simulation are indicated in Figure 8.30, where the voltage across the probing source is plotted as a function current. The zero intercept gives the Thevenin equivalent voltage and the slope of the line is equal to the Thevenin impedance. As expected, these values agree with the results of (8.44).

The next circuit that we examine comes from Example 8.6 and is redrawn with the schematics editor in Figure 8.31. This time we will use a probing voltage source and a dc sweep analysis with PSpice to determine the Norton equivalent nonideal current source for the resistances given in the figure. The dc source parameters for this example are $V_{s1} = 2$ V, $V_{s3} = 5$ V, $A = 3$, and $R = 10\ \Omega$ (these values are defined

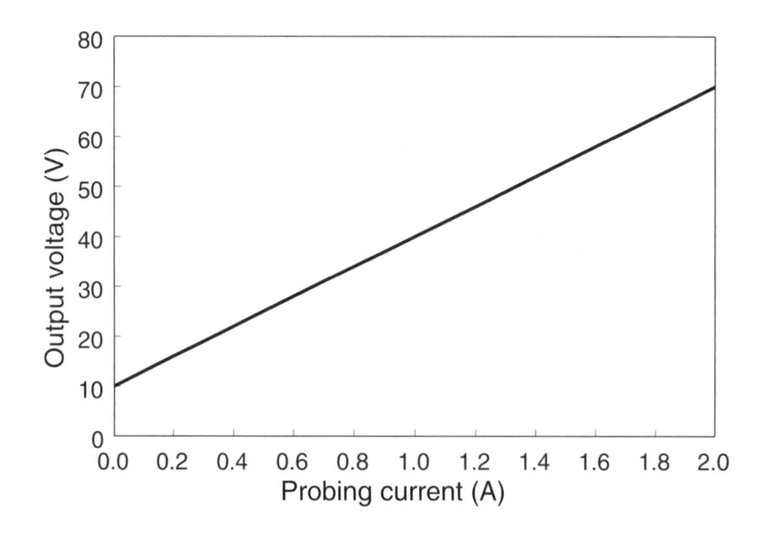

Figure 8.30: The simulated output voltage as a function of probing current for the circuit of Example 8.5.

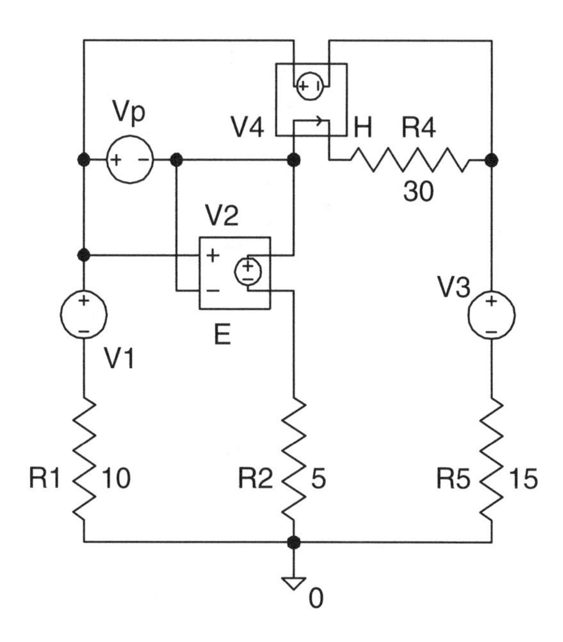

Figure 8.31: The schematic drawing of the circuit of Example 8.6.

in Figure 8.16). The part name for the current-controlled voltage source is "H" and the model, contained in the "analog.slb" library, is once again a four-port device. The direction of the controlling current is indicated by the direction of the arrow which is connected between the two "sensing" terminals. (Note: the current-controlled current source, which is not considered in either of these examples, has the part name "F" and is in the same library as the other dependent sources.)

In the "DC Sweep" dialog box we select "Voltage source" for the "Swept Var. Type" and "Linear" for the "Sweep Type." We enter "Vp" under "Name" and select a start value of zero, an end value of ten, and an increment of 0.2. The current through the probing source, as calculated by PSpice, is plotted in Figure 8.32 as a function probe voltage. According to equation (8.48), the zero intercept gives the Norton equivalent current and the negative of the slope of the line is equal to the Norton admittance. The circuit values given above, when inserted in equations (8.49) and (8.50), result in the Norton parameters $I_{NO} = 0.29$ A and $Y_{NO} = 0.43$ ℧. Again, these values are virtually identical to those indicated on the PSpice output graph.

The schematic for a noninverting amplifier circuit, with the nonideal op-amp circuit shown in Figure 8.19, is indicated in Figure 8.33a. The model for the op-amp includes the voltage-controlled voltage source Av with a gain of 10,000 and the two resistors Ri and Ro. An ac voltage source (part name = "VSRC") with an amplitude of 1 V is the input. The gain of an ideal noninverting amplifier is $A_f = 1 + \text{Rf\_a} / \text{R1\_a} = 11$. An "AC Sweep" analysis is selected with 21 points per decade calculated from 100 Hz to 100 kHz. The output voltage magnitude over that range of frequencies, indicated by the solid line in Figure 8.34, is about 0.12% below the ideal value of

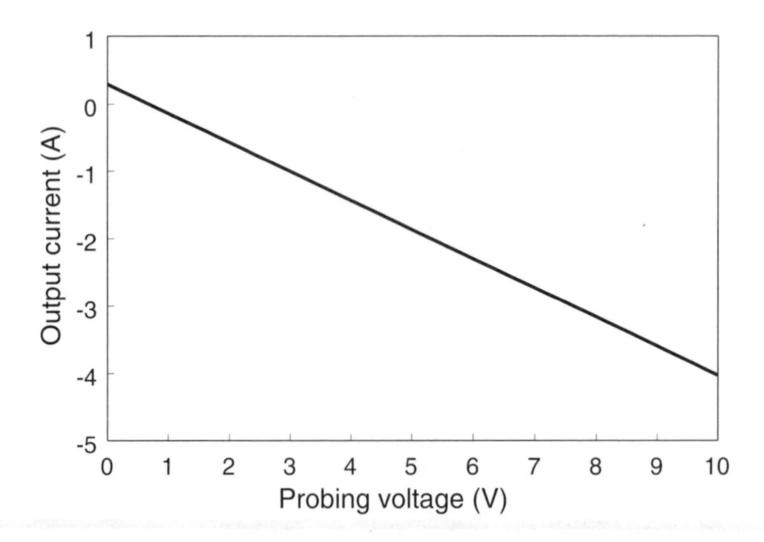

Figure 8.32: The simulated output current as a function of probing voltage for the circuit of Example 8.6.

11 V. Our simple model of an op-amp is independent of frequency, of course, so the voltage curve is just a line with zero slope.

Since practical circuits often contain many op-amps, it would be useful if the set of elements that make up an op-amp model (Av, Ri, and Ro for our simple model) could be combined into one block that could be inserted into a schematic as a single entity. PSpice has the means to do this by defining a "subcircuit." In the windows version of PSpice, subcircuits can be created by the "Create Subcircuit" line in the

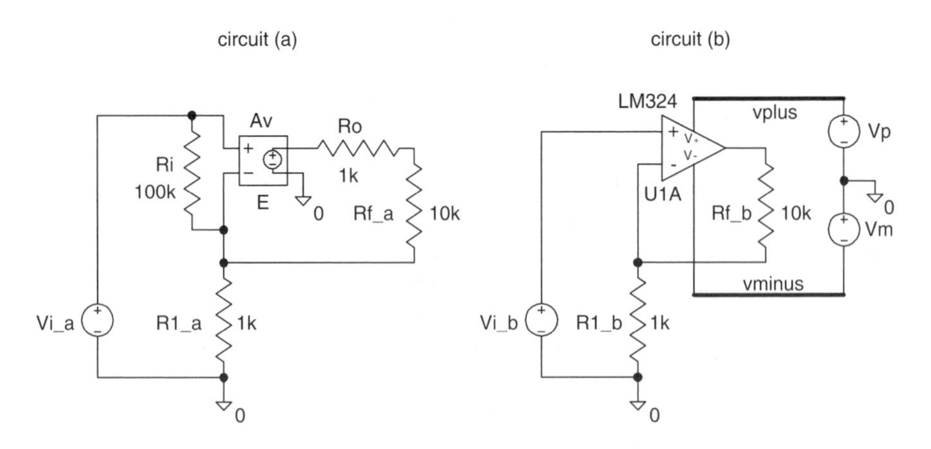

Figure 8.33: Noninverting amplifier circuits: (a) nonideal op-amp model and (b) LM 324 op-amp model.

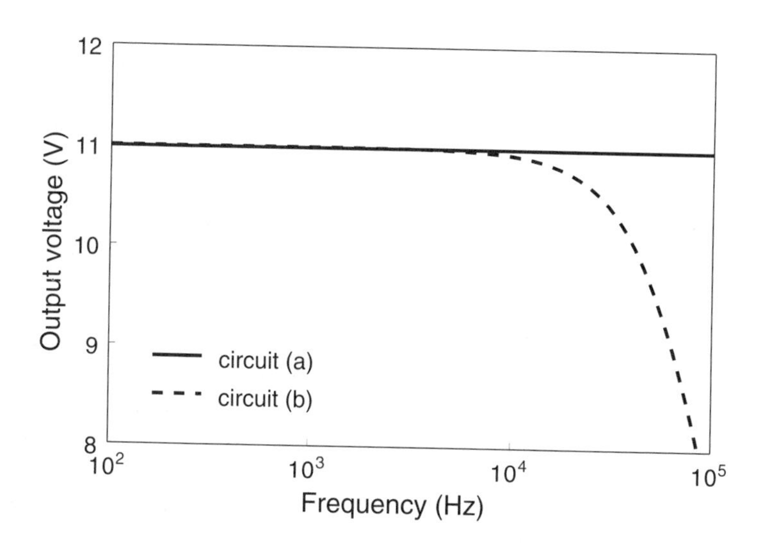

Figure 8.34: The frequency dependence of the output voltage.

"Tools" drop-down menu. Subcircuit generation is beyond the scope of this text, however, and the interested reader is referred to the references given in Appendix C.

We will instead use one of the op-amp models that is provided in the "eval.slb" library. The part name for this commonly used op-amp is "LM324" and it can be obtained from the "Get New Part" line of the "Draw" drop-down menu in the usual way. This model is considerably more sophisticated than the nonideal model pictured in Figure 8.33a and is based on the actual makeup of the device (transistors, resistors, etc., and their interconnections). A thorough analysis of the operation of this op-amp is a subject for a course on electronics; here we will just mention two general properties of real op-amps that should be understood prior to their introduction into PSpice schematics. First, they require connections to external power supplies which provide, among other things, the energy necessary to amplify input signals. Often, two power supplies are used with equal voltage magnitudes but opposite signs (± 12 V are widely used for many applications). This is required, for example, to produce an amplified sinusoidal signal with zero offset voltage, because the range of possible output voltages is necessarily contained between the two source voltages. Second, op-amps exhibit frequency dependences due to either capacitors which are deliberately placed in the op-amp circuit or capacitances that are a consequence of transistor structure. Typically, these capacitances result in a decrease in the open-loop gain as the drive frequency is increased.

The noninverting circuit with the LM324 op-amp is shown in Figure 8.33b. The op-amp requires five connections. Three of the connections are the same as for our simple model: the inverting terminal ($-$), the noninverting terminal ($+$), and the output terminal (not marked). The other two terminals are for the power supply connections. Vp and Vm are dc power supplies of $+12$ and -12 V, respectively. They are connected to the V+ and V− terminals of the op-amp via "bus" lines. Bus lines are typically used in a circuit when many connections are required to the same point and can be generated by the "Bus" line of the "Draw" drop-down menu. The output voltage terminals of power supplies are excellent examples of bus lines. The gain of this circuit, which ideally should be the same as that in the circuit of Figure 8.33a, is given by the dashed line in Figure 8.34. The low-frequency gain of the op-amp circuit is slightly higher than in the previous case, but the gain drops off sharply as the frequency goes above 10 kHz. Note that the two circuits are placed in the same schematic drawing and are simulated at the same time (hence with the same analysis specifications) in PSpice.

The next op-amp circuit that we will investigate is the integrating circuit shown in Figure 8.35. The supply voltages are ± 12 V and the resistance and capacitance are taken so that R1Cf = 1. The voltage source to be integrated uses a piecewise linear model and has the part name "VPWL" in the "source.slb" library. The time dependence of the input voltage is indicated by the solid line in Figure 8.36. This voltage rises from zero to 6 volts in 1 second, remains at 6 volts for 2 seconds, falls back to zero in 1 second, and then decreases to -12 V in the final second. Two cases are simulated via a parametric analysis: Vo = 0 V and Vo = 5 V.

A transient analysis was conducted for 5 seconds and the results for both initial conditions are plotted in Figure 8.36. The nonzero initial condition case is plotted with the dot-dashed line. As expected, the output is parabolic when the input is changing linearly and varies linearly when the input is constant. The output decreases for positive input voltages due to the inverting nature of the integrator. The zero initial condition case output is indicated by the dashed line. The circuit clearly does not

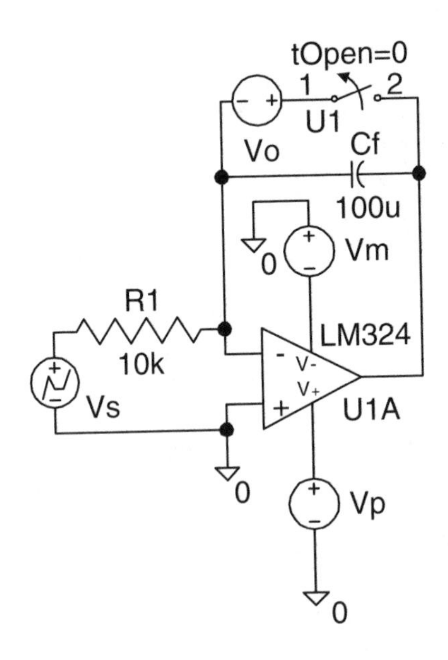

Figure 8.35: An integrator circuit with an LM324 op-amp.

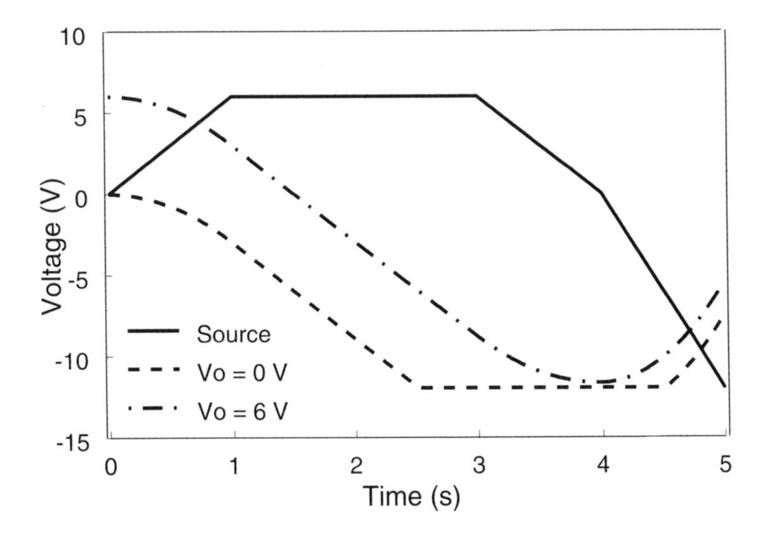

Figure 8.36: Input and output voltages of the integrator circuit.

integrate the input signal after about 2.5 seconds. The problem arises because the output voltage attempts to go below the minimum allowable value of Vm = −12 V. We say that the output has "saturated" during this time. It is interesting to note that there is a delay after the input goes negative before the output again tracks the (negative of the) integral of the input signal. The interested reader can find an explanation for this phenomenon by plotting the voltage at the inverting terminal. After the output saturates, it is no longer necessarily true that V+ = V− and the output does not begin to function properly again until this condition is restored.

The final PSpice simulation presented in this section involves the harmonic generator op-amp circuit of Example 8.8. This circuit, with parameters selected to generate an angular frequency of 10 rad/s, is shown in Figure 8.37. The power supplies

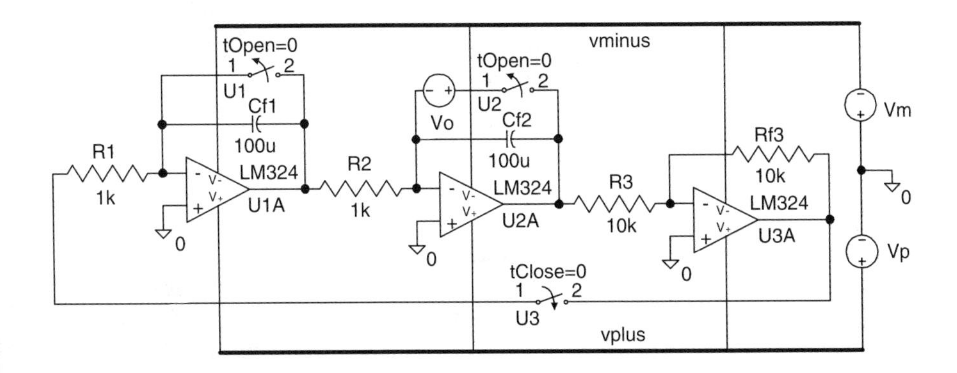

Figure 8.37: The schematic drawing of the circuit of Example 8.8.

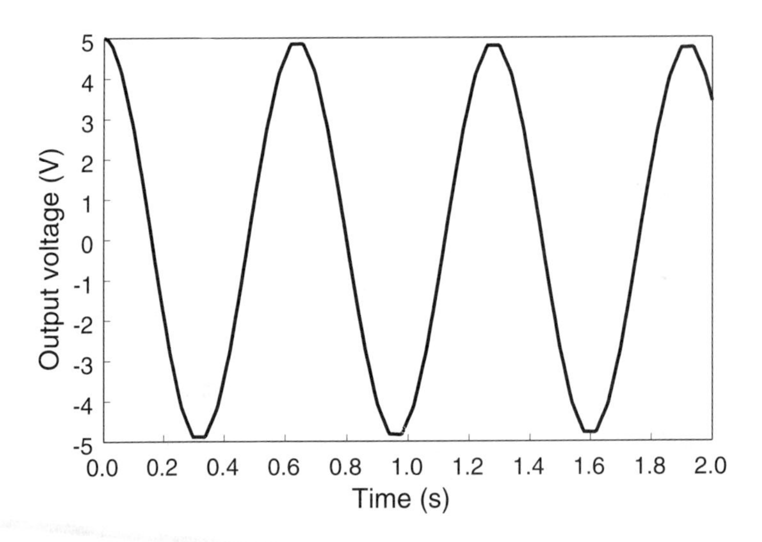

Figure 8.38: The output voltage of the circuit of Example 8.8.

are again ± 12 V and the initial voltage is Vo = 5 V. A transient analysis is selected with a time step of 5 ms and a final time of 2 s. The time dependence of the output voltage is given in Figure 8.38 and clearly indicates the sinusoidal nature of the output with the proper initial condition and a frequency of $f = 10/(2\pi)$ Hz.

8.6 Summary

The main results of this chapter can be summarized as follows:

- Four types of dependent sources have been introduced. They are the voltage-controlled voltage source (VCVS), current-controlled voltage source (ICVS), voltage-controlled current source (VCIS), and current-controlled current source (ICIS). It has been demonstrated that dependent sources usually appear as **linear models for transistors** and semiconductor devices.

- The nodal analysis and mesh analysis techniques for circuits with dependent sources have been presented. These techniques involve the following steps: (a) Write the nodal potential or mesh current equations as if all the sources were independent. (b) Express the dependent sources in terms of the unknown variables. For nodal analysis, this means to express the dependent sources in terms of nodal potentials. For mesh current analysis, the dependent sources should be expressed in terms of mesh currents. (c) Substitute the above expressions for the dependent sources in terms of unknown variables into the original equations. Then, modify the coefficient matrix by moving these expressions to the left-hand side of the equations.

- The application of Thevenin's theorem to the analysis of electric circuits with dependent sources has been discussed. It has been pointed out that the calculation of the open-circuit voltage and the short-circuit current with respect to the terminals of the branch in question is required in order to find the parameters of the Thevenin equivalent circuit. These calculations are substantially simplified when the branch in question is a **controlling branch**. The "probing" technique for the calculation of Thevenin and Norton equivalent circuits has been presented as well.

- An ideal operational amplifier has been defined as a voltage-controlled voltage source with infinite gain, infinite input resistance, and zero output resistance. It has been demonstrated that operational amplifiers can be used to design various circuits such as buffer amplifiers (voltage follower), noninverting and inverting amplifiers, adder circuits, integrators, and differentiators. It has been shown that the above operational amplifier circuits can be utilized to design analog computers and sine wave generators.

8.7 Problems

1. Write, but do not solve, the nodal equations for the circuit shown in Figure P8-1.

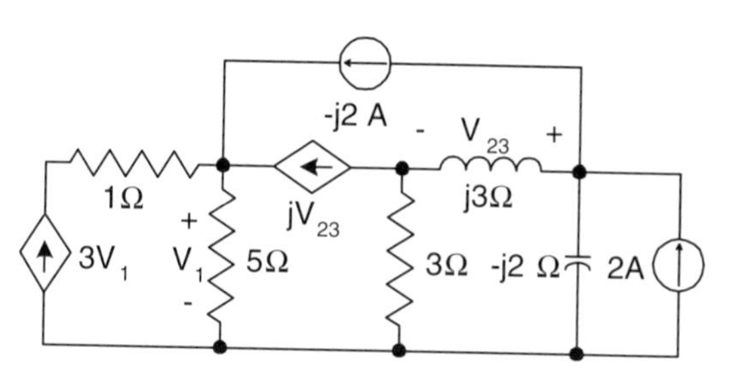

Figure P8-1

2. Find the voltages of all the nodes in the circuit shown in Figure P8-2.

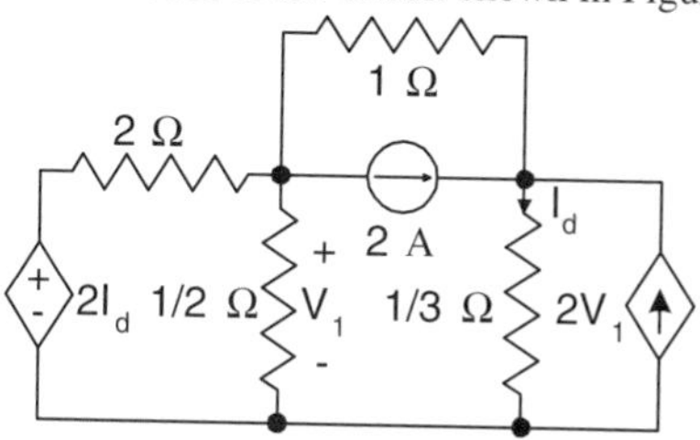

Figure P8-2

3. Determine the value of the resistance R required to achieve a 2 V potential difference across the 1 Ω resistor in the circuit shown in Figure P8-3.

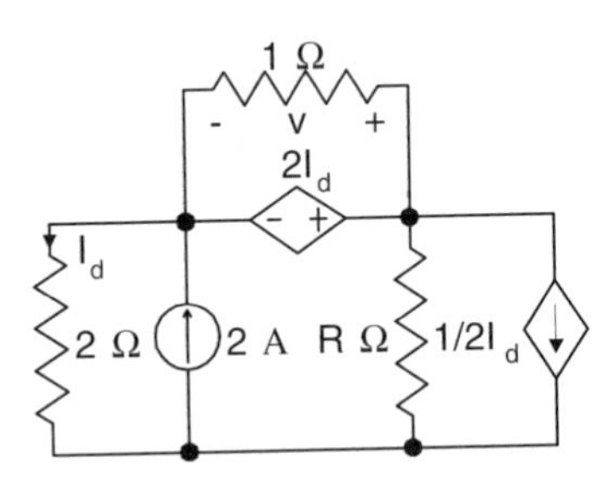

Figure P8-3

In Problems 4 and 5, assume the sources all have an angular frequency of ω.

4. Write, but do not solve, the nodal equations for the circuit shown in Figure P8-4.

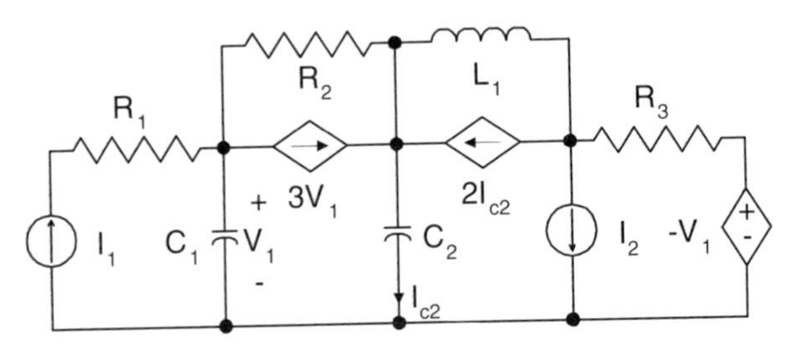

Figure P8-4

5. Write, but do not solve, the mesh current equations for the circuit shown in Figure P8-5.

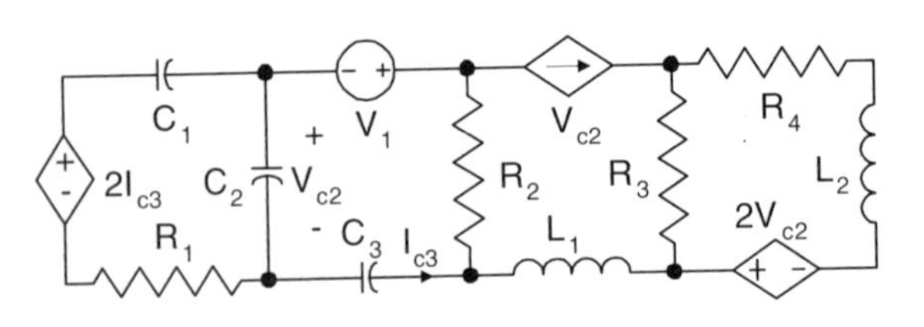

Figure P8-5

6. Determine the value of the resistance R which results in a current of 1 A emerging from the positive terminal of the dependent voltage source in the circuit shown in Figure P8-6.

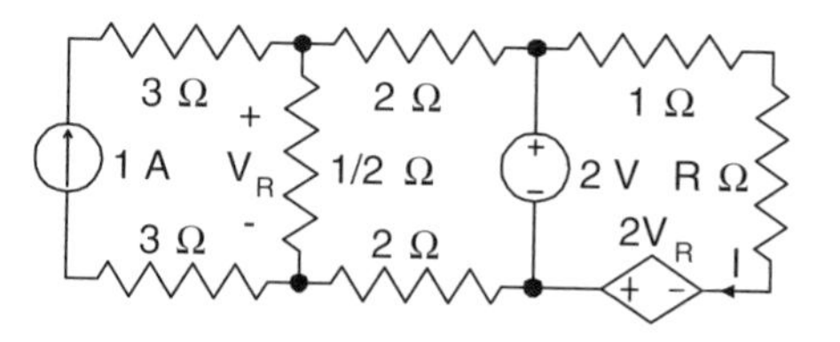

Figure P8-6

7. Find all of the mesh currents in the circuit shown in Figure P8-7.
8. Write, but do not solve, the mesh current equations for the circuit shown in Figure P8-8.

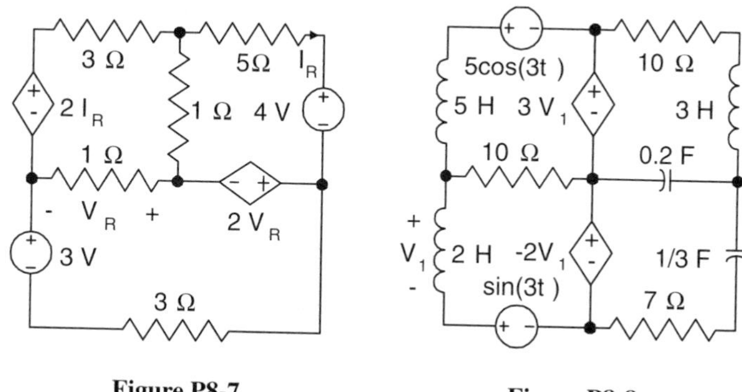

Figure P8-7

Figure P8-8

9. Remove the 1/3 Ω resistor from the circuit shown in Figure P8-2 and find the Thevenin equivalent circuit with respect to the open terminals.
10. Remove the 2 Ω resistor from the circuit shown in Figure P8-3 and find the Norton equivalent circuit with respect to the open terminals. Let $R = 1\ \Omega$.
11. Remove the 1/2 Ω resistor from the circuit shown in Figure P8-6 and find the Norton equivalent circuit with respect to the open terminals. Let $R = 1\ \Omega$.
12. Remove the 5 Ω resistor from the circuit shown in Figure P8-7 and find the Thevenin equivalent circuit with respect to the open terminals.
13. Consider the circuit shown in Figure P8-13. Solve for all the node voltages.

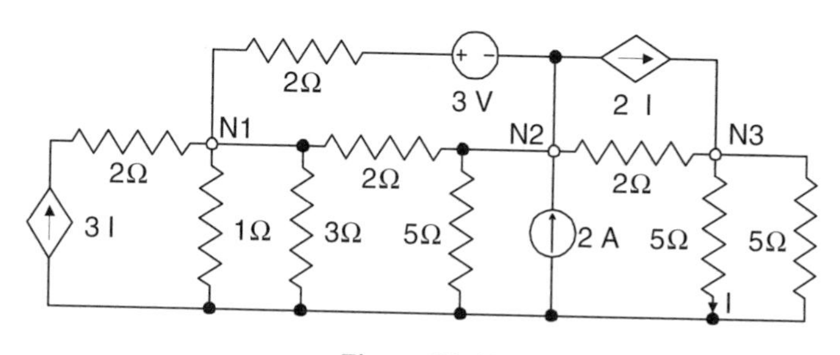

Figure P8-13

14. Consider the circuit shown in Figure P8-14. Solve for all the node voltages.

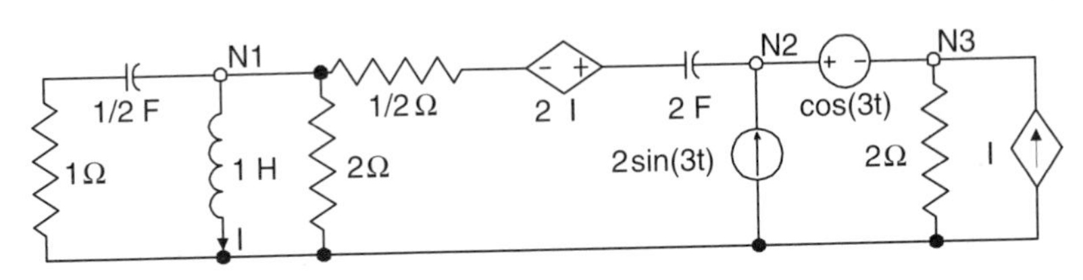

Figure P8-14

15. Consider the circuit shown in Figure P8-15. Identify and label clearly all the meshes in the circuit and solve for all the corresponding mesh currents.

16. Consider the circuit shown in Figure P8-16. Solve for all the phasor mesh currents.

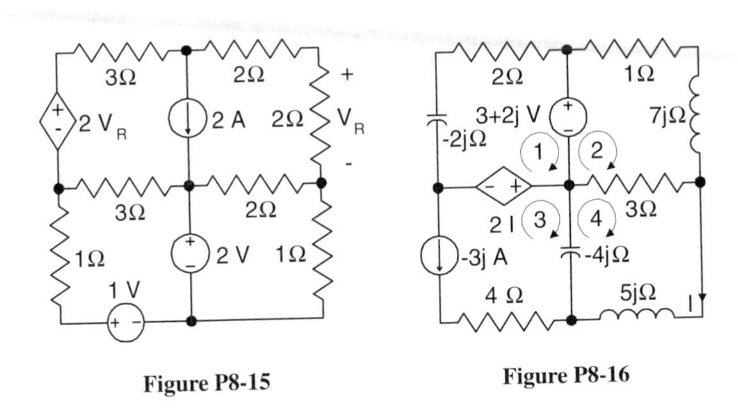

Figure P8-15 **Figure P8-16**

17. Given the circuit shown in Figure P8-17, find the Norton equivalent circuit at the open terminals.

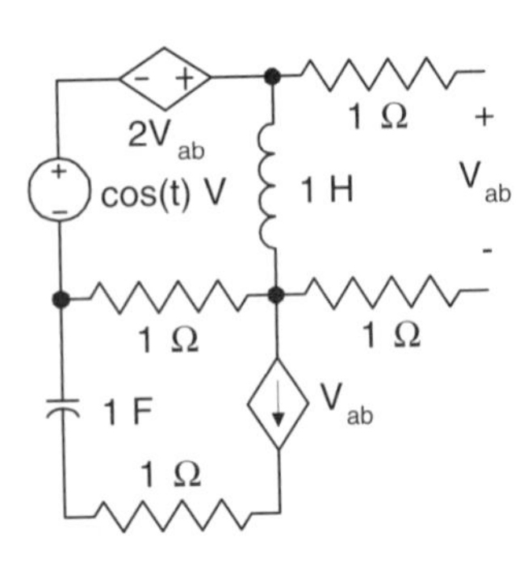

Figure P8-17

18. Design an ideal op-amp circuit which produces an output voltage of $v_{out}(t) = 1000v_3 - 100v_2 + 10v_1 + v_o$, where v_o, v_1, v_2, v_3 are input voltages. Use only resistors in the range 100 Ω to 10 kΩ and minimize the number of op-amps.

19. Consider the circuit shown in Figure P8-19. Find the output voltage in terms of the input voltage as a function of v_1, v_2.

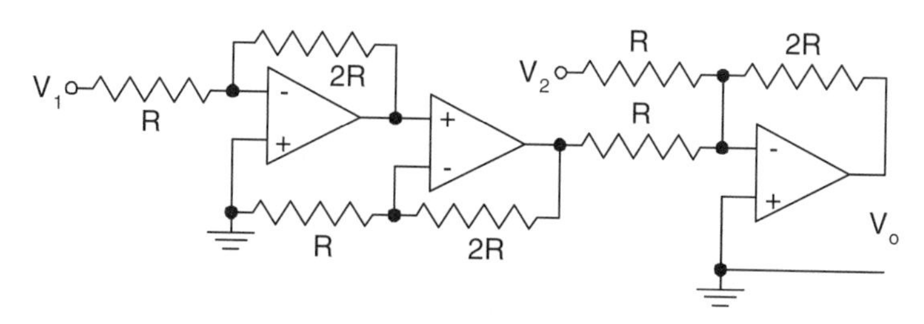

Figure P8-19

20. Analyze the circuit shown in Figure P8-20 to find the output voltage as a function of $v_1, v_2, \ldots v_n$.

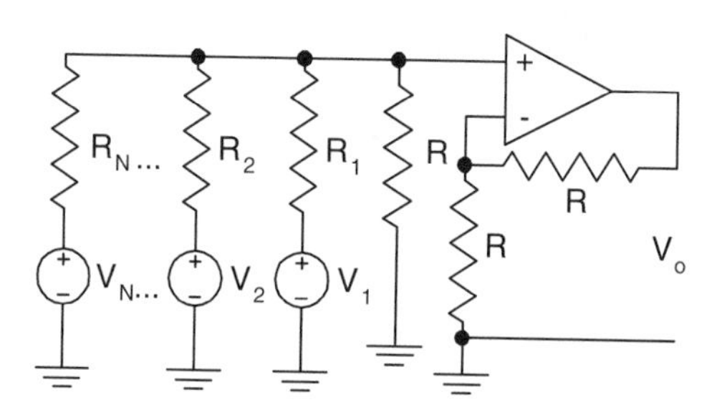

Figure P8-20

21. Design an ideal op-amp circuit which produces an output voltage of $v_{out}(t) = 22v_1 + 49v_2 - 17v_3$, where v_1, v_2, and v_3 are input voltages. The maximum resistance value you may use is 10 kΩ.

22. Design an op-amp circuit that produces an output voltage of $v_o(t) = 5t^4 + 12t^2 + 6$ given an input voltage source of $v_s(t) = t^6/6 + t^4 + 7t$. You can use any combination of capacitors, batteries, and switches, but you may only use 10 kΩ resistors.

23. Design an op-amp circuit that produces an output voltage of $v_o(t) = 5t^4 + 12t^2 + 6$ given an input voltage source of $v_s(t) = 10t^3 + 12t$. You can use any combination of capacitors, batteries, and switches, but you may only use 10 kΩ resistors.

24. Design an op-amp circuit that solves the differential equation:

$$3\frac{d^2v}{dt^2} + 4\frac{dv}{dt} + 2v(t) = 2t.$$

Assume $v(0) = 0$ and $v'(0) = 2$ V. You can use any combination of capacitors, batteries, switches, and resistors (up to 10 kΩ).

25. Design an op-amp circuit that can generate the function $v(t) = e^{-t}(\cos 2t + 1/2 \sin 2t)$. You can use any combination of capacitors, batteries, switches, and resistors (up to 10 kΩ).

26. Using only switches, 10 μF capacitors, resistors with 100 Ω $\leq R \leq$ 10 kΩ, and four op-amps, design (and draw) a circuit to realize the following equation, where v_1, v_2, v_3, v_4 are input voltage sources:

$$v_o = 10v_1 - 100v_2 + 4\frac{dv_3}{dt} - 7\int_0^t v_4(\tau)\, d\tau.$$

27. Design an op-amp circuit that can generate the function $v(t) = te^{-2t}u(t)$. You can use any combination of capacitors, batteries, switches, and resistors (up to 10 kΩ).

28. Analyze the op-amp circuit shown in Figure P8-28 to determine the time dependence of the output voltage v_o on the input voltages v_1 and v_2.

29. Analyze the op-amp circuit shown in Figure P8-29 to determine the time dependence of the output voltage v_o on the input voltages v_1 and v_2.

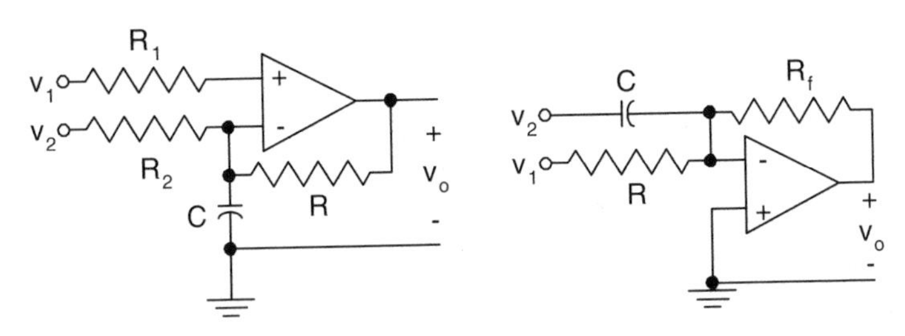

Figure P8-28

Figure P8-29

30. Analyze the op-amp circuit shown in Figure P8-30 to determine the time dependence of the output voltage v_o on the input voltages v_1, v_2, and v_3.

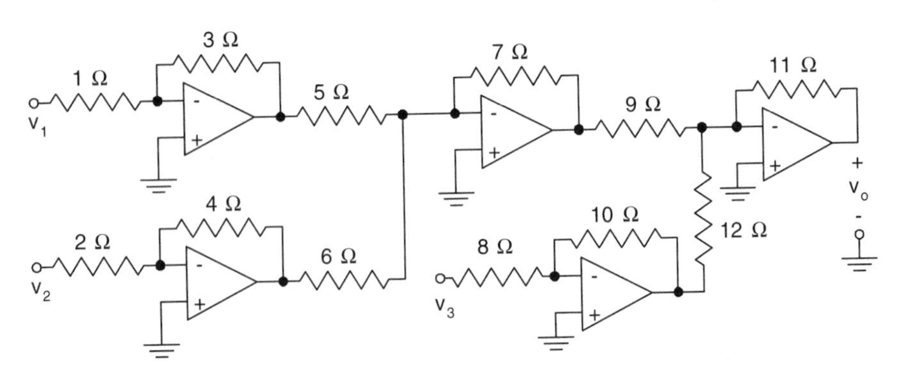

Figure P8-30

31. The op-amp circuit shown in Figure P8-31 (for $t > 0$) has been constructed to solve a specific differential equation. Find the differential equation which the output voltage v_o satisfies. (The circuit is shown for time $t > 0$; do not attempt to find the initial conditions.)

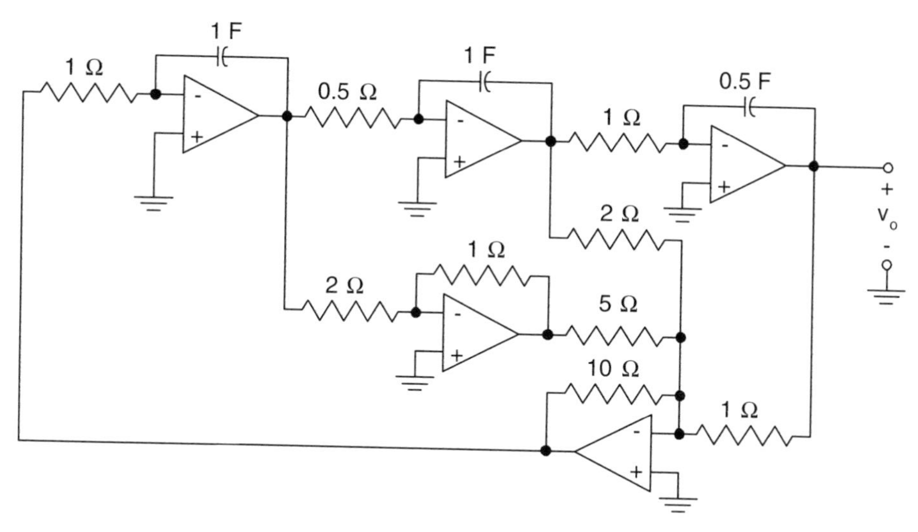

Figure P8-31

Use PSpice to solve Problems 32–41.

32. Find the voltages of all the nodes in the circuit shown in Figure P8-2.

33. Determine the value of the resistance R which results in a current of 1 A emerging from the positive terminal of the dependent voltage source in the circuit shown in Figure P8-6.

34. Remove the 2 Ω resistor from the circuit shown in Figure P8-3 and find the Norton equivalent circuit with respect to the open terminals. Let $R = 1\ \Omega$.

35. Remove the 5 Ω resistor from the circuit shown in Figure P8-7 and find the Thevenin equivalent circuit with respect to the open terminals.

36. Consider the circuit shown in Figure P8-14. Plot the node voltages N1, N2, and N3 as a function of time.

37. Design an op-amp circuit that solves the differential equation:

$$3\frac{d^2v}{dt^2} + 4\frac{dv}{dt} + 2v(t) = 2t.$$

You can use any combination of capacitors, batteries, switches, and resistors (up to 10 kΩ). Plot the output voltage and the analytic result for $v(t)$ over an appropriate time interval.

38. Design an op-amp circuit that can generate the function $v(t) = e^{-t}(\cos 2t + 1/2 \sin 2t)$. You can use any combination of capacitors, batteries, switches, and resistors (up to 10 kΩ). Plot the output voltage and the analytic result for $v(t)$ over an appropriate time interval.

39. Design an op-amp circuit that can generate the function $v(t) = te^{-2t}u(t)$. Plot the output voltage and the analytic result for $v(t)$ from $t = 0$ to $t = 5$ s.

40. Plot the output voltage of the op-amp circuit shown in Figure P8-29 when $v_1 = \cos(10t)$ V, $v_2 = 2\sin(5t)$ V, $R_f = 10$ kΩ, $R = 5$ kΩ, and $C = 100$ μF.

41. Plot the output voltage $v_o(t)$ for the op-amp circuit shown in Figure P8-31 over an appropriate time interval.

Chapter 9

Frequency Characteristics of Electric Circuits

9.1 Introduction

In earlier chapters, we discussed electric circuits that were driven by sinusoidal voltage and current sources. It was always assumed in our previous discussions that the frequencies of these sources were given and fixed. Now, we consider how responses of electric circuits vary with changing frequencies.

First, we focus on a simple circuit with an ac voltage source, resistor, inductor, and capacitor all connected in series. We demonstrate that this circuit may exhibit an interesting phenomenon called resonance. The resonance condition (that is, when the current is in phase with the source voltage) is discussed in detail. We then introduce the notion of a *filter* and demonstrate the properties of four different filter types by evaluating the voltage signals across the various passive components as a function of frequency. An approximate method enabling the quick plotting of the transfer function amplitude and phase as functions of frequency is presented. This method is based on the logarithmic graphs that are known as Bode plots.

Active filters, which utilize op-amp circuits that contain only resistors and capacitors, are introduced. We demonstrate that each of the four filter types can be realized by some combination of op-amp circuits. Here the fact that the input and output stages of op-amp circuits are decoupled from each other is exploited to design cascaded circuits. The chapter is closed with the development of a procedure for the synthesis of transfer functions by using active-*RC* circuits. The procedure utilizes poles and zeros of transfer functions. Generic op-amp circuits for the realization of pole factors and zero-factors are introduced, and the synthesis of the transfer functions is accomplished by cascading these generic circuits.

9.2 Resonance

We shall study the simple *RLC* circuit shown in Figure 9.1. We assume that $v_s(t)$ is an ac voltage source:

$$v_s(t) = V_{ms}\cos(\omega t + \phi_s), \tag{9.1}$$

and we shall be interested in the response of this circuit as a function of frequency ω. The current $i(t)$ produced by the above voltage source is sinusoidal as well:

$$i(t) = I_m\cos(\omega t + \phi_I). \tag{9.2}$$

However, the peak value, I_m, and initial phase, ϕ_I, of the current depend on the value of ω. The relationship between these quantities and ω is easily found by using the phasor technique:

$$v_s(t) \rightarrow \hat{V}_s = V_{ms}e^{j\phi_s}, \tag{9.3}$$

$$i(t) \rightarrow \hat{I}(\omega) = I_m(\omega)e^{j\phi_I(\omega)}, \tag{9.4}$$

$$Z(\omega) = R + j\left(\omega L - \frac{1}{\omega C}\right) = R + jX(\omega). \tag{9.5}$$

By using the impedance-type relationship between voltage and current, we find that:

$$\hat{I}(\omega) = \frac{\hat{V}_s}{Z(\omega)}, \tag{9.6}$$

$$I_m(\omega) = \frac{V_{ms}}{|Z(\omega)|} = \frac{V_{ms}}{\sqrt{R^2 + (\omega L - \frac{1}{\omega C})^2}}, \tag{9.7}$$

$$\phi_I(\omega) = \phi_s - \phi(\omega), \quad \tan\phi(\omega) = \frac{X(\omega)}{R}. \tag{9.8}$$

The frequency dependence of I_m and ϕ_I is clear from equations (9.7) and (9.8).

We are interested in finding the particular frequency ω_0 which causes the polar angle of the impedance to be equal to zero. Mathematically, we seek such values of

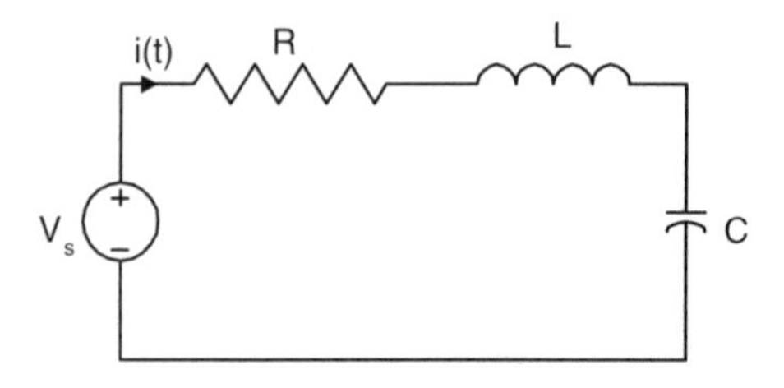

Figure 9.1: A simple *RLC* circuit.

ω_0 that satisfy:

$$\phi(\omega_0) = 0. \tag{9.9}$$

Equations (9.9) and (9.8) imply that the reactance must be equal to zero:

$$X(\omega_0) = 0. \tag{9.10}$$

From (9.5) and (9.10), we find:

$$\omega_0 L - \frac{1}{\omega_0 C} = 0, \tag{9.11}$$

$$\omega_0 = \frac{1}{\sqrt{LC}}. \tag{9.12}$$

Thus, at the frequency ω_0, the circuit behaves as if it contains only a resistor. The contributions of the inductance and the capacitance are canceled out. As a result, the initial phase of the current is equal to the initial phase of the voltage source:

$$\phi_I(\omega_0) = \phi_s. \tag{9.13}$$

In other words, the voltage and current are in phase. This phenomenon is called *resonance*. The frequency ω_0 at which resonance occurs is called the *resonant frequency*.

A circuit possesses a few interesting properties when at resonance. First, the impedance achieves its minimum absolute value at the resonant frequency:

$$|Z(\omega_0)| = \min_\omega |Z(\omega)| = R. \tag{9.14}$$

A direct result of equation (9.14) is that the peak value of the current achieves its maximum value at the resonant frequency:

$$I_m(\omega_0) = \max_\omega I_m(\omega) = \frac{V_{ms}}{R}. \tag{9.15}$$

Second, the voltages across the inductor and the capacitor have an interesting relationship at resonance. To see this, we recall the following expressions for these voltages:

$$\hat{V}_L(\omega) = j\omega L\hat{I}(\omega), \tag{9.16}$$

$$\hat{V}_C(\omega) = -\frac{j}{\omega C}\hat{I}(\omega). \tag{9.17}$$

By combining (9.11) with (9.16) and (9.17), we arrive at the following relationship:

$$\hat{V}_L(\omega_0) = j\omega_0 L\hat{I}(\omega_0) = \frac{j}{\omega_0 C}\hat{I}(\omega_0) = -\hat{V}_C(\omega_0), \tag{9.18}$$

$$\hat{V}_L(\omega_0) = -\hat{V}_C(\omega_0). \tag{9.19}$$

This means that at resonance the phase difference between these voltages is 180°, while the peak values are the same. As a result, the sum of the voltages across the inductor and the capacitor is equal to zero at any instant of time. Thus, we can conclude that the voltage across the resistor is equal to the source voltage:

$$\hat{V}_R(\omega_0) = \hat{V}_s = R\hat{I}(\omega_0). \tag{9.20}$$

This leads to some interesting consequences. In particular, we examine the ratio of the peak values of the voltages across the energy storage elements to the peak value of the voltage source:

$$\frac{V_{Lm}(\omega_0)}{V_{ms}} = \frac{V_{Cm}(\omega_0)}{V_{ms}} = \frac{\omega_0 L I_m(\omega_0)}{R I_m(\omega_0)} = \frac{\omega_0 L}{R}. \tag{9.21}$$

Thus, we come to what may seem to be a startling conclusion: if $\omega_0 L \gg R$, then $V_{Lm}(\omega_0) \gg V_{ms}$. In other words, *the voltage across the energy storage elements could attain peak values that are far greater than the source voltage itself*. This "amplification" depends solely on the values of the circuit elements and the frequency of the voltage source. This result suggests that it is unwise to touch exposed ac circuits even if the voltage source is known to be relatively small. Voltages across energy storage elements in the circuit may be much larger than the source voltage and may cause serious harm.

Another property of a circuit at resonance is that all supplied power is consumed; none of it is stored. This is because the circuit behaves as a pure resistor. As far as the inductor and capacitor are concerned, at resonance they continuously *exchange* the electromagnetic energy without involving the source in this process.

When a circuit is not operating at its resonant frequency ω_0, the reactive part of the impedance will act as a capacitance or as an inductance depending on whether the operating frequency ω is less than or greater than ω_0.
If $\omega < \omega_0$,

$$\omega L - \frac{1}{\omega C} < 0, \tag{9.22}$$

$$X(\omega) < 0, \tag{9.23}$$

and the current leads the voltage and we deal with "capacitive" reactance.
If $\omega > \omega_0$,

$$\omega L - \frac{1}{\omega C} > 0, \tag{9.24}$$

$$X(\omega) > 0, \tag{9.25}$$

and the current lags behind the voltage and we deal with "inductive" reactance.

EXAMPLE 9.1 For the circuit shown in Figure 9.1, we assume $R = 10\ \Omega$, $L = 1$ mH, $C = 470$ nF, $V_{ms} = 10$ V, and $\phi_s = 0$. We intend to calculate the circuit's resonant frequency, the maximum voltage across the inductor at resonance, and the magnitude and phase of the current at $\omega_0/2$, ω_0, and $3\omega_0$.

Table 9.1: Current magnitude and phase values.

Frequency ω	Current magnitude $I_m(\omega)$	Current phase $\phi_I(\omega)$
$\omega_0/2$	143 mA	$+81.78°$
ω_0	1 A	$0°$
$3\omega_0$	81 mA	$-85.34°$

The resonant frequency is given by equation (9.12) as:

$$\omega_0 = 1/\sqrt{(4.7 \times 10^{-7})(1 \times 10^{-3})} = 4.61 \times 10^4 \text{ rad/s,} \tag{9.26}$$

The maximum voltage across the inductor is given by expression (9.21) as:

$$V_{Lm}(\omega_0) = V_{ms}\omega_0 L/R = 10 \times 4.61 \times 10^4 \times 10^{-3}/10 = 46.1 \text{ V!} \tag{9.27}$$

We can rewrite the expressions for the magnitude and phase of the current via equations (9.7), (9.8), and (9.12) as follows:

$$I_m(\omega) = \frac{V_{ms}/R}{\sqrt{1 + \left[\omega L/R\right]^2 \left[1 - (\omega_0/\omega)^2\right]^2}} \tag{9.28}$$

and

$$\phi_I(\omega) = -\tan^{-1}\left\{\frac{\omega L}{R}\left[1 - (\omega_0/\omega)^2\right]\right\}. \tag{9.29}$$

With these two equations we can quickly fill out the entries in Table 9.1, which completes this problem. ■

9.3 Passive Filters

The behavior of a circuit can be changed appreciably by varying the frequency of excitation. By exploiting this fact, a circuit can be used as a *filter*. In general, a filter is a device designed to selectively let signals at some frequencies pass through while suppressing signals at other frequencies. When applied to electric circuits, a filter is a circuit that lets some voltage or current signals pass through while attenuating other signals based on the frequency of those signals. We consider below four types of filters: high-pass filters, low-pass filters, band-pass filters, and band-notch filters. A high-pass filter allows only high-frequency signals to pass through, while a low-pass filter permits the passage of only low-frequency signals. A band-pass filter limits the passage of signals to a somewhat narrow band of frequencies around the resonant frequency, while a band-notch filter attenuates only the signals near the resonant frequency. Each of these filters can be realized by the same circuit shown in Figure 9.1.

9.3.1 High-Pass Filter

To demonstrate the high-pass filtering properties of the above circuit, we consider specifically the voltage across the terminals of the inductor. By using the familiar impedance-type relationship between voltage and current phasors, we can express this voltage in phasor form as follows:

$$\hat{V}_L(\omega) = j\omega L\hat{I}(\omega) = \frac{j\omega L\hat{V}_s}{R + j(\omega L - \frac{1}{\omega C})}. \tag{9.30}$$

We now define a *transfer function* $H_L(\omega)$ as the ratio of $\hat{V}_L$ to $\hat{V}_s$:

$$\hat{H}_L(\omega) = \frac{\hat{V}_L(\omega)}{\hat{V}_s} = \frac{j\omega L}{R + j(\omega L - \frac{1}{\omega C})}. \tag{9.31}$$

Equation (9.31) defines a complex transfer function whose absolute value is:

$$|\hat{H}_L(\omega)| = \frac{\omega L}{\sqrt{R^2 + (\omega L - \frac{1}{\omega C})^2}} = \frac{\omega^2 LC}{\sqrt{\omega^2 R^2 C^2 + (\omega^2 LC - 1)^2}}. \tag{9.32}$$

It is clear from (9.32) that $|\hat{H}_L(\omega)|$ is equal to zero when $\omega = 0$. As ω is increased, $|\hat{H}_L(\omega)|$ is increased as well and reaches its maximum value. With further increase of ω, $|\hat{H}_L(\omega)|$ is somewhat decreased and asymptotically reaches the value that is equal to 1. The plot of $|\hat{H}_L(\omega)|$ is shown in Figure 9.2. It is clear from this plot that if we measure the voltage across inductor terminals, then the output signals with low frequencies will be suppressed, while the output signals at high frequencies will have almost the same peak values as the input signals. For this reason, if the inductor terminals are used as the output terminals, then the circuit shown in Figure 9.1 acts as a high-pass filter. We shall further investigate the function $|\hat{H}_L(\omega)|$. Namely, we shall find the answer to the following question. What is the maximum value of $|\hat{H}_L(\omega)|$ and at what frequency (relative to ω_0) does it occur?

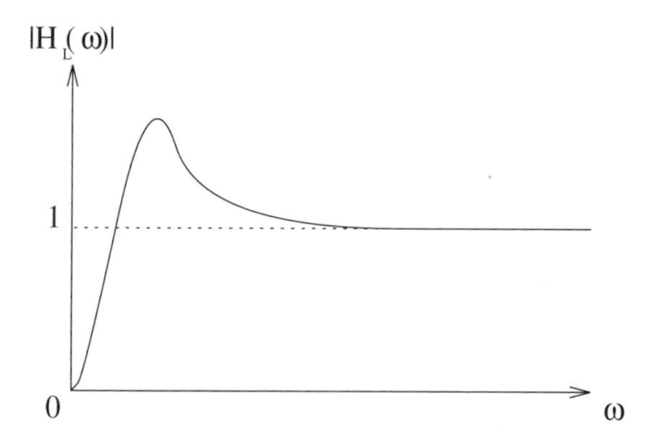

Figure 9.2: Plot of $|H_L(\omega)|$ vs. ω.

If we insert equation (9.12) into (9.32) and define the damping factor ζ by $2\zeta/\omega_0 = RC$, then we obtain (with the abbreviation $x = (\omega/\omega_0)^2$):

$$|H_L(x)| = \frac{x}{\sqrt{4\zeta^2 x + (x-1)^2}}. \tag{9.33}$$

We find the maximum by solving for the zero of the derivative of $|\hat{H}_L(x)|$ with respect to x:

$$\omega_{max}/\omega_0 = 1/\sqrt{1-2\zeta^2} = 1/\sqrt{1-(\omega_0 RC)^2/2} \tag{9.34}$$

and

$$|\hat{H}_L(\omega_{max})| = \frac{1}{\sqrt{(\omega_0 RC)^2 - (\omega_0 RC)^4/4}}. \tag{9.35}$$

There are two things to note here. First, the maximum occurs at the resonant frequency only when $R = 0$, in which case the maximum is infinite. Otherwise, the maximum frequency exceeds the resonant frequency. Second, if the resistance is so large that $R > \sqrt{2L/C}$, then there is no local maximum and the transfer function magnitude is a monotonically increasing function of frequency which asymptotically approaches 1. For the parameters of the previous example ($R = 10\ \Omega$, $L = 1$ mH, $C = 470$ nF), we find that $\zeta = 0.108$, $\omega_{max}/\omega_0 = 1.012$, and $|H_L(\omega_{max})| = 4.64$. This represents a peak voltage of 46.4 V, which is only slightly higher than the calculated voltage at resonance.

9.3.2 Low-Pass Filter

To arrive at the low-pass filter, we examine the voltage across the terminals of the capacitor:

$$\hat{V}_C(\omega) = \frac{\hat{I}(\omega)}{j\omega C} = \hat{V}_s \frac{1}{j\omega C[R + j(\omega L - \frac{1}{\omega C})]}. \tag{9.36}$$

The transfer function, $\hat{H}_C(\omega)$, defined as the ratio of $\hat{V}_C$ to $\hat{V}_s$, is:

$$\hat{H}_C(\omega) = \frac{\hat{V}_C(\omega)}{\hat{V}_s} = \frac{1}{j\omega C[R + j(\omega L - \frac{1}{\omega C})]}. \tag{9.37}$$

The absolute value of $\hat{H}_C(\omega)$ is given by:

$$|\hat{H}_C(\omega)| = \frac{1}{\omega C\sqrt{R^2 + (\omega L - \frac{1}{\omega C})^2}} = \frac{1}{\sqrt{\omega^2 R^2 C^2 + (\omega^2 LC - 1)^2}}. \tag{9.38}$$

It is clear from (9.38) that $|H_C(\omega)|$ is equal to 1 when $\omega = 0$. As ω is increased, $|\hat{H}_C(\omega)|$ is increased as well and reaches its maximum value. With further increase

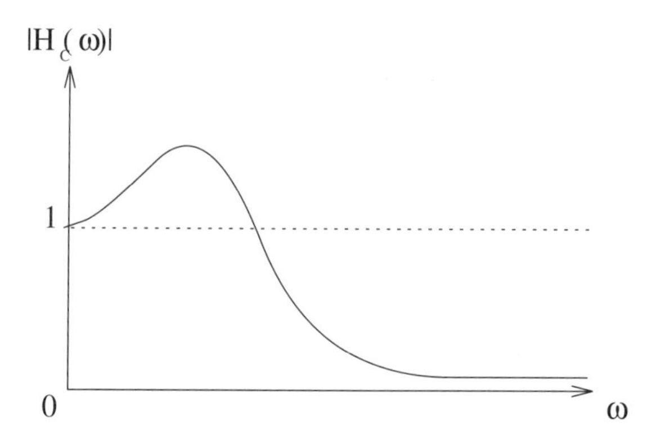

Figure 9.3: A plot of $|H_c(\omega)|$ vs. ω.

of ω, $|\hat{H}_C(\omega)|$ monotonically decreases and asymptotically reaches zero. A typical plot of $|\hat{H}_C(\omega)|$ is shown in Figure 9.3. It is evident from this plot that if the capacitor terminals are viewed as output terminals, then the output signal at high frequencies will be suppressed, while the output signal at low frequencies will have the same (or even somewhat larger) peak values as the input signal. Thus, if the capacitor terminals are used as the output terminals, then the circuit shown in Figure 9.1 acts as a low-pass filter.

Next, we consider the following question. What is the maximum value of $|H_C(\omega)|$ and at what frequency (relative to ω_0) does it occur?

If we insert (9.12) into (9.38) and define $2\zeta/\omega_0 = RC$ and $x = (\omega/\omega_0)^2$, then we obtain:

$$|\hat{H}_C(x)| = \frac{1}{\sqrt{4\zeta^2 x + (x-1)^2}}. \tag{9.39}$$

We find the maximum by solving for the zero of the derivative of $|\hat{H}_C(x)|$ with respect to x:

$$\omega_{\max}/\omega_0 = \sqrt{1-2\zeta^2} = \sqrt{1-(\omega_0 RC)^2/2} \tag{9.40}$$

and

$$|\hat{H}_C(\omega_{\max})| = \frac{1}{\sqrt{(\omega_0 RC)^2 - (\omega_0 RC)^4/4}}. \tag{9.41}$$

Again, the maximum occurs at the resonant frequency only when $R = 0$, in which case the maximum is infinite. Otherwise, the maximum frequency is below the resonant frequency. Furthermore, if the resistance is so large that $R > \sqrt{2L/C}$, then there is no local maximum and the transfer function magnitude is a monotonically decreasing function of frequency. For the parameters of the previous example ($R = 10\ \Omega$, $L = 1$ mH, $C = 470$ nF), we still have $\zeta = 0.108$ and $|\hat{H}_C(\omega_{\max})| = 4.64$, but now $\omega_{\max}/\omega_0 = 0.988$.

9.3.3 Band-Pass Filter

To arrive at the band-pass filter, we consider the voltage across the terminals of the resistor:

$$\hat{V}_R(\omega) = R\hat{I}(\omega) = \hat{V}_s \frac{R}{R + j(\omega L - \frac{1}{\omega C})}. \tag{9.42}$$

The transfer function, $\hat{H}_R(\omega)$, defined as the ratio of $\hat{V}_R$ to $\hat{V}_s$, is:

$$\hat{H}_R(\omega) = \frac{\hat{V}_R(\omega)}{\hat{V}_s} = \frac{R}{R + j(\omega L - \frac{1}{\omega C})}. \tag{9.43}$$

The absolute value of $|\hat{H}_R(\omega)|$ is given by:

$$|\hat{H}_R(\omega)| = \frac{R}{\sqrt{R^2 + (\omega L - \frac{1}{\omega C})^2}} = \frac{\omega RC}{\sqrt{\omega^2 R^2 C^2 + (\omega^2 LC - 1)^2}}. \tag{9.44}$$

It is clear from (9.44) that $|\hat{H}_R(\omega)|$ is equal to zero when $\omega = 0$. As ω is increased, $|\hat{H}_R(\omega)|$ is increased as well and reaches its maximum value of one at the resonance frequency ω_0 given by (9.12). As the frequency ω is further increased, $|\hat{H}_R(\omega)|$ is montonically decreased and asymptotically approaches zero. The plot of $|\hat{H}_R(\omega)|$ is shown in Figure 9.4. It is evident from this plot that if the resistor terminals are viewed as the output terminals, then the output signals of low and high frequencies will be suppressed, while the output signals at frequencies centered around ω_0 will have approximately the same peak values as the input signals. Thus, if the resistor terminals are used as the output terminals, the circuit shown in Figure 9.1 acts as a band-pass filter.

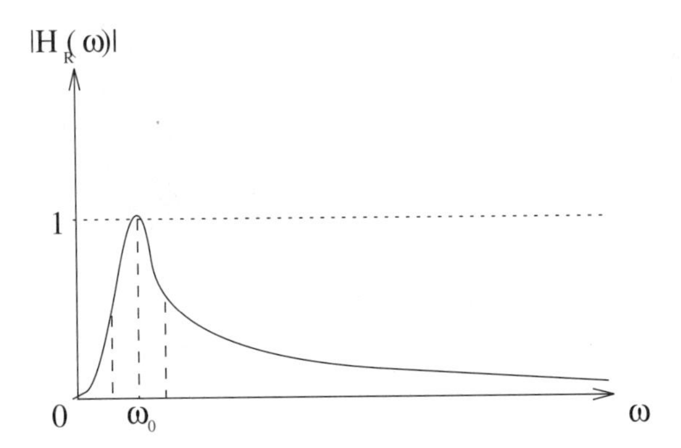

Figure 9.4: A plot of $|H_R(\omega)|$ vs. ω.

9.3.4 Band-Notch Filter

This filter comes from considering the net output voltage across the inductor and the capacitor. According to KVL, this voltage is equal to the source voltage minus the resistor voltage. Thus, from equation (9.42) we get:

$$\hat{H}_{LC}(\omega) = \hat{V}_{LC}(\omega)/\hat{V}_s = \frac{j(\omega L - 1/\omega C)}{R + j(\omega L - 1/\omega C)} \tag{9.45}$$

which results in the following expression for the magnitude of the transfer function:

$$|\hat{H}_{LC}(\omega)| = \frac{|\omega^2 LC - 1|}{\sqrt{(\omega RC)^2 + (\omega^2 LC - 1)^2}}. \tag{9.46}$$

Comparing this equation to (9.44), we see that $|\hat{H}_{LC}(\omega)| = \sqrt{1 - |\hat{H}_R(\omega)|^2}$. Consequently, this transfer function has a minimum of zero at the resonant frequency and tends toward a maximum of one as the frequency heads toward either zero or infinity. This circuit arrangement is called a band-notch filter because it allows signals to pass at all frequencies except for a small band of frequencies near the circuit resonance.

9.4 Bode Plots

Bode plots are piecewise straight-line approximations for the frequency dependence of the magnitude and phase of a transfer function. They will be demonstrated through the discussion of several examples.

EXAMPLE 9.2 This example deals with the circuit shown in Figure 9.5. From the voltage divider rule, we get,

$$\hat{H}(\omega) = \frac{\hat{V}_0}{\hat{V}_s} = \frac{1/j\omega C}{R + 1/j\omega C} = \frac{1}{1 + j\omega RC} = \frac{1}{1 + j\omega/\omega_0} \tag{9.47}$$

where $\omega_0 = 1/RC$. This transfer function has a simple pole in the complex plane. The magnitude of the transfer function is $|\hat{H}(\omega)| = 1/\sqrt{1 + (\omega/\omega_0)^2}$, while the phase is given by $\phi = -\tan^{-1}(\omega/\omega_0)$. We convert the magnitude to the decibel (dB) scale by taking the logarithm of the transfer function magnitude and multiplying the

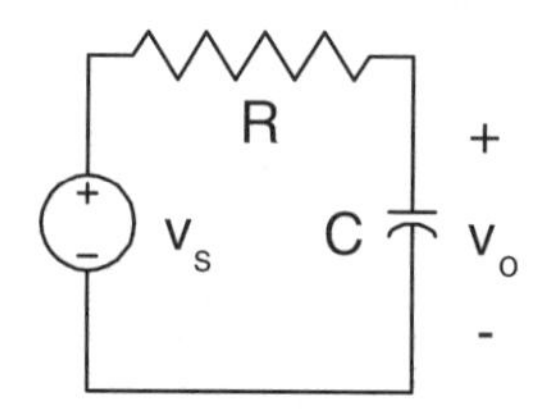

Figure 9.5: Example of a low-pass filter.

result by 20:

$$|\hat{H}(\omega)|(\text{dB}) = 20\log_{10}\left[\frac{1}{\sqrt{1+(\omega/\omega_0)^2}}\right] = -20\log_{10}\left[\sqrt{1+(\omega/\omega_0)^2}\right]$$
$$= -10\log_{10}\left(1+(\omega/\omega_0)^2\right). \tag{9.48}$$

The last two equalities result from standard logarithmic manipulations which are used here to get the answer in a simple form. The factor of 20 is used (rather than 10) because decibels are usually taken as an indication of relative power rather than relative voltage.

Notice that in the low-frequency limit ($\omega << \omega_0$), we get that $|\hat{H}|(\text{dB}) \approx -10\log_{10}(1) = 0$ dB. In contrast, we can neglect the constant in the high-frequency limit ($\omega >> \omega_0$) to approximate the transfer function magnitude as

$$|\hat{H}|(\text{dB}) \approx -10\log_{10}((\omega/\omega_0)^2) = -20\log_{10}(\omega/\omega_0)\text{ dB}.$$

On a plot of $|\hat{H}|$ (dB) versus frequency, where ω is plotted on a logarithmic scale, this high-frequency approximation represents a straight line which decreases by 20 dB for every order of magnitude increase in ω. We approximate the actual magnitude by combining the two straight lines corresponding to the low- and high-frequency approximations. These two straight lines intersect at $\omega = \omega_0$, which is sometimes referred to as the breakpoint (or corner) frequency. A plot of the actual magnitude and its approximation is shown in Figure 9.6. We can see that the greatest discrepancy between the two curves occurs at the breakpoint frequency. The approximate curve estimates the transfer function magnitude to be 0 dB but the exact result is $|\hat{H}|(\text{dB}) = -10\log_{10}(2) \approx -3$ dB. Therefore, the two curves agree to within 3 dB

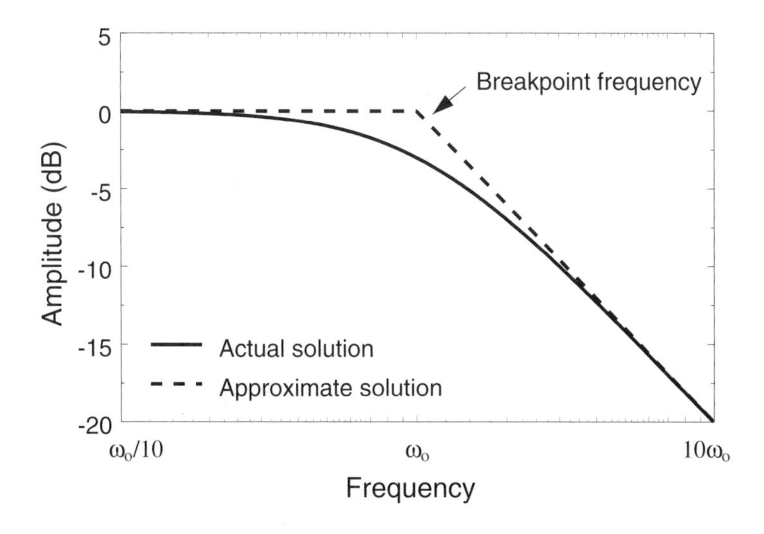

Figure 9.6: Bode plot of the magnitude of the transfer function.

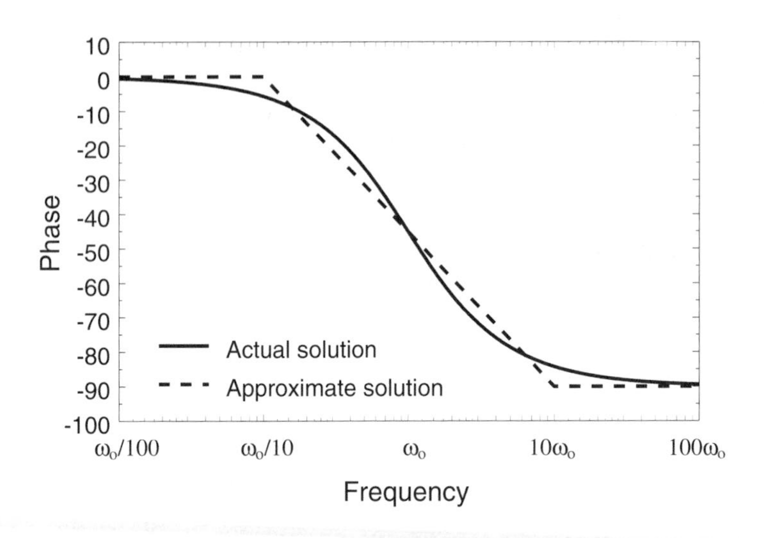

Figure 9.7: Bode plot of the phase of the transfer function.

everywhere. Because of this, the breakpoint (corner) frequency is also often referred to as the 3 dB point.

A plot of the phase of $\hat{H}(\omega)$ as a function of frequency is given in Figure 9.7. The curve reaches $0°$ as the frequency approaches zero and asymptotically approaches $-90°$ as the frequency tends to infinity. This curve is usually approximated by a piecewise linear function that is equal to $0°$ for frequencies up to a decade below the corner frequency and is equal to $-90°$ for all frequencies exceeding the corner frequency by an order of magnitude. These two lines are connected by a third line with a slope of $-45°$ per decade which goes through the breakpoint frequency at $-45°$. The rationale for the above approximate plot is not as firmly established as for the transfer function magnitude approximation; however, it is a fairly simple and reasonably accurate approximation nonetheless. ■

EXAMPLE 9.3 We revisit the second-order high-pass transfer function defined by equation (9.31). Recall that after we defined the resonant frequency $\omega_0 = 1/\sqrt{LC}$ and the damping factor $2\zeta/\omega_0 = RC$, we got the transfer function magnitude in the form:

$$|\hat{H}(\omega)| = \frac{(\omega/\omega_0)^2}{\sqrt{4\zeta^2(\omega/\omega_0)^2 + ((\omega/\omega_0)^2 - 1)^2}}.$$

When we convert this magnitude to the dB scale we obtain two distinct terms:

$$|\hat{H}(\omega)|(\text{ dB}) = +40\log_{10}(\omega/\omega_0) - 10\log_{10}\{4\zeta^2(\omega/\omega_0)^2 + [(\omega/\omega_0)^2 - 1]^2\}.$$

This clearly reveals the advantage of using a logarithmic scale. Namely, we can construct an approximation for the transfer function magnitude by graphically combining the two individual terms. The first term needs no approximation at all. It generates a simple straight line with a slope of +40 dB/decade which passes through 0 dB at ω_0. The second term can be approximated on the basis of its behavior in the high- and low-frequency limits. In the low-frequency limit ($\omega << \omega_0$), we find the second term is equal to $-10\log_{10}(1) = 0$ dB. In the high-frequency limit ($\omega >> \omega_0$) we approximate the second term as $-10\log_{10}((\omega/\omega_0)^4) = -40\log_{10}(\omega/\omega_0)$ dB. The two approximations intersect at ω_0. In Figure 9.8 we plot the first term, the piecewise straight-line approximation of the second term, their sum (dashed line), and the actual function (solid line). When $\omega < \omega_0$, the second term is approximated as zero, so the net result is just the line with the +40 dB slope. When $\omega > \omega_0$, the two adding terms cancel one another, resulting in a flat line at 0 dB. Note that this high-pass filter rolls off at twice the rate of the previous example. This is because in the previous example we dealt with a first-order pole, while in this example we have a second-order pole. To appreciate the level of discrepancy between the piecewise straight-line approximation and actual curve, we note that at resonance ($\omega = \omega_0$) the approximation estimates a gain of 0 dB, while the actual value depends on ζ: $|\hat{H}|(\text{dB}) = -20\log_{10}(2\zeta)$. Exact curves for $\zeta = 1$ and $\zeta = 1/4$ are given in the figure.

The frequency dependence of the phase of the transfer function is given by

$$\phi = \tan^{-1}\left\{\frac{2\zeta\omega/\omega_0}{\left[(\omega/\omega_0)^2 - 1\right]}\right\}$$

and is plotted in Figure 9.9. The curve reaches 180° as the frequency approaches zero and it asymptotically approaches 0° as the frequency tends to infinity. A straight-line approximation is given by a piecewise linear function that is equal to 180° for frequencies up to a decade below the breakpoint and is equal to 0° for all frequencies exceeding the breakpoint by an order of magnitude. These two lines are connected

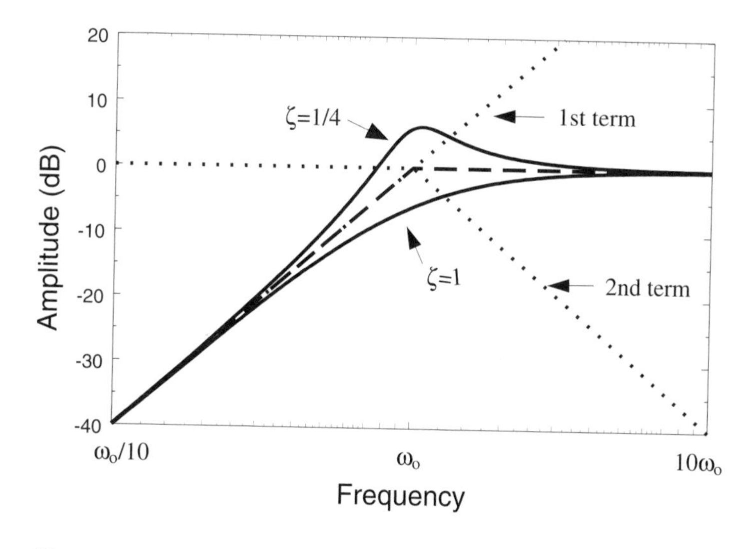

Figure 9.8: Bode plot magnitude of a second-order high-pass filter.

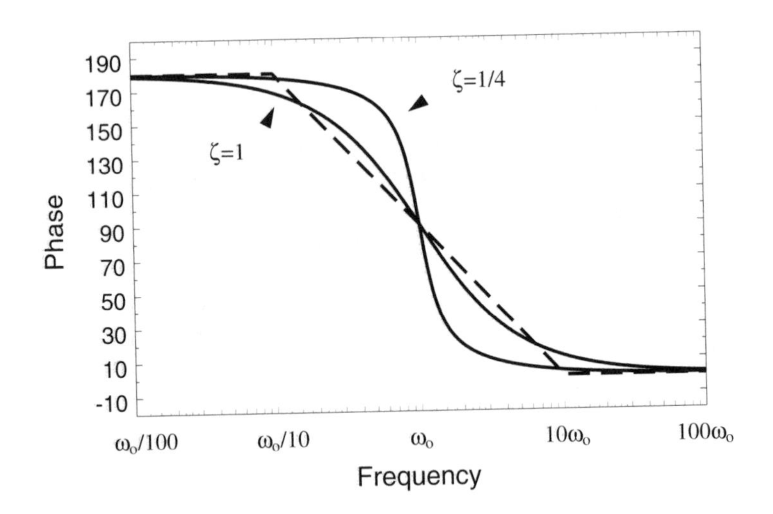

Figure 9.9: Bode plot phase of a second-order high-pass filter.

by a third line with a slope of $-90°$ per decade which goes through the breakpoint frequency at $90°$. Exact curves for $\zeta = 1$ and $\zeta = 1/4$ are given in the figure. ■

EXAMPLE 9.4 Consider the circuit shown in Figure 9.10. To find the transfer function $H(\omega) = \hat{V}_0/\hat{V}_s$, we use the standard mesh analysis with the three meshes indicated in the figure. The matrix equation is found to be:

$$\begin{bmatrix} 1 - j/\omega & j/\omega & 0 \\ j/\omega & 2j(\omega - 1/\omega) & j/\omega \\ 0 & j/\omega & 1 - j/\omega \end{bmatrix} \begin{bmatrix} \hat{\mathcal{I}}_1 \\ \hat{\mathcal{I}}_2 \\ \hat{\mathcal{I}}_3 \end{bmatrix} = \begin{bmatrix} \hat{V}_s \\ 0 \\ 0 \end{bmatrix}. \tag{9.49}$$

Because the output is the voltage across a 1 Ω resistor, we find:

$$\hat{H}(\omega) = \frac{\hat{\mathcal{I}}_3}{\hat{V}_s} = \frac{1/2}{(1 + j\omega)[1 + j\omega + (j\omega)^2]}.$$

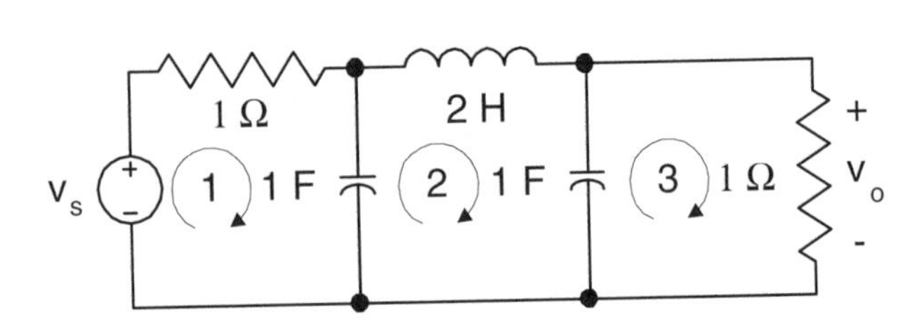

Figure 9.10: Example of a Butterworth circuit.

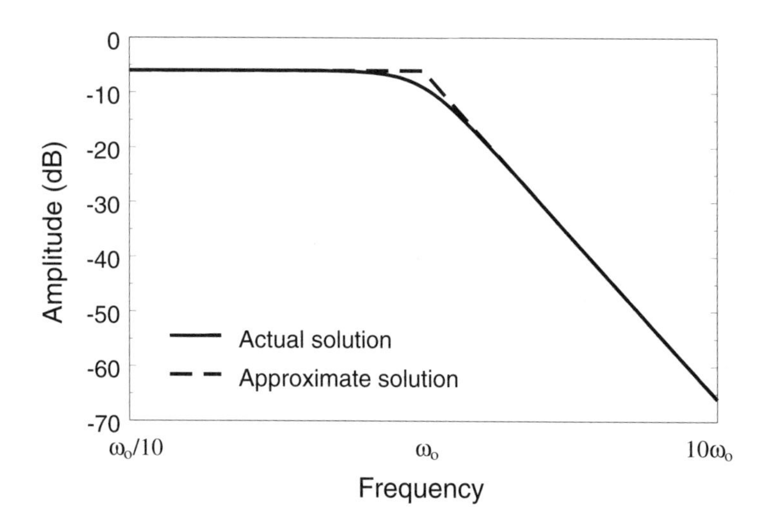

Figure 9.11: Bode plot magnitude of a third-order Butterworth filter.

The last result is obtained by applying Gaussian elimination to the above matrix equation. We will discuss the Bode plot only for the magnitude and leave the Bode plot for the phase as an exercise for the reader. The magnitude of the transfer function is: $|\hat{H}(\omega)| = 1/(2\sqrt{1+\omega^6})$. This function is a third-order low-pass filter. The circuit is called a third-order Butterworth filter and has the property that $|\hat{H}(\omega)|$ is maximally flat, i.e., has many derivatives equal to zero at $\omega = 0$. The approximate and exact plots are given in Figure 9.11. The constant line is at -6 dB as a result of the factor of 1/2 in the numerator. The breakpoint frequency ω_0 is 1 rad/s, where the two curves differ by 3 dB. The actual plot is asymptotic to the approximate line with a slope of -60 dB/decade. ■

EXAMPLE 9.5 Assume we have a circuit whose transfer function is given by:

$$\hat{H}(\omega) = \frac{j\omega(1 + j\omega/100)}{[1 + j\omega/10 + (j\omega/10)^2](1 + j\omega/1000)^2} = \frac{\hat{H}_1(\omega)\hat{H}_2(\omega)}{\hat{H}_3(\omega)\hat{H}_4(\omega)}. \tag{9.50}$$

Let us construct Bode plots for both the magnitude and the phase of this function. Because of the properties of logarithms, we can write:

$$|\hat{H}(\omega)|(\text{dB}) = |\hat{H}_1(\omega)|(\text{dB}) + |\hat{H}_2(\omega)|(\text{dB}) - |\hat{H}_3(\omega)|(\text{dB}) - |\hat{H}_4(\omega)|(\text{dB}).$$

Likewise, for the phase we have:

$$\arg(\hat{H}) = \arg(\hat{H}_1) + \arg(\hat{H}_2) - \arg(\hat{H}_3) - \arg(\hat{H}_4).$$

Thus, we can approximate each individual term separately and add them up at the end. Figure 9.12 shows the individual magnitude contributions and Figure 9.13 gives the phase contributions. $\hat{H}_1(\omega) = j\omega$ is similar to a factor we considered while

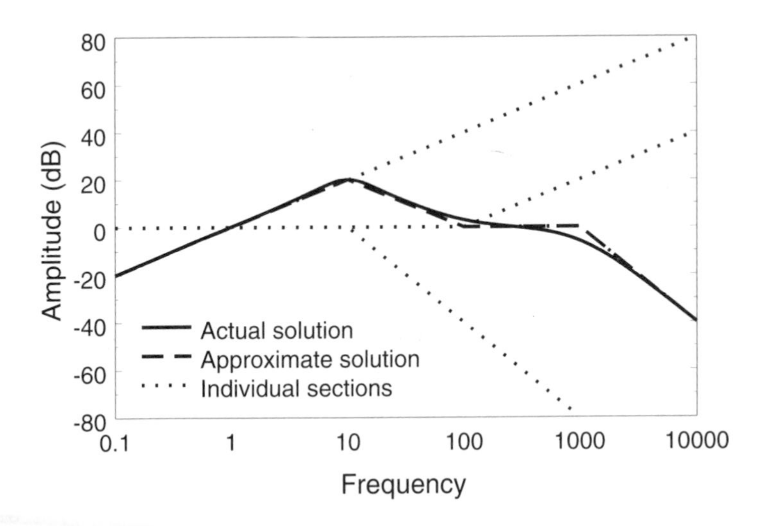

Figure 9.12: Bode plot magnitude of the transfer function given in equation (9.50).

investigating the high-pass filter and results in a straight line of slope 20 dB/decade which goes through 0 dB at 1 rad/s. The phase is a constant 90°. $\hat{H}_2(\omega) = 1 + j\omega/100$ has a simple zero with a breakpoint of $\omega_0 = 100$ rad/s. The magnitude Bode plot is equal to 0 dB until the breakpoint, then increases by 20 dB/decade. The phase begins at 0° and then increases to 90°. The two flat sections are linked by a line with a slope of 45° which passes through 45° at the breakpoint. $\hat{H}_3(\omega)$ is similar in form to the "second-order pole" transfer function we discussed in Example 9.3 with

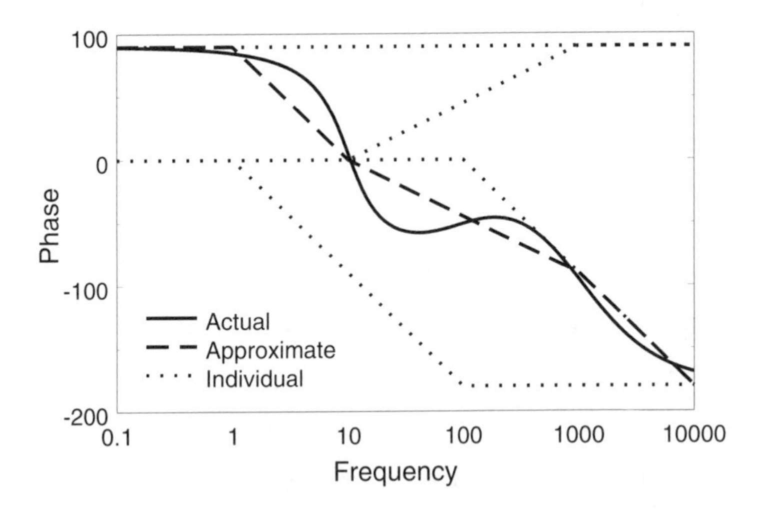

Figure 9.13: Bode plot phase of the transfer function given in equation (9.50).

$\zeta = 1/2$ and $\omega_0 = 10$. Thus, $-\hat{H}_3$ has a magnitude Bode plot that is flat at 0 dB until 10 rad/s, then decreases at -40 dB/decade. The phase ($-\arg(H_3)$) starts at 0°, then decreases to $-180°$ via a line with a slope of $-90°$ which passes through $-90°$ when $\omega = 10$ rad/s. The final part, $-\hat{H}_4$, is really the product of two "first-order pole" transfer functions. Therefore, the magnitude Bode plot is flat at 0 dB until the breakpoint frequency of 1000 rad/s. Afterwards it decreases at a rate of -40 dB/decade. The phase stays at 0° until $\omega = 100$ rad/s and decreases to $-180°$ by the time the frequency reaches 10,000 rad/s. The resulting Bode plots are also shown in the figure, as are the actual plots. It is quite clear that the approximation is fairly accurate and is considerably easier to reproduce than the actual function. ■

The following conclusion can be drawn from the above discussion. Bode plots are piecewise linear approximations for transfer function magnitudes and phases as functions of frequency on logarithmic scale. **The logarithmic scale allows one to represent Bode plots for complicated transfer functions as sums of very simple piecewise straight-line plots which are determined by the zeros and poles of the transfer functions.**

9.5 Active-*RC* Filters

The filters discussed in the previous sections have been realized by using circuits that contain only passive elements (resistors, inductors, and capacitors). For this reason, these circuits are called passive filters. The main deficiency of passive filters is the use of inductors. Inductors are "bulky" circuit elements which are difficult to miniaturize and which are not compatible with integrated circuit technology. Thus, it can be concluded that it is desirable to replace passive filters with some other circuits which will perform the same functions without using inductors. It is also desirable to combine filtering properties of electric circuits with amplification. It turns out that this can be accomplished by using resistors, capacitors, and operational amplifiers. As a result, a new class of electric circuits called active-*RC* filters emerges. *The main idea of the design of these filters is to come up with active-RC circuits which realize the transfer functions with the same frequency dependence as in the case of transfer functions* $H_L(\omega)$, $H_C(\omega)$, $H_R(\omega)$, and $H_{LC}(\omega)$.

For the sake of generality (and notational simplicity) we again introduce the complex frequencies. For time harmonic signals, the relationship is given by

$$s = j\omega. \tag{9.51}$$

Now, expressions (9.31), (9.37), (9.43), and (9.45) can be written as follows:

$$H_L(s) = \frac{sL}{R + sL + \frac{1}{sC}} = \frac{s^2LC}{s^2LC + sRC + 1}, \tag{9.52}$$

$$H_C(s) = \frac{1}{sC(R + sL + \frac{1}{sC})} = \frac{1}{s^2LC + sRC + 1}, \tag{9.53}$$

$$H_R(s) = \frac{R}{R + sL + \frac{1}{sC}} = \frac{sRC}{s^2LC + sRC + 1}, \tag{9.54}$$

$$H_{LC}(s) = \frac{s^2LC + 1}{s^2LC + sRC + 1}. \tag{9.55}$$

Transfer functions for active-*RC* high-pass, low-pass, band-pass, and band-notch filters should mimic the transfer functions (9.52), (9.53), (9.54), and (9.55), respectively. This means that they should have the following functional dependence on s:

$$H_{hp}(s) = \frac{s^2 a}{s^2 + sb + c}, \tag{9.56}$$

$$H_{lp}(s) = \frac{d}{s^2 + sb + c}, \tag{9.57}$$

$$H_{bp}(s) = \frac{sg}{s^2 + sb + c}, \tag{9.58}$$

$$H_{bn}(s) = \frac{s^2 e + f}{s^2 + sb + c}, \tag{9.59}$$

where the subscripts "hp," "lp," "bp," and "bn" stand for "high-pass," "low-pass," "band-pass," and "band-notch," respectively. Now, we demonstrate that the low-pass transfer function (9.57) can be realized by the active-*RC* circuit shown in Figure 9.14. This demonstration is also useful as an example of the analysis of electric circuits with operational amplifiers.

We begin our analysis of the circuit shown in Figure 9.14 by writing the nodal equation for node number 3, and we shall use the version of nodal equations developed for the circuits which contain voltage sources in series with admittances. According to this version we have:

$$\left(sC_1 + \frac{1}{R_1} + \frac{1}{R_2} + \frac{1}{R_3}\right)\hat{V}_3 - \frac{1}{R_2}\hat{V}_1 - \frac{1}{R_3}\hat{V}_o = \frac{1}{R_1}\hat{V}_{in}. \tag{9.60}$$

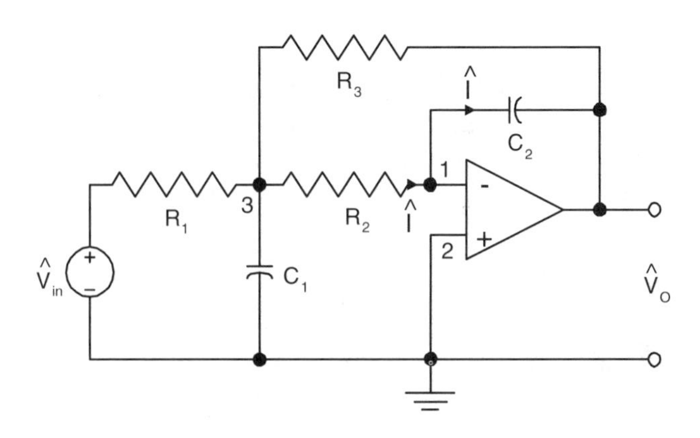

Figure 9.14: Active-*RC* low-pass filter.

Since the input impedance of operational amplifier is assumed to be infinite, the same current $\hat{I}$ flows through the resistance R_2 and capacitance C_2. This results in the following equation:

$$\hat{I} = \frac{\hat{V}_3 - \hat{V}_1}{R_2} = sC_2(\hat{V}_1 - \hat{V}_o). \tag{9.61}$$

Since the gain of the operational amplifier is assumed to be infinite, nodes 1 and 2 have equal potentials. Taking into account that noninverting terminal 2 is grounded, we obtain:

$$\hat{V}_1 = \hat{V}_2 = 0. \tag{9.62}$$

By substituting (9.62) into (9.61), we obtain:

$$\frac{\hat{V}_3}{R_2} = -sC_2\hat{V}_o, \tag{9.63}$$

which is tantamount to:

$$\hat{V}_3 = -sR_2C_2\hat{V}_o. \tag{9.64}$$

By substituting (9.62) and (9.64) into (9.60), we arrive at the following equation for $\hat{V}_o$:

$$-sR_1R_2C_2\left(sC_1 + \frac{1}{R_1} + \frac{1}{R_2} + \frac{1}{R_3}\right)\hat{V}_o - \frac{R_1}{R_3}\hat{V}_o = \hat{V}_{in}. \tag{9.65}$$

From (9.65), we derive the following expression for the transfer function:

$$\hat{H}(s) = \frac{\hat{V}_o}{\hat{V}_{in}} = \frac{-1}{s^2R_1R_2C_1C_2 + sR_1R_2C_2\left(\frac{1}{R_1} + \frac{1}{R_2} + \frac{1}{R_3}\right) + \frac{R_1}{R_3}}, \tag{9.66}$$

which can be further transformed as follows:

$$\hat{H}(s) = \frac{-\frac{1}{R_1R_2C_1C_2}}{s^2 + s\left(\frac{1}{R_1C_1} + \frac{1}{R_2C_1} + \frac{1}{R_3C_1}\right) + \frac{1}{R_2R_3C_1C_2}}. \tag{9.67}$$

Now, we easily observe that the transfer function (9.67) has the same functional dependence on s as the transfer function (9.57) for a low-pass filter. This proves that the circuit shown in Figure 9.14 is an active-*RC* low-pass filter. By using this circuit, we can also design active-*RC* band-pass and high-pass filters. An active-*RC* circuit which accomplishes this task is shown in Figure 9.15. It is clear that the left block of this circuit, which contains the operational amplifier 1, is identical to the circuit shown in Figure 9.14. For this reason:

$$\frac{\hat{V}_o^{(1)}}{\hat{V}_{in}} = \hat{H}_{lp}(s) = \frac{-\frac{1}{R_1R_2C_1C_2}}{s^2 + s\left(\frac{1}{R_1C_1} + \frac{1}{R_2C_1} + \frac{1}{R_3C_1}\right) + \frac{1}{R_2R_3C_1C_2}}. \tag{9.68}$$

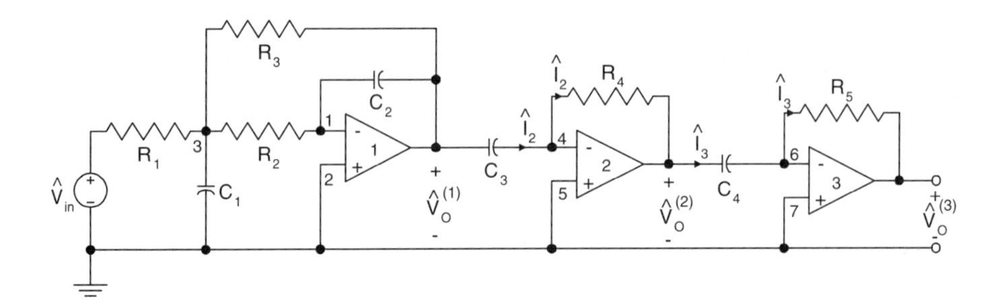

Figure 9.15: Active-*RC* filter circuit.

Next, we find the relation between $\hat{V}_o^{(1)}$ and $\hat{V}_o^{(2)}$. Since the input impedance of operational amplifier 2 is assumed to be infinite, we find that the same current $\hat{I}_2$ flows through capacitor C_3 and resistance R_4. Consequently, we have:

$$\hat{I}_2 = sC_3(\hat{V}_o^{(1)} - \hat{V}_4) = \frac{\hat{V}_4 - \hat{V}_o^{(2)}}{R_4}. \tag{9.69}$$

Next, we recall that

$$\hat{V}_4 = \hat{V}_5 = 0. \tag{9.70}$$

By substituting (9.70) into (9.69), we obtain:

$$sC_3\hat{V}_o^{(1)} = -\frac{\hat{V}_o^{(2)}}{R_4}, \tag{9.71}$$

which is tantamount to:

$$\frac{\hat{V}_o^{(2)}}{\hat{V}_o^{(1)}} = -sR_4C_3. \tag{9.72}$$

From (9.68) and (9.72), we derive:

$$\frac{\hat{V}_o^{(2)}}{\hat{V}_{in}} = \hat{H}_{bp}(s) = \frac{s\frac{R_4C_3}{R_1R_2C_1C_2}}{s^2 + s\left(\frac{1}{R_1C_1} + \frac{1}{R_2C_1} + \frac{1}{R_3C_1}\right) + \frac{1}{R_2R_3C_1C_2}}. \tag{9.73}$$

Now, we easily observe that the transfer function (9.73) has the same functional dependence on s as the transfer function (9.58) for a band-pass filter. This proves that if voltage $\hat{V}_o^{(2)}$ is taken as the output voltage, then the circuit shown in Figure 9.15 acts as a band-pass filter.

Next, we relate $\hat{V}_o^{(2)}$ to $\hat{V}_o^{(3)}$. By using the same line of reasoning as before, we find:

$$\hat{I}_3 = sC_4(\hat{V}_o^{(2)} - \hat{V}_6) = \frac{\hat{V}_6 - \hat{V}_o^{(3)}}{R_5}, \tag{9.74}$$

and

$$\hat{V}_6 = \hat{V}_7 = 0. \tag{9.75}$$

By substituting (9.75) into (9.74), we derive:

$$\frac{\hat{V}_o^{(3)}}{\hat{V}_o^{(2)}} = -sR_5C_4. \tag{9.76}$$

By multiplying (9.73) by (9.76), we obtain:

$$\frac{\hat{V}_o^{(3)}}{\hat{V}_{in}} = \hat{H}_{hp}(s) = \frac{-s^2 \frac{R_4R_5C_3C_4}{R_1R_2C_1C_2}}{s^2 + s\left(\frac{1}{R_1C_1} + \frac{1}{R_2C_1} + \frac{1}{R_3C_1}\right) + \frac{1}{R_2R_3C_1C_2}}. \tag{9.77}$$

Now, we easily observe that the transfer function (9.77) has the same functional dependence on s as the transfer function (9.56) for a high-pass filter. This proves that if voltage $\hat{V}_o^{(3)}$ is taken as the output voltage, then the circuit shown in Figure 9.15 acts as a high-pass filter.

The construction of an active-*RC* band-notch filter is somewhat more difficult than the previous cases. For this reason, we outline only the general idea. Recall that the band-notch filter arose from the *RLC* circuit by taking the difference between the source voltage and the voltage across the resistance. In other words, the band-notch filter was constructed by "subtracting" a band-pass filter from the source voltage. In a similar manner, we can generate a band-notch filter by introducing an inverter and a summer circuit to the end of a band-pass filter. We leave the task of determining the resistance values necessary to achieve the proper cancellation to the reader.

9.6 Synthesis of Transfer Functions with Active-*RC* Circuits

In the previous section we exploited the idea of *cascading* op-amp circuits. Namely, we have constructed the active-*RC* circuit shown in Figure 9.15 as a cascade of the active-*RC* low-pass filter circuit shown in Figure 9.14 and two active circuits with similar transfer functions which can be generically described as:

$$\hat{H}(s) = sd. \tag{9.78}$$

Now, we shall further exploit the idea of cascading. Let us suppose that we want to design a circuit realization for a given transfer function $\hat{H}(s)$. Let us also suppose that this function can be represented as a product of several (or many) presumably simple functions $\hat{H}_k(s) \quad (k = 1, 2, ..., n)$:

$$\hat{H}(s) = \hat{H}_1(s)\hat{H}_2(s)\cdots\hat{H}_n(s). \tag{9.79}$$

Such a representation is called factorization.

Now, if we can find a circuit realization for each factor in (9.79), then a circuit realization for $\hat{H}(s)$ can be designed as the cascade of circuits whose transfer functions

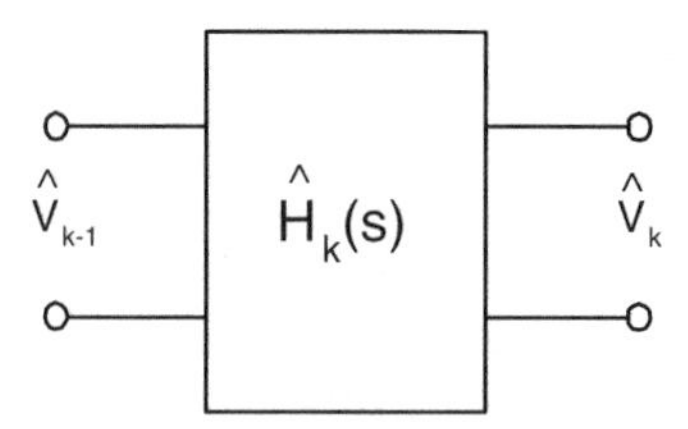

Figure 9.16: Notation for a circuit with transfer function $\hat{H}_k(s)$.

are the factors $\hat{H}_k(s)$ in (9.79). This statement can be proved as follows. Figure 9.16 presents the notation for a circuit, which provides a realization of the factor $\hat{H}_k(s)$. This means that the input $\hat{V}_{k-1}$ and output $\hat{V}_k$ of this circuit are related to one another as follows:

$$\frac{\hat{V}_k(s)}{\hat{V}_{k-1}(s)} = \hat{H}_k(s). \tag{9.80}$$

Consider the cascade of the above circuits (see Figure 9.17). According to (9.80) we have:

$$\frac{\hat{V}_1(s)}{\hat{V}_{in}(s)} = \hat{H}_1(s), \quad \frac{\hat{V}_2(s)}{\hat{V}_1(s)} = \hat{H}_2(s), \cdots \frac{\hat{V}_{n-1}(s)}{\hat{V}_{n-2}(s)} = \hat{H}_{n-1}(s), \quad \frac{\hat{V}_o(s)}{\hat{V}_{n-1}(s)} = \hat{H}_n(s). \tag{9.81}$$

By multiplying the last equalities, we end up with:

$$\frac{\hat{V}_1(s)\hat{V}_2(s)\cdots\hat{V}_{n-1}(s)\hat{V}_o(s)}{\hat{V}_{in}(s)\hat{V}_1(s)\cdots\hat{V}_{n-2}(s)\hat{V}_{n-1}(s)} = \hat{H}_1(s)\hat{H}_2(s)\cdots\hat{H}_{n-1}(s)\hat{H}_n(s). \tag{9.82}$$

After obvious cancellations in (9.82), we obtain:

$$\frac{\hat{V}_o(s)}{\hat{V}_{in}(s)} = \hat{H}(s) = \hat{H}_1(s)\hat{H}_2(s)\cdots\hat{H}_{n-1}(s)\hat{H}_n(s). \tag{9.83}$$

This proves that the cascade shown in Figure 9.17 indeed provides the realization for the given transfer function $\hat{H}(s)$.

The above discussion underlines the practical utility of factorization (9.79). It shows that if a complicated transfer function, $\hat{H}(s)$, can be represented as a product of

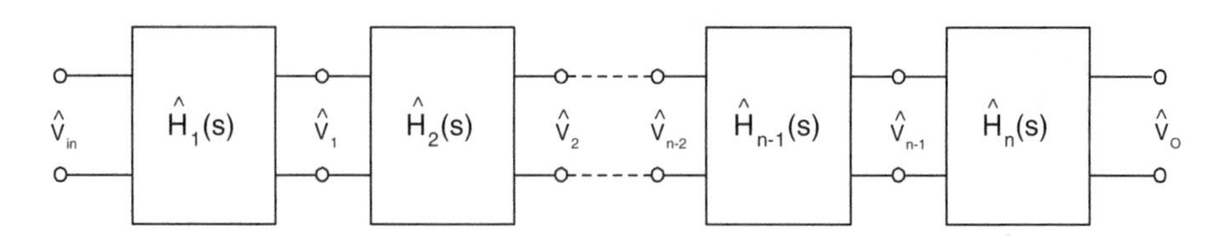

Figure 9.17: Cascade of circuits with transfer functions $\hat{H}_k(s)$.

simple transfer functions $\hat{H}_k(s)$ which can be separately realized by simple circuits, then $\hat{H}(s)$ can be realized as the cascade of these simple circuits. *Thus, cascading is a way to build complicated circuits by using simple ones as the main building blocks.*

To utilize the idea of cascading, we have to find a way to factorize transfer functions. Here, we note that a transfer function can be represented (or approximated) as a ratio of two polynomials:

$$\hat{H}(s) = \frac{N_n(s)}{D_m(s)} = \frac{a_n s^n + a_{n-1}s^{n-1} + \cdots + a_1 s + a_0}{b_m s^m + b_{m-1}s^{m-1} + \cdots + b_1 s + b_0}. \tag{9.84}$$

We next consider the case when "numerator" polynomial $N_m(s)$ and "denominator" polynomial $D_m(s)$ have only real and negative roots. Roots of $N_n(s)$ are zeros of transfer functions and we shall use the following notation for them:

$$z_1, z_2, \cdots z_n. \tag{9.85}$$

Roots of $D_m(s)$ are poles of transfer functions and they will be denoted as:

$$p_1, p_2, \cdots p_m. \tag{9.86}$$

It is well known that polynomials can be factorized by using their roots. These factorizations for $N_n(s)$ and $D_m(s)$ can be written as follows:

$$N_n(s) = a_n(s - z_1)(s - z_2)\cdots(s - z_n), \tag{9.87}$$

$$D_m(s) = b_m(s - p_1)(s - p_2)\cdots(s - p_m). \tag{9.88}$$

By substituting (9.87) and (9.88) into (9.84), we obtain:

$$\hat{H}(s) = d\,\frac{(s - z_1)(s - z_2)\cdots(s - z_n)}{(s - p_1)(s - p_2)\cdots(s - p_m)}, \tag{9.89}$$

where $d = a_n/b_m$. Now, the factorization of transfer functions is readily available. Indeed, expression (9.89) can be rewritten as follows:

$$\hat{H}(s) = (s - z_1)(s - z_2)\cdots(s - z_n)\frac{d_1}{s - p_1} \cdot \frac{d_2}{s - p_2} \cdots \frac{d_m}{s - p_m}, \tag{9.90}$$

where $d = d_1 d_2 \cdots d_m$.

Thus, the whole problem of circuit realization of transfer functions is reduced to the design of generic circuits which will realize the factors $(s - z_k)$ and $d_k/(s - p_k)$. As soon as the design of these circuits is accomplished, the transfer function, $\hat{H}(s)$, can be realized as a cascade of these generic circuits.

To arrive at the above generic circuits, let us consider the circuit shown in Figure 9.18. Since the input impedance of the operational amplifier is assumed to be infinite, we find:

$$\hat{I} = \frac{\hat{V}_{in} - \hat{V}_1}{Z(s)} = \frac{\hat{V}_1 - \hat{V}_o}{Z_f(s)}. \tag{9.91}$$

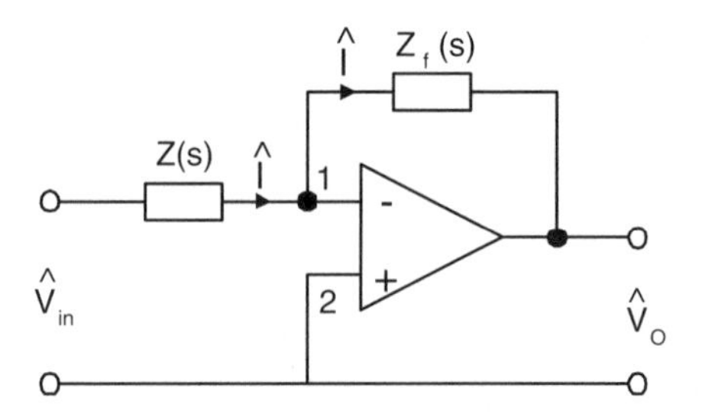

Figure 9.18: Generic circuit.

It is also clear from the above circuit that:

$$\hat{V}_1 = \hat{V}_2 = 0. \tag{9.92}$$

By substituting (9.92) into (9.91), we obtain:

$$\hat{H}(s) = \frac{\hat{V}_o}{\hat{V}_{in}} = -\frac{Z_f(s)}{Z(s)}. \tag{9.93}$$

Now, consider the circuit shown in Figure 9.19. This circuit is a particular case of the circuit shown in Figure 9.18. For this circuit, we have:

$$Z(s) = R, \tag{9.94}$$

$$Z_f(s) = \frac{R_f \cdot \frac{1}{sC_f}}{R_f + \frac{1}{sC_f}} = \frac{\frac{1}{C_f}}{s + \frac{1}{R_f C_f}}. \tag{9.95}$$

By substituting (9.94) and (9.95) into (9.93), we derive:

$$\hat{H}(s) = \frac{-\frac{1}{RC_f}}{s + \frac{1}{R_f C_f}}. \tag{9.96}$$

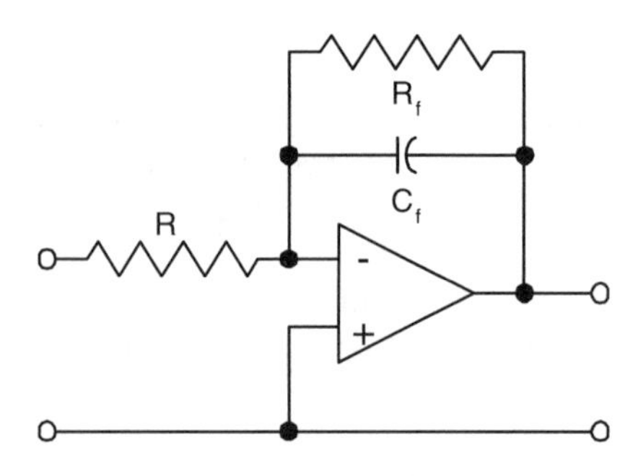

Figure 9.19: Active-*RC* circuit with a prescribed pole.

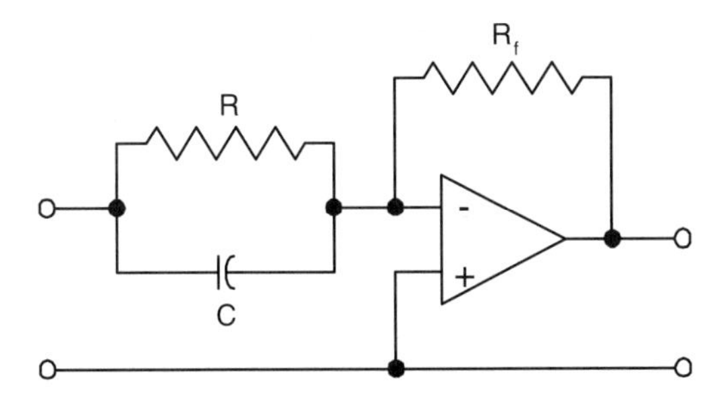

Figure 9.20: Active-RC circuit with a prescribed zero.

From (9.96), we conclude that the circuit shown in Figure 9.19 has a transfer function with a prescribed pole:

$$p = -\frac{1}{R_f C_f}. \tag{9.97}$$

This means that this circuit can be used for realization of factors $d_k/(s-p_k)$ in (9.90).

Next, consider the circuit shown in Figure 9.20. This circuit is also a particular case of the circuit shown in Figure 9.18. For this circuit, we have:

$$Z(s) = \frac{R \cdot \frac{1}{sC}}{R + \frac{1}{sC}} = \frac{\frac{1}{C}}{s + \frac{1}{RC}}, \tag{9.98}$$

$$Z_f(s) = R_f. \tag{9.99}$$

By substituting (9.98) and (9.99) into (9.93), we derive:

$$\hat{H}(s) = -R_f C \left(s + \frac{1}{RC} \right). \tag{9.100}$$

From (9.100), we observe that the circuit shown in Figure 9.20 has a transfer function with a prescribed zero:

$$z = -\frac{1}{RC}. \tag{9.101}$$

This means that this is a generic circuit for realization of factors $(s - z_k)$ in (9.90).

Thus, we can conclude that the circuit realization of a transfer function (9.90) with real and negative poles and zeros can be achieved by cascading generic circuits shown in Figures 9.19 and 9.20.

We shall illustrate the previous discussion by the following examples.

EXAMPLE 9.6 We would like to design an active-RC circuit with a transfer function given by the following expression:

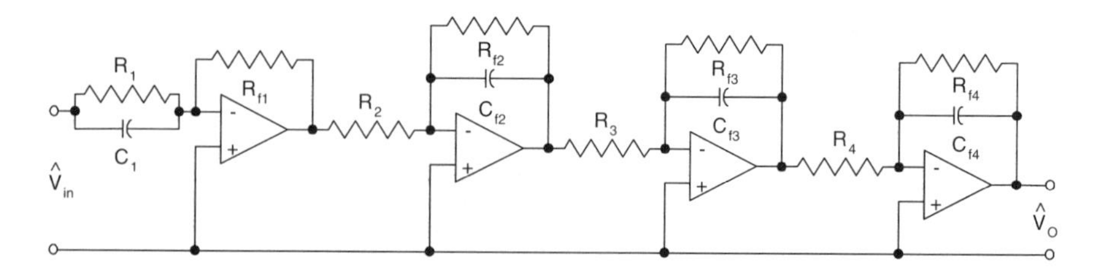

Figure 9.21: Active-*RC* circuit realization of the example transfer function.

$$H(s) = \frac{s+1}{(s+2)(s+3)(s+4)}. \tag{9.102}$$

This transfer function has one zero

$$z_1 = -1 \tag{9.103}$$

and three poles

$$p_1 = -2, \quad p_2 = -3, \quad \text{and} \quad p_3 = -4. \tag{9.104}$$

By using the technique described above, we can realize this transfer function by cascading one circuit of the type shown in Figure 9.20 and three circuits of the type shown in Figure 9.19. The resulting circuit is shown in Figure 9.21.

To guarantee the correct poles and zero of the transfer function (9.102), resistances and capacitances of the above circuit can be chosen as follows:

$$R_1C_1 = 1, \quad R_{f1}C_1 = 1, \tag{9.105}$$

$$R_{f2}C_{f2} = \frac{1}{2}, \quad R_2C_{f2} = 1, \tag{9.106}$$

$$R_{f3}C_{f3} = \frac{1}{3}, \quad R_3C_{f3} = 1, \tag{9.107}$$

$$R_{f4}C_{f4} = \frac{1}{4}, \quad R_4C_{f4} = 1. \tag{9.108}$$

Expressions (9.105)–(9.108) follow from (9.100) and (9.96). ■

EXAMPLE 9.7 We will synthesize a normalized, third-order Butterworth filter by using active-*RC* filter circuits. The transfer function for this filter is given by:

$$H(s) = \frac{1/2}{(s^2+s+1)(s+1)}. \tag{9.109}$$

Therefore, we will need to use op-amp circuits cascaded in the arrangement shown in Figure 9.22. The leftmost op-amp circuit will generate the second-order polynomial in the denominator and the rightmost circuit will generate the first-order pole. There are a few extra degrees of freedom in this problem, so let us make a few arbitrary

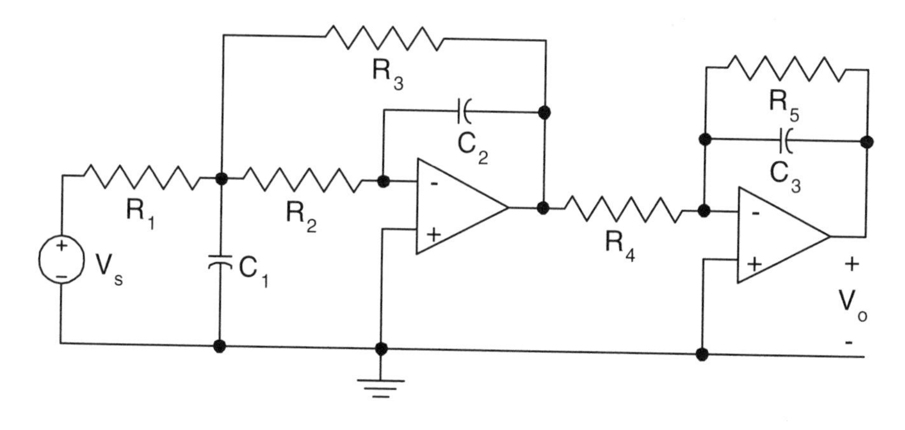

Figure 9.22: Active circuit realization of a third-order Butterworth filter.

assignments. First, we will assume that $R_1 = R_2 = R_3 = 1\ \Omega$. Comparing equation (9.67) with the Butterworth transfer function, we see that we will need to set $C_1 = 3$ F and $C_2 = 1/3$ F. This gives us the proper form for the denominator of the first part of the transfer function and leaves the numerator at -1. The second op-amp circuit will then have to yield the proper pole and have a numerator equal to $-1/2$. Again there is some flexibility, so we arbitrarily pick $C_3 = 1$ F. Equation (9.96) then requires that $R_5 = 1\ \Omega$ and $R_4 = 2\ \Omega$. This completely specifies one possible design of the third-order Butterworth filter. ■

In conclusion, it is important to mention that cascading is possible due to very small (almost zero) output impedances of operational amplifiers. Because of that property, any load (circuit) which is placed across the output terminals of operational amplifier does not affect the output voltage of this amplifier, and, consequently, it does not affect the transfer function of the circuit to which this amplifier belongs.

9.7 MicroSim PSpice Simulations

The first circuit that we will analyze with PSpice is the resonant circuit shown in Figure 9.1 and drawn by the PSpice schematic generator in Figure 9.23. The values of the passive elements are indicated in the figure and correspond to the parameters of Example 9.1. The voltage source uses the "VSRC" model and has an ac magnitude of 10 V. An "AC Sweep" is selected in the "Analysis Setup" dialog box. In Example 9.1 the resonant frequency was calculated to be fo = 7337 Hz, so we sweep in frequency from 100 Hz to 100 kHz.

The results of the simulation for the voltages across each of the passive elements is plotted on a log-log scale in Figure 9.24. The voltage across the inductor is indicated by the solid line. For frequencies far above the resonance frequency the voltage is nearly constant and equal to the source voltage of 10 V. For frequencies far below

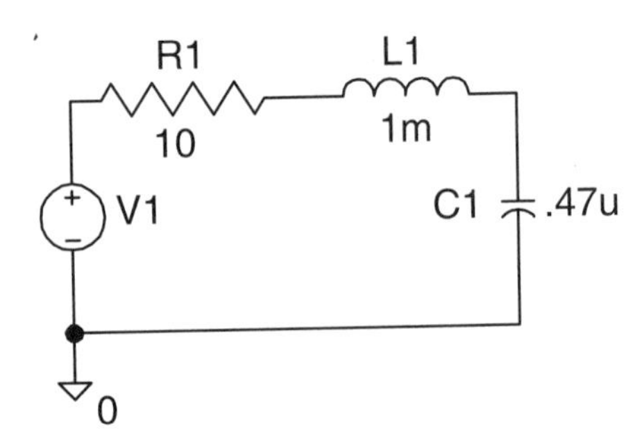

Figure 9.23: The PSpice schematic for the circuit in Example 9.1.

resonance, the voltage decreases at about a rate of 20 dB per decade in voltage (40 dB/decade in power). These are exactly the properties expected of a second-order high-pass filter. At resonance, the analytic calculation predicted the inductor voltage to exceed the source voltage by a factor of 4.64 = 6.67 dB. This is indeed the case, as indicated in the figure.

The voltage across the capacitor is indicated in the figure by the dot-dashed line. This curve displays the classic behavior of a second-order low-pass filter as expected. The overshoot of the voltage at resonance matches that of the inductor (recall that at resonance the inductor and capacitor voltages are equal in magnitude but out of phase by 180°). The dashed line traces the voltage across the resistor as a function of frequency. The resistor acts as a band-pass filter, with a maximum voltage at

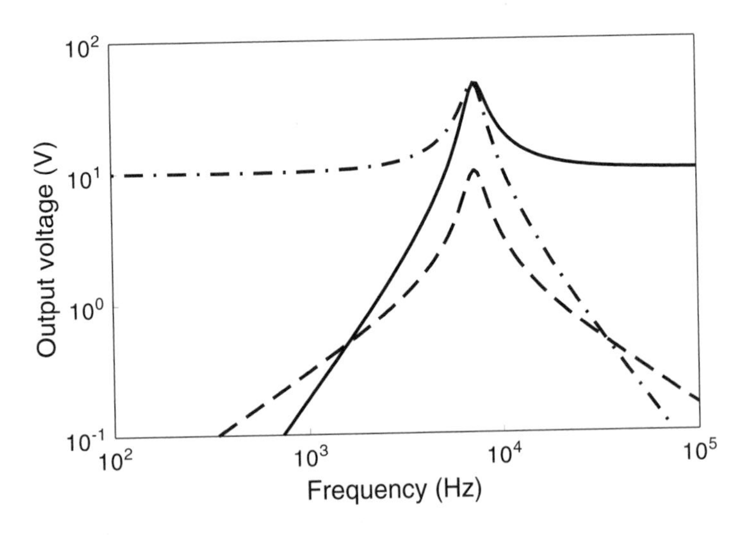

Figure 9.24: The simulated voltages across the passive elements for the circuit shown in Figure 9.23. The inductor voltage is given by the solid line, the resistor voltage by the dashed line, and the capacitor voltage by the dot-dashed line.

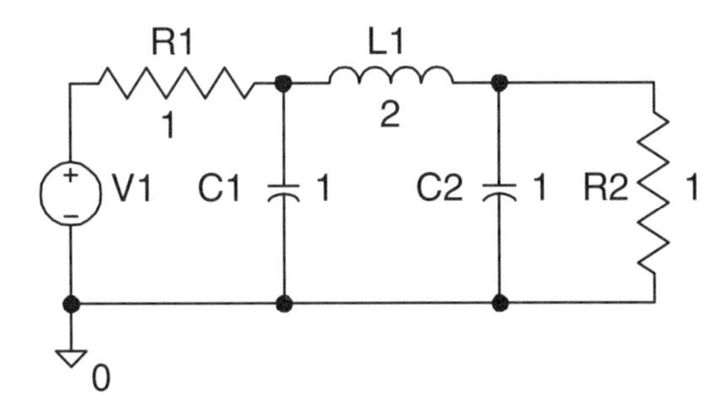

Figure 9.25: The PSpice schematic for the circuit in Example 9.4.

resonance equal to the source voltage and a drop-off away from resonance of about 10 dB per decade (in voltage) on both sides of the maximum.

For the second PSpice simulation we investigate the third-order, low-pass Butterworth filter circuit of Example 9.4. The circuit is redrawn by the schematic editor in Figure 9.25. The passive component values are the same as in the example and are shown in the figure. Again we use the "VSRC" part to model the source, but we set the ac magnitude to 5 V. The breakpoint frequency is calculated in be $\omega = 1$ rad/s (f = 0.159 Hz), so the "AC Sweep" is taken from 0.001 Hz to 100 Hz.

The results of the simulation for the output voltage across the resistor are plotted in Figure 9.26. The voltage at low frequencies is one-half the source voltage as predicted in the example. The effect of the "maximally flat" property of this filter is evident in this figure by the extremely flat output voltage in the passband of

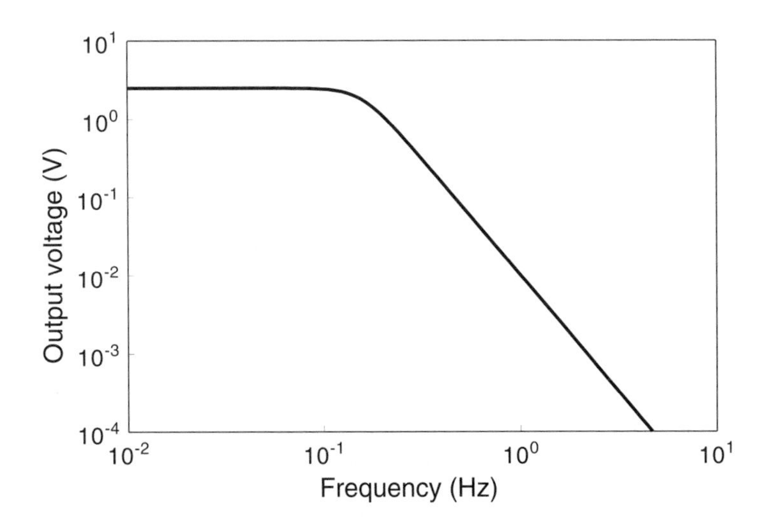

Figure 9.26: The simulated output voltage of the third-order Butterworth filter.

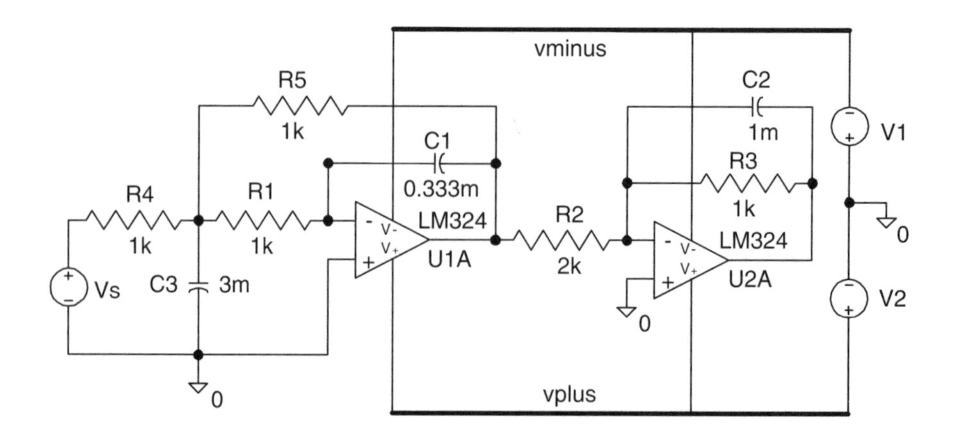

Figure 9.27: The PSpice schematic for the circuit in Example 9.7.

the circuit. The output voltage drops off far above the breakpoint frequency at the expected third-order rate of 30 dB/decade.

The final PSpice simulation examines the active version of the third-order low-pass Butterworth filter. This circuit was analyzed in Example 9.7 and is redrawn by the circuit simulator in Figure 9.27. The resistors have been scaled by a factor of 1000 ($\Omega \rightarrow k\Omega$) and the capacitors have been reduced by a factor of 1000 (F $\rightarrow$ mF) to yield somewhat more realistic component values. However, since any resistance in the transfer function is always multiplied by a capacitance, the net transfer function

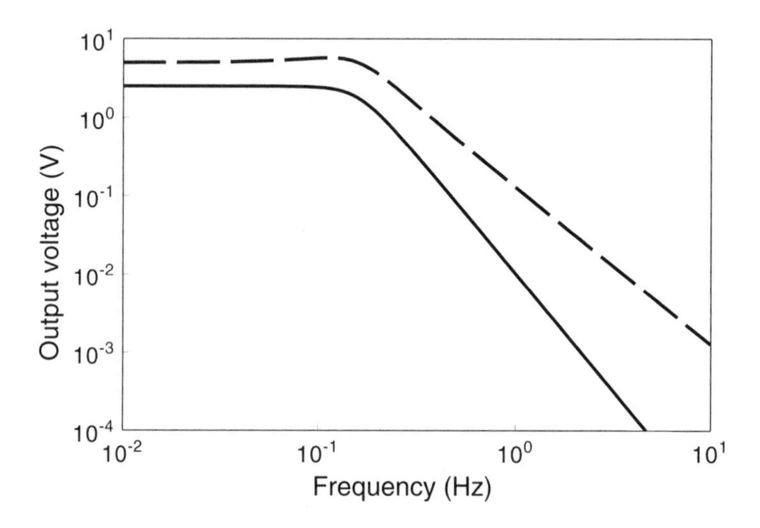

Figure 9.28: The simulated voltage at the output of each stage of the third-order Butterworth filter. The dashed line gives the output of the first stage and the solid line indicates the output of the final stage.

of this circuit is identical to the one in the example. The two op-amps use the LM324 five-connection subcircuit model from the "eval.slb" library with all the standard default values. This model was described previously in Section 8.5. V1 and V2 are the op-amp dc supply voltages of minus and plus 12 V, respectively. Vs is the source voltage, which has an ac magnitude of 5 V to match that of the previous example. All sources are "VSRC" parts.

The voltage at the output of each of the op-amp stages is shown in Figure 9.28. Recall that this filter was synthesized by combining a second-order low-pass filter and a first-order low-pass filter. The dashed line indicates the output of the second-order filter. The low-frequency gain is unity and the voltage drop-off at high frequency is 20 dB/decade. Note that there is a slight overshoot in the voltage at the resonance frequency. This overshoot helps to compensate for the knee of the first-order low-pass filter of the second stage to produce the "maximally flat" characteristic of the filter. The second stage has a low-frequency gain of 1/2 and the same breakpoint frequency as the first stage. The net output voltage of the filter is indicated in the figure by the solid line. This curve is virtually identical to the output dependence of the passive version of this filter, as expected.

9.8 Summary

In this chapter, we have discussed the frequency characteristics of electric circuits. The main results of this chapter can be briefly summarized as follows:

- It has been demonstrated that a simple *RLC* circuit exhibits an interesting phenomenon called resonance. At resonance, the current through the circuit is **in phase** with the source voltage in spite of the presence of energy storage elements. This occurs at a resonant frequency $\omega_0 = 1/\sqrt{LC}$. At resonance, the impedance as a function of frequency assumes its minimum value, while the current as a function of frequency assumes its maximum value. Voltages across the capacitor and inductor have a phase difference of 180°, while their peak values are the same. These peak values may significantly exceed the peak value of the voltage source.

- It has been demonstrated that the *RLC* circuit can be utilized as a filter. If the inductor terminals are used as the output terminals, then the circuit acts as a **high-pass** filter. This means that output signals with low frequencies will be suppressed, while the output signals at high frequencies will have almost the same peak values as the input signals. If the capacitor terminals are used as the output terminals, then the circuit acts as a **low-pass** filter. This means that the output signals at high frequencies will be suppressed, while the output signals at low frequencies will have almost the same peak values as the input signals. If the resistor terminals are used as the output terminals, then the circuit acts as a **band-pass** filter. This means that the output signals will be suppressed at low and high frequencies, while the output signals at frequencies centered around the resonant frequency ω_0 will have approximately the same peak values as the input signals. If the net voltage across the inductor and capacitor is used as the output voltage, then the circuit acts as a **band-notch** filter. This means that the output signals at frequencies centered around ω_0 will be suppressed, while the output signals at low and high frequencies will have almost the same peak values as the input signals.

- Bode plots for transfer functions have been presented. **These plots are piecewise straight-line approximations for transfer function magnitudes and phases as functions of frequency on a logarithmic scale**. The logarithmic scale allows one to represent Bode plots for complicated transfer functions as sums of very simple straight-line plots which are determined by the zeros and poles of the transfer functions.
- Active-RC filters have been discussed. These filters contain resistors, capacitors, and operational amplifiers, but they do not contain inductors (which are incompatible with integrated circuit technology). The main idea of the design of an active-RC filter is to come up with active-RC circuits which have the same transfer functions as high-pass, low-pass, band-pass, and band-notch filters. Examples of the design and analysis of active-RC filters have been given.
- The synthesis of transfer functions by using active-RC circuits has been presented. The main idea of this synthesis procedure is to represent a complicated transfer function as a product of simple transfer functions which can be separately realized with simple active-RC circuits. Then, the original transfer function can be realized as the cascade of these simple circuits. **Thus, cascading is a way to build complicated circuits by using simple ones as the main building blocks**. In the chapter, poles and zeros have been used in the factorization of transfer functions. Generic op-amp circuits for realization of pole-factors and zero-factors have been designed and the synthesis of the transfer functions has been accomplished by cascading these generic circuits.

9.9 Problems

In Problems 1–6, find the transfer function indicated in the appropriate figure and identify the type of filter it represents (low-pass, high-pass, band-pass, or band-notch). Sketch on a graph the magnitude of the transfer function from $\omega_0/10$ to $10\omega_0$.

1. $H(s) = v_o/v_s$ for the circuit shown in Figure P9-1.
2. $H(s) = v_o/v_s$ for the circuit shown in Figure P9-2.

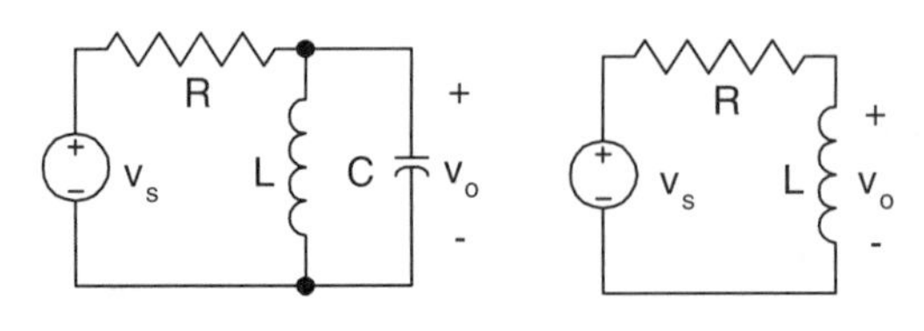

Figure P9-1 Figure P9-2

3. $H(s) = i_o/i_s$ for the circuit shown in Figure P9-3.

4. $H(s) = v_o/v_s$ for the circuit shown in Figure P9-4.

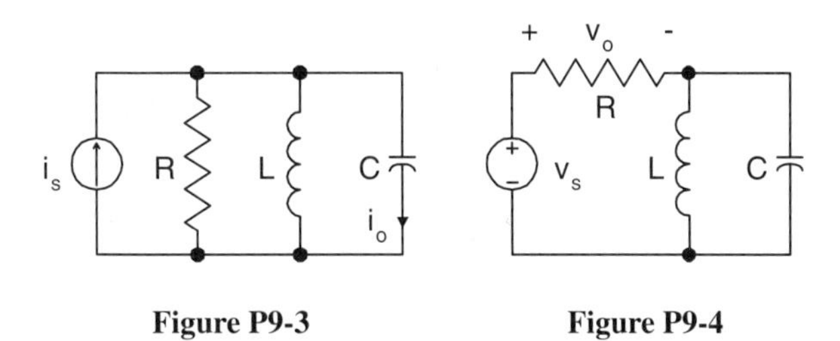

Figure P9-3 **Figure P9-4**

5. $H(s) = v_o/v_s$ for the circuit shown in Figure P9-5.

6. $H(s) = v_o/v_s$ for the circuit shown in Figure P9-6.

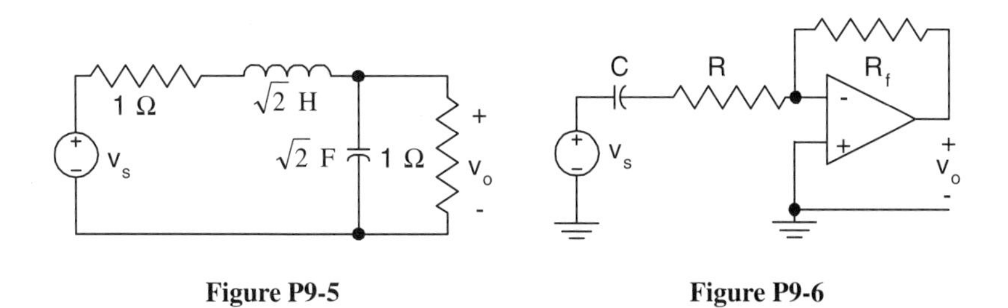

Figure P9-5 **Figure P9-6**

In Problems 7–12, find the indicated transfer function and construct Bode plots for the magnitude and phase dependence on frequency.

7. $H(s) = v_o/v_s$ for the circuit shown in Figure P9-7.

8. $H(s) = v_o/v_s$ for the circuit shown in Figure P9-8.

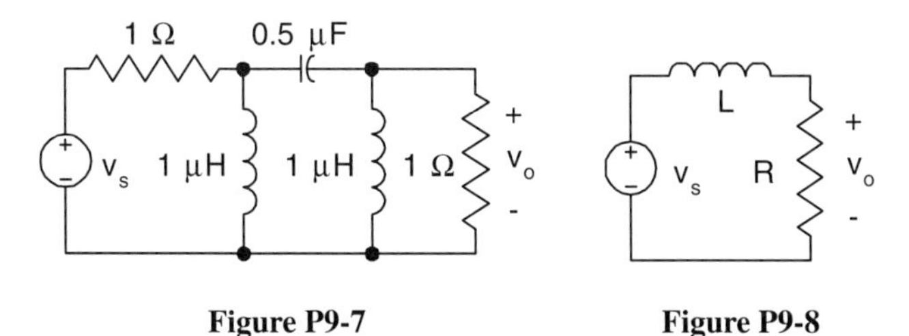

Figure P9-7 **Figure P9-8**

9. $H(s) = v_o/v_s$ for the circuit shown in Figure P9-9.

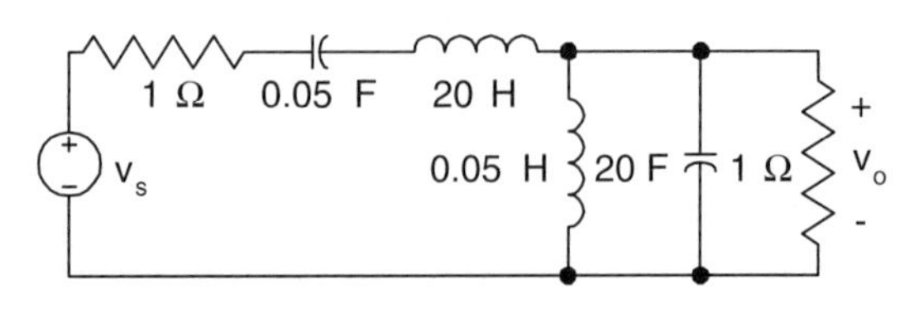

Figure P9-9

10. $H(s) = v_o/v_s$ for the circuit shown in Figure P9-10.

11. $H(s) = v_o/v_s$ for the circuit shown in Figure P9-11.

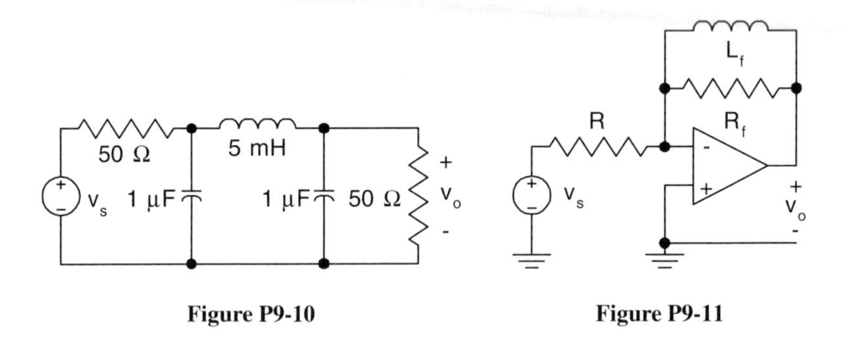

Figure P9-10 **Figure P9-11**

12. $H(s) = v_o/v_s$ for the circuit shown in Figure P9-12.

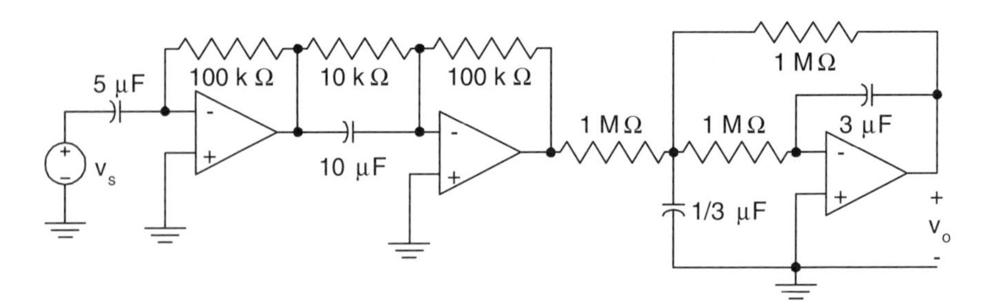

Figure P9-12

In Problems 13–18, construct Bode plots for the magnitude and phase dependence on frequency of the transfer functions given in Table 9.2.

Table 9.2: Sample transfer functions.

Transfer functions
$H_1(s) = \dfrac{1 + s/10}{(1 + s)(1 + s/100)}$
$H_2(s) = \dfrac{(s + 10)(s + 1000)}{(s + 1)(s + 100)^2}$
$H_3(s) = \dfrac{100s(1 + s/100)}{(1 + s/10)(1 + s/1000)}$
$H_4(s) = \dfrac{10s}{(1 + 5s + s^2)(1 + s/10)}$
$H_5(s) = \dfrac{(s + 5)s}{(s^2 + 3s + 1)^2(s + 50)}$
$H_6(s) = \dfrac{2000(s + 1)s^3}{(s + 100)^2(s + 10)^2}$

13. Transfer function H_1.

14. Transfer function H_2.

15. Transfer function H_3.

16. Transfer function H_4.

17. Transfer function H_5.

18. Transfer function H_6.

In Problems 19-24, design active *RC* circuits which realize the transfer functions given in Table 9.2. Use only 10 μF capacitors. Use resistance values $R \leq 1$ MΩ and as few op-amps as possible.

19. Transfer function H_1.

20. Transfer function H_2.

21. Transfer function H_3.

22. Transfer function H_4.

23. Transfer function H_5.

24. Transfer function H_6.

In Problems 25–34, use a frequency analysis in PSpice to plot the frequency response of the transfer function magnitudes indicated. Plot at least one decade above and below any corner frequencies.

25. $H(\omega) = i_o/i_s$ for the circuit shown in Figure P9-3. Let $R = 2$ kΩ, $C = 1$ mF, and $L_f = 1$ mH.

26. $H(\omega) = v_o/v_s$ for the circuit shown in Figure P9-5.

27. $H(\omega) = v_o/v_s$ for the circuit shown in Figure P9-6. Let $R = 3R_f = 24$ kΩ and $C = 47$ μF.

28. $H(\omega) = v_o/v_s$ for the circuit shown in Figure P9-8. Plot the Bode estimate on the same graph.

29. $H(\omega) = v_o/v_s$ for the circuit shown in Figure P9-9. Plot the Bode estimate on the same graph.

30. $H(\omega) = v_o/v_s$ for the circuit shown in Figure P9-10. Plot the Bode estimate on the same graph.

31. $H(\omega) = v_o/v_s$ for the circuit shown in Figure P9-12. Plot the Bode estimate on the same graph.

32. $H(\omega) = v_o/v_s$ for the circuit shown in Figure P9-11. Let $2R = R_f = 10$ kΩ and $L_f = 1$ mH.

33. Design an active RC circuit with the transfer function given by H_1 in Table 9.2. Plot the simulated transfer function magnitude and the Bode plot estimate on the same graph.

34. Design an active RC circuit with the transfer function given by H_5 in Table 9.2. Plot the simulated transfer function magnitude and the Bode plot estimate on the same graph.

Chapter 10

Magnetically Coupled Circuits and Two-Port Elements

10.1 Introduction

In our previous discussions, we have dealt with circuits in which circuit elements have been coupled electrically. This has been realized by using electric wires. It turns out that circuit elements (and circuits) can also be coupled magnetically, that is, without direct electric contacts. This coupling occurs because time-varying electric currents in one circuit produce time-varying magnetic fields and magnetic field lines. These field lines may link an adjacent electric circuit and, according to Faraday's law, induce a voltage in this circuit. In this way, the time-varying electric currents in one circuit may affect (may be coupled with) the currents in the adjacent circuit.

In some cases, magnetic coupling can be useful and, for this reason, it can be utilized and even deliberately enhanced in the design of certain electric devices. Examples include transformers and induction motors. In other cases, magnetic coupling can be detrimental to design purposes and should be suppressed. An important example of this is high-density interconnects of VLSI circuits where "cross-talk" is becoming prohibitively large due to wiring proliferation on silicon chips.

In this chapter, we consider magnetically coupled circuits. We begin our discussion with the definition of mutual inductance and establish a very important inequality between the mutual inductance and the product of self-inductances. We then use the concept of mutual inductance to derive the coupled circuit equations. These equations constitute the foundation for the discussion of a transformer. We first consider the theory of an ideal transformer when many small parameters and imperfections are neglected. The presentation of the theory of a nonideal transformer then follows. The importance of such small parameters as leakage reactances is exposed and stressed, and the coupled circuit equations are modified in order to account explicitly for these leakage reactances. This modification leads directly to the equivalent circuit for the transformer.

The second part of the chapter deals with the theory of two-port elements. It is emphasized that the two-port elements serve a twofold purpose: they can be used as equivalent replacements for complicated circuits and, more importantly, they can be employed as circuit models for real devices such as transistors, transformers, and transmission lines. The linear and homogeneous equations which describe terminal voltage–current relations for two-port elements can be written in different forms which result in different representations of two-port elements. In the chapter, these various representations are studied in detail along with their experimental identifications and circuit realizations.

10.2 Mutual Inductance and Coupled Circuit Equations

In circuit theory, magnetic coupling between electric circuits can be described by using the concept of mutual inductance. To define the mutual inductance, consider the two coils shown in Figure 10.1. The first coil has N_1 turns, while the second coil has N_2 turns. First, we assume that the first coil is energized while the current through the second coil is equal to zero:

$$i_1 \neq 0, \qquad i_2 = 0. \tag{10.1}$$

The current through the first coil creates the magnetic field around this coil, which can be represented by magnetic field lines (see Figure 10.1a). Some of these field lines may link the second coil resulting in the flux linkages, ψ_{21}. Here the order of subscripts indicates that ψ_{21} is the flux linkages of the second coil due to the current through the first coil. The mutual inductance M_{12} between the first and the second coils is defined as the following ratio:

$$M_{12} = \frac{\psi_{21}}{i_1}. \tag{10.2}$$

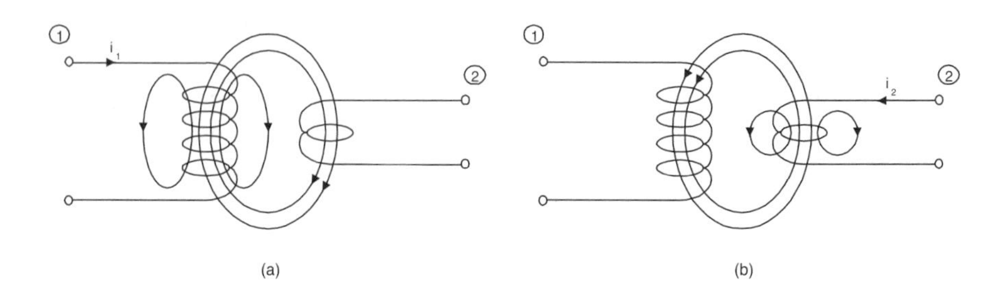

Figure 10.1: Two magnetically coupled coils.

Next, consider the case when the second coil is energized while the current through the first coil is equal to zero:

$$i_1 = 0, \qquad i_2 \neq 0. \tag{10.3}$$

In this case, the current through the second coil creates the magnetic field around this coil which can be represented by magnetic field lines (see Figure 10.1b). Some of these field lines may link the first coil resulting in the flux linkages ψ_{12}. Here, the order of subscripts indicates that ψ_{12} is the flux linkages of the first coil due to the current through the second coil. The mutual inductance M_{21} between the second and the first coils is defined as

$$M_{21} = \frac{\psi_{12}}{i_2}. \tag{10.4}$$

It turns out that mutual inductances M_{12} and M_{21} are the same:

$$M_{21} = M_{12} = M. \tag{10.5}$$

This fact is rigorously proved in electromagnetic field theory and it is known as the reciprocity principle. The physical meaning of the reciprocity principle is apparent from the definitions of M_{12} and M_{21} given above. Namely, this principle states that the **interchange of excitation and response** does not affect the result. Indeed, in the case of M_{12}, the first coil is excited while we consider the response (flux linkages) of the second coil. In the case of M_{21}, the second coil is excited while we consider the response (flux linkages) of the first coil. It is clear from (10.2), (10.4), and (10.5) that if the excitations are the same, then the responses will be the same, which is in compliance with the meaning of the reciprocity principle. Reciprocity will be further discussed in this chapter when we study the theory of two-port elements.

By using equality (10.5) in formulas (10.2) and (10.4), we can rewrite them as follows:

$$M = \frac{\psi_{21}}{i_1} = \frac{\psi_{12}}{i_2}. \tag{10.6}$$

The last expression is tantamount to:

$$\psi_{21}(t) = Mi_1(t), \tag{10.7}$$

$$\psi_{12}(t) = Mi_2(t). \tag{10.8}$$

It is clear from expressions (10.7) and (10.8) that the mutual inductance, M, has the physical meaning of the measure of magnetic coupling between two coils. Indeed, the more the mutual inductance between two coils, the more the flux linkages of the second coil for the same current through the first coil and, consequently, the stronger the magnetic coupling between two coils.

The mutual inductance was defined as the ratio of flux linkages to currents. However, the mutual inductance does not depend on flux linkages or currents. It is rather the coefficient of proportionality between flux linkages and currents as clearly suggested by formulas (10.7) and (10.8). The mutual inductance depends on relative locations of two coils with respect to one another, their geometric dimensions, the physical properties of the media in the vicinity of the two coils, and the numbers of turns of the coils. Actually, the mutual inductance is proportional to the product of the numbers of turns:

$$M \sim N_1 N_2. \tag{10.9}$$

The calculation of mutual inductance is a difficult problem which belongs to electromagnetic field theory. In electric circuit theory, it is always assumed that the mutual inductance is known. The MKSA unit of mutual inductance is the henry, that is the same as that of inductance (which it is now more proper to call self-inductance).

The given definition of mutual inductance may create the impression that the mutual inductance is always smaller than the self-inductance. However, this impression is erroneous. To elucidate this issue, let us recall the definition of self-inductance given in Chapter 1. The self-inductances of the first and second coils shown in Figure 10.1 can be defined as follows:

$$L_1 = \frac{\psi_{11}}{i_1}, \tag{10.10}$$

$$L_2 = \frac{\psi_{22}}{i_2}. \tag{10.11}$$

Here, ψ_{11} is the flux linkages of the first coil when only this coil is energized, and ψ_{22} has a similar physical meaning. It is clear that ψ_{11} and ψ_{22} can be called self-flux linkages.

Since ψ_{11} is the flux linkages due to all magnetic field lines and ψ_{21} is the flux linkages due to only some part of these field lines, the impression may be created that ψ_{11} is always larger than ψ_{21} and thus, according to (10.6) and (10.10), L_1 is always larger than M. However, this is not true because the flux linkages are determined not only by the number of field lines which link the coil but by the number of turns in the coil as well. Although only a small part of magnetic field lines may link the turns of the second coil, the flux linkages ψ_{21} can be made arbitrarily large by increasing the number N_2 of turns of this coil. In this way, we can make M larger than L_1. This suggests that there is no direct and universally true inequality between M and L_1 (or between M and L_2, for that matter). However, there is an important and universally true inequality between M and the product of L_1 and L_2, which is given below:

$$M \leq \sqrt{L_1 L_2}. \tag{10.12}$$

We shall prove this inequality for the practically important case when the turns of each of the coils are closely spaced. In this case, all turns of the same coil are linked by the same number of magnetic field lines and, consequently, by the same flux. By using this fact, we can write:

$$\psi_{21} = N_2 \phi_{21}, \tag{10.13}$$

where ϕ_{21} is the flux which is created by the current through the first coil and which links all the turns of the second coil (Figure 10.1a).

It is clear that

$$\phi_{21} \leq \phi_{11}, \tag{10.14}$$

where ϕ_{11} is the flux which is created by the current through the first coil and which links all the turns of the same coil.

By substituting (10.14) into (10.13), after simple transformations we find:

$$\psi_{21} \leq N_2\phi_{11} = \frac{N_2}{N_1}N_1\phi_{11}. \tag{10.15}$$

It is clear that:

$$\psi_{11} = N_1\phi_{11}. \tag{10.16}$$

By using formula (10.16) in (10.15), we obtain:

$$\psi_{21} \leq \frac{N_2}{N_1}\psi_{11}. \tag{10.17}$$

By dividing both sides of (10.17) by i_1 and by using the definitions (10.6) and (10.10), we derive:

$$M \leq \frac{N_2}{N_1}L_{11}. \tag{10.18}$$

Next, consider the case shown in Figure 10.1b. It is clear that:

$$\psi_{12} = N_1\phi_{12}, \tag{10.19}$$

where ϕ_{12} is the flux which is created by the current through the second coil and which links all the turns of the first coil.

It is also clear that ϕ_{12} is only some part of the flux ϕ_{22} which is created by i_2 and links all the turns of the second coil. Thus:

$$\phi_{12} \leq \phi_{22}. \tag{10.20}$$

By using inequality (10.20) in expression (10.19), after simple transformations we find:

$$\psi_{12} \leq N_1\phi_{22} = \frac{N_1}{N_2}N_2\phi_{22}. \tag{10.21}$$

Since

$$\psi_{22} = N_2\phi_{22},$$

inequality (10.21) can be written as follows:

$$\psi_{12} \leq \frac{N_1}{N_2}\psi_{22}.$$

By dividing both sides of the last formula by i_2 and by recalling the definitions (10.6) and (10.11), we derive:

$$M \leq \frac{N_1}{N_2}L_2. \tag{10.22}$$

Now, by multiplying inequalities (10.18) and (10.22), we obtain:

$$M^2 \leq L_1 L_2, \tag{10.23}$$

which is tantamount to (10.12).

Inequality (10.12) suggests the following definition of a very important quantity k called the coupling coefficient:

$$k = \frac{M}{\sqrt{L_1 L_2}}. \tag{10.24}$$

It is clear from inequality (10.12) that

$$0 \leq k \leq 1. \tag{10.25}$$

It is also clear that k does not depend on the numbers N_1 and N_2 of turns of the first and second coils. This is because the self-inductances L_1 and L_2 are proportional to the squares of the numbers of turns of their coils:

$$L_1 \sim N_1^2, \qquad L_2 \sim N_2^2, \tag{10.26}$$

while for the mutual inductance M the expression (10.9) is valid. By using formulas (10.26) and (10.9) in the definition (10.24), we conclude that k does not depend on N_1 and N_2. This assertion suggests that the coupling coefficient can be used as a very important measure of magnetic coupling between two coils, a measure which does not depend on the numbers of turns of these coils. The coupling coefficient depends only on relative locations of two coils and the physical properties of the media in the vicinity of these coils. The latter point is very important and it is used in practice in order to enhance the magnetic coupling between two coils. To achieve this, the coils are usually placed on the same iron (or ferrite) core of high magnetic permeability (see Figure 10.2). Because of its high magnetic permeability, the iron core is an excellent

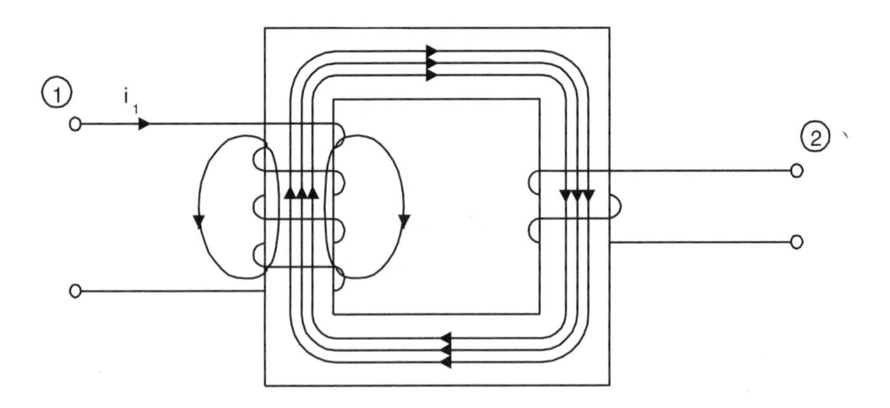

Figure 10.2: The use of an iron core to enhance the magnetic coupling between two coils.

guide for magnetic field lines. This means that most of the magnetic field lines lie entirely within the iron core, while only a very small fraction of them (attributed to leakage) partially go through air. As a result, the first and second coils are always linked by almost the same flux. This suggests that inequalities (10.14) and (10.20) are very close to equalities, which, in turn, suggests that the coupling coefficient is very close to unity.

Having defined and discussed the concept of mutual inductance, we shall next consider how this concept can be used in electric circuit theory. Our ultimate goal is to find the relationship between the terminal voltages $v_1(t)$ and $v_2(t)$ and terminal currents $i_1(t)$ and $i_2(t)$ for two magnetically coupled coils. To this end, consider the case when the two coils shown in Figure 10.1 are simultaneously energized. By using the superposition principle, we can claim that the total flux linkage of the first coil is the algebraic sum of self-flux linkage $\psi_{11}(t)$, which are due to the current $i_1(t)$ through the first coil, and mutual flux linkage $\psi_{12}(t)$, which are due to the current $i_2(t)$ through the second coil:

$$\psi_1(t) = \psi_{11}(t) \pm \psi_{12}(t). \tag{10.27}$$

Similarly, for the total flux linkage of the second coil we have:

$$\psi_2(t) = \psi_{22}(t) \pm \psi_{21}(t). \tag{10.28}$$

We remark that the "$\pm$" signs in equations (10.27) and (10.28) indicate that mutual flux linkages may add to or subtract from self-flux linkages. Which of these two eventualities occurs depends on the relative directions of two currents as well as the relative winding directions of two coils. This matter will be discussed in detail a bit later.

By using expressions (10.7), (10.8), (10.10), and (10.11) in formulas (10.27) and (10.28), we can express the total flux linkages in terms of terminal currents as follows:

$$\psi_1(t) = L_1 i_1(t) \pm M i_2(t), \tag{10.29}$$

$$\psi_2(t) = L_2 i_2(t) \pm M i_1(t). \tag{10.30}$$

According to Faraday's law, terminal voltages can be related to total flux linkages in the manner specified below:

$$v_1(t) = \frac{d\psi_1(t)}{dt}, \qquad v_2(t) = \frac{d\psi_2(t)}{dt}. \tag{10.31}$$

By substituting expressions (10.29) and (10.30) into equations (10.31) and by assuming that self-inductance and mutual inductance are time invariant, we obtain:

$$v_1(t) = L_1 \frac{di_1(t)}{dt} \pm M \frac{di_2(t)}{dt}, \tag{10.32}$$

$$v_2(t) = L_2 \frac{di_2(t)}{dt} \pm M \frac{di_1(t)}{dt}. \tag{10.33}$$

Now, we shall address the issue of $\pm$ signs in equations (10.32) and (10.33) (as well as in previous equations). It has been remarked before that the sign ambiguity appears because two factors are simultaneously involved: relative current directions and relative coil winding directions. To illustrate the second factor, two sets of magnically coupled coils are shown in Figures 10.3a and b. These two sets of coupled coils are identical in all respects except that the secondary coils have opposite winding directions. As a consequence, the same directions of secondary currents will result in opposite directions of magnetic field lines and in opposite signs in equations (10.32) and (10.33). This sign ambiguity can be removed by the preliminary calibration of two magnetically coupled coils and by establishing the **dot convention** on the basis of this calibration. The calibration can be accomplished experimentally without any prior knowledge of the relative coil winding directions or it can be accomplished theoretically by using the **right-hand rule** which relates the directions of currents to the directions of magnetic field lines produced by these currents. In any case, it is assumed in circuit theory that this calibration has been performed and its results are represented by the dot convention. The essence of the dot convention can be stated as follows: an increasing (in time) current entering the dotted terminal of one coil results in such an open-circuit voltage at the terminals of the other coil that the dotted terminal is positive.

The dot convention is indicated in Figures 10.4a and b along with relative (reference) directions of electric currents. The coupled circuit equations (10.32) and (10.33) for the above two cases can be written, respectively, as follows. For case (a):

$$v_1(t) = L_1 \frac{di_1(t)}{dt} + M \frac{di_2(t)}{dt}, \tag{10.34}$$

$$v_2(t) = L_2 \frac{di_2(t)}{dt} + M \frac{di_1(t)}{dt}, \tag{10.35}$$

while for case (b):

$$v_1(t) = L_1 \frac{di_1(t)}{dt} - M \frac{di_2(t)}{dt}, \tag{10.36}$$

$$v_2(t) = L_2 \frac{di_2(t)}{dt} - M \frac{di_1(t)}{dt}. \tag{10.37}$$

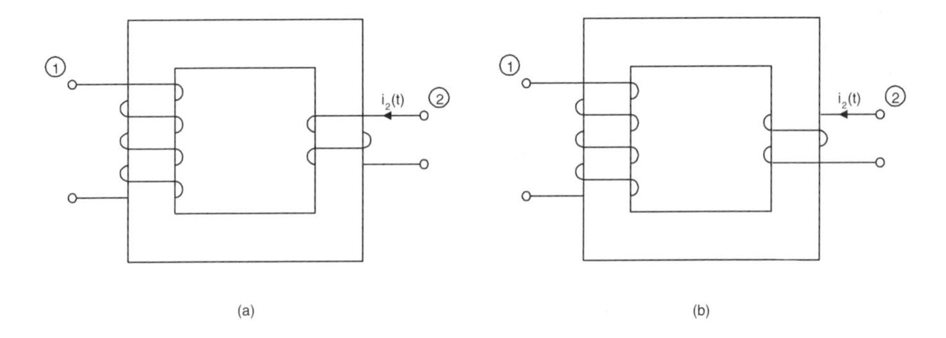

Figure 10.3: The case when the secondary coils have opposite winding directions.

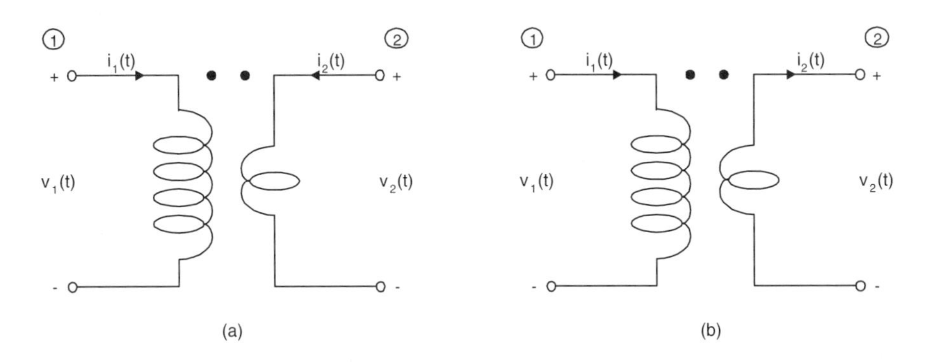

Figure 10.4: Dot convention.

The coupled circuit equations in the form (10.36)–(10.37) are sometimes preferable from the physical point of view. This is the case when the second coil does not have an independent source of excitation and is excited only due to the magnetic coupling with the first coil. Then, according to Lenz's law, the current $i_2(t)$, in the second coil is always induced in such a way as to counteract the cause of induction. The primary cause of induction of $i_2(t)$ is the voltage $v_1(t)$, and the minus sign in (10.36) reflects the counter action of this cause of induction. The described situation is typical for transformers and induction motors. For this reason, coupled circuit equations in the form (10.36)–(10.37) will be mostly used in our subsequent discussion. These equations can be easily generalized in order to account for finite resistances of the coupled coils. These resistances result in additional drops of voltages which lead to the following modification of coupled circuit equations:

$$v_1(t) = R_1 i_1(t) + L_1 \frac{di_1(t)}{dt} - M \frac{di_2(t)}{dt}, \tag{10.38}$$

$$v_2(t) = R_2 i_2(t) + L_2 \frac{di_2(t)}{dt} - M \frac{di_1(t)}{dt}. \tag{10.39}$$

In the case of ac steady state, the above equations can be written in the phasor form as follows:

$$\hat{V}_1 = R_1 \hat{I}_1 + jx_{11} \hat{I}_1 - jx_{12} \hat{I}_2, \tag{10.40}$$

$$\hat{V}_2 = R_2 \hat{I}_2 + jx_{22} \hat{I}_2 - jx_{12} \hat{I}_1, \tag{10.41}$$

where x_{11}, x_{22}, and x_{12} are the self-reactances and mutual reactance, which are related to self-inductances and mutual inductance by the expressions:

$$x_{11} = \omega L_1, \qquad x_{22} = \omega L_2, \tag{10.42}$$

$$x_{12} = \omega M. \tag{10.43}$$

The above coupled circuit equations can be also written in the impedance form:

$$\hat{V}_1 = z_{11}\hat{I}_1 + z_{12}\hat{I}_2, \tag{10.44}$$

$$\hat{V}_2 = z_{12}\hat{I}_1 + z_{22}\hat{I}_2, \tag{10.45}$$

where:

$$z_{11} = R_1 + jx_{11}, \qquad z_{22} = R_2 + jx_{22}, \qquad z_{12} = -jx_{12}. \tag{10.46}$$

It is appropriate to stress here that the mutual inductance is associated with two pairs of terminals or with two ports. In this respect, two inductively coupled coils can be treated as a two-port element (see Section 10.4). The coupled circuit equations (10.44)–(10.45) provide a complete terminal description of this two-port element. These equations can be used to complement the KVL and KCL equations written for circuits connected to the first and second coils, respectively. In this way, we can obtain the complete set of equations for magnetically coupled circuits.

It is clear from the previous discussion that magnetic coupling between two coils is realized for time-varying currents in these coils. This follows from Faraday's law (see formula (10.31)), which leads to induced voltages only for time-varying flux linkages. It is instructive to note that there is an interesting and important exception to the above assertion. This exception holds for superconducting coils. This is because for superconducting coils their total flux linkages are "frozen." They do not change with time and they remain equal to their initial values, that is, the values at the instance of superconducting transition. According to equations (10.29) and (10.30), this leads to the following constraints imposed on currents in the magnetically coupled superconducting coils:

$$\psi_1(t) = L_1 i_1(t) \pm M i_2(t) = \text{ constant}, \tag{10.47}$$

$$\psi_2(t) = L_2 i_2(t) \pm M i_1(t) = \text{ constant}. \tag{10.48}$$

It is apparent from equations (10.47) and (10.48) that even very slow variations in one current result in immediate adjustments in another current in order to keep the total flux linkages the same. This suggests that there is "static" (dc) coupling between superconducting coils. In practice, this coupling prohibits independent (separate) excitation of superconducting coils.

10.3 Transformers

In this section, we shall use the coupled circuit equations in order to discuss the operation and performance of a very important electric device known as a transformer. The transformer is used to step up or step down ac voltages. For this reason, the transformer is an indispensable link in electric power systems. The transformer can also be used to electrically isolate a source from a load or for impedance matching purposes. This explains why the transformer is also a vital component in many low-power applications.

The transformer can be defined as a device in which two (or more) stationary coils (called windings) are coupled magnetically. One of the windings, known as the primary, receives power at a specified voltage and frequency from the source, while the other winding, known as the secondary, delivers power to the load at a different voltage but the same frequency. To enhance electromagnetic coupling between the primary and secondary windings, they are placed on the same iron core (see Figure 10.5). This iron core is subject to a time-varying magnetic flux which links the primary and secondary windings. Since the iron core has a finite (nonzero) conductivity, this time-varying flux induces eddy currents in the iron core. These eddy currents may produce substantial power losses called eddy current losses. To reduce eddy current losses, the iron core is laminated. This means that the iron core is assembled from many very thin steel laminations which are electrically isolated from one another by very thin oxidation (or varnish) layers. The steel used for transformer laminations is called transformer steel. It is usually deliberately doped with silicon in order to reduce its intrinsic conductivity without appreciably affecting its high magnetic permeability. This reduction in steel conductivity further diminishes eddy current losses.

To better understand the principle of operation of the transformer, we shall first consider the ideal transformer. In the case of the ideal transformer, we neglect the small resistances R_1 and R_2 of the primary and secondary windings, respectively. We also neglect leakage flux ψ_ℓ. In other words, we assume that all magnetic field lines are entirely confined to the iron core. We also assume that the magnetic permeability, μ_c, of the iron core is infinite, while the conductivity, σ_c, of the same core is equal to zero. All the mentioned assumptions are summarized below:

$$R_1 = R_2 = 0, \qquad \psi_\ell = 0, \qquad \mu_c = \infty, \qquad \sigma_c = 0. \tag{10.49}$$

Since we neglect the leakage flux, it means that we assume that all turns of the primary and the secondary windings are linked with the same flux $\phi(t)$ which is formed by the magnetic field lines entirely confined to the iron core. This means that the flux linkages of the primary and secondary windings can be expressed as follows:

$$\psi_1(t) = N_1\phi(t), \qquad \psi_2(t) = N_2\phi(t). \tag{10.50}$$

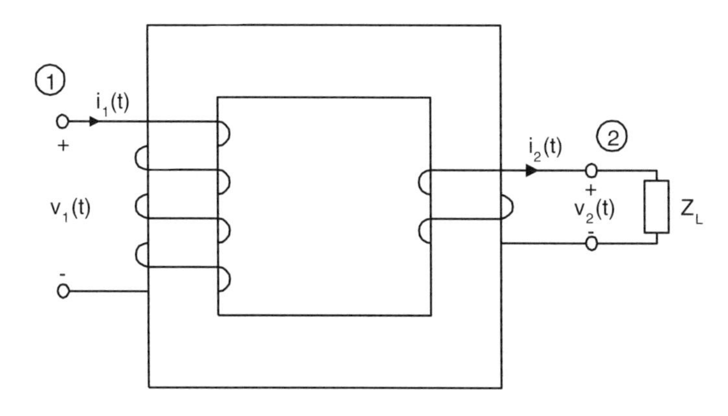

Figure 10.5: Transformer.

Now, by using Faraday's law and expressions (10.50), we find:

$$v_1(t) = \frac{d\psi_1(t)}{dt} = N_1\frac{d\phi(t)}{dt}, \tag{10.51}$$

$$v_2(t) = \frac{d\psi_2(t)}{dt} = N_2\frac{d\phi(t)}{dt}. \tag{10.52}$$

By dividing expression (10.51) by expression (10.52), we derive:

$$\frac{v_1(t)}{v_2(t)} = a, \tag{10.53}$$

where a stands for the turns ratio:

$$a = \frac{N_1}{N_2}. \tag{10.54}$$

If the applied (source) voltage, $v_1(t)$, of the primary winding is sinusoidal, then, according to (10.53), the secondary (induced) voltage, $v_2(t)$, is sinusoidal as well. It is also clear from (10.53) that the phasors of the primary and secondary voltages are related to one another as follows:

$$\frac{\hat{V}_1}{\hat{V}_2} = a. \tag{10.55}$$

Expressions (10.53) and (10.55) clearly elucidate the principle of operation of the transformer. They suggest that by manipulation of the turns ratio we can easily achieve the desired secondary voltage.

Next, we shall derive the expression for the ratio of primary and secondary currents. To this end, we shall exploit our assumptions that $R_1 = R_2 = 0$ and $\sigma_c = 0$. These assumptions mean that there are no power losses in the ideal transformer. Consequently, the primary power, $p_1(t)$, delivered to the terminals of the primary winding is equal to the secondary power, $p_2(t)$, delivered to the load:

$$p_1(t) = p_2(t). \tag{10.56}$$

Recalling the expressions for instantaneous power, we have:

$$p_1(t) = v_1(t)i_1(t), \qquad p_2(t) = v_2(t)i_2(t). \tag{10.57}$$

By substituting expressions (10.57) into equality (10.56), we end up with:

$$v_1(t)i_1(t) = v_2(t)i_2(t). \tag{10.58}$$

By dividing both parts of expression (10.58) by the product $v_1(t)i_2(t)$, we find:

$$\frac{i_1(t)}{i_2(t)} = \frac{v_2(t)}{v_1(t)}. \tag{10.59}$$

Now, by recalling expression (10.53), we finally derive:

$$\frac{i_1(t)}{i_2(t)} = \frac{1}{a}. \tag{10.60}$$

The last formula can be written in the phasor form as follows:

$$\frac{\hat{I}_1}{\hat{I}_2} = \frac{1}{a}. \tag{10.61}$$

Expressions (10.55) and (10.61) completely define the ideal transformer as a two-port element of electric circuits. It is sometimes more convenient to write these equations in the form:

$$\hat{V}_1 = a\hat{V}_2, \tag{10.62}$$

$$\hat{I}_1 = \frac{1}{a}\hat{I}_2, \tag{10.63}$$

which provide complete "terminal" characterization of this two-port element. These terminal relations should be combined with KVL and KCL equations for the rest of electric circuits in order to analyze the electric circuits with ideal transformers.

To conclude our discussion of the ideal transformer, consider its input impedance. We recall that the input impedance is defined as:

$$Z_{in} = \frac{\hat{V}_1}{\hat{I}_1}. \tag{10.64}$$

By using expressions (10.62) and (10.63) in formula (10.64), we obtain:

$$Z_{in} = a^2\frac{\hat{V}_2}{\hat{I}_2}. \tag{10.65}$$

However, the ratio of the secondary voltage $\hat{V}_2$ to the secondary current $\hat{I}_2$ is the load impedance:

$$Z_L = \frac{\hat{V}_2}{\hat{I}_2}. \tag{10.66}$$

By combining expressions (10.65) and (10.66), we derive:

$$Z_{in} = a^2 Z_L. \tag{10.67}$$

Thus, the load impedance Z_L viewed from the primary terminals is equal to $a^2 Z_L$. This fact suggests that the ideal (and real) transformer can be used for impedance matching purposes, and this is actually the case in some applications. It is also clear from expression (10.67) that, with respect to the primary terminals, the ideal transformer can be represented by the equivalent circuit shown in Figure 10.6. The equivalence here is understood in the sense that, as far as the relationship between the primary voltage $\hat{V}_1$ and primary current $\hat{I}_1$ is concerned, the ideal transformer and the circuit shown in Figure 10.6 are indistinguishable. It is a very powerful idea to

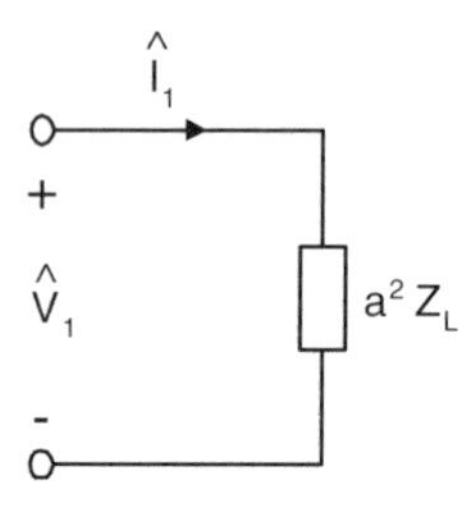

Figure 10.6: Equivalent circuit for the ideal transformer.

replace a complicated device by a simple equivalent electric circuit which provides the same relationship for terminal voltages and currents as we have for the actual device. This idea permeates many different areas of electrical engineering.

Next, we proceed to the discussion of the transformer theory by removing the first three assumptions in (10.49). The only assumption that still will be in place is that $\sigma_c = 0$, which is tantamount to the neglect of eddy current losses. These losses will be taken into account at the very end of our discussion, albeit in a somewhat ad hoc manner.

By removing the first three assumptions in (10.49), we can now base our analysis of the transformer on the coupled circuit equations (10.40)–(10.41). All the quantities in these equations marked by subscript 1 are related to the primary winding, while all the quantities marked by subscript 2 are related to the secondary winding, and x_{12} is the mutual reactance between the primary and secondary windings.

First, we use equation (10.40) and solve it for $\hat{I}_1$:

$$\hat{I}_1 = \frac{\hat{V}_1}{R_1 + jx_{11}} + \frac{jx_{12}}{R_1 + jx_{11}} \hat{I}_2. \tag{10.68}$$

Next, we substitute expression (10.68) into equation (10.41) and, after simple transformations, we obtain:

$$\hat{V}_2 = \frac{-jx_{12}}{R_1 + jx_{11}} \hat{V}_1 + (R_2 + jx_{22}) \left[1 - \frac{(jx_{12})^2}{(R_1 + jx_{11})(R_2 + jx_{22})} \right] \hat{I}_2. \tag{10.69}$$

Expression (10.69) for $\hat{V}_2$ has two distinct terms. The first term

$$\hat{V}_2^{\text{ind}} = \frac{-jx_{12}}{R_1 + jx_{11}} \hat{V}_1 \tag{10.70}$$

is the secondary voltage in the case when $\hat{I}_2 = 0$. In other words, this is an open-circuit secondary voltage which can be physically interpreted as the voltage induced in the secondary winding by the magnetic flux created by the current in the primary winding. This explains why we use superscript "ind" for $\hat{V}_2$ in (10.70).

The second term

$$\hat{V}_2^{\text{drop}} = (R_2 + jx_{22}) \left[1 - \frac{(jx_{12})^2}{(R_1 + jx_{11})(R_2 + jx_{22})} \right] \hat{I}_2 \tag{10.71}$$

has the physical meaning of the drop of the voltage due to the finite (nonzero) load current $\hat{I}_2$. It is highly desirable to have this term as small as possible in order to keep the secondary voltage $\hat{V}_2$ (which is the voltage across the load terminals) as constant as possible in the face of continuously (and quite often unpredictably) changing load. It is clear that the smaller $\hat{V}_2^{\text{drop}}$ the better the quality of the transformer.

Next, we consider another important characteristic of the transformer, namely, its secondary short-circuit current $\hat{I}_2^{sc}$. This current occurs when the secondary winding is accidently short-circuited, that is, when:

$$\hat{V}_2 = 0.$$

From the last expression and (10.69), we find:

$$\hat{I}_2^{sc} = \hat{V}_1 \frac{jx_{12}}{(R_1 + jx_{11})(R_2 + jx_{22}) \left[1 - \frac{(jx_{12})^2}{(R_1 + jx_{11})(R_2 + jx_{22})}\right]}. \tag{10.72}$$

We can see that the same quantity

$$D = 1 - \frac{(jx_{12})^2}{(R_1 + jx_{11})(R_2 + jx_{22})} \tag{10.73}$$

appears in the expressions (10.71) and (10.72) for $\hat{V}_2^{\text{drop}}$ and $\hat{I}_2^{sc}$, respectively. The smaller D, the better the quality of the transformer with respect to its ability to maintain more or less constant voltage across the load terminals in the face of changing load. However, small D results in a large short-circuit current $\hat{I}_2^{sc}$ which is undesirable. This suggests that the successful design of the transformer should carefully balance out these competing requirements.

The previous discussion clearly reveals the importance of the quantity D. For this reason, we shall carefully analyze this quantity. Usually, the primary and secondary resistances are much smaller than the primary and secondary reactances:

$$R_1 \ll x_{11}, \qquad R_2 \ll x_{22}. \tag{10.74}$$

For this reason, expression (10.73) for D can be simplified as follows:

$$D \approx 1 - \frac{x_{12}^2}{x_{11}x_{22}}. \tag{10.75}$$

By recalling formulas (10.42), (10.43), and (10.24), expression for D can be further transformed as follows:

$$D = 1 - \frac{M^2}{L_1 L_2} = 1 - k^2. \tag{10.76}$$

It has been pointed out in the previous section that, for strongly coupled coils (which is certainly the case for the primary and secondary transformer windings), the coupling coefficient k is very close to unity. This suggests that D is a small quantity. We shall next show that this quantity is fully determined by leakage parameters.

Consider the self-flux linkages ψ_{11} of the primary winding. These flux linkages are equal to the sum of flux linkages due to the magnetic flux ϕ, which is formed by the magnetic field lines entirely confined to the iron core, and the flux linkages ψ_1^{ℓ}, which are due to leakage field lines which partially or entirely lie outside the core (see Figure 10.2). Thus, the flux linkages ψ_{11} can be written as follows:

$$\psi_{11} = N_1\phi + \psi_1^{\ell}. \tag{10.77}$$

By recalling expression (10.10) for self-inductance, from formula (10.77) we find:

$$L_1 = \frac{N_1\phi}{i_1} + \frac{\psi_1^{\ell}}{i_1}. \tag{10.78}$$

Thus, we can see that the self-inductance consists of two distinct components. The first component

$$L_1^m = \frac{N_1\phi}{i_1} \tag{10.79}$$

can be interpreted as the main self-inductance. The term "main" emphasizes the fact that L_1^m is the predominant part of L_1.

The second term

$$L_1^{\ell} = \frac{\psi_1^{\ell}}{i_1} \tag{10.80}$$

can be interpreted as the "leakage" self-inductance.

Thus:

$$L_1 = L_1^m + L_1^{\ell}. \tag{10.81}$$

Next, we shall establish the connection between the mutual inductance M and the main self-inductance L_1^m. First, we recall that:

$$M = \frac{\psi_{21}}{i_1}. \tag{10.82}$$

It is also clear that:

$$\psi_{21} = N_2\phi. \tag{10.83}$$

By substituting expression (10.83) into formula (10.82), after simple transformation we obtain:

$$M = \frac{N_2\phi}{i} = \frac{N_2}{N_1}\frac{N_1\phi}{i_1}. \tag{10.84}$$

By using expression (10.79) in (10.84), we derive:

$$M = \frac{N_2}{N_1}L_1^m. \tag{10.85}$$

By literally repeating the same line of reasoning as was used in derivation of (10.81) and (10.85), we can establish that:

$$L_2 = L_2^m + L_2^{\ell}, \tag{10.86}$$

and

$$M = \frac{N_1}{N_2} L_2^m. \tag{10.87}$$

By substituting expressions (10.85) and (10.87) into formulas (10.81) and (10.86), respectively, we find:

$$\frac{N_1}{N_2} M = L_1 - L_1^{\ell}, \tag{10.88}$$

$$\frac{N_2}{N_1} M = L_2 - L_2^{\ell}. \tag{10.89}$$

By multiplying the expressions (10.88) and (10.89), we obtain:

$$M^2 = (L_1 - L_1^{\ell})(L_2 - L_2^{\ell}) = L_1 L_2 - L_1 L_2^{\ell} - L_2 L_1^{\ell} + L_1^{\ell} L_2^{\ell}. \tag{10.90}$$

By dividing both sides of expression (10.90) by the product $L_1 L_2$, we arrive at:

$$\frac{M^2}{L_1 L_2} = 1 - \frac{L_1^{\ell}}{L_1} - \frac{L_2^{\ell}}{L_2} + \frac{L_1^{\ell} L_2^{\ell}}{L_1 L_2}. \tag{10.91}$$

By substituting the last expression into formula (10.76), we finally derive:

$$D = \frac{L_1^{\ell}}{L_1} + \frac{L_2^{\ell}}{L_2} - \frac{L_1^{\ell} L_2^{\ell}}{L_1 L_2}. \tag{10.92}$$

The last expression clearly reveals the importance of leakage inductances. These inductances determine the value of D, which, in turn, determines the ability of the transformer to maintain more or less constant voltage across the load terminals as well as the vulnerability of the transformer with respect to the occurrence of accidental shorts of the load terminals.

The importance of the leakage inductances can also be demonstrated from the purely mathematical point of view. Consider coupled circuit equations (10.40) and (10.41) as a set of two linear simultaneous equations with respect to $\hat{I}_1$ and $\hat{I}_2$. It is easy to see that the determinant, Δ, of these equations is equal to:

$$\Delta = (R_1 + jx_{11})(R_2 + jx_{22}) - (jx_{12})^2 = (R_1 + jx_{11})(R_2 + jx_{22})D. \tag{10.93}$$

It is evident from expressions (10.93) and (10.92) that this determinant is small and it becomes equal to zero when we neglect the leakage inductances (along with small resistances R_1 and R_2). In other words, the set of coupled circuit equations (10.40)–(10.41) becomes degenerate (singular) if we neglect the leakage inductances. In mathematics, the problems in which the neglect of small parameters leads to degeneracy (singularity) are called **singularly perturbed** problems. Thus, the coupled circuit equations (10.40)–(10.41), which describe the operation of the transformer, are singularly perturbed. It has long been understood in mathematics that small parameters in singularly perturbed problems are very important. This is because it is the small parameters which make the singularly pertrubed problems well defined (nonsingular). The presented discussion brings other (mathematical) evidence for the importance of the leakage inductances.

Thus, we have established that the leakage inductances are very important. However, the coupled circuit equations (10.40)–(10.41) do not contain the leakage reactances explicitly. These reactances are absorbed by the total reactances x_{11} and x_{22} and their significance is masked and lost. This suggests that it is highly desirable to modify the coupled circuit equations (10.40)–(10.41) in such a way that the leakage reactances will be **explicitly accounted for**. As is typical in singularly perturbed problems, the small parameters—leakage reactances—can be exposed as a result of the appropriate **scaling**. For the coupled circuit equations (10.40)–(10.41), this scaling is accomplished by introducing scaled (or primed) secondary voltage and secondary current:

$$\hat{V}_2' = a\hat{V}_2, \qquad \hat{I}_2' = \frac{\hat{I}_2}{a}, \tag{10.94}$$

where, as before, a is the turns ratio.

The coupled circuit equations (10.40)–(10.41) can be written in terms of $\hat{V}_2'$ and $\hat{I}_2'$ as follows:

$$\hat{V}_1 = R_1\hat{I}_1 + jx_{11}\hat{I}_1 - jax_{12}\hat{I}_2', \tag{10.95}$$

$$\hat{V}_2' = a^2R_2\hat{I}_2' + ja^2x_{22}\hat{I}_2' - jax_{12}\hat{I}_1. \tag{10.96}$$

Now, we introduce the scaled secondary resistance and secondary reactance:

$$R_2' = a^2R_2, \qquad x_{22}' = a^2x_{22}, \tag{10.97}$$

and rewrite the previous equations in the form given below:

$$\hat{I}_1 = R_1\hat{I}_1 + jx_{11}\hat{I}_1 - jax_{12}\hat{I}_2', \tag{10.98}$$

$$\hat{V}_2' = R_2'\hat{I}_2' + jx_{22}'\hat{I}_2' - jax_{12}\hat{I}_1. \tag{10.99}$$

Next, we shall make the following equivalent transformations of the last two equations: we shall add and subtract the terms $jax_{12}\hat{I}_1$ and $jax_{12}\hat{I}_2'$ in equations (10.98) and (10.99), respectively:

$$\hat{V}_1 = R_1\hat{I}_1 + j(x_{11} - ax_{12})\hat{I}_1 + jax_{12}(\hat{I}_1 - \hat{I}_2'), \tag{10.100}$$

$$\hat{V}_2' = R_2'\hat{I}_2' + j(x_{22}' - ax_{12})\hat{I}_2' + jax_{12}(\hat{I}_2' - \hat{I}_1). \tag{10.101}$$

Now, consider the physical meaning of the coefficients in equations (10.100)–(10.101). First, by using formulas (10.43), (10.54), and (10.85), we derive:

$$ax_{12} = \frac{N_1}{N_2}\omega M = \frac{N_1}{N_2}\omega\frac{N_2}{N_1}L_1^m = \omega L_1^m = x_{11}^m, \tag{10.102}$$

where x_{11}^m has the physical meaning of the main reactance of the primary winding. Next, by using formulas (10.42), (10.43), and (10.81), we find:

$$x_{11} - ax_{12} = \omega L_1 - \omega L_1^m = \omega L_1^\ell = x_1^\ell, \tag{10.103}$$

where x_1^ℓ stands for the leakage reactance of the primary winding.

Finally, by using formulas (10.42), (10.43), (10.86), (10.87), and (10.97), we obtain:

$$x'_{22} - ax_{12} = a^2 x_{22} - ax_{12} = a^2 \left(x_{22} - \frac{1}{a} x_{12} \right), \tag{10.104}$$

$$x_{22} - \frac{1}{a} x_{12} = \omega L_2 - \frac{N_2}{N_1} \omega M = \omega (L_2 - L_2^m) = \omega L_2^\ell = x_2^\ell, \tag{10.105}$$

$$x'_{22} - ax_{12} = a^2 x_2^\ell = (x_2^\ell)', \tag{10.106}$$

where x_2^ℓ stands for the leakage reactance of the secondary winding and $(x_2^\ell)'$ is the scaled value of this reactance.

By substituting expressions (10.102), (10.103), and (10.106) into equations (10.100)–(10.101), we derive:

$$\hat{V}_1 = R_1 \hat{I}_1 + j x_1^\ell \hat{I}_1 + j x_{11}^m (\hat{I}_1 - \hat{I}'_2), \tag{10.107}$$

$$\hat{V}'_2 = R'_2 \hat{I}'_2 + j (x_2^\ell)' \hat{I}'_2 + j x_{11}^m (\hat{I}'_2 - \hat{I}_1). \tag{10.108}$$

Thus, by means of equivalent mathematical transformations we have represented the coupled circuit equations (10.40) and (10.41) in the form (10.107)–(10.108) in which the leakage reactances are explicitly accounted for. The useful by-product of the above equivalent transformations is the fact that equations (10.107) and (10.108) can be interpreted as KVL equations for the electric circuit shown in Figure 10.7. This circuit can be considered the equivalent electric circuit for the transformer. In other words, this circuit and the transformer are described by mathematically identical sets of equations and, consequently, the transformer and this circuit are indistinguishable as far as the relationship between terminal voltages and terminal currents are concerned. The load impedance of the transformer is modeled in the equivalent circuit by its scaled value $a^2 Z_L$. This immediately follows from (10.94) and the following brief derivation:

$$\frac{\hat{V}'_2}{\hat{I}'_2} = a^2 \frac{\hat{V}_2}{\hat{I}_2} = a^2 Z_L. \tag{10.109}$$

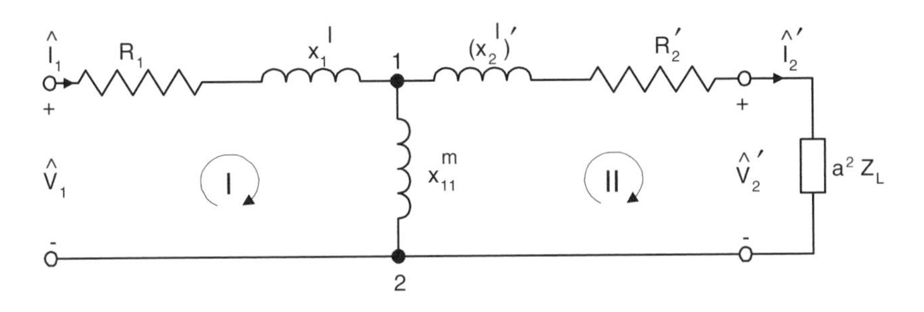

Figure 10.7: Equivalent circuit for the transformer without eddy current losses.

In our discussions, we have so far neglected eddy current losses in the transformer core. Now, we shall try to take them into account in the equivalent circuit for the transformer. It can be shown (and this is beyond the scope of this text) that eddy current losses are directly proportional to the square of the magnetic flux through the iron core. By using Faraday's law, it is easy to see that the rms value of the voltage induced in the primary winding by the core flux is directly proportional to this flux. Thus, we can conclude that the eddy current losses are proportional to the square of the voltage induced in the primary winding by the core flux. In the equivalent circuit, this voltage is the voltage, V_{12}, across the terminals of the reactance x_{11}^{m}. Thus, we can write that

$$P_e \sim V_{12}^2, \tag{10.110}$$

where P_e is the eddy current loss.

Expression (10.110) leads to the idea of modeling the actual eddy current losses P_e in the iron core of the transformer by the ohmic losses P_{Re} in resistor R_e connected across the terminals of x_{11}^{m} (see Figure 10.8). The rationale behind this idea is the fact that the ohmic losses in R_e are also proportional to the square of the same voltage:

$$P_{Re} = \frac{V_{12}^2}{R_e}. \tag{10.111}$$

The resistor, R_e, can be chosen from the condition that losses P_e and P_{R_e} should be the same:

$$P_e = P_{Re}. \tag{10.112}$$

By using expression (10.112) in formula (10.111), we easily derive:

$$R_e = \frac{V_{12}^2}{P_e}. \tag{10.113}$$

The electric circuit shown in Figure 10.8 is the complete equivalent circuit for the transformer. One of the important features of this equivalent circuit is that the important small parameters such as the leakage reactances are explicitly accounted

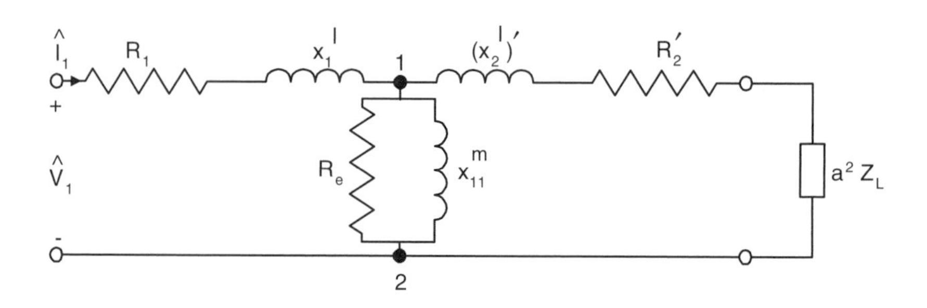

Figure 10.8: Complete equivalent circuit for the transformer.

for. It is interesting to note that, in the case of the ideal transformer, the equivalent circuit shown in Figure 10.8 is reduced to the equivalent circuit shown in Figure 10.6. Indeed, in the case of the ideal transformer we assume that primary and secondary resistances R_1 and R_2 are equal to zero. We also assume that leakage flux linkages are equal to zero, which is tantamount to the assumption that leakage reactances x_1^ℓ and $(x_2^\ell)'$ are zeros. The above assumptions mean that we should replace the corresponding circuit elements by short-circuit branches. In the case of the ideal transformer, we also assume that $\sigma_c = 0$ and, consequently, $P_e = 0$, which according to (10.113) leads to the infinite value of R_e. It is also assumed that $\mu_c = \infty$, which leads to the infinite value of x_{11}^m. Thus, the last two assumptions suggest that the corresponding circuit elements should be replaced by open-circuit branches. By introducing the above-mentioned short-circuit and open-circuit branches, we reduce the equivalent circuit shown in Figure 10.8 to the equivalent circuit shown in Figure 10.6.

EXAMPLE 10.1 Consider the circuit shown in Figure 10.9a. The parameters of the transformer are $L_1^m = 1$ mH, $L_2^m = 25$ mH, $L_1^\ell = 40\ \mu$H, $L_2^\ell = 1$ mH, $R_1 = 0.04\ \Omega$, $R_2 = 1.0\ \Omega$, and $a = 0.2$. The voltage source is given by $v_1(t) = 100\sin(2500t)$ V. Find the voltage across the load resistor.

First note that if the transformer were ideal, the output voltage would be $v_o(t) = 500\sin(2500t)$ V. To solve the problem with the nonideal transformer, we first need to find the equivalent transformer parameters indicated in Figure 10.7. Then we can use mesh analysis to compute the scaled output voltage. Finally, we can use the scaling relation (10.94) to convert the scaled result to the actual output voltage.

From (10.102)–(10.106) and the parameters listed above, we find $x_{11}^m = 2.5\ \Omega$, $x_1^\ell = 0.1\ \Omega$, and $(x_2^\ell)' = 0.1\ \Omega$. From (10.97) we find $R_2' = 0.04\ \Omega$ and from (10.67) we calculate the scaled load impedance to be $a^2 R_L = 4\ \Omega$. The equivalent circuit is shown in Figure 10.9b.

The generalized mesh equations are:

$$\begin{bmatrix} 0.04 + 2.6j & -2.5j \\ -2.5j & 4.04 + 2.6j \end{bmatrix} \begin{bmatrix} \hat{\mathcal{I}}_1 \\ \hat{\mathcal{I}}_2' \end{bmatrix} = \begin{bmatrix} -100j \\ 0 \end{bmatrix} \text{V}$$

from which we can find the scaled secondary voltage

$$\hat{V}_2' = 4\hat{\mathcal{I}}_2' = \frac{-1000j}{10.608 + 0.3484j} = -94.2je^{-j0.0328}\ \text{V}.$$

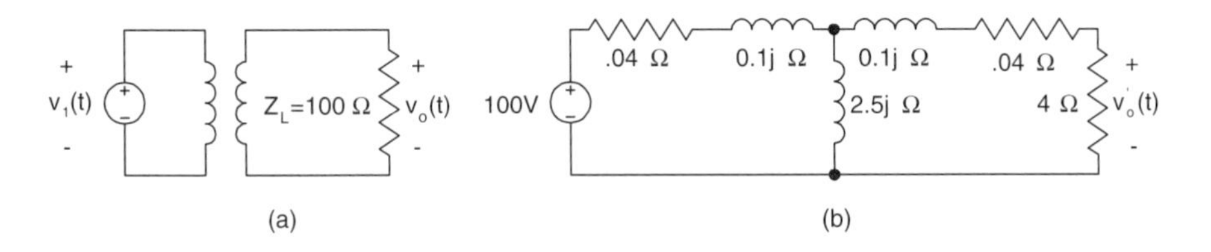

Figure 10.9: (a) Transformer circuit and (b) equivalent circuit with scaled parameters.

The actual voltage is $\hat{V}_2 = \hat{V}_2'/a$, which can be represented in the time domain as follows:

$$v_2(t) = 471 \sin(2500t - 1.88°) \text{ V}. \qquad \blacksquare$$

In the conclusion of this section, it is appropriate to point out that the theory developed here for the transformer is quite general in nature and is applicable to many electric devices which contain **strongly** coupled coils (windings). For these devices, the corresponding coupled circuit equations are singularly perturbed and the leakage inductances (or leakage reactances) play a very important role which suggests that they should be explicitly accounted for. This, for instance, explains why the theory of induction motors has many features similar to the theory of the transformer.

10.4 Theory of Two-Port Elements

A two-port element is an element of an electric circuit which is associated with two pairs of terminals (see Figure 10.10). Two left terminals will be called the first port and the voltage and current associated with these terminals will be marked by subscript "1." Similarly, two right terminals will be called the second port and the voltage and current associated with these terminals will be marked by subscript "2."

In circuit theory, two-port elements serve a twofold purpose. First, they can be used as equivalent replacements for complicated electric circuits. In this way, the use of the two-port elements can simplify (and facilitate) the analysis of electric circuits. Second (and more important), the two-port elements can be employed as circuit models for complicated devices such as transistors, transformers, and transmission lines. The interior dynamics of these devices is usually too complex to be fully described by electric circuit theory. However, in the analysis of electric circuits with these devices, quite often one does not need to know all the details of their behavior; the mathematical relationships between device terminal voltages and currents are quite sufficient. This is exactly the purpose that the theory of two-port elements fulfills. The two-port theory provides terminal voltage-current characterizations of the devices. These terminal voltage-current relations can then be coupled with KVL

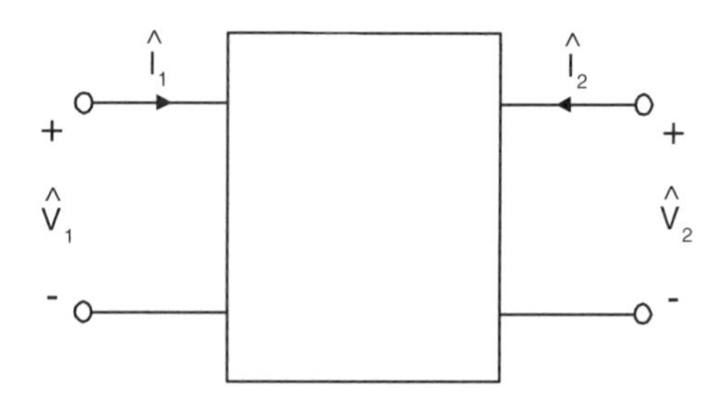

Figure 10.10: The two-port element.

and KCL equations and used for the analysis of electric circuits with complicated devices.

In this section, we shall consider linear and passive two-port elements. The terminal voltage-current relations for these two-port elements are described by **linear** and **homogeneous** equations. These equations can be written in many different forms which result in different representations of two-port elements.

We begin with the z-representation of two-port elements. In this representation, the terminal voltages are expressed as linear and homogeneous functions of terminal currents:

$$\hat{V}_1 = z_{11}\hat{I}_1 + z_{12}\hat{I}_2, \tag{10.114}$$

$$\hat{V}_2 = z_{21}\hat{I}_1 + z_{22}\hat{I}_2. \tag{10.115}$$

The linearity of equations (10.114) and (10.115) is a consequence of linearity of the two-port element. Indeed, the two-port element can be viewed as being excited by two current sources $\hat{I}_{s1} = \hat{I}_1$ and $\hat{I}_{s2} = \hat{I}_2$ connected to the first and second ports, respectively. Then, according to the superposition principle, the terminal voltages should be linear combinations of these current sources. This is exactly what is expressed by equations (10.114) and (10.115). The homogeneity of these equations is a consequence of the passivity of the two-port element. In other words, the homogeneity of the above equations comes from the fact that the two-port element does not contain any independent sources; as a result, terminal voltages $\hat{V}_1$ and $\hat{V}_2$ should be equal to zero when terminal source currents $\hat{I}_1$ and $\hat{I}_2$ are set to zero.

Equations (10.114) and (10.115) can be written in the matrix form:

$$\vec{V} = \tilde{Z}\vec{I}, \tag{10.116}$$

where vectors $\vec{V}, \vec{I}$ and matrix $\tilde{Z}$ are defined as follows:

$$\vec{V} = \begin{pmatrix} \hat{V}_1 \\ \hat{V}_2 \end{pmatrix}, \quad \vec{I} = \begin{pmatrix} \hat{I}_1 \\ \hat{I}_2 \end{pmatrix}, \quad \tilde{Z} = \begin{pmatrix} z_{11} & z_{12} \\ z_{21} & z_{22} \end{pmatrix}. \tag{10.117}$$

Next, we show that the z-parameters in the above equation can be determined by using the two open-circuit tests shown in Figure 10.11. In test (a), the current source

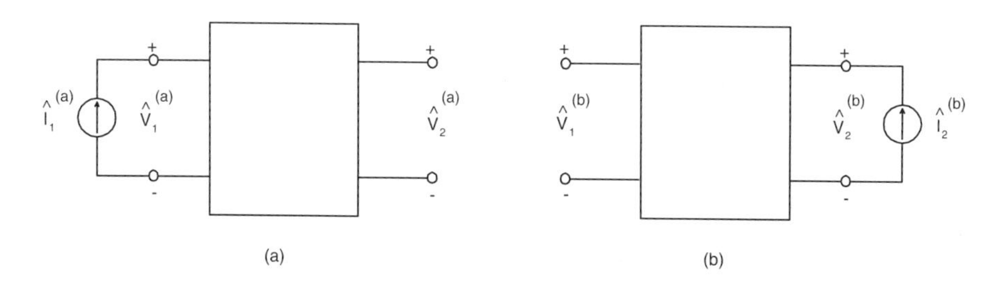

Figure 10.11: Two open-circuit tests for identification of the z-representation.

$\hat{I}_1^{(a)}$ is connected to the terminals of the first port while the terminals of the second port are kept open. The latter means that:

$$\hat{I}_2^{(a)} = 0. \tag{10.118}$$

In this test, voltages $\hat{V}_1^{(a)}$ and $\hat{V}_2^{(a)}$ are measured. From equations (10.114), (10.115), and (10.118), we find:

$$\hat{V}_1^{(a)} = z_{11}\hat{I}_1^{(a)}, \qquad \hat{V}_2^{(a)} = z_{12}\hat{I}_1^{(a)}. \tag{10.119}$$

Consequently,

$$z_{11} = \frac{\hat{V}_1^{(a)}}{\hat{I}_1^{(a)}}, \qquad z_{21} = \frac{\hat{V}_1^{(a)}}{\hat{I}_1^{(a)}}. \tag{10.120}$$

The last two expressions are quite often written in the forms:

$$z_{11} = \frac{\hat{V}_1}{\hat{I}_1}\Big|_{\hat{I}_2=0}, \qquad z_{21} = \frac{\hat{V}_2}{\hat{I}_1}\Big|_{\hat{I}_2=0}, \tag{10.121}$$

which explicitly reflect the condition of the test (a).

In the case of test (b), the current source $\hat{I}_2^{(b)}$ is connected across the second port, while the terminals of the first port are kept open. The latter means that

$$\hat{I}_1^{(b)} = 0. \tag{10.122}$$

In this test, voltages $\hat{V}_1^{(b)}$ and $\hat{V}_2^{(b)}$ are measured. From equations (10.114), (10.115), and (10.122), we find:

$$\hat{V}_1^{(b)} = z_{12}\hat{I}_2^{(b)}, \qquad \hat{V}_2^{(b)} = z_{22}\hat{I}_2^{(b)}. \tag{10.123}$$

From expressions (10.123), we derive:

$$z_{12} = \frac{\hat{V}_1^{(b)}}{\hat{I}_2^{(b)}}, \qquad z_{22} = \frac{\hat{V}_2^{(b)}}{\hat{I}_2^{(b)}}, \tag{10.124}$$

which can also be written as:

$$z_{12} = \frac{\hat{V}_1}{\hat{I}_2}\Big|_{\hat{I}_1=0}, \qquad z_{22} = \frac{\hat{V}_2}{\hat{I}_2}\Big|_{\hat{I}_1=0}. \tag{10.125}$$

The tests (a) and (b) can be used to find out if the two-port element is reciprocal. In the case of the reciprocal two-port element, the equality of excitations

$$\hat{I}_1^{(a)} = \hat{I}_2^{(b)} \tag{10.126}$$

results in the following equality of responses:

$$\hat{V}_1^{(b)} = \hat{V}_2^{(a)} \tag{10.127}$$

In other words, reciprocity means that measurement results are invariant with respect to the interchange of excitation and response terminals. In the case of the reciprocal two-port element, from expressions (10.120), (10.124), (10.126), and (10.127) we derive:

$$z_{12} = z_{21}. \tag{10.128}$$

This means that for reciprocal two-port elements $\tilde{Z}$-matrices are symmetric.

It can be shown that, when two-port elements are used as equivalent replacements for passive electric circuits without dependent sources, then these two-port elements are always reciprocal. The formal proof is based on the symmetry of matrices for mesh current equations and Cramer's rule. This proof is omitted.

When two-port elements are used as circuit models for devices, they are not necessarily reciprocal. For instance, the two-port elements used as the models for transformers are reciprocal (see the previous section), while the two-port elements used as the models for transistors are not reciprocal.

The tests (a) and (b) described above can be used for experimental determination of z-parameters. In this sense, these two tests constitute the experimental procedure for the identification of the two-port elements as circuit models for real devices. The above two tests can be also employed for computational determination of z-parameters of the two-port elements when these two-port elements are used as equivalent replacements for electric circuits. We illustrate this statement by the following example.

EXAMPLE 10.2 Consider the two-port circuit shown in Figure 10.12. We want to find z-parameters of the equivalent two-port element.

It is clear from Figure 10.11 and formulas (10.120) and (10.124) that z_{11} and z_{22} can be construed as the input impedances of circuits shown in Figures 10.13a and b, respectively. In the case of the circuit shown in Figure 10.13a, we can clearly see that impedances Z_4 and Z_5 are connected in series and together they connected in parallel with Z_3. This group of three impedances is connected in series with Z_2 and the group of the above four impedances is connected in parallel with Z_5. Thus, z_{11} as the input impedance of the circuit shown in Figure 10.13a is given by:

$$z_{11} = \frac{Z_1\left(Z_2 + \frac{Z_3(Z_4+Z_5)}{Z_3+Z_4+Z_5}\right)}{Z_1 + Z_2 + \frac{Z_3(Z_4+Z_5)}{Z_3+Z_4+Z_5}}. \tag{10.129}$$

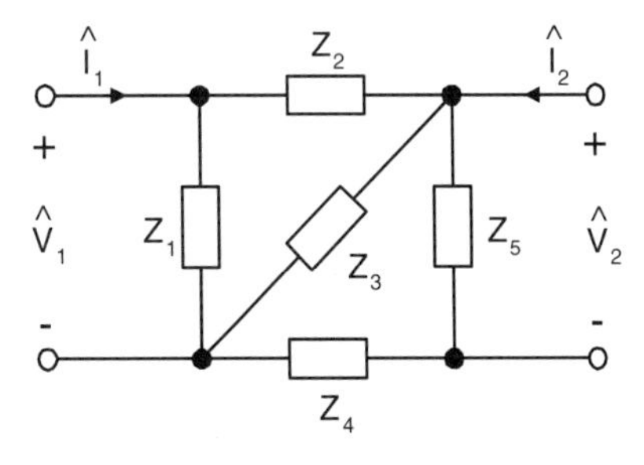

Figure 10.12: A two-port electric circuit.

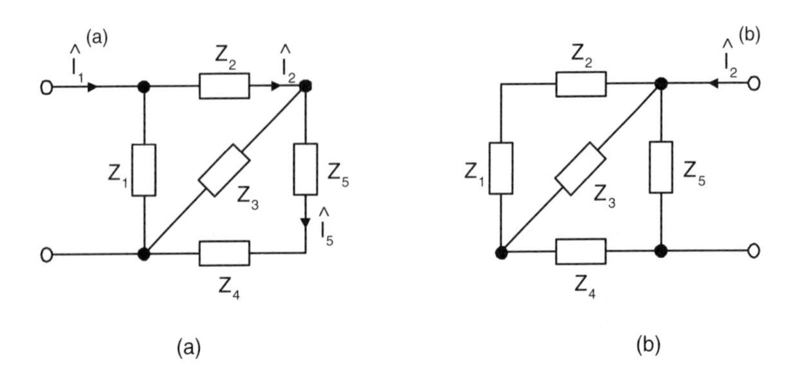

Figure 10.13: Two circuits used for the determination of z-parameters.

By using Figure 10.13b and the same line of reasoning, we find:

$$z_{22} = \frac{Z_5(Z_4 + \frac{Z_3(Z_1+Z_2)}{Z_3+Z_1+Z_2})}{Z_5 + Z_4 + \frac{Z_3(Z_1+Z_2)}{Z_3+Z_1+Z_2}}. \tag{10.130}$$

To find z_{21} we shall use the second formula given in (10.119). It is clear that in our case, voltage $\hat{V}_2^{(a)}$ is equal to the voltage, $\hat{V}_5$, across impedance Z_5:

$$\hat{V}_2^{(a)} = \hat{V}_5. \tag{10.131}$$

By consecutively using the current divider rule for the circuit shown in Figure 10.13a, we find:

$$\hat{I}_2 = \hat{I}_1^{(a)} \frac{Z_1}{Z_1 + Z_2 + \frac{Z_3(Z_4+Z_5)}{Z_3+Z_4+Z_5}}, \tag{10.132}$$

$$\hat{I}_5 = \hat{I}_2 \frac{Z_3}{Z_3 + Z_4 + Z_5}. \tag{10.133}$$

From formula (10.131), we conclude that:

$$\hat{V}_2^{(a)} = \hat{I}_5 Z_5. \tag{10.134}$$

By combining expressions (10.119), (10.132), (10.133), and (10.134), we obtain:

$$z_{21} = \frac{Z_1 Z_3 Z_5}{(Z_3 + Z_4 + Z_5)(Z_1 + Z_2 + \frac{Z_3(Z_4+Z_5)}{Z_3+Z_4+Z_5})}. \tag{10.135}$$

The expression for z_{12} is identical to that for z_{21} because in our problem the equivalent two-port element is reciprocal. However, it is suggested that the reader verify the reciprocity by the direct analysis of the circuit shown in Figure 10.13b. ■

Thus, we have seen that two-port elements can be used as equivalent replacements for electric circuits. It is interesting to consider the inverse problem. Given a two-port element defined by equations (10.114)–(10.115), we want to find a simple circuit realization for this two-port element. In other words, we want to find a simple two-port electric circuit with terminal voltage-current relations identical to those described by formulas (10.114) and (10.115).

First, we shall consider the case of reciprocal two-port elements. In this case, it is natural to look for circuit realizations without dependent sources. To this end, consider the circuit shown in Figure 10.14. By writing KVL equations for loops I and II, we find:

$$\hat{V}_1 = Z_a\hat{I}_1 + Z_c(\hat{I}_1 + \hat{I}_2), \tag{10.136}$$

$$\hat{V}_2 = Z_b\hat{I}_2 + Z_c(\hat{I}_1 + \hat{I}_2). \tag{10.137}$$

By combining the similar terms in the above equations, we obtain:

$$\hat{V}_1 = (Z_a + Z_c)\hat{I}_1 + Z_c\hat{I}_2, \tag{10.138}$$

$$\hat{V}_2 = Z_c\hat{I}_1 + (Z_b + Z_c)\hat{I}_2. \tag{10.139}$$

By comparing the last two equations with equations (10.114) and (10.115), we conclude that these pairs of equations will be identical under the conditions:

$$Z_a + Z_c = z_{11}, \quad Z_b + Z_c = z_{22}, \quad Z_c = z_{12} = z_{21}. \tag{10.140}$$

By solving equations (10.140) for Z_a, Z_b, and Z_c, we find:

$$Z_a = z_{11} - z_{12}, \quad Z_b = z_{22} - z_{12}, \qquad Z_c = z_{12}. \tag{10.141}$$

Thus, for impedances defined by expression (10.141), the electric circuit shown in Figure 10.14 will have terminal voltage-current relations identical to those of the two-port element described by equations (10.114)–(10.115). In this sense, we have arrived at the circuit realization of the two-port element. It is important to keep in mind that impedances Z_a, Z_b, and Z_c may not always be physically realizable. This will be the case when real parts of these impedances are negative. To circumvent this difficulty as well as to include nonreciprocal two-port elements, circuit realizations with dependent sources can be used. One of these realizations which contains two

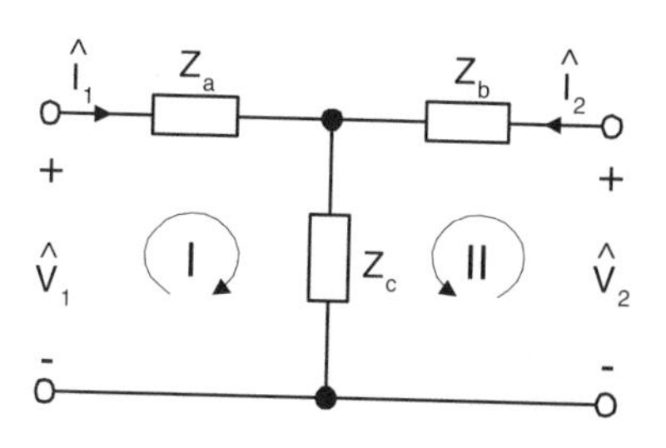

Figure 10.14: A circuit realization of the z-representation.

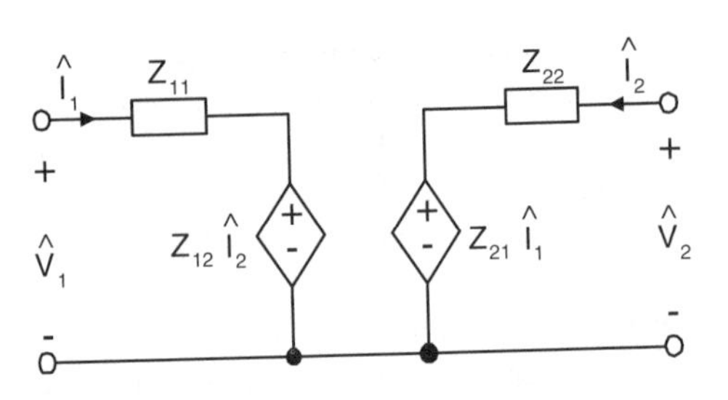

Figure 10.15: A circuit realization of the z-representation with two dependent sources.

dependent voltage sources is shown in Figure 10.15. It is apparent that the two KVL equations for this circuit coincide with equations (10.114)–(10.115), which means that Figure 10.15 provides a circuit realization for the two-port element. Another circuit realization for the two-port elements is given in the following example.

EXAMPLE 10.3 Consider the circuit shown in Figure 10.16. We want to show that it provides a circuit realization for the two-port element.

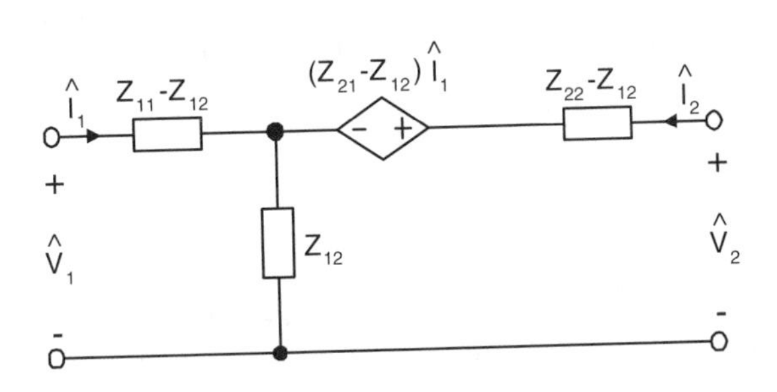

Figure 10.16: A circuit realization of the z-representation with one dependent source.

First, we write two KVL equations:

$$\hat{V}_1 = (z_{11} - z_{12})\hat{I}_1 + z_{12}(\hat{I}_1 + \hat{I}_2), \tag{10.142}$$

$$\hat{V}_2 = (z_{22} - z_{12})\hat{I}_2 + (z_{21} - z_{12})\hat{I}_1 + z_{12}(\hat{I}_1 + \hat{I}_2). \tag{10.143}$$

Be performing cancellation of the identical terms in equations (10.142) and (10.143), we end up with:

$$\hat{V}_1 = z_{11}\hat{I}_1 + z_{12}\hat{I}_2, \tag{10.144}$$

$$\hat{V}_2 = z_{21}\hat{I}_1 + z_{22}\hat{I}_2. \tag{10.145}$$

The last two equations are the same as equations (10.114) and (10.115). ■

The z-representation of two-port elements is very convenient for the analysis of series connection of two-port elements. This connection is shown in Figure 10.17.

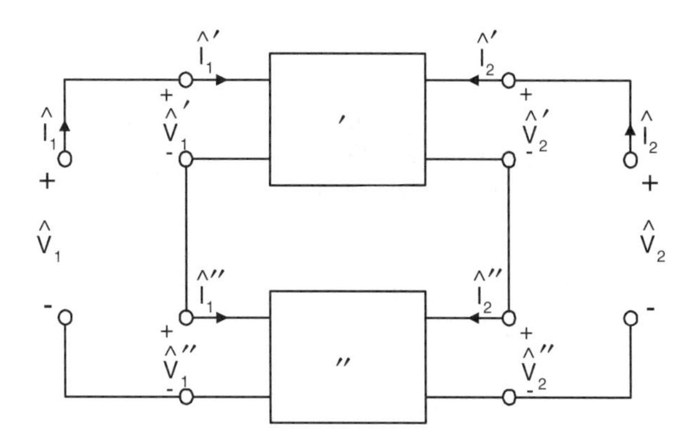

Figure 10.17: Series connection of two-port elements.

Our intent is to find the z-representation for the resulting two-port element. We start our derivation by writing z-representations for the "primed" and "double-primed" two-port elements:

$$\begin{pmatrix} \hat{V}_1' \\ \hat{V}_2' \end{pmatrix} = \begin{pmatrix} \hat{z}_{11}' & z_{12}' \\ z_{21}' & z_{22}' \end{pmatrix} \begin{pmatrix} \hat{I}_1' \\ \hat{I}_2' \end{pmatrix}, \tag{10.146}$$

$$\begin{pmatrix} \hat{V}_1'' \\ \hat{V}_2'' \end{pmatrix} = \begin{pmatrix} z_{11}'' & z_{12}'' \\ z_{27}'' & z_{22}'' \end{pmatrix} \begin{pmatrix} \hat{I}_1'' \\ \hat{I}_2'' \end{pmatrix}. \tag{10.147}$$

Due to the series connection of the above two-port elements, we have:

$$\hat{V}_1 = \hat{V}_1' + \hat{V}_1'', \qquad \hat{V}_2 = \hat{V}_2' + \hat{V}_2'', \tag{10.148}$$

$$\hat{I}_1 = \hat{I}_1' = \hat{I}_1'', \qquad \hat{I}_2 = \hat{I}_2' = \hat{I}_2''. \tag{10.149}$$

By adding equations (10.146) and (10.147) and then by taking into account expressions (10.148) and (10.149), we derive:

$$\begin{pmatrix} \hat{V}_1 \\ \hat{V}_2 \end{pmatrix} = \begin{pmatrix} z_{11}' + z_{11}'' & z_{12}' + z_{12}'' \\ z_{21}' + z_{22}'' & z_{22}' + z_{22}'' \end{pmatrix} \begin{pmatrix} \hat{I}_1 \\ \hat{I}_2 \end{pmatrix}, \tag{10.150}$$

which is the z-representation of the resulting two-port element. The last result can be written in concise matrix form as follows:

$$\tilde{Z} = \tilde{Z}' + \tilde{Z}''. \tag{10.151}$$

It is very convenient to couple the z-representation of the two-port element with the mesh current analysis technique. Indeed, the mesh currents can be introduced in such a way that branch currents $\hat{I}_1$ and $\hat{I}_2$ (see Figure 10.10) will coincide with mesh currents. Then, two-port element equations (10.114) and (10.115) can be directly combined with mesh current equations for the circuits connected to the first and second ports. In this way, we end up with a complete set of mesh current equations for the overall circuit.

In the case of the nodal analysis technique, the y-representation of the two-port elements is preferable. In this representation, the terminal currents are expressed as linear and homogeneous functions of terminal voltages:

$$\hat{I}_1 = y_{11}\hat{V}_1 + y_{12}\hat{V}_2, \tag{10.152}$$

$$\hat{I}_2 = y_{21}\hat{V}_1 + y_{22}\hat{V}_2. \tag{10.153}$$

These equations can be written in the matrix form as follows:

$$\vec{I} = \tilde{Y}\vec{V}, \tag{10.154}$$

where vectors $\vec{I}$ and $\vec{V}$ are defined in formula (10.117), while the matrix $\tilde{Y}$ is given by:

$$\tilde{Y} = \begin{pmatrix} y_{11} & y_{12} \\ y_{21} & y_{22} \end{pmatrix}. \tag{10.155}$$

By comparing expressions (10.116) and (10.154), we find the following connection between $\tilde{Z}$ and $\tilde{Y}$ matrices:

$$\tilde{Y} = \tilde{Z}^{-1}, \qquad \tilde{Z} = \tilde{Y}^{-1}. \tag{10.156}$$

To find y-parameters, the two short-circuit tests shown in Figure 10.18 can be used. In test (a), the voltage source is connected to the terminals of the first port, while the terminals of the second port are short-circuited. The latter means that:

$$\hat{V}_2 = 0. \tag{10.157}$$

By using condition (10.157) in equations (10.152) and (10.153), we derive:

$$y_{11} = \frac{\hat{I}_1}{\hat{V}_1}\Big|_{\hat{V}_2=0}, \qquad y_{21} = \frac{\hat{I}_2}{\hat{V}_1}\Big|_{\hat{V}_2=0}. \tag{10.158}$$

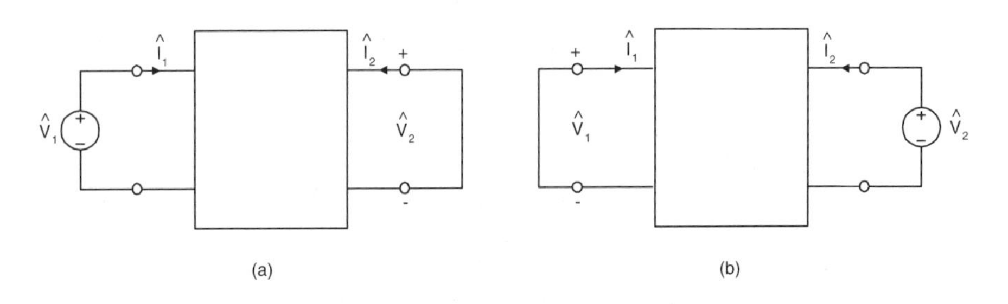

Figure 10.18: Two short-circuit tests used for identification of the y-representation.

In test (b), the voltage source is connected to the terminals of the second port, while the terminals of the first port are short-circuited, which means that:

$$\hat{V}_1 = 0. \tag{10.159}$$

By using condition (10.159) in equations (10.152) and (10.153), we derive:

$$y_{12} = \frac{\hat{I}_1}{\hat{V}_2}|_{\hat{V}_1=0}, \qquad y_{22} = \frac{\hat{I}_2}{\hat{V}_2}|_{\hat{V}_2=0}. \tag{10.160}$$

It is clear that for reciprocal two-port elements we have:

$$y_{12} = y_{21}. \tag{10.161}$$

Next, we consider circuit realizations of two-port elements defined by equations (10.152)–(10.153). As before, we want to find a simple two-port electric circuit with terminal voltage-current relations identical to those described by equations (10.152)–(10.153). First, we discuss the case of reciprocal two-port elements. In this case, it is natural to look for circuit realizations without dependent sources. To this end, consider the circuit shown in Figure 10.19. By using KCL equations for the two upper nodes, we find:

$$\hat{I}_1 = Y_a\hat{V}_1 + Y_c(\hat{V}_1 - \hat{V}_2), \tag{10.162}$$

$$\hat{I}_2 = Y_b\hat{V}_2 - Y_c(\hat{V}_1 - \hat{V}_2). \tag{10.163}$$

By rearranging the terms in the last two equations, we obtain:

$$\hat{I}_1 = (Y_a + Y_c)\hat{V}_1 - Y_c\hat{V}_2, \tag{10.164}$$

$$\hat{I}_2 = -Y_c\hat{V}_1 + (Y_b + Y_c)\hat{V}_2. \tag{10.165}$$

By comparing the last two equations with equations (10.152)–(10.153), we conclude that these pairs of equations will be identical under the following conditions:

$$Y_a + Y_c = y_{11}, \qquad Y_b + Y_c = y_{22}, \qquad Y_c = -y_{12}. \tag{10.166}$$

By solving equations (10.166) for Y_a, Y_b and Y_c, we obtain:

$$Y_a = y_{11} + y_{12}, \quad Y_b = y_{22} + y_{12}, \quad Y_c = -y_{12}. \tag{10.167}$$

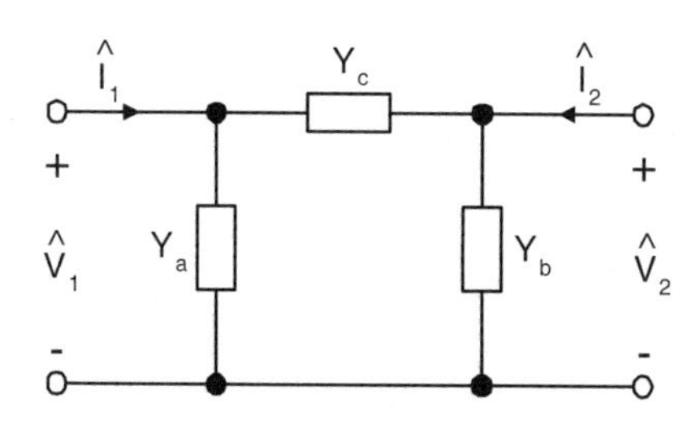

Figure 10.19: A circuit realization of the y-representation.

Thus, for admittances defined by expressions (10.167), the electric circuit shown in Figure 10.19 has the same terminal voltage-current relations as the two-port element described by equations (10.152)–(10.153). The admittances given by formula (10.167) may not be physically realizable if their real parts are negative. To circumvent this difficulty and to extend circuit realizations to nonreciprocal two-port elements, electric circuits with dependent sources can be used. This is demonstrated in the following two examples.

EXAMPLE 10.4 Consider the circuit shown in Figure 10.20 and demonstrate that it is a circuit realization for the two-port element described by equations (10.152) and (10.153).

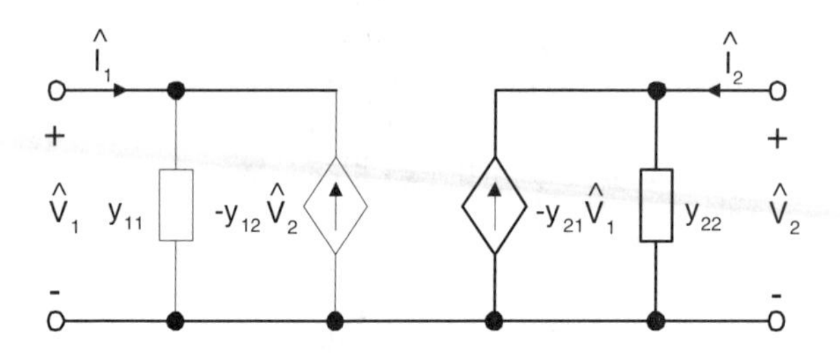

Figure 10.20: A circuit realization of the y-representation with two dependent sources.

By using KCL equations, we easily derive the following expressions for $\hat{I}_1$ and $\hat{I}_2$:

$$\hat{I}_1 = y_{11}\hat{V}_1 + y_{12}\hat{V}_2, \tag{10.168}$$

$$\hat{I}_2 = y_{21}\hat{V}_1 + y_{22}\hat{V}_2, \tag{10.169}$$

which are identical to (10.152)–(10.153). ■

EXAMPLE 10.5 Consider the circuit shown in Figure 10.21. This circuit contains only one dependent current source. We want to show that it provides a circuit realization for the two-port element defined by equations (10.152)–(10.153).

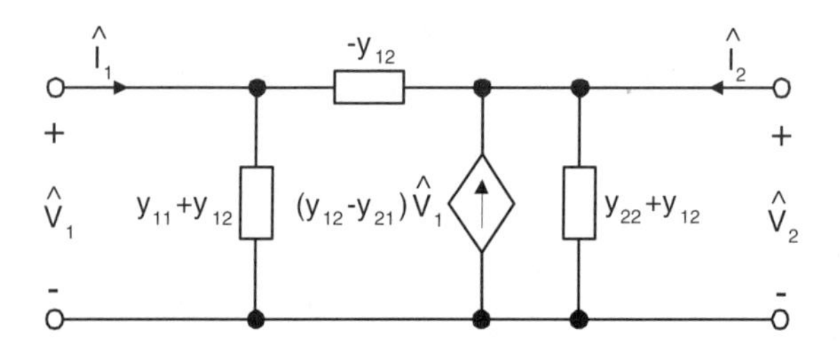

Figure 10.21: A circuit realization of the y-representation with one dependent source.

By using KCL equations for two upper nodes of the above circuit, we obtain:

$$\hat{I}_1 = (y_{11} + y_{12})\hat{V}_1 - y_{12}(\hat{V}_1 - \hat{V}_2), \tag{10.170}$$

$$\hat{I}_2 = (y_{22} + y_{12})\hat{V}_2 - (y_{12} - y_{21})\hat{V}_1 + y_{12}(\hat{V}_1 - \hat{V}_2). \tag{10.171}$$

By performing cancellation of the identical terms in the last two equations, we derive:

$$\hat{I}_1 = y_{11}\hat{V}_1 + y_{12}\hat{V}_2, \tag{10.172}$$

$$\hat{I}_2 = y_{21}\hat{V}_1 + y_{22}\hat{V}_2. \tag{10.173}$$

The equations obtained are the same as equations (10.152)–(10.153). ∎

We conclude our discussion of the y-representation with the analysis of parallel connection of two-port elements. This connection is shown in Figure 10.22. Our intent is to find the y-representation for the resulting two-port element. We begin our reasoning by writing y-representations for the primed and double-primed two-port elements:

$$\begin{pmatrix} \hat{I}'_1 \\ \hat{I}'_2 \end{pmatrix} = \begin{pmatrix} y'_{11} & y'_{12} \\ y'_{21} & y'_{22} \end{pmatrix} \begin{pmatrix} \hat{V}'_1 \\ \hat{V}'_2 \end{pmatrix}, \tag{10.174}$$

$$\begin{pmatrix} \hat{I}''_1 \\ \hat{I}''_2 \end{pmatrix} = \begin{pmatrix} y''_{11} & y''_{12} \\ y''_{21} & y''_{22} \end{pmatrix} \begin{pmatrix} \hat{V}''_1 \\ \hat{V}''_2 \end{pmatrix}. \tag{10.175}$$

The parallel connection of the two-port elements results in the following equations for terminal voltages and terminal currents:

$$\hat{V}'_1 = \hat{V}''_1 = \hat{V}_1, \qquad \hat{V}'_2 = \hat{V}''_2 = \hat{V}_2, \tag{10.176}$$

$$\hat{I}_1 = \hat{I}'_1 + \hat{I}''_1, \qquad \hat{I}_2 = \hat{I}'_2 + \hat{I}''_2. \tag{10.177}$$

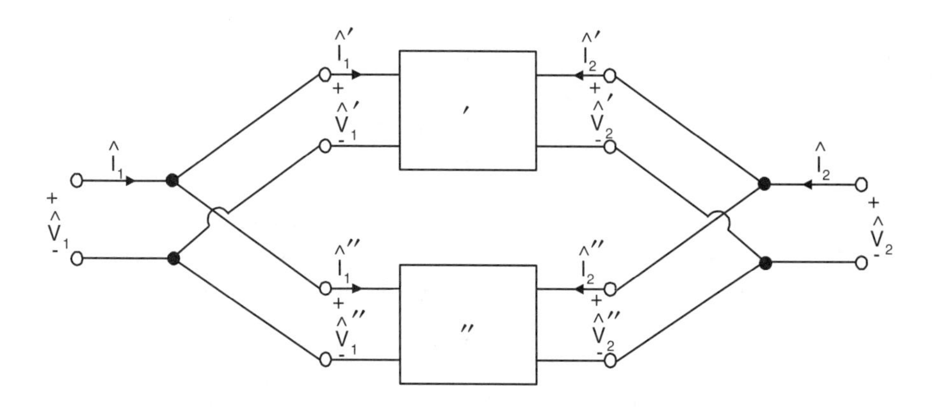

Figure 10.22: Parallel connection of two-port elements.

By adding equations (10.174) and (10.175) and then by taking into account expressions (10.176) and (10.177), after simple transformations we derive:

$$\begin{pmatrix} \hat{I}_1 \\ \hat{I}_2 \end{pmatrix} = \begin{pmatrix} y'_{11} + y''_{11} & y'_{12} + y''_{12} \\ y'_{21} + y''_{21} & y'_{22} + y''_{22} \end{pmatrix} \begin{pmatrix} \hat{V}_1 \\ \hat{V}_2 \end{pmatrix}. \tag{10.178}$$

The last result can be written in concise matrix form as follows:

$$\tilde{Y} = \tilde{Y}' + \tilde{Y}''. \tag{10.179}$$

It turns out that not all two-port elements have z- or y-representations. For instance, the ideal transformer discussed in the previous section is the two-port element with the following terminal relations[1]:

$$\hat{V}_2 = a\hat{V}_1, \tag{10.180}$$

$$\hat{I}_2 = -\frac{1}{a}\hat{I}_1 \tag{10.181}$$

which cannot be framed in terms of z- or y-representations.

This suggests the importance of hybrid representations for the two-port elements and such representations are indeed useful in various applications. There are two hybrid representations for the two-port elements: h-representation and g-representation. In the case of h-representation, the terminal voltage $\hat{V}_1$ and the terminal current $\hat{I}_2$ are expressed as linear and homogeneous functions of $\hat{I}_1$ and $\hat{V}_2$:

$$\hat{V}_1 = h_{11}\hat{I}_1 + h_{12}\hat{V}_2, \tag{10.182}$$

$$\hat{I}_2 = h_{21}\hat{I}_1 + h_{22}\hat{V}_2. \tag{10.183}$$

The above equations can also be written in the matrix form:

$$\begin{pmatrix} \hat{V}_1 \\ \hat{I}_2 \end{pmatrix} = \tilde{H} \begin{pmatrix} \hat{I}_1 \\ \hat{V}_2 \end{pmatrix}, \tag{10.184}$$

where the matrix $\tilde{H}$ is defined as follows:

$$\tilde{H} = \begin{pmatrix} h_{11} & h_{12} \\ h_{21} & h_{22} \end{pmatrix}. \tag{10.185}$$

To find h-parameters, the two tests shown in Figure 10.23 can be used. In test (a), the current source is connected to the terminals of the first port, while the terminals of the second port are short-circuited. The latter means that

$$\hat{V}_2 = 0. \tag{10.186}$$

---

[1]Please note the appearance of minus sign in equation (10.181) in comparison with equation (10.63). This is because the reference direction for $\hat{I}_2$ in the two-port element (see Figure 10.10) is opposite to the reference direction of $\hat{I}_2$ in the transformer (see Figure 10.5).

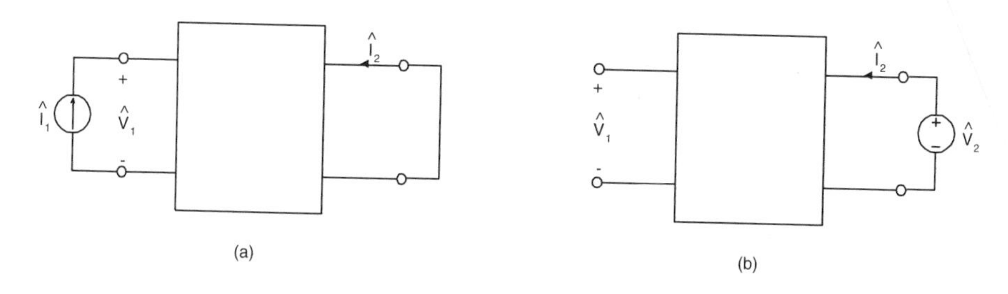

Figure 10.23: Two tests used for identification of the h-representation.

By using condition (10.186) in equations (10.182) and (10.183), we derive:

$$h_{11} = \frac{\hat{V}_1}{\hat{I}_1}\Big|_{\hat{V}_2=0}, \qquad h_{21} = \frac{\hat{I}_2}{\hat{I}_1}\Big|_{\hat{V}_2=0}. \tag{10.187}$$

In test (b), the voltage source is connected to the terminals of the second port, while the terminals of the first port are kept open. The latter means that:

$$\hat{I}_1 = 0. \tag{10.188}$$

By using the last condition in equations (10.182) and (10.183), we derive:

$$h_{12} = \frac{\hat{V}_1}{\hat{V}_2}\Big|_{\hat{I}_1=0}, \qquad h_{22} = \frac{\hat{I}_2}{\hat{V}_2}\Big|_{\hat{I}_1=0}. \tag{10.189}$$

It is instructive to note that the h-representation is extensively used for the characterization of bipolar junction transistors and the following terminology is adopted for h-parameters:

$h_{11} = h_i$ — short circuit input impedance,
$h_{12} = h_r$ — open circuit reverse voltage gain,
$h_{21} = h_f$ — short circuit forward current gain,
$h_{22} = h_o$ — open circuit output admittance.

The above terminology is consistent with expressions (10.187) and (10.189).

In the case of g-representation, the terminal current $\hat{I}_1$ and terminal voltage $\hat{V}_2$ are expressed as linear and homogeneous functions of $\hat{V}_1$ and $\hat{I}_2$:

$$\hat{I}_1 = g_{11}\hat{V}_1 + g_{12}\hat{I}_2, \tag{10.190}$$

$$\hat{V}_2 = g_{21}\hat{V}_1 + g_{22}\hat{I}_2. \tag{10.191}$$

The above equations can also be written in the matrix form:

$$\begin{pmatrix} \hat{I}_1 \\ \hat{V}_2 \end{pmatrix} = \tilde{G} \begin{pmatrix} \hat{V}_1 \\ \hat{I}_2 \end{pmatrix}, \tag{10.192}$$

where the matrix $\tilde{G}$ is defined as follows:

$$\tilde{G} = \begin{pmatrix} g_{11} & g_{12} \\ g_{21} & g_{22} \end{pmatrix}. \tag{10.193}$$

By comparing equations (10.184) and (10.192), we conclude that matrices $\tilde{H}$ and $\tilde{G}$ are related to one another by the expressions:

$$\tilde{H} = \tilde{G}^{-1}, \qquad \tilde{G} = \tilde{H}^{-1}. \tag{10.194}$$

The identification of g-parameters can be accomplished by using the two tests shown in Figure 10.24. It is clear that in test (a):

$$\hat{I}_2 = 0, \tag{10.195}$$

which, according to equations (10.190) and (10.191), leads to the following formulas for g_{11} and g_{21}:

$$g_{11} = \frac{\hat{I}_1}{\hat{V}_1}\Big|_{\hat{I}_2=0}, \qquad g_{21} = \frac{\hat{V}_2}{\hat{V}_1}\Big|_{\hat{I}_2=0}. \tag{10.196}$$

In the case of test (b), we have:

$$\hat{V}_1 = 0, \tag{10.197}$$

which, according to equations (10.190) and (10.191), leads to the following expressions for g_{12} and g_{22}:

$$g_{12} = \frac{\hat{I}_1}{\hat{I}_2}\Big|_{\hat{V}_1=0}, \qquad g_{22} = \frac{\hat{V}_2}{\hat{I}_2}\Big|_{\hat{V}_1=0}. \tag{10.198}$$

It is interesting to establish connections between the y-representation and g-representation. To this end, we solve equation (10.153) for $\hat{V}_2$:

$$\hat{V}_2 = \frac{-y_{21}}{y_{22}}\hat{V}_1 + \frac{1}{y_{22}}\hat{I}_2. \tag{10.199}$$

By substituting expression (10.199) into equation (10.152), after simple transformations we find:

$$\hat{I}_1 = \frac{\det \tilde{Y}}{y_{22}}\hat{V}_1 + \frac{y_{12}}{y_{22}}\hat{I}_2. \tag{10.200}$$

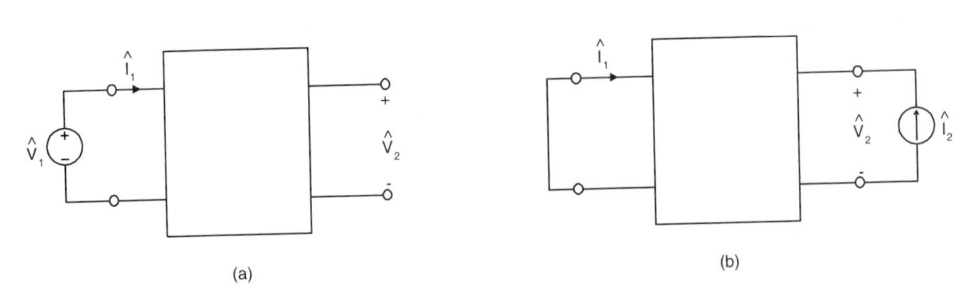

Figure 10.24: Two tests used for the identification of g-representation.

By comparing equations (10.190) and (10.191) with equations (10.200) and (10.199), respectively, we derive:

$$g_{11} = \frac{\det \tilde{Y}}{y_{22}}, \quad g_{12} = \frac{y_{12}}{y_{22}}, \quad g_{21} = \frac{-y_{21}}{y_{22}}, \quad g_{22} = \frac{1}{y_{22}}. \tag{10.201}$$

From the above formulas and symmetry condition (10.161), we find that for the reciprocal two-port elements the following equality holds:

$$g_{21} = -g_{12}. \tag{10.202}$$

From the last expression and formula (10.194), it can be derived that reciprocity imposes the following constraint on h-parameters:

$$h_{21} = -h_{12}. \tag{10.203}$$

Next, we consider some examples related to h- and g-representations.

EXAMPLE 10.6 We consider how h- and g-representations can be applied to the ideal transformer.

By comparing equations (10.180)–(10.181) with equations (10.182)–(10.183), we find that matrix $\tilde{H}$ has the form:

$$\tilde{H} = \begin{pmatrix} 0 & \frac{1}{a} \\ -\frac{1}{a} & 0 \end{pmatrix}. \tag{10.204}$$

Similarly, by comparing equations (10.180)–(10.181) with equations (10.190)–(10.191), we conclude that matrix $\tilde{G}$ can be written as follows:

$$\tilde{G} = \begin{pmatrix} 0 & -a \\ a & 0 \end{pmatrix}. \tag{10.205}$$

By using expressions (10.204) and (10.205), we easily verify that:

$$\tilde{H}\tilde{G} = \tilde{G}\tilde{H} = \begin{pmatrix} 1 & 0 \\ 0 & 1 \end{pmatrix}, \tag{10.206}$$

which is consistent with expressions (10.194). ■

EXAMPLE 10.7 Consider the circuit shown in Figure 10.25 and demonstrate that it gives a circuit realization for h-representation.

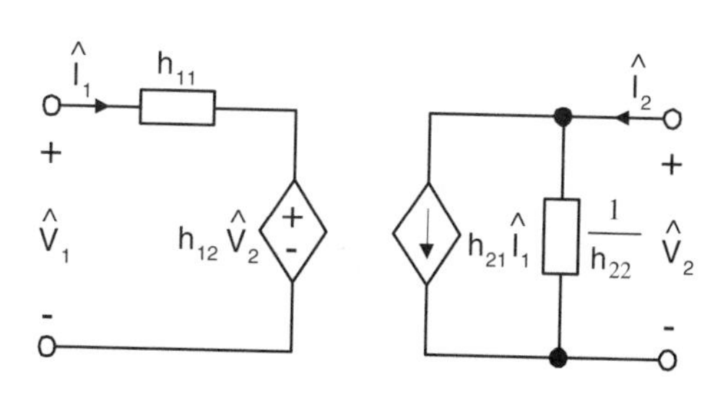

Figure 10.25: A circuit realization of the h-representation.

By applying KVL to the left part of the circuit and KCL to the right part of the same circuit, we find:

$$\hat{V}_1 = h_{11}\hat{I}_1 + h_{12}\hat{V}_2, \tag{10.207}$$

$$\hat{I}_2 = h_{21}\hat{I}_1 + h_{22}V_2. \tag{10.208}$$

The last "terminal" equations are identical to those which define the h-representation of the two-port element. ■

EXAMPLE 10.8 Consider the circuit shown in Figure 10.26 and demonstrate that it gives a circuit realization for g-representation.

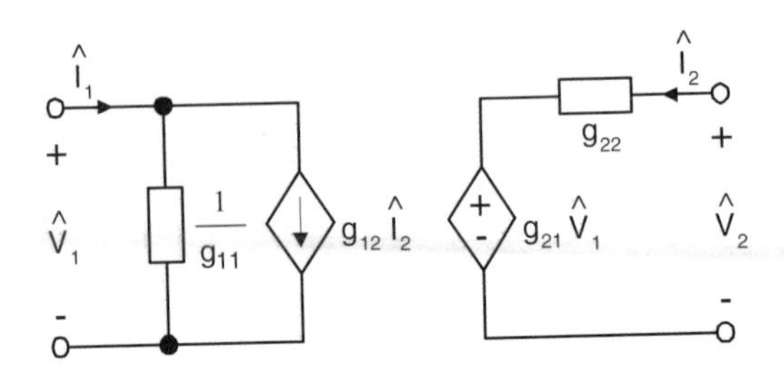

Figure 10.26: A circuit realization of the g-representation.

By using KCL for the left part of the circuit and KVL for the right part of the same circuit, we derive:

$$\hat{I}_1 = g_{11}\hat{V}_1 + g_{12}\hat{I}_2, \tag{10.209}$$

$$\hat{V}_2 = g_{21}\hat{V}_1 + g_{22}\hat{V}_2. \tag{10.210}$$

The last "terminal" equations are identical to those which define the g-representation of the two-port element. ■

We shall conclude this section with the discussion of $ABCD$-representation of the two-port elements. This representation is also called the transmission representation and it is very convenient for the analysis of cascading of two-port elements. In the case of $ABCD$-representation, $\hat{V}_1$ and $\hat{I}_1$ are expressed as linear and homogeneous functions of $\hat{V}_2$ and $\hat{I}_2$:

$$\hat{V}_1 = A\hat{V}_2 - B\hat{I}_2, \tag{10.211}$$

$$\hat{I}_1 = C\hat{V}_2 - D\hat{I}_2. \tag{10.212}$$

The last equations can be written in the matrix form:

$$\begin{pmatrix} \hat{V}_1 \\ \hat{I}_1 \end{pmatrix} = \tilde{T} \begin{pmatrix} \hat{V}_2 \\ -\hat{I}_2 \end{pmatrix}, \tag{10.213}$$

where the matrix $\tilde{T}$ is defined as follows:

$$\tilde{T} = \begin{pmatrix} A & B \\ C & D \end{pmatrix}. \tag{10.214}$$

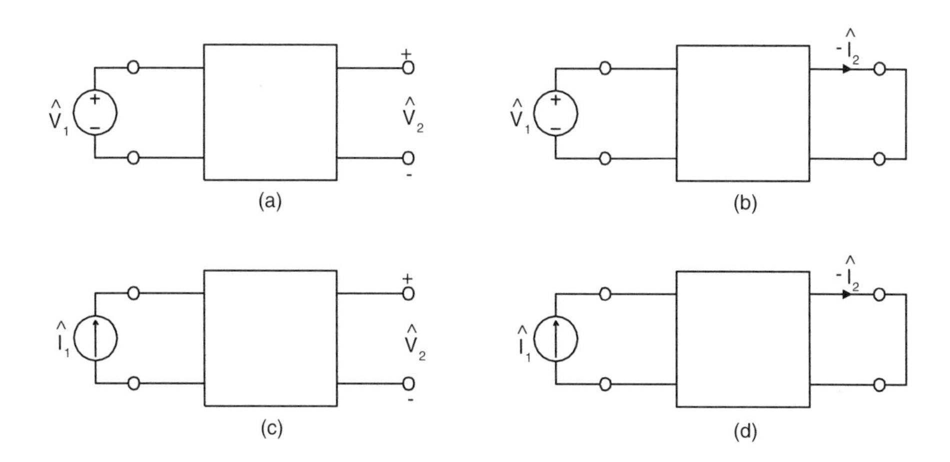

Figure 10.27: Tests used for identification of the *ABCD*-representation.

The identification of the parameters *A, B, C, D* can be performed by using the four tests shown in Figure 10.27. By employing the same line of reasoning as before, it is easy to show that the following expressions are valid.

$$A = \frac{\hat{V}_1}{\hat{V}_2}|_{\hat{I}_2=0}, \quad B = \frac{\hat{V}_1}{-\hat{I}_2}|_{\hat{V}_2=0}, \quad C = \frac{\hat{I}_1}{\hat{V}_2}|_{\hat{I}_2=0}, \quad D = \frac{\hat{I}_1}{-\hat{I}_2}|_{\hat{V}_2=0}. \tag{10.215}$$

EXAMPLE 10.9 Find the transmission representation for the ideal transformer.

By comparing equations (10.180)–(10.181) with equations (10.211)–(10.212), we conclude that matrix $\tilde{T}$ has the form:

$$\tilde{T} = \begin{pmatrix} \frac{1}{a} & 0 \\ 0 & a \end{pmatrix}. \tag{10.216}$$

It is easy to see from expression (10.216) that

$$\det \tilde{T} = 1. \tag{10.217}$$

It can be shown that formula (10.217) is generally true for any reciprocal two-port element. ■

Next, consider cascading of two-port elements shown in Figure 10.28. Our goal is to find the $\tilde{T}$-matrix for the resulting two-port element. To this end, we shall first remark that the following terminal relations are valid due to the very nature of the cascading connection:

$$\hat{I}_1'' = -\hat{I}_2', \qquad \hat{V}_1'' = \hat{V}_2'. \tag{10.218}$$

Now, we shall write the transmission representations for the primed and double-

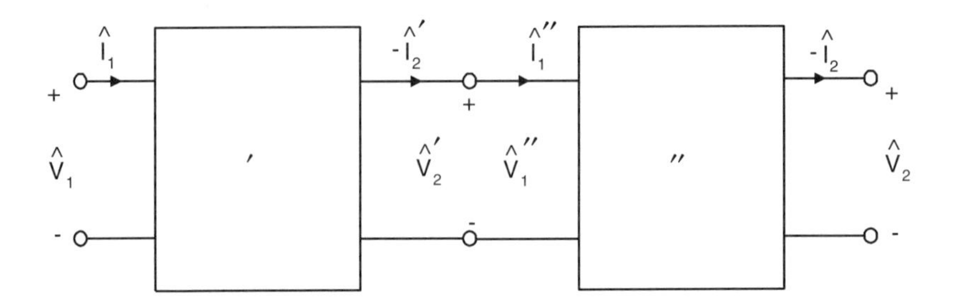

Figure 10.28: Cascading of two-port elements.

primed two-port elements and, in doing so, we shall take into account formulas (10.218):

$$\begin{pmatrix} \hat{V}_1 \\ \hat{I}_1 \end{pmatrix} = \begin{pmatrix} A' & B' \\ C' & D' \end{pmatrix} \begin{pmatrix} \hat{V}_2' \\ -\hat{I}_2' \end{pmatrix} = \begin{pmatrix} A' & B' \\ C' & D' \end{pmatrix} \begin{pmatrix} \hat{V}_1' \\ \hat{I}_1' \end{pmatrix}, \tag{10.219}$$

$$\begin{pmatrix} \hat{V}_1' \\ \hat{I}_1' \end{pmatrix} = \begin{pmatrix} A'' & B'' \\ C'' & D'' \end{pmatrix} \begin{pmatrix} \hat{V}_2 \\ -\hat{I}_2 \end{pmatrix}. \tag{10.220}$$

By substituting (10.220) into (10.219), we derive:

$$\begin{pmatrix} \hat{V}_1 \\ \hat{I}_1 \end{pmatrix} = \begin{pmatrix} A' & B' \\ C' & B' \end{pmatrix} \begin{pmatrix} A'' & B'' \\ C'' & D'' \end{pmatrix} \begin{pmatrix} \hat{V}_2 \\ -\hat{I}_2 \end{pmatrix}. \tag{10.221}$$

The last result can be expressed succinctly in the matrix form as follows:

$$\tilde{T} = \tilde{T}'\tilde{T}''. \tag{10.222}$$

Finally, we shall remark that the same formalism of *ABCD*-matrices is extensively used (albeit with a different physical connotation) in laser courses for ray tracing and transformation of Gaussian beams.

10.5 MicroSim PSpice Simulations

Consider the transformer circuit which was analyzed in Example 10.1 and is pictured as circuit (a) in Figure 10.29 (as drawn by the schematics editor). The voltage source is the "VSIN" type with an amplitude of "VAMPL=100," a frequency of

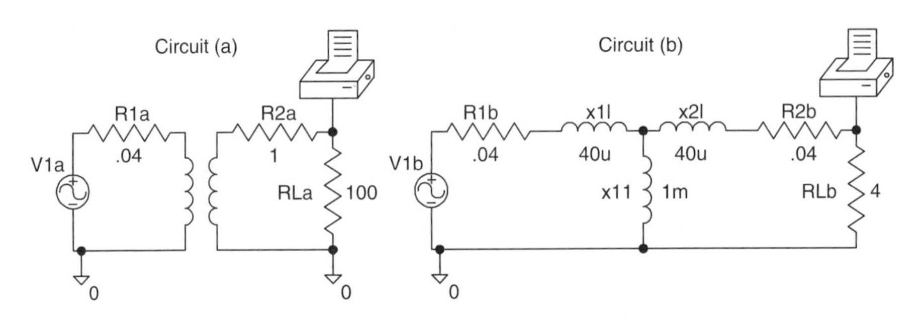

Figure 10.29: Circuits from Example 10.1.

"FREQ=397.887," a phase of "PHASE=0," and an offset voltage of "VOFF"=0. Two grounds are required because the two coils of the transformer are electrically isolated. The part name of the transformer is "XFRM_LINEAR" and it resides in the "analog" library. Double-clicking on the transformer brings up the dialog box where the primary inductance, secondary inductance, and coupling coefficient can be defined. From the example, we have "L1_value=1m," "L2_value=25m," and "COUPLING=0.9583." The latter value comes from (10.91), which can be rewritten:

$$k = \frac{M}{\sqrt{L_1 L_2}} = \left(1 - \frac{L_1^\ell}{L_1} - \frac{L_2^\ell}{L_2} - \frac{L_1^\ell L_2^\ell}{L_1 L_2}\right)^{1/2} . \tag{10.223}$$

The winding resistances are not included in the PSpice model, so they are explicitly included as $R1a$ and $R2a$.

The scaled equivalent circuit is shown in circuit (b) of Figure 10.29. Note that $x2l$, $R2b$, and RLb have been multiplied by $a^2 = 0.04$, but all other quantities are unchanged. The printer symbols come from the "VPRINT1" device in the "special" library. Attaching them to the nodes indicated will cause PSpice to print the corresponding node voltages in the output file. The default printing occurs for transient analyses, but double-clicking on the printer symbol brings up a dialog box that can be used to select other types of analyses.

First, we chose a transient analysis to compare the time dependences to the analytic result. In the "Transient" dialog box select a "Print Step" of 20 us and a "Final Time" of 4 ms. The voltage across the load resistor is given in Figure 10.30. The solid line gives the output voltage of circuit (a). The dashed line indicates the actual voltage across the load resistor as computed for circuit (b). This voltage is five times ($=1/a$) the voltage across the scaled 4 Ω resistor. The dotted line gives the

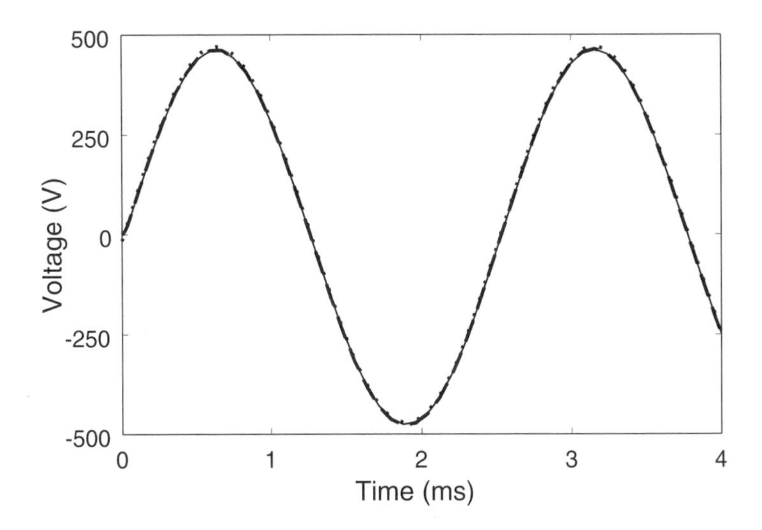

Figure 10.30: Time dependence of the output voltages.

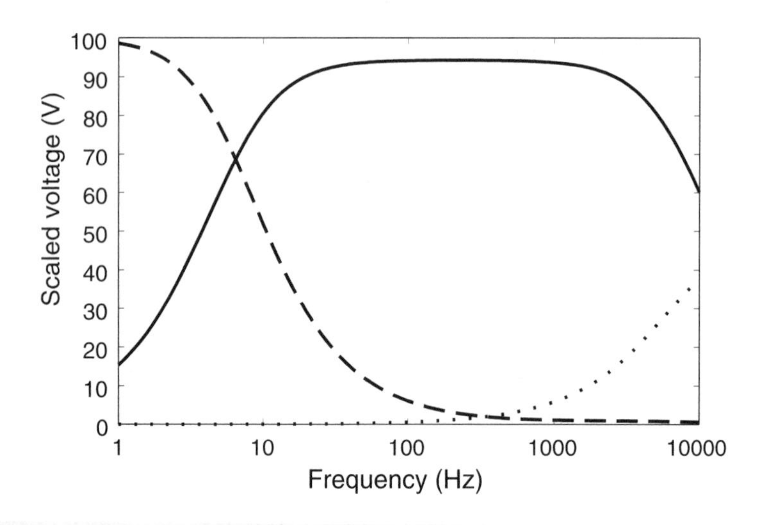

Figure 10.31: Frequency dependence of circuit voltages.

analytic result as computed in Example 10.1. All three results are virtually identical, as expected.

It is instructive to analyze the output voltage as a function of frequency. This can be easily done in PSpice by making two changes to the previous example. First, edit both voltage sources and add the ac voltage "AC=100." In the "Analysis Setup" dialog box, switch to an ac sweep. Set the "AC Sweep Type" to "Decade," the "Pts/Decade" to 51, the "Start Freq." to 1, and the "End Freq." to 10k. The results of the simulation are shown in Figure 10.31. The scaled voltage across the load resistor is given by the solid line. The response is quite flat from 100 Hz to 1 kHz but falls off at lower and higher frequencies. At low frequencies, the primary winding draws too much current and a large voltage drop occurs due to the primary winding resistance. The voltage across $R1b$ is plotted with a dashed line in the figure. At high frequencies, the drop is caused by the leakage inductance. The scaled voltage across the secondary leakage inductance is given by the dotted line.

10.6 Summary

The main results of this chapter can be briefly summarized as follows:

- The magnetic coupling of two coils can be characterized by the mutual inductance M. By definition, M is equal to the ratio of the flux linkages of one coil to the current flowing through the other coil (assuming zero current in the first coil):

$$M = \psi_{21}(t)/i_1(t) = \psi_{12}(t)/i_2(t). \tag{10.224}$$

- The mutual inductance is always less than the geometric mean of the self-inductances of the two coils:

$$M \leq \sqrt{L_1 L_2}. \tag{10.225}$$

- The coupling coefficient k is defined as

$$k = M/\sqrt{L_1 L_2} \tag{10.226}$$

  and it depends on the relative locations of the two coils, their geometric dimensions, and the properties of the media in the vicinity of the two coils.

- The coupled circuit equations for two magnetically coupled coils can be written as follows:

$$\hat{V}_1 = R_1 \hat{I}_1 + jx_{11}\hat{I}_1 - jx_{12}\hat{I}_2, \tag{10.227}$$

$$\hat{V}_2 = R_2 \hat{I}_2 + jx_{22}\hat{I}_2 - jx_{12}\hat{I}_1. \tag{10.228}$$

- An ideal transformer has a coupling coefficient of unity, zero resistances of the primary and secondary windings, and no eddy currents in the iron core.

- For an ideal transformer, the ratio of secondary voltage (current) to the primary voltage (current) is directly (inversely) proportional to the turns ratio. Thus, a load impedance Z_L has an effective value of $a^2 Z_L$ when viewed from the primary terminals. For this reason, transformers are frequently used to match load impedances to source impedances.

- Leakage inductances result from imperfectly coupled ($k < 1$) transformer windings and they diminish the transformer's ability to maintain a constant voltage across the secondary terminals in the face of varying loads.

- Leakage reactances are very important and should be explicitly accounted for in transformer models.

- A nonideal transformer can be modeled as a T-network.

- Two-port elements are often used as models for complicated devices and also as equivalent replacements for complex circuits.

- The "input" and "output" variables for the various two-port representations are given in Table 10.1.

- Simple open- and short-circuit tests can be performed to determine the matrix elements of the various representations.

- If a two-port element contains only passive components, then the $\tilde{Z}$ and $\tilde{Y}$ matrices are symmetric.

- z-parameters are useful for modeling series connections of two-port elements.

Table 10.1: Summary of two-port representations.

Representation	Matrix	"Input" variables	"Output" variables
Impedance	$\tilde{Z}$	I_1, I_2	V_1, V_2
Admittance	$\tilde{Y}$	V_1, V_2	I_1, I_2
Hybrid	$\tilde{H}$	I_1, V_2	I_2, V_1
Hybrid	$\tilde{G}$	I_2, V_1	I_1, V_2
Transmission	$\tilde{T}$	$-I_2$, V_2	I_1, V_1

- y-parameters are useful for modeling parallel connections of two-port elements.
- $ABCD$-parameters are useful for the analysis of cascading of two-port elements.

10.7 Problems

1. Two inductors in a circuit are magnetically coupled so that when 30 mA is flowing in the first inductor, L_1, a net flux of 20 μWb links the second coil, L_2, when it is open-circuited. (a) What is the mutual inductance of this arrangement? (b) How much flux will link an open-circuited L_1 if 200 μA is made to flow in L_2?

2. Two sets of measurements were made on a pair of magnetically coupled coils and the results are indicated in the following table. Calculate the self-inductances of the two coils, the mutual inductance, and the coupling coefficient.

Test #	i_1 (A)	i_2 (A)	ψ_1 (μWb)	ψ_2 (μWb)
1	1	2	25	3
2	1	3	20	7

3. Two identical inductors are magnetically coupled so that when 1 A flows through the first coil and 2 A flows through the second coil, a net flux of 50 mWb links the first coil while 200 mWb links the second coil. Calculate the mutual and self-inductances of this arrangement as well as the coupling coefficient.

4. Two identical inductors are magnetically coupled so that when 3 mA flows through the first coil and 5 mA flows through the second coil, a net flux of 125 μWb links the first coil while 150 μWb links the second coil. Calculate the mutual and self-inductances of this arrangement as well as the coupling coefficient.

5. Two inductors are magnetically coupled. When the 5 μH inductor is energized with 2 A, it produces a flux linkage of 6.2 μWb in the open-circuited 2 μH inductor. Calculate the coupling coefficient.

6. Two nonideal inductors are magnetically coupled. Find the voltages across the inductors if the driving currents are $i_1(t) = 20\cos(377t - 45°)$ mA and $i_2(t) = 30\sin(377t + 45°)$ mA and the inductor parameters are $L1 = 9$ mH, $L2 = 4$ mH, and $M = 6$ mH.

7. Two magnetically coupled nonideal inductors have inductances and winding resistances of $L_1 = 2$ mH, $R_1 = 0.1\ \Omega$, and $L_2 = 5$ mH, $R_2 = 0.2\ \Omega$. The coupling coefficient is $k = 0.5$ and the voltages across the coils are found to be $v_1(t) = 5\cos(377t + 30°)$ and $v_2(t) = 4\cos(377t - 45°)$. Find the time dependences of the currents flowing through the two coils.

8. A transformer has a primary self-inductance of $L_1 = 2$ mH and a leakage inductance of 27.7 μH. The secondary winding has a primary self-inductance of $L_2 = 72$ mH and a leakage inductance of 1 mH. The turns ratio is $a = 1/6$. Calculate the scaled inductances of the transformer's equivalent circuit, x_1^{ℓ}, $x_2^{\ell'}$, and x_{11}^m, if the transformer is driven at $\omega = 377$ rad/s.

9. Consider the circuit shown in Figure P10-9. The parameters of the nonideal transformer are as follows: $L_1^m = 100\ \mu$H, $L_1^{\ell} = 2\ \mu$H, $L_2^m = 10$ mH, $L_2^{\ell} = 200\ \mu$H, $R_1 = 0.05\ \Omega$, and $R_2 = 10\ \Omega$. Find an expression for the voltage across the load resistor $R_L = 2$ kΩ if the source voltage is given by: $v_s(t) = 25\cos(10^6 t)$ kV. The turns ratio is $a = 0.1$.

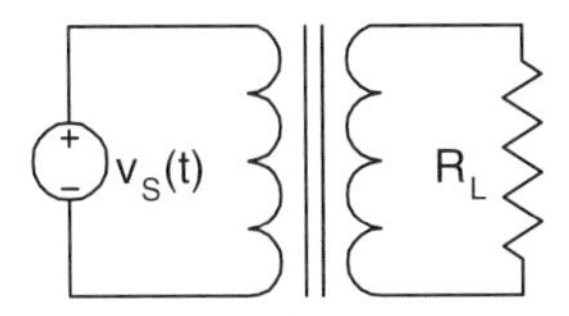

Figure P10-9

10. A nonideal transformer has the following properties: $L_1^m = 25\ \mu$H, $L_1^{\ell} = 1\ \mu$H, $L_2^m = 10$ mh, $L_2^{\ell} = 400\ \mu$H, $R_1 = 0.05\ \Omega$, and $R_2 = 2.5\ \Omega$. Find the z-representation of the transformer's scaled equivalent circuit at a frequency of $\omega = 2000$ rad/s. The turns ratio is $a = 0.05$.

In Problems 11–34, the two leftmost terminals in the circuit indicated constitute the first port and the two rightmost terminals constitute port 2.

In Problems 11–16, calculate the z-parameters of the circuit shown in the figure indicated.

11. Figure P10-11.

12. Figure P10-12.

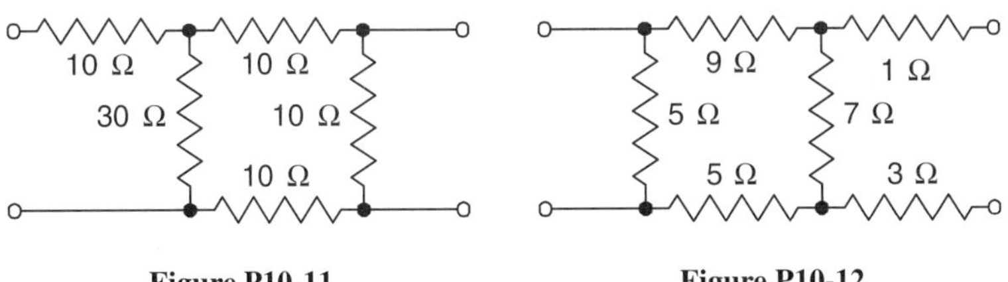

Figure P10-11 **Figure P10-12**

13. Figure P10-13, $\omega = 5000$ rad/s.

14. Figure P10-14, $\omega = 2 \times 10^6$ rad/s.

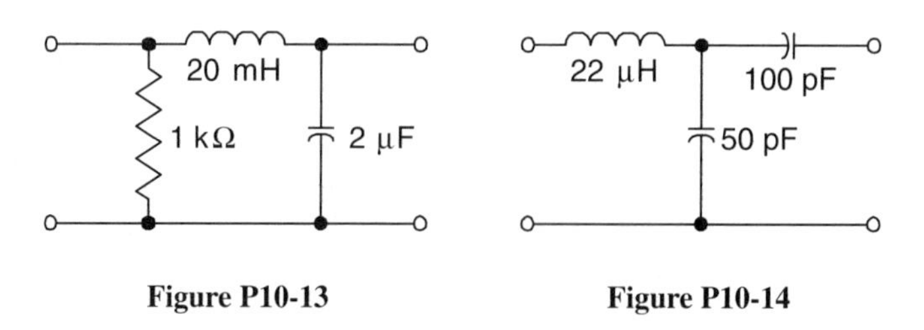

Figure P10-13 **Figure P10-14**

15. Figure P10-15.

16. Figure P10-16.

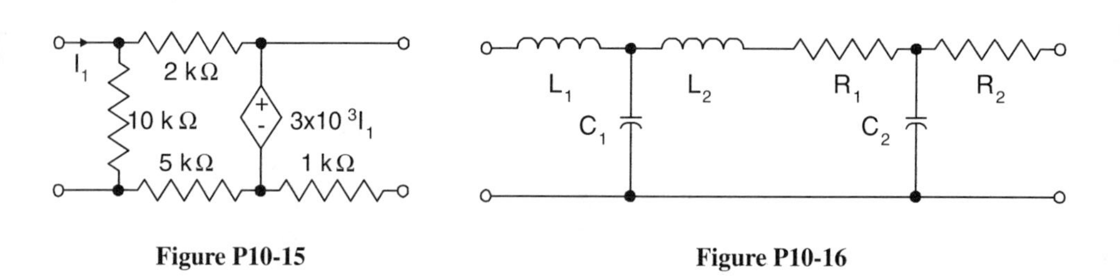

Figure P10-15 **Figure P10-16**

17. Consider the series combination of two two-port networks in Figure P10-17 where the z-parameters of network 1 are given by:

$$\begin{pmatrix} 125 & 50 \\ 50 & 100 \end{pmatrix}.$$

Find the z-parameters of the total network.

18. Consider the circuit shown in Figure P10-18 where the z-parameters of network 1 are given by:

$$\begin{pmatrix} 69 & 22 \\ 22 & 64 \end{pmatrix}.$$

Use mesh analysis to find the current in the 10 Ω resistor.

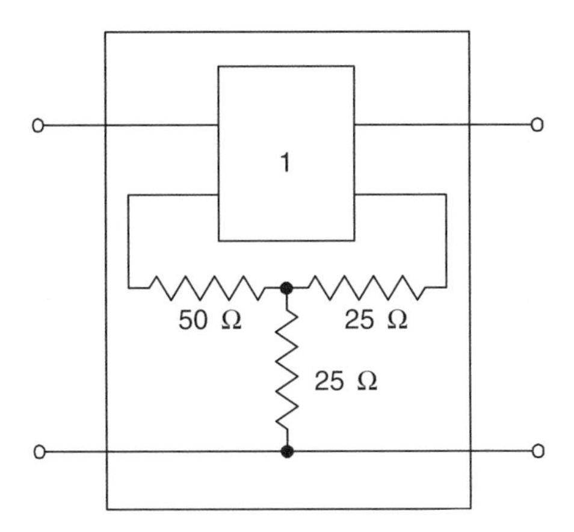

Figure P10-17

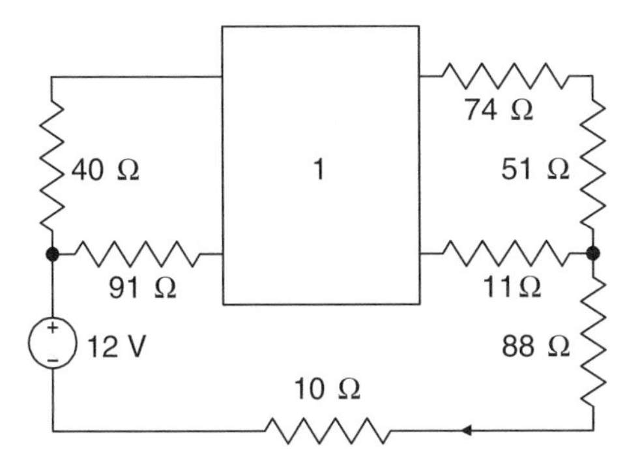

Figure P10-18

19. Design a circuit which realizes each set of z-parameters given in the following table. Use dependent sources in the design only if necessary.

Case	z_{11} (Ω)	z_{12} (Ω)	z_{21} (Ω)	z_{22} (Ω)	ω (rad/s)
a	75	50	50	125	0
b	25	50	50	75	0
c	$50 - j75$	50	75	$50 + j50$	400

20. Design a circuit which realizes each set of y-parameters given in the following table. Use dependent sources in the design only if necessary.

Case	y_{11} (℧)	y_{12} (℧)	y_{21} (℧)	y_{22} (℧)	ω (rad/s)
a	75	-50	-50	100	0
b	$50 - j10$	$j25$	$j25$	$-j100$	377
c	$20 + j20$	$j25$	$15 + j25$	$10 - j40$	5000

21. Consider the series combination of two two-port networks in Figure P10-21 where the y-parameters of network 1 are given by:

$$\begin{pmatrix} 0.3 & -0.1 \\ -0.1 & 0.2 \end{pmatrix}.$$

Find the y-parameters of the total network.

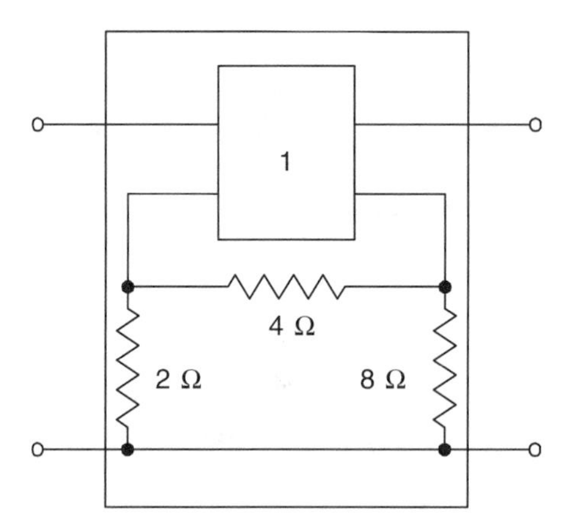

Figure P10-21

In Problems 22–24 calculate the y-parameters of the circuit shown in the figure indicated.

22. Figure P10-11.

23. Figure P10-13, $\omega = 7000$ rad/s.

24. Figure P10-24.

In Problems 25–27 calculate the g-parameters of the circuit shown in the figure indicated.

25. Figure P10-12.

26. Figure P10-14, $\omega = 3 \times 10^6$ rad/s.

27. Figure P10-27.

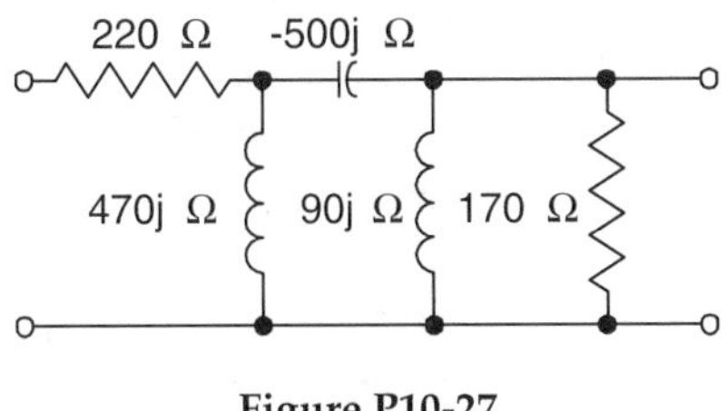

Figure P10-27

In Problems 28–30 calculate the h-parameters of the circuit shown in the figure indicated. The notation N_1:N_2 for an ideal transformer indicates that the turns ratio is $a = N_1/N_2$ and the secondary winding is the rightmost inductor.

28. Figure P10-15.

29. Figure P10-24.

30. Figure P10-30.

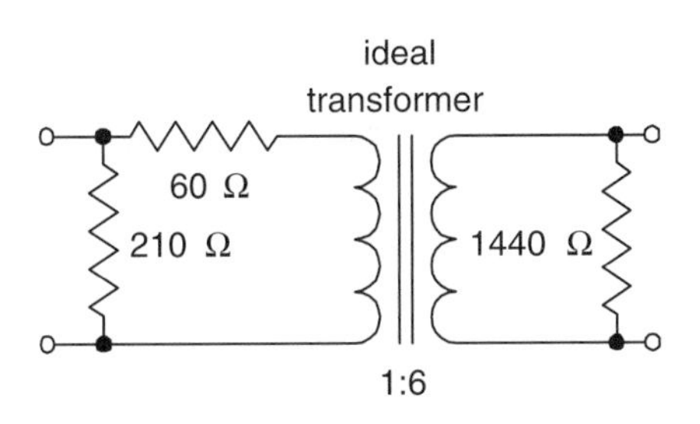

Figure P10-30

In Problems 31–33 calculate the ABCD-parameters of the circuit shown in the figure indicated.

31. Figure P10-16.

32. Figure P10-27.

33. Figure P10-30.

34. Consider the cascaded combination of two-port networks in Figure P10-34 where the z-parameters of network 1 are given by:

$$\begin{pmatrix} 6 & 3 \\ 3 & 7 \end{pmatrix}.$$

Find the ABCD-parameters of the total network.

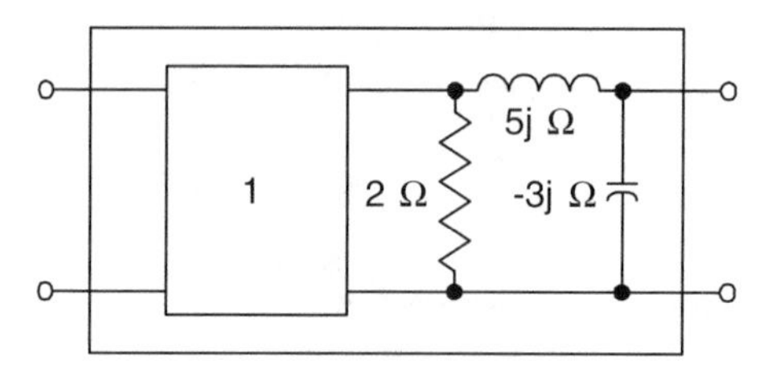

Figure P10-34

35. The y-representation of a two-port network is given by:

$$\begin{pmatrix} 5 & -3 \\ -3 & 7 \end{pmatrix}.$$

Find the g- and $ABCD$-representations of the same network.

36. The z-representation of a two-port network is given by:

$$\begin{pmatrix} z_{11} & z_{12} \\ z_{21} & z_{22} \end{pmatrix}.$$

Find the h-representation of the same network.

37. The h-representation of a two-port network is given by:

$$\begin{pmatrix} 2 & 3 \\ 5 & 7 \end{pmatrix}.$$

Find the g- and $ABCD$-representations of the same network.

38. A transformer has a primary self-inductance of $L_1 = 2$ mH and a leakage inductance of 27.7 μH. The secondary winding has a primary self-inductance of $L_2 = 72$ mH and a leakage inductance of 1 mH. The transformer is driven with a 12 volt sinusoidal signal at $\omega = 377$ rad/s and is connected to a load resistance of 1 kΩ (see Figure P10-9). Use PSpice to find the time dependence of the voltage across the load.

39. Consider the circuit shown in Figure P10-9. The parameters of the nonideal transformer are as follows: $L_1^m = 100\ \mu$H, $L_1^\ell = 2\ \mu$H, $L_2^m = 10$ mH, $L_2^\ell = 200\ \mu$H, $R_1 = 0.05\ \Omega$, and $R_2 = 10\ \Omega$. Use PSpice to plot the expression for the voltage across the load resistor $R_L = 2$ kΩ if the source voltage is given by: $v_s(t) = 25\cos(10^6 t)$ kV.

40. A nonideal transformer has the following properties: $L_1^m = 25\ \mu$H, $L_1^\ell = 1\ \mu$H, $L_2^m = 10$ mH, $L_2^\ell = 400\ \mu$H, $R_1 = 0.05\ \Omega$, and $R_2 = 2.5\ \Omega$. The transformer has an equivalent core loss of 1 Ω and is connected to a 1 kΩ load as in Figure P10-9. Use PSpice to plot the ratio of the voltage across the load to the input voltage from 10 Hz to 10 kHz.

Appendix A

Complex Numbers

To become fluent with phasors and phasor manipulations, one should have a solid background in the area of complex numbers. There are many good introductory textbooks on complex numbers; one such book is *Complex Variables and Applications* by R. V. Churchill, J. W. Brown, and R. F. Verhey (McGraw-Hill, 1976). Here, we briefly summarize a few important facts concerning the complex numbers that are relevant to the complex algebra required for basic ac circuit analysis.

The cornerstone of the concept of an imaginary number comes from the equation

$$j^2 = -1, \tag{A.1}$$

for which there is no real number solution. By accepting equation (A.1) as the definition of the number j, we can form an imaginary number line by multiplying j by any real number. A general complex number z is composed of both real and imaginary parts: $z = x + jy$, where x and y are both real numbers. This form of complex number is referred to as the algebraic or Cartesian form. It is a convenient form when we need to add or subtract complex numbers.

It is often useful to plot complex numbers on a two-dimensional plane as indicated in Figure A.1. This graphical representation leads us to consider a polar form for z. The magnitude of z is clearly $|z| = \sqrt{x^2 + y^2}$ and the polar angle of z is given by $\tan\theta = y/x$. The graphical representation of complex numbers is extremely useful for ensuring the selection of the proper quadrant for θ.

There is a third way to express a complex number known as the exponential or exponential-polar form. This form is the most relevant to the phasor technique and it is defined as $z = |z|e^{j\theta}$, where $|z|$ and θ are the same magnitude and phase as in the polar form, respectively. To understand the basis for this representation, consider the Taylor series expansion for e^x:

$$e^x = 1 + x + x^2/2 + x^3/3! + \cdots + x^n/n! + \cdots. \tag{A.2}$$

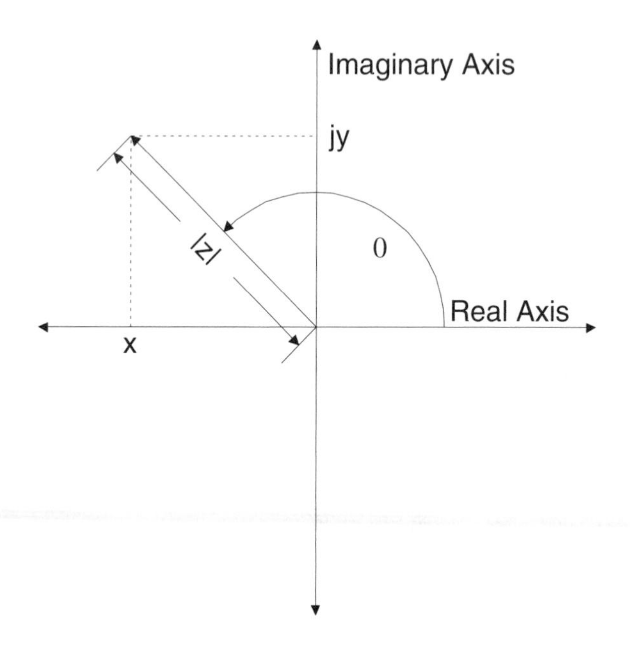

Figure A.1: The complex plane.

Recall that $j^2 = -1$, so that $j^3 = -j$ and $j^4 = 1$. If we replace x with $j\theta$ in (A.2) and group the terms into real and imaginary parts, we arrive at:

$$e^{j\theta} = [1-\theta^2/2+\theta^4/4!-\theta^6/6!+\cdots]+j[\theta-\theta^3/3!+\theta^5/5!-\theta^7/7!+\cdots]. \quad \text{(A.3)}$$

The first term in brackets is just the Taylor series expansion for $\cos\theta$ and the second bracketed term is just $\sin\theta$. The relation we have just derived is known as Euler's equation:

$$e^{j\theta} = \cos\theta + j\sin\theta. \quad \text{(A.4)}$$

The exponential form is quite convenient for multiplication and division of complex numbers. Indeed:

$$z = z_1 z_2 = |z_1|e^{j\theta_1}|z_2|e^{j\theta_2} = |z_1||z_2|e^{j(\theta_1+\theta_2)}, \quad \text{(A.5)}$$

$$z = \frac{z_1}{z_2} = \frac{|z_1|}{|z_2|}\frac{e^{j\theta_1}}{e^{j\theta_2}} = \frac{|z_1|}{|z_2|}e^{j(\theta_1-\theta_2)}. \quad \text{(A.6)}$$

There are several useful manipulations on complex numbers that we will use in this text. We will often take the real and imaginary parts of a complex number: $x = \text{Re}(z)$ and $y = \text{Im}(z)$. Care must be used with these two operations. If $z_1 = x_1 + jy_1$ and $z_2 = x_2 + jy_2$ then in general

$$\text{Re}(z_1)\,\text{Re}(z_2) \neq \text{Re}(z_1 z_2). \quad \text{(A.7)}$$

The proof comes directly from the multiplication result:

$$z_1 z_2 = (x_1 x_2 - y_1 y_2) + j(x_1 y_2 + x_2 y_1). \tag{A.8}$$

Thus,

$$\text{Re}\,(z_1 z_2) = x_1 x_2 - y_1 y_2, \tag{A.9}$$

$$= \text{Re}\,(z_1)\,\text{Re}\,(z_2) - \text{Im}\,(z_1)\,\text{Im}\,(z_2). \tag{A.10}$$

By examining the imaginary part of (A.8), one can also easily see that

$$\text{Im}\,(z_1 z_2) \neq \text{Im}\,(z_1)\,\text{Im}(z_2). \tag{A.11}$$

A direct consequence of (A.9) is that you can "pull" real numbers in and out of the Re( ) operator. If $y_1 = 0$, for example, then $z_1 = x_1$ and $\text{Re}\,(x_1 z_2) = x_1 x_2 = x_1\,\text{Re}\,(z_2)$.

We will denote the complex conjugate of z as z^* and define it as a number equal in magnitude to z but possessing the polar angle of opposite sign. Thus, if $z = |z|e^{j\theta} = x + jy$, then $z^* = |z|e^{-j\theta} = x - jy$. One use of this definition is to note that

$$zz^* = |z|e^{j\theta}|z|e^{-j\theta} = |z|^2 = x^2 + y^2. \tag{A.12}$$

Let us consider a concrete example to gain some familiarity with the above concepts. Let $z_1 = -3 - j4$ and $z_2 = 2e^{j45^\circ}$ as indicated in Figure A.2. Then

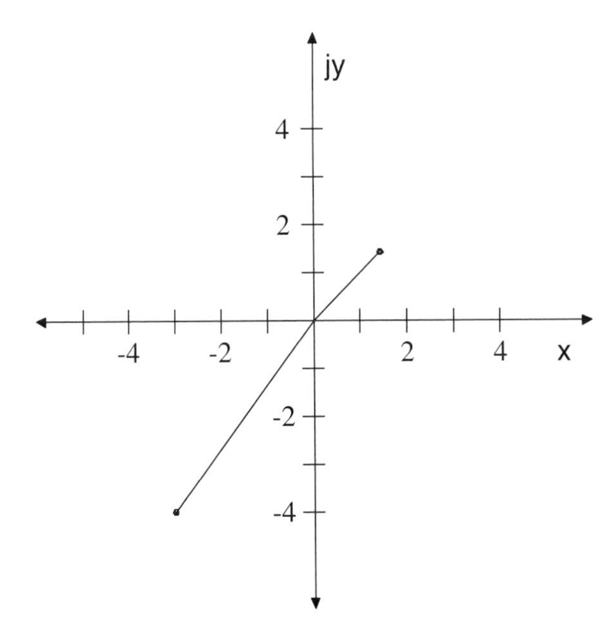

Figure A.2: Graphical representation of two complex numbers.

Re $(z_1) = -3$ and Im $(z_1) = -4$. The complex conjugate $z_1^* = -3 + j4$. To convert into exponential form, we find $|z_1| = \sqrt{x_1^2 + y_1^2} = 5$ and $\theta = \tan^{-1}(y_1/x_1) = \tan^{-1}(4/3)$. We can use a calculator, and take into account the fact that z_1 is in the third quadrant, to get $\theta = -126.9°$. From Euler's equation we see that $z_2 = 2(\cos 45° + j \sin 45°) = \sqrt{2} + j\sqrt{2}$. We subtract by using the Cartesian form: $z_1^* - z_2 = -3 + j4 - (\sqrt{2} + j\sqrt{2}) = -(3 + \sqrt{2}) + j(4 - \sqrt{2}) = -4.414 + j2.586$. Finally, we divide by using the polar form: $z_1/z_2 = (5/2)e^{j[-126.9-45°]} = 2.5e^{-j171.9°}$.

Drill Exercises

1. Convert the following numbers to exponential form:
(a) $5 + j2$, (b) $12 - j5$, (c) $-3 + \sqrt{7}j$, (d) $-1 - 5j$.

2. Convert the following numbers to Cartesian form:
(a) $3e^{j\pi/6}$, (b) $7e^{j122°}$, (c) $25e^{4.2\pi j}$, (d) $-e^{-j/7}$.

3. Plot the following numbers on the complex plane:
(a) $6\sqrt{2}e^{j125°}$, (b) $6 - 5j$, (c) $10 + 24j$, (d) $13e^{-j1.7\pi}$.

4. Perform the specified operations and express the answer in Cartesian form:
(a) $(5 + 3j) * 2e^{j45°}$, (b) $3e^{j\pi/6} + 4e^{-j\pi/7}$, (c) $7e^{j100°}/(1 + j)$, (d) $(7 + 5j) - 5e^{j120°}$.

5. Perform the specified operations and express the answer in polar form:
(a) $(1 + j) * (-2 + 3j)$, (b) $7j + 4e^{j\pi}$, (c) $(-1 + j)/(3 + 4j)$, (d) $4 - e^{j90°}$.

Appendix B

Gaussian Elimination

The study of linear algebraic equations belongs to linear algebra, which is a broad area of mathematics that is far beyond the scope of this text. There are many good references on the subject, an example of which is *Linear Algebra and Its Applications*, by G. Strang (Academic Press, 1976).

In this text, the reader is exposed to a very specific subset of matrix operations related to the solution of general nodal and mesh equations. In these problems, we are always equating a known source vector with the product of a known $n \times n$ matrix and an unknown column vector (an $n \times 1$ matrix):

$$\begin{pmatrix} M_{11} & M_{12} & \cdots & M_{1n} \\ M_{21} & \ddots & & \\ \vdots & \ddots & \ddots & \\ M_{N1} & & & M_{nn} \end{pmatrix} \begin{pmatrix} U_1 \\ U_2 \\ \vdots \\ U_n \end{pmatrix} = \begin{pmatrix} S_1 \\ S_2 \\ \vdots \\ S_n \end{pmatrix}. \tag{B.1}$$

We can write this more compactly as $\tilde{M} \cdot \vec{U} = \vec{S}$. Circuit analysis usually requires that we find one or more of the elements in the unknown vector.

The complete solution for $\vec{U}$ is given symbolically by

$$\vec{U} = \tilde{M}^{-1} \cdot \vec{S} \tag{B.2}$$

where $\tilde{M}^{-1}$ is the inverse matrix of $\tilde{M}$, defined by

$$\tilde{M} \cdot \tilde{M}^{-1} = \tilde{M}^{-1} \cdot \tilde{M} = \tilde{I}. \tag{B.3}$$

Here $\tilde{I}$, known as the identity matrix, is the matrix equivalent of unity. It is an $n \times n$ matrix whose diagonal elements are one and whose off-diagonal elements are zero:

$$\tilde{I} = \begin{pmatrix} 1 & 0 & \cdots & 0 \\ 0 & 1 & & \\ \vdots & & \ddots & \\ 0 & & & 1 \end{pmatrix}. \tag{B.4}$$

The product of an identity matrix with any matrix $\tilde{M}$ of the same dimensions is just equal to $\tilde{M}$ itself. This fact, coupled with equations (B.1) and (B.3), can be used to prove that (B.2) is the solution. Consequently, if we can find the inverse matrix for the circuit problem we are attempting to solve, simple matrix multiplication will yield the unknowns. Many analytic and numeric techniques for inverting matrices are described in the relevant literature. One particularly powerful method is known as *Gaussian elimination with back-substitution.* For this reason, Gaussian elimination is the only technique described in this appendix. Readers who are interested in the other methods are referred to the literature.

The basic idea behind Gaussian elimination is described below. Recall that we want to solve the matrix equation (B.1) for the elements of the unknown column vector. To do this, we will transform the given matrix equation into an equivalent problem where the $n \times n$ matrix is upper-triangular (i.e., a matrix for which $i > l$ implies that $M_{il} = 0$):

$$\begin{pmatrix} M'_{11} & M'_{12} & \cdots & M'_{1n} \\ 0 & M'_{22} & & \\ \vdots & \ddots & & \\ 0 & \cdots & 0 & M'_{nn} \end{pmatrix} \begin{pmatrix} U_1 \\ U_2 \\ \vdots \\ U_n \end{pmatrix} = \begin{pmatrix} S'_1 \\ S'_2 \\ \vdots \\ S'_n \end{pmatrix}. \tag{B.5}$$

If we can find this equivalent problem, a simple algorithm can be used to compute the unknowns. This procedure is known as back-substitution and works as follows. The final row of the equivalent matrix equation (B.5) is:

$$M'_{nn} U_n = S'_n, \tag{B.6}$$

so that the final unknown element is given by

$$U_n = S'_n / M'_{nn}. \tag{B.7}$$

The second-to-last row can be written:

$$M'_{n-1,n-1} U_{n-1} + M'_{n-1,n} U_n = S'_n \tag{B.8}$$

which can be solved for U_{n-1}:

$$U_{n-1} = (S'_{n-1} - M'_{n-1,n} U_n) / M'_{n-1,n-1}. \tag{B.9}$$

All the terms on the right-hand side of equation (B.9) are known, so U_{n-1} can be calculated. This procedure can be repeated to find all of the unknowns. The general expression for U_j is ($1 \le j \le n-1$):

$$U_j = \left(S'_j - \sum_{i=j+1}^{n} M'_{ji} U_i \right) / M'_{jj}. \tag{B.10}$$

Equations (B.7) and (B.10) fully describe the back-substitution technique.

To validate this technique, we return to the question of producing the equivalent problem with an upper-triangular matrix. This procedure is known as Gaussian elimination and is described below. Let us assume that the matrix equation in (B.1) represents a collection of n linearly independent equations. It is known from linear algebra that if we multiply any of the equations by a nonzero constant and add the result to any other equation, we still have a set of n linearly independent equations with the same solution.

Let's assume that $M_{11} \neq 0 \neq M_{12}$. If we multiply the first row in (B.1) by $-M_{12}/M_{11}$ and add the result to the second row we obtain:

$$\begin{pmatrix} M_{11} & M_{12} & \cdots & M_{1n} \\ 0 & M_{22} - M_{12}M_{21}/M_{11} & \cdots & M_{2n} - M_{1n}M_{21}/M_{11} \\ M_{31} & \vdots & & \\ \vdots & & \vdots & \\ M_{n1} & & M_{nn} & \end{pmatrix} \begin{pmatrix} U_1 \\ U_2 \\ \vdots \\ U_n \end{pmatrix} = \begin{pmatrix} S_1 \\ S_2 - S_1 \frac{M_{21}}{M_{11}} \\ \vdots \\ S_n \end{pmatrix}. \tag{B.11}$$

If we continue in this fashion (i.e., we subtract from the jth row M_{j1}/M_{11} times the first row) we will get

$$\begin{pmatrix} M_{11} & M_{12} & \cdots & M_{1n} \\ 0 & M_{22} - \frac{M_{12}M_{21}}{M_{11}} & & \\ \vdots & \vdots & & \\ 0 & M_{n1} - \frac{M_{12}M_{n1}}{M_{11}} & \cdots & M_{nn} - \frac{M_{1n}M_{n1}}{M_{11}} \end{pmatrix} \begin{pmatrix} U_1 \\ U_2 \\ \vdots \\ U_n \end{pmatrix} = \begin{pmatrix} S_1 \\ S_2 - S_1 \frac{M_{21}}{M_{11}} \\ \vdots \\ S_n - S_1 \frac{M_{n1}}{M_{11}} \end{pmatrix}. \tag{B.12}$$

The first column is in the proper form, so we turn our attention to the second column. By multiplying the second row by appropriate factors and adding it to the third through the nth row, we can zero the lower elements of the second column. Repeating this procedure for the third through the $(n-1)$st row, we can zero the lower elements of the third through the $(n-1)$st columns, respectively, and arrive at the desired result.

Let's consider the specific case of a two-mesh problem. The matrix equation is:

$$\begin{pmatrix} Z_{11} & Z_{12} \\ Z_{21} & Z_{22} \end{pmatrix} \begin{pmatrix} \hat{I}_1 \\ \hat{I}_2 \end{pmatrix} = \begin{pmatrix} \hat{V}_{S_1} \\ \hat{V}_{S_2} \end{pmatrix}. \tag{B.13}$$

Multiplying the first row by $-Z_{21}/Z_{11}$ and adding it to the second row yields:

$$\begin{pmatrix} Z_{11} & Z_{12}^1 \\ 0 & Z_{22} - Z_{12}Z_{21}/Z_{11} \end{pmatrix} \begin{pmatrix} \hat{I}_1 \\ \hat{I}_2 \end{pmatrix} = \begin{pmatrix} \hat{V}_{S_1} \\ \hat{V}_{S_2} - \hat{V}_{S_1}Z_{21}/Z_{11} \end{pmatrix}. \tag{B.14}$$

The second row is used to find $\hat{I}_2$:

$$\hat{I}_2 = \frac{-Z_{21}\hat{V}_{S_1} + Z_{11}\hat{V}_{S_2}}{Z_{11}Z_{22} - Z_{12}Z_{21}}. \tag{B.15}$$

An application of (B.10) gives:

$$\hat{I}_1 = (\hat{V}_{S_1} - Z_{12}\hat{I}_2)/Z_{11} = \frac{Z_{22}\hat{V}_{S_1} - Z_{12}\hat{V}_{S_2}}{Z_{11}Z_{22} - Z_{12}Z_{21}}. \tag{B.16}$$

As a second example, suppose that we have a four-node circuit that results in the following phasor problem

$$\begin{pmatrix} 1+j & -j & 0 \\ -j & 1-j & -1 \\ 0 & -1 & 2 \end{pmatrix} \begin{pmatrix} \hat{V}_1 \\ \hat{V}_2 \\ \hat{V}_3 \end{pmatrix} = \begin{pmatrix} 1 \\ 0 \\ -j \end{pmatrix}. \tag{B.17}$$

Let us compute all three node potentials. To zero Y_{21}, we multiply the first row by $j/(1+j) = (1+j)/2$ and add it to the second row. The resulting equation is:

$$\begin{pmatrix} 1+j & j & 0 \\ 0 & \frac{3}{2}(1-j) & -1 \\ 0 & -1 & 2 \end{pmatrix} \begin{pmatrix} \hat{V}_1 \\ \hat{V}_2 \\ \hat{V}_3 \end{pmatrix} = \begin{pmatrix} 1 \\ \frac{1+j}{2} \\ -j \end{pmatrix}. \tag{B.18}$$

The first column is now zeroed because we had $Y_{31} = 0$ to begin with, so we move on to the second column. We need to multiply the second row by $\frac{2}{3}/(1-j) = (1+j)/3$ and add it to the third row to get:

$$\begin{pmatrix} 1+j & -j & 0 \\ 0 & \frac{3}{2}(1-j) & -1 \\ 0 & 0 & (5-j)/3 \end{pmatrix} \begin{pmatrix} \hat{V}_1 \\ \hat{V}_2 \\ \hat{V}_3 \end{pmatrix} = \begin{pmatrix} 1 \\ (1+j)/2 \\ -2j/3 \end{pmatrix}. \tag{B.19}$$

We use equation (B.7) to find $\hat{V}_3$:

$$\hat{V}_3 = \frac{-2j}{3} \Big/ \frac{(5-j)}{3} = (1-5j)/13. \tag{B.20}$$

From (B.10) we find that:

$$\hat{V}_2 = \left(\frac{1+j}{2} + \hat{V}_3\right)\left(\frac{1+j}{3}\right) = (2+3j)/13 \tag{B.21}$$

and

$$\hat{V}_1 = (1 + j\hat{V}_2)/(1+j) = (6-4j)/13. \tag{B.22}$$

While equations (B.20)–(B.22) form the answer to this example and the problem is completed, it is often prudent to check for algebraic mistakes by plugging the answer back into the original problem (B.17). Performing that exercise with this example reveals that the answer is consistent, so we conclude that we in fact have found the correct solution.

Drill Problems

In Problems 1–5, use Gaussian elimination and back-substitution to find all the unknowns.

1. $\begin{pmatrix} 1 & 2 \\ 3 & 5 \end{pmatrix} \begin{pmatrix} \hat{V}_1 \\ \hat{V}_2 \end{pmatrix} = \begin{pmatrix} 1 \\ -1 \end{pmatrix}$

2. $$\begin{pmatrix} 1 & j \\ j & (2+j) \end{pmatrix} \begin{pmatrix} \hat{I}_1 \\ \hat{I}_2 \end{pmatrix} = \begin{pmatrix} 2-j \\ 1 \end{pmatrix}$$

3. $$\begin{pmatrix} 0 & 3 & 0 \\ 2 & 1 & -1 \\ 1 & 5 & 4 \end{pmatrix} \begin{pmatrix} \hat{I}_1 \\ \hat{I}_2 \\ \hat{I}_3 \end{pmatrix} = \begin{pmatrix} 2 \\ -3 \\ 7 \end{pmatrix}$$

4. $$\begin{pmatrix} 1+j & -j & 3+j \\ -j & 2 & 0 \\ 3+j & 0 & 2+j \end{pmatrix} \begin{pmatrix} \hat{V}_1 \\ \hat{V}_2 \\ \hat{V}_3 \end{pmatrix} = \begin{pmatrix} 1 \\ -j \\ 3-j \end{pmatrix}$$

5. $$\begin{pmatrix} 1 & -1/2 & -2 & 0 \\ -1/2 & 2 & -1 & -1/3 \\ -2 & -1 & 3 & -1 \\ 0 & -1/3 & -1 & 4 \end{pmatrix} \begin{pmatrix} \hat{V}_1 \\ \hat{V}_2 \\ \hat{V}_3 \\ \hat{V}_4 \end{pmatrix} = \begin{pmatrix} 1 \\ 0 \\ 0 \\ -2 \end{pmatrix}$$

Appendix C

MicroSim PSpice References

Conant, Roger, *Engineering Circuit Analysis with PSpice and Probe*. McGraw-Hill, New York, 1993.

Goody, Roy, *PSpice for Windows*. Prentice Hall, Englewood Cliffs, NJ, 1995.

Herniter, Marc, *Schematic Capture with MicroSim PSpice*, 2nd Edition. Prentice Hall, Englewood Cliffs, NJ, 1996.

MicroSim Corporation, *Circuit Analysis Users Guide*. Irvine, CA, 1992.

Nilsson, James, *Introduction to Spice*. Addison-Wesley, New York, 1990.

Index